Agroecological Approaches for Sustainable Soil Management

Agroecological Approaches for Sustainable Soil Management

Edited by

Majeti Narasimha Vara Prasad
University of Hyderabad (An Institution of Eminence)
Hyderabad, India

Chitranjan Kumar
Amity Institute of Organic Agriculture
Amity University, Noida, India

WILEY

This edition first published 2023

Registered Offices
John Wiley & Sons, Inc., 111 River Street, Hoboken, NJ 07030, USA
John Wiley & Sons Ltd, The Atrium, Southern Gate, Chichester, West Sussex, PO19 8SQ, UK

For details of our global editorial offices, customer services, and more information about Wiley products visit us at www.wiley.com.

Wiley also publishes its books in a variety of electronic formats and by print-on-demand. Some content that appears in standard print versions of this book may not be available in other formats.

Library of Congress Cataloging-in-Publication Data
Names: Prasad, M. N. V. (Majeti Narasimha Vara), 1953– editor. | Kumar, Chitranjan, editor.
Title: Agroecological approaches for sustainable soil management / edited by Majeti Narasimha Vara Prasad and Chitranjan Kumar.
Description: Hoboken, NJ: Wiley, 2023. | Includes bibliographical references and index.
Identifiers: LCCN 2023007310 (print) | LCCN 2023007311 (ebook) | ISBN 9781119911968 (hardback) | ISBN 9781119911975 (adobe pdf) | ISBN 9781119911982 (epub)
Subjects: LCSH: Soil management. | Land degradation. | Agricultural ecology.
Classification: LCC S591.A59173 2023 (print) | LCC S591 (ebook) | DDC 631.4–dc23/eng/20230512
LC record available at https://lccn.loc.gov/2023007310
LC ebook record available at https://lccn.loc.gov/2023007311

Cover Design: Wiley
Cover Image: Courtesy of Dr. Chitranjan Kumar

Set in 9.5/12.5pt STIXTwoText by Straive, Pondicherry, India

Contents

List of Contributors

Yassine Aallam
Laboratory of Agro-Industrial and Medical Biotechnologies
Faculty of Science and Techniques
University of Sultan Moulay Slimane
Beni Mellal, Morocco

Sanggono Adisasmito
Department of Chemical Engineering
Institut Teknologi Bandung
Bandung, Indonesia

Lander de Jesus Alves
Postgraduate Program in Biology and Biotechnology of Microorganisms,
State University of Santa Cruz (UESC)
Ilhéus, Bahia, Brazil

Soumia Amir
Centre d'Agrobiotechnologie et Bioingénierie
Unité de Recherche Labellisée CNRST
Cadi Ayyad University, Marrakech

Hossein Beyrami
National Salinity Research Center (NSRC)
Agricultural Research
Education and Extension Organization
(AREEO), Yazd, Iran

Wassila Bouta
Laboratory of Agro-Industrial and Medical Biotechnologies
Faculty of Science and Techniques
University of Sultan Moulay Slimane
Beni Mellal, Morocco; Office de la formation professionnelle et de la promotion du travail (Ofppt)
Pole agriculture, La Cité des Métiers et des Compétences de la région Béni Mellal-Khénifra
Béni Mellal, Morocco

Piyapatr Busababodhin
Isan Saline Soil Research Unit (ISSRU)
Faculty of Science
Mahasarakham University
Khamriang, Kantarawichi
Mahasarakham, Thailand

Cláudia Cseko Nolasco de Carvalho
Department of Biology
State University of Alagoas
Santana do Ipanema, Alagoas, Brazil

A. J. Palace Carvalho
Chemistry and Biochemistry Department
School of Sciences and Technology
University of Évora
Évora, Portugal;
LAQV-REQUIMTE
School of Sciences and Technology
University of Évora
Évora, Portugal

Wutthisat Chokkuea
Isan Saline Soil Research Unit (ISSRU)
Faculty of Science
Mahasarakham University
Khamriang, Kantarawichi
Mahasarakham, Thailand

David E. Clay
Department of Plant Sciences
South Dakota State University, USA;
American Society of Agronomy
Madison, USA

Cláudia Marques-dos-Santos
School of Agriculture, University of Lisbon
CEF—Forest Research Centre
Tapada da Ajuda
Lisbon, Portugal

A. V. Dordio
Chemistry and Biochemistry Department
School of Sciences and Technology
University of Évora
Évora, Portugal;
MARE – Marine and Environmental
Sciences Centre
School of Sciences and Technology
University of Évora
Évora, Portugal

Nawal K. Dubey
Laboratory of Herbal Pesticides
Centre of Advanced Study (CAS) in Botany
Institute of Science
Banaras Hindu University
Varanasi, India

Winya Dungkaew
Isan Saline Soil Research Unit (ISSRU)
Faculty of Science
Mahasarakham University
Khamriang, Kantarawichi
Mahasarakham, Thailand

Jorge M.S. Faria
Mediterranean Institute for Agriculture
Environment and Development
MED & Global Change and Sustainability
Institute, CHANGE
Institute for Advanced Studies and
Research
University of Évora
Évora, Portugal;
National Institute for Agrarian and
Veterinarian Research, INIAV
I. P., Quinta do Marquês
Oeiras, Portugal

Mohamed Farissi
Laboratory of Biotechnology and
Sustainable Development of Natural
Resources, Polydisciplinary Faculty, USMS
Beni Mellal, Morocco

Tino Fauk
Department Water Environment
Civil Engineering and Safety
Magdeburg-Stendal University of Applied
Sciences, Breitsscheidstr
Magdeburg, Germany

Ana Fonseca
School of Agriculture, University of Lisbon
CEF—Forest Research Centre
Tapada da Ajuda
Lisbon, Portugal

V. Girijaveni
Central Research Institute for Dryland
Agriculture, Santoshnagar
Hyderabad, India

Yolanda González-García
Departamento de Botánica
Universidad Autónoma Agraria
Antonio Narro
Saltillo, México

Abdelmajid Haddioui
Laboratory of Agro-Industrial and
Medical Biotechnologies
Faculty of Science and Techniques
University of Sultan Moulay Slimane
Beni Mellal, Morocco

Hanane Hamdali
Laboratory of Agro-Industrial and
Medical Biotechnologies

Faculty of Science and Techniques
University of Sultan Moulay Slimane
Beni Mellal, Morocco

Ayoub Haouas
Centre d'Agrobiotechnologie
et Bioingénierie
Unité de Recherche Labellisée CNRST
(Centre AgroBiotech-URL-CNRST-05)
Cadi Ayyad University Marrakech, Morocco

Antonius Indarto
Department of Chemical Engineering
Institut Teknologi Bandung
Bandung, Indonesia

Ponlakit Jitto
Faculty of Environment and
Resource Studies
Mahasarakham University, Khamriang
Kantarawichi District
Maha Sarakham, Thailand

Deepak R. Joshi
Department of Plant Sciences
South Dakota State University, USA

Antonio Juárez-Maldonado
Departamento de Botánica
Universidad Autónoma Agraria
Antonio Narro
Saltillo, México

Piyanutt Khanema
Isan Saline Soil Research Unit (ISSRU)
Faculty of Science
Mahasarakham University
Khamriang, Kantarawichi
Mahasarakham, Thailand

Surasak Khankhum
Isan Saline Soil Research Unit (ISSRU)
Faculty of Science
Mahasarakham University
Khamriang, Kantarawichi
Mahasarakham, Thailand

Chitranjan Kumar
Amity Institute of Organic Agriculture
Amity University, Uttar Pradesh
Noida, India

Mehdi Mahbod
Department of Water Sciences
and Engineering, College of Agriculture
Jahrom University, Jahrom, Iran

Joana Marinheiro
School of Agriculture
University of Lisbon
CEF—Forest Research Centre
Tapada da Ajuda
Lisbon, Portugal

Florin-Constantin Mihai
CERNESIM Environmental Research Center
Institute of Interdisciplinary Research
"Alexandru Ioan Cuza" University of Iasi
Bulevardul Carol I
Iași, Romania

Chanchal K. Mitra
School of Life Sciences
University of Hyderabad
Hyderabad, India

Ahmed El Moukhtari
Laboratory of Ecology and Environment
Faculty of Sciences Ben Msik
Hassan II University
PO 7955 Sidi Othmane
Casablanca, Morocco;
Laboratory of Biotechnology and
Sustainable Development of Natural
Resources
Polydisciplinary Faculty, USMS
Beni Mellal, Morocco

Rachid Mrabet
Institut National de la Recherche
Agronomique (INRA)
Avenue de la Victoire
Morocco

Ruttanakorn Munjit
Isan Saline Soil Research Unit (ISSRU)
Faculty of Science
Mahasarakham University
Khamriang, Kantarawichi
Mahasarakham, Thailand

Woranan Nakbanpote
Isan Saline Soil Research Unit (ISSRU)
Faculty of Science
Mahasarakham University
Khamriang, Kantarawichi
Mahasarakham, Thailand

Fábio Carvalho Nunes
Academic Department
Federal Institute Baiano (IF BAIANO)
Santa Inês, Bahia, Brazil

Bülent OKUR
Faculty of Agriculture
Soil Science and Plant Nutrition
Department, Ege University
Bornova, İzmir, Turkey

Nur OKUR
Faculty of Agriculture
Soil Science and Plant Nutrition Department
Ege University, Bornova
İzmir, Turkey

Nesrin ÖRÇEN
Faculty of Agriculture, Field Crop Science
Ege University, Bornova
İzmir, Turkey

Abdallah Oukarroum
High Throughput Multidisciplinary
Research Laboratory University
Mohammed VI Polytechnic (UM6P)
Ben Guerir, Morocco

Marek Pająk
Department of Ecological Engineering
and Forest Hydrology
Faculty of Forestry
University of Agriculture in Krakow
Krakow, Poland

Sangeeta Pandey
Amity Institute of Organic Agriculture
Amity University, Uttar Pradesh
Noida, India

Amir Parnian
National Salinity Research Center (NSRC)
Agricultural Research
Education and Extension
Organization (AREEO)
Yazd, Iran

Marcin Pietrzykowski
Department of Ecological Engineering
and Forest Hydrology
Faculty of Forestry
University of Agriculture in Krakow
Krakow, Poland

Ana Paula Pinto
Mediterranean Institute for Agriculture,
Environment and Development
MED & Global Change and Sustainability
Institute, CHANGE
Institute for Advanced Studies and Research
University of Évora,
Évora, Portugal;
Chemistry and Biochemistry Department
School of Sciences and Technology
University of Évora,
Évora, Portugal

Daniel Pramudita
Department of Chemical Engineering
Institut Teknologi Bandung
Bandung, Indonesia;
Department of Food Engineering
Institut Teknologi Bandung

J. V. N. S. Prasad
Central Research Institute for Dryland
Agriculture, Santoshnagar
Hyderabad, India

Majeti Narasimha Vara Prasad
School of Life Sciences
University of Hyderabad
Hyderabad, India

Ronny Purwadi
Department of Chemical Engineering
Institut Teknologi Bandung
Bandung
Indonesia; Department of Food Engineering
Institut Teknologi Bandung
Jl. Let. Jend. Purn. Dr. (HC) Mashudi No.1
Sumedang, Indonesia

Prabhat Kumar Rai
Department of Environmental Science
School of Earth Sciences and Natural Resource Management
Mizoram University, Aizawl
Mizoram, India

Taoufik El Rasafi
High Throughput Multidisciplinary Research Laboratory
University Mohammed VI Polytechnic (UM6P)
Ben Guerir, Morocco;
Laboratory of Health and Environment
Hassan II University
Faculty of Sciences Ain Chock
Casablanca, Morocco

K. Sammi Reddy
Central Research Institute for Dryland Agriculture, Santoshnagar
Hyderabad, India

Petra Schneider
Department Water Environment
Civil Engineering and Safety
Magdeburg-Stendal University of Applied Sciences, Breitsscheidstr
Magdeburg, Germany

João Serra
School of Agriculture University of Lisbon
CEF—Forest Research Centre
Tapada da Ajuda, Lisbon, Portugal

Tarun Sharma
Department of Agronomy
CSK Himachal Pradesh Agricultural University
Palampur, India

Akashdeep Singh
Department of Agronomy
CSK Himachal Pradesh Agricultural University
Palampur, India

Anil K. Singh
Shri Murli Manohar Town Post Graduate College
Ballia, India

V. K. Singh
Central Research Institute for Dryland Agriculture, Santoshnagar
Hyderabad, India

Khanitta Somtrakoon
Isan Saline Soil Research Unit (ISSRU)
Faculty of Science
Mahasarakham University
Khamriang, Kantarawichi
Mahasarakham, Thailand

Pranee Srihaban
Land Development Regional Office 5
Department of Land, Development
Ministry of Agriculture and Cooperatives
Khon Kaen, Thailand

Ágnes Szepesi
Department of Plant Biology
Institute of Biology
Faculty of Science and Informatics
University of Szeged
Szeged, Hungary

Anas Tallou
Centre d'Agrobiotechnologie et Bioingénierie
Unité de Recherche Labellisée CNRST
Cadi Ayyad University, Marrakech

Uraiwan Taya
Land Development Regional Office 5
Department of Land, Development
Ministry of Agriculture and Cooperatives
Khon Kaen, Thailand

Shikha Tiwari
Department of Botany
S.S.S.V. S. Government Post Graduate College, Chunar, Mirzapur, India

Ajay Tomar
Amity Institute of Organic Agriculture
Amity University, Uttar Pradesh
Noida, India

Diana Cota-Ungson
Doctorado en Agricultura Protegida
Universidad Autónoma Agraria Antonio Narro, Saltillo, México

Bartłomiej Woś
Department of Ecological Engineering and Forest Hydrology, Faculty of Forestry
University of Agriculture in Krakow
Krakow, Poland

Preface

Majeti Narasimha Vara Prasad and Chitranjan Kumar

Agroecological Approaches for Sustainable Soil Management

Agroecology is a significant part of the global productivity system, and it involves interactions between biotic and abiotic in many different ways. In order to meet the demands of the growing population, it provides a variety of ecosystem services such as halting climate change, carbon sequestration, soil conservation, and preserving productivity and output. The management and sustainability of agroecosystems are under danger due to the increase in global food demand. The adoption of modern technology has resulted in increased resource usage, which is slowly increasing the ecological footprints day by day. The food production system is also hampered by problems with food shortages, security, and overproduction beyond the ecosystem's biocapacity. As a result, the agroecosystem and food production system require urgently adequate management.

All nations on the planet rely heavily on agriculture, hence appropriate management of it is urgently needed. The book discusses a thorough strategy for managing ecological footprints in the agroecosystem effectively. The book also examines key elements such as energy, carbon, nitrogen, climate, water, land, and livestock footprints using a systematic and scientific approach that paves the way for their efficient management. The present book has also addressed specific strategies, planning, and policy formulation toward mitigating the various forms of ecological footprint based on research and developmental activities within the agroecosystem.

The advancement of agriculture is essential to the progress of humanity as a whole and is essential to the timely achievement of numerous sustainable development goals. Nevertheless, agriculture accounts for 34% of the world's yearly emissions of greenhouse gases and is the greatest nonpoint source of pollution due to the widespread use of agrochemicals. Agricultural activities account for 70% of the freshwater withdrawn from the planet each year, making them one of the main causes of the loss of biodiversity globally. The Food and Agricultural Organization of the United Nations estimates that agricultural production must quadruple by the year 2050 in order to feed a population of 9.7 billion people worldwide. However, this intensification needs to be carried out sustainably (not at the cost of planetary resilience).

However, this intensification must be carried out sustainably (not at the expense of planetary resilience), and fresh approaches must be developed to highlight agriculture's vital

Figure 1 A word cloud map was generated using the titles and keywords of papers of this book.

contribution to the achievement of the global goals. In order to improve human well-being while protecting the life-supporting systems of a highly polluted, overexploited, and resource-constrained world, the concept of "planet friendly agriculture," where food production must be performed within the planetary bounds, is gaining global prominence. The goal of this graphic review was to illustrate different environmentally friendly farming techniques based on resource replenishment and conservation to improve food and nutrition security for the present and future generations while lowering pollution, greenhouse gas emissions, biodiversity loss, and water footprint, even in the face of changing climatic conditions.

Agroecologists are working hard to provide sustainable food systems for the world's expanding population while also lowering their environmental carbon impact. Any agricultural system's sustainability largely depends on soil sustainability, which meets current demand without jeopardizing the ability of future generations to satisfy their own requirements with sustainable soil resources. Scientists studying soil are concerned about an ongoing rise in soil deterioration from prehistoric times.

We choose to incorporate many seasoned as well as aspiring agroecologists and soil scientists working throughout the globe in our dream project in response to a widespread desire for a visionary research experience to close a significant gap between plant ecology and soil biology.

As the greatest hazard of the twenty-first century, soil degradation caused by alien invading plants, management of the soil via carbon sequestration, restoration of resilience, and sustainable usage ecological restoration via protected restoration and organic farming, biochar incorporation and amendment-assisted restoration, sustainable control of soil-borne plant diseases, rooftop vertical hydroponic and aeroponic farming systems, and other agroecological transformation techniques saline soil remediation for sustained crop production, aggravation-conservation agriculture based on agroecology, cover crops, tree cover dynamics, and intercropping ecology industries and oil-palm plantations.

Reclaiming urban brownfield and industrial areas, and Internet of things in soil management for smart agriculture, in the structured format of (i) Introduction, (ii) Need for agroecology: Justification, (iii) Restoration of degraded lands, (iv) Agroecology for sustainable food systems, (v) Ecosystem services and conservation agriculture, and (vi) Case studies for restoration of degraded lands.

Editors are indebted to all contributors for their valuable research and work as a chapter for this book. We express our sincere gratitude to Dr. Sakeena Quraishi and Dr. Gudrun Walter, the editors of the Wiley publisher team for their keen interest, cooperation, and prompt efforts. We acknowledge the quality work of the production team of John Wiley & Sons, UK, who helped us to bring this book in its final form.

About the Editors

Majeti Narasimha Vara Prasad is currently Emeritus Professor, School of Life Sciences, University of Hyderabad, Hyderabad, India.

Formerly Dean, School of Life Sciences

Formerly Head, Dept of Plant Sciences

Formerly Coordinator, Biotechnology Program

Formerly Coordinator of PG Diploma in Environmental education and management

M.Sc. (Botany) from Andhra University, Waltair 1973–1975; Ph.D (Botany) Lucknow University, Lucknow 1975–1979 (Research conducted at Birbal Sahni Institute of Palaeosciences, An autonomous Inst. Under the Dept. of Sci & Tech, Govt. of India).

Professional experience Lecturer, June 1980–85, Dept. of Botany,

North Eastern Hill University, Shillon, Meghalaya

Lecturer, 1985–86, Lecturer (Senior Scale) 1986–90 **University of Hyderabad,** School of Life Science; Reader, 1990–98, and Professor since 1998 to May 10, 2018, in the Department of Plant Sciences, University of Hyderabad.

Made significant contributions to the field of plant-metal nteractions, bioremediation, and bioeconomy. Published 221 research articles in peer-reviewed Journals, 141 book chapters. Edited Books 34: Elsevier [15], JohnWiley [6], Springer [6]; One each by Fizmatlit Russia, Kluwer Academic, Ministry of Environment and Forests, GoI, Marcel Dekker, Narosa, Russian Academy of Sciences and CRCPress, Taylor & Francis

Academic Honors

1) XIX Int. Bot. Congress Excellent Scholar Awardee, July 23–29, 2017, Shenzhen, China
2) Pitamber Pant National Environment Fellow 2007 awarded by the Ministry of Environment, Forests and Climate Change, Government of India
3) Recipient of Prof. KS Bilgrami award – 2015 by the Soc for Plant Research, India
4) Served as COST action 859 (Phytotechnology) working group member, ESF
5) Elected Fellow – Linnean Society of London, UK
   Elected Fellow – National Institute of Ecology, New Delhi

Global Recognition

Based on data compiled and studied by John Ioannidis of Stanford University, Elsevier and SciTech Strategies, MNV Prasad **figured in top 2% consecutively for three years (2020–2022)**.

Visiting Assignments in Various Universities – Widely Traveled

1) University Quebec INRS-Eau, ***Canada***, NSERC foreign research awardee *1994*
2) Dept of Plant Physiology and Biochemistry, Jagiellonian University, Krakow, ***Poland***, *1996*
3) University of Coimbra, ***Portugal***, *1999*
4) Stockholm University, Institute of Botany, ***Sweden***, *2000*
5) University of Oulu, Oulu by Finnish Academy ***Finland***, *2002*
6) University of South Australia, Adelaide, ***Australia***, *2005*
7) Al-Farabi Kazakh National University, Department. of Botany, Almaty, ***Kazakhstan***, *2006*
8) Ural Federal University, Ekaterinburg, ***Russia***, *2007–2015*
9) Ghent University Faculty of Bioscience Engineering, Gent, ***Belgium***, *2011*
10) Mahasarakham University, ***Thailand***, *2013–2014*
11) University of Santa Cruz, Ilheus-Bahia, ***Brazil***, *2015*
12) Govt. of India deputation to Asian Institute of Technology, ***Thailand*** for one semester *2017*

Dr. Chitranjan Kumar (M.Sc. Gold Medalist NET D.Phil.), is Assistant Professor-III and Program Leader, Amity Institute of Organic Agriculture, Amity University, Noida, India; Associate Editor, Agronomy Journal, American Society of Agronomy, USA. He is Formerly Technical Editor, The National Academy of Sciences India, Prayagraj. Dr. Chitranjan Kumar is working in the field of phyto-bio-remediation/agroecological management of sewage-irrigated soils/organic fertigation. He is serving Agronomy Journal, American Society of Agronomy, USA as Associate Editor continuously for the last 10 years. He has published 45 research/ review articles in journals of national and international repute. He has published five-books from reputed publishers including John Wiley & Sons, Germany, and Lap Lambert Academic Publishing, Germany. He has developed an understanding of enhanced phytoremediation of heavy metals through plant species, oilcake, vermicompost, and microbial strains of *Pseudomonas sp.*, *Thiobacillus sp.*, and *Glomus sp.* as their individual as well as their combinatorial treatments. He presented a systematic collection of data on the Ganges River pollution, its scientific analyses, and its relationship with 6Ps (namely population, poverty, pollution, precipitation, plantation, and periodicity). He presented a noble integrated micro-biochemical approach through microbe-nutrient-ornamental plant-assisted phytoremediation in sewage-contaminated soils. Studied long-term integrated nutrient management and C-sequestration in soils. The top 12 international papers were published in reputed journals, like Ecotoxicology and Environmental Safety, Land Degradation and Development, Journal of Soil Science and Plant Nutrition, Agronomy Journal, 3 Biotech, International Journal of Environmental Science and Technology, Environmental Monitoring and Assessment, International Journal of Phytoremediation, Bulletin of Environmental Contamination and Toxicology, and Agronomy Journal. Edited 12-Annual Reports of The National Academy of Sciences, India (NASI, Prayagraj) during 2009-2020. Also edited >1060 Springer articles, 24-Wiley Articles, 12-Year Books of NASI as Technical Editor, and reviewed > 240 international articles from 40 international journals.

World Scientist Ranking: One of the top rankings in the field of Agriculture (Organic Agriculture | Environment | Phytoremediation | Bioremediation | Heavy Metal Interaction) as per the link: https://www.adscientificindex.com/scientist/chitranjan-kumar-sharma/352588

1

Soil Degradation: A Major Challenge in the Twenty-First Century

Fábio Carvalho Nunes[1], *Cláudia Cseko Nolasco de Carvalho*[2], *Lander de Jesus Alves*[3], *and Majeti Narasimha Vara Prasad*[4]

[1] *Academic Department, Federal Institute Baiano (IF BAIANO), Santa Inês, Bahia, Brazil*
[2] *Department of Biology, State University of Alagoas, Santana do Ipanema, Alagoas, Brazil*
[3] *Postgraduate Program in Biology and Biotechnology of Microorganisms, State University of Santa Cruz (UESC), Ilhéus, Bahia, Brazil*
[4] *School of Life Sciences, University of Hyderabad, Hyderabad, India*

1.1 Introduction

Soils are the basis for food production, and, in addition, they perform essential functions for the balance of ecosystems. Although the functions related to material aspects are more emphasized, such as provision, support, and regulation, the environmental functions performed by the soil are much more comprehensive and also include the immaterial dimension, such as information, culture, leisure, and religion (Nunes et al. 2020) (Table 1.1).

Despite the importance of soil, it is often neglected as a non-renewable and indispensable resource for human and non-human life. Its degradation has been unfolding year after year due to a combination of processes that range from natural to anthropic factors. According to the Food and Agriculture Organization of the United Nations (FAO 2021) about 34% of soil globally is moderately or highly degraded.

Soil degradation consists of the reduction or loss of its biological or economic productivity due to a process or combination of processes, among which water or wind erosion and deterioration of physical, chemical, and/or biological soil properties stand out. Such processes can be triggered by climate change and direct human actions such as deforestation, fires, mining, irrigation, increased use of machines, plastic deposition, use of fertilizers and pesticides, and such actions can cause erosion, desertification, sandization, salinization, pollution, and soil contamination.

According to FAO (2021), 24 billion tons of fertile soil/year are lost through degradation processes, displacing about 135 million people, mainly from arid, semi-arid, and sub-humid to drought areas. Proper soil management is a major challenge in the twenty-first century, as both an abrupt change in the current production model and its maintenance can cause a collapse that compromises nutritional and food security on a global scale.

Agroecological Approaches for Sustainable Soil Management, First Edition.
Edited by Majeti Narasimha Vara Prasad and Chitranjan Kumar.

Table 1.1 Soil functions: regulation and/or support, provision, *information, culture, leisure,* and *religion*

Regulation and/or support functions	References
Recharge of aquifers, control, and storage of water	Ghaley et al. (2018), van Leeuwen et al. (2017)
Water purification, assimilation, and recycling of pollutants	Ghaley et al. (2018), van Leeuwen et al. (2017)
Regulation of floods	FAO (2015)
Provide refuge, nurseries, and habitats for organisms	Ghaley et al. (2018), Schulte et al. (2015)
Support engineering works	FAO (2015)
Support bacterial culture for antibiotics production	Ling et al. (2015)[a]
Support raising livestock	Schulte et al. (2015), Volchko et al. (2014)
Nutrient cycling	Ghaley et al. (2018), Schulte et al. (2015)
Climate regulation	Ghaley et al. (2018), Volchko et al. (2014)
Provision functions	
Natural production of foods, fibers, and medicines	Ghaley et al. (2018), FAO (2015)
Crops for food, fiber, and crop production in cultivated areas	Ghaley et al. (2018), FAO (2015)
Crops for the production of energy	FAO (2015)
Materials for ornaments, handicrafts, and household items	Alves et al. (2005)[b]
Genetic resources	Schaefer (2011)[c], FAO (2015)
Materials for construction	FAO (2015)
Materials for pharmaceuticals and cosmetics	FAO (2015)
Functions of information, culture, leisure, and religion	
Plaeoenvironmental information	Inda, Fink and Santos (2018), Ladeira (2010)
Cultural heritage	Capra et al. (2017), Alves et al. (2005)
Recreation and leisure	Sindelar (2015)
Religious rituals	Barrera-Bassols and Zinck (2003), Minami (2009)[d]
Educational activities	Hayhoe et al. (2016), Hartemink et al. (2014)

[a] The authors do not mention soil functioning in support of bacterial culture for the production of antibiotics, but do reference soil as a culture medium in the production of antibiotics.
[b] The authors make no reference to soil functioning as a medium for the provision of materials for ornaments, handicrafts, and domestic utensils, but present a study that documented indigenous communities using soil for such purposes,
[c] The authors make no reference to soil functioning as a means for acquiring genetic resources, but emphasizes that soil has such resources.
[d] The authors make no reference to soil functioning as a means of performing religious rituals, but show that soil is used for such purposes.
Source: Adapted Nunes et al. (2020).

Furthermore, as the soil is a complex ecological system, therefore interdependent and full of non-linear responses (Nunes et al. 2021), its degradation can cause negative impacts on the hydrosphere, atmosphere, lithosphere, and biosphere. Figure 1.1 shows, in a non-exhaustive way, examples of interactions between these systems, responsible for soil fertility and fundamental for the balance of the planet and, therefore, for humanity; hence, the importance of conservation and recovery of pedological covers.

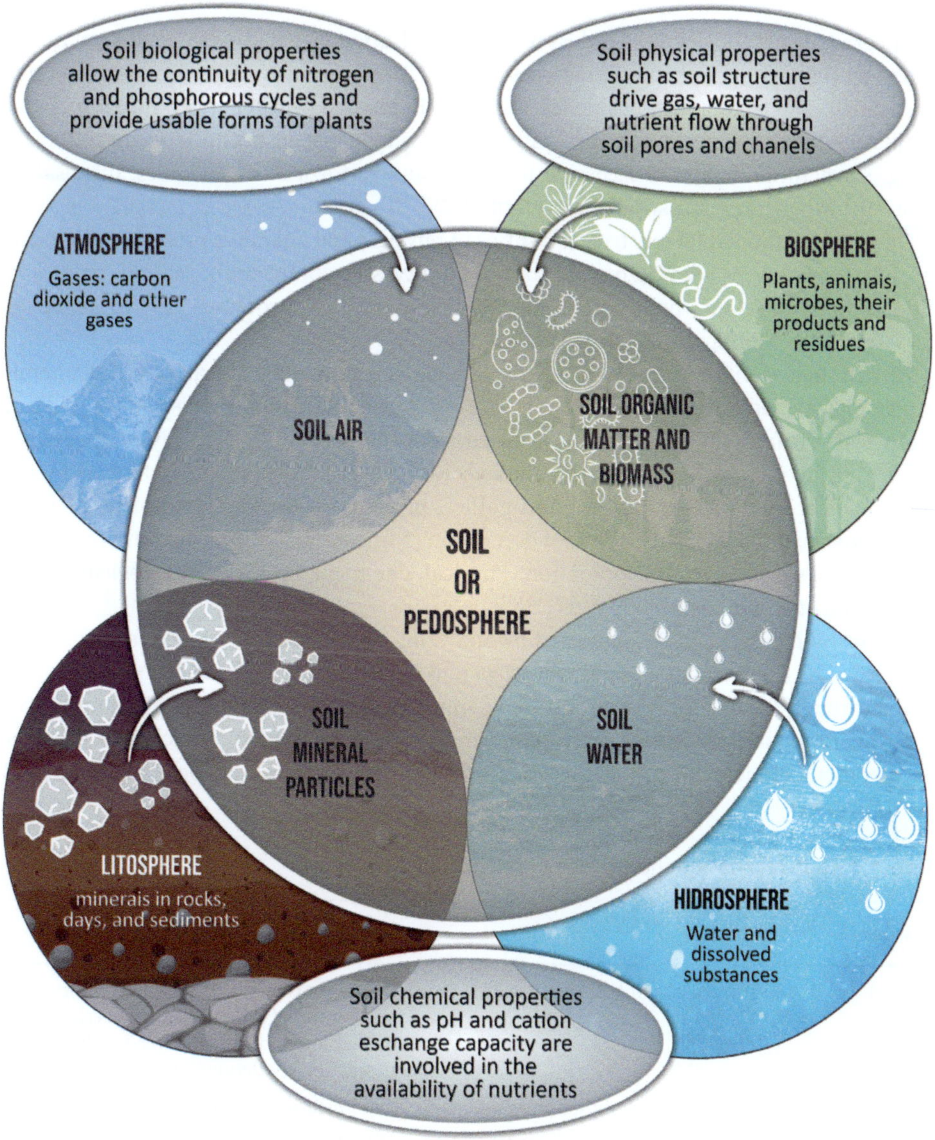

Figure 1.1 Interactions between terrestrial systems and soils responsible for their fertility. *Source:* Adapted from Weil and Brady (2017).

A more gradual change in the productive system associated with research for the recovery of degraded areas and for increasing productivity using green technologies seems to be a good path. Numerous studies have been carried out on the subject in recent decades and the results are promising, however, the application of green techniques on a large scale is still a challenge, as demonstrated by the work carried out by Prasad et al. (2021).

1.2 Soil Degradation: Start and Consequences

Soil degradation is the decrease or loss of its physical, chemical, and/or biological qualities, either by natural or anthropic processes, which affects its ecosystem functions (Nunes et al. 2020). The factors responsible for soil degradation can be categorized into direct and facilitating factors, such as torrential rains, strong winds, use of machinery, animal trampling, shortening of fallow, excessive irrigation and fertilization, pesticides, disposal of domestic, urban, and industrial waste (Table 1.2).

The aforementioned factors trigger a series of physical and chemical processes that favor erosion, depletion of organic matter, loss of biodiversity, compaction, sealing, point and diffuse contamination, pollution, and salinization. These processes are evidence of degraded soils (Montanarella 2007).

We can suggest different indicators for soil degradation, namely visual, physical, chemical, biological, and integrative (Ribeiro et al. 2009; Lal 2018), which in this text we will call ecological. Visual indicators can be obtained from field observation, analysis of satellite images, radar, or aerial photographs (Figures 1.2–1.4). Examples include changes in soil color, evidence of furrows, ravines and gullies, presence of weeds, monitoring of plant development, transport, and sediment deposition (Nunes et al. 2020), which represent the visual aspects arising from degradation. Physical, chemical, biological, ecological, or the integration of several factors (Figure 1.5).

Physical indicators can be measured by analyzing the disposition of solid soil fractions (Ribeiro et al. 2009), such as sand, silt, clay, and organic matter. These are mainly manifested through the decline of aggregation of solid fractions, plasma-skeletal disjunction of the soil, formation of crust, compaction, and increase in density, which hinder plant development, reduce infiltration, increase surface runoff, and favor erosion. It is important

Table 1.2 Natural and anthropic factors of soil degradation.

Factors	Natural	Anthropics
Direct	Torrential rains, strong winds	Use of machinery, animal trampling, shortening of fallow, excessive irrigation or inefficient drainage, over-fertilization, pesticides, inadequate waste disposal
Facilitators	Topography, granulometry, and chemical and mineralogical composition of the soil, vegetation cover, water regime	Deforestation, overgrazing, overuse of vegetation, cut slopes

Figure 1.2 Field evidence of soils in the process of degradation. (a) Shallow, stony soils with little vegetation cover in a semi-arid environment. Municipality of Gentio do Ouro, Bahia, Brazil. (b) Shallow soils, with loss of surface horizons and exposed due to suppression of vegetation. Semi-arid region, municipality of Gentio do Ouro, Bahia, Brazil.

Figure 1.3 Vegetation suppression and susceptibility to soil degradation. Municipality of Igaporã, Bahia, Brazil.

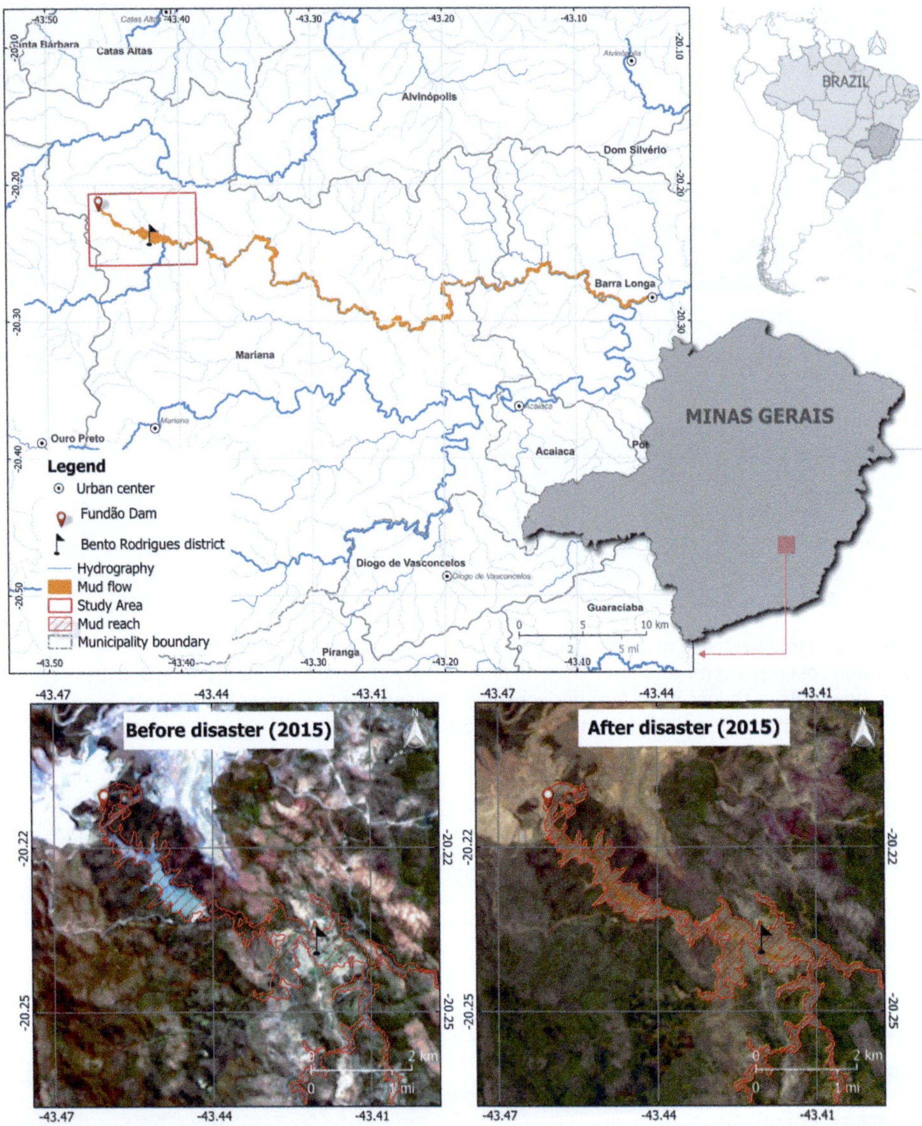

Figure 1.4 Satellite images used to ascertain the extent of the Mariana red mud disaster, Minas Gerais, Brazil. *Source:* Adapted from Nunes et al. (2022).

to emphasize that the physical degradation of the soil can be evidenced from the microscopic to the macroscopic scale (Figure 1.6) and ends up affecting the chemical, biological, ecological attributes, and vice versa, as suggested in Figure 1.5.

Chemical indicators can be measured by monitoring soil pH, salinity, organic matter content, cation, and anion exchange capacity, nutrient cycling and the presence of toxic or radioactive elements (Figure 1.7), while biological indicators can include measurements the presence or absence of macro and microorganisms, as well as their activities and by-products (Figure 1.8).

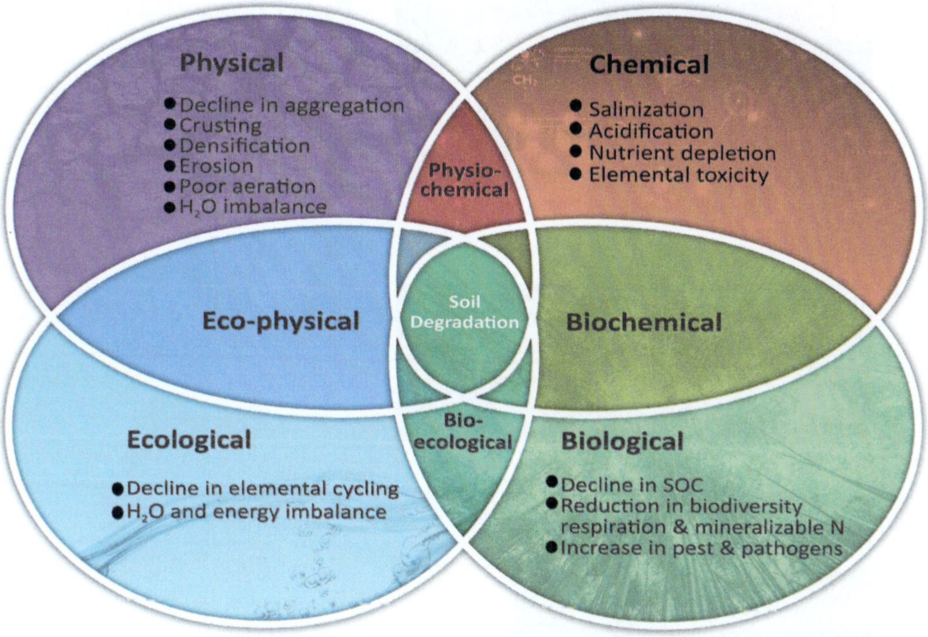

Figure 1.5 Processes and types of soil degradation. *Source:* Adapted from Lal (2018), Elsevier, CC BY-NC-ND 4.0.

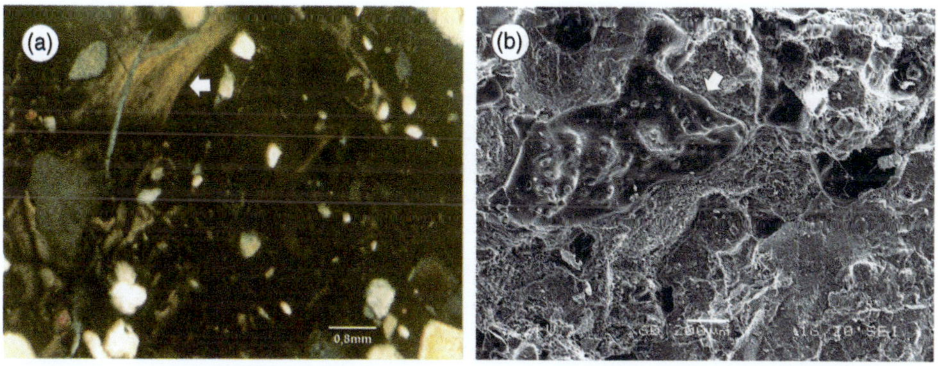

Figure 1.6 Photos using the Petrographic Microscope (A) and Scanning Electronic Microscope (SEM) of soil that is undergoing plasma-skeleton disjunction, making it more susceptible to erosion. Arrows indicate clay slurry (cutãs) in subsurface soil horizons. Photomicrograph (B) was adapted from Nunes et al. (2020) Caminhos de Geografia, CC BY-NC-ND 4.0.

The abusive use of chemical substances in agriculture, domestic, and industrial waste and plastics accumulated year after years in soils are worldwide problems. Chapman (1996) defines pollution as alterations that are capable of promoting the degradation of an ecosystem, through the removal or introduction, directly or indirectly, of substances resulting from human activities with deleterious effects on biological resources and risk to human health (Figure 1.9). Generally, in environmental pollution there is the presence of complex mixtures that trigger toxic effects different from those known for isolated

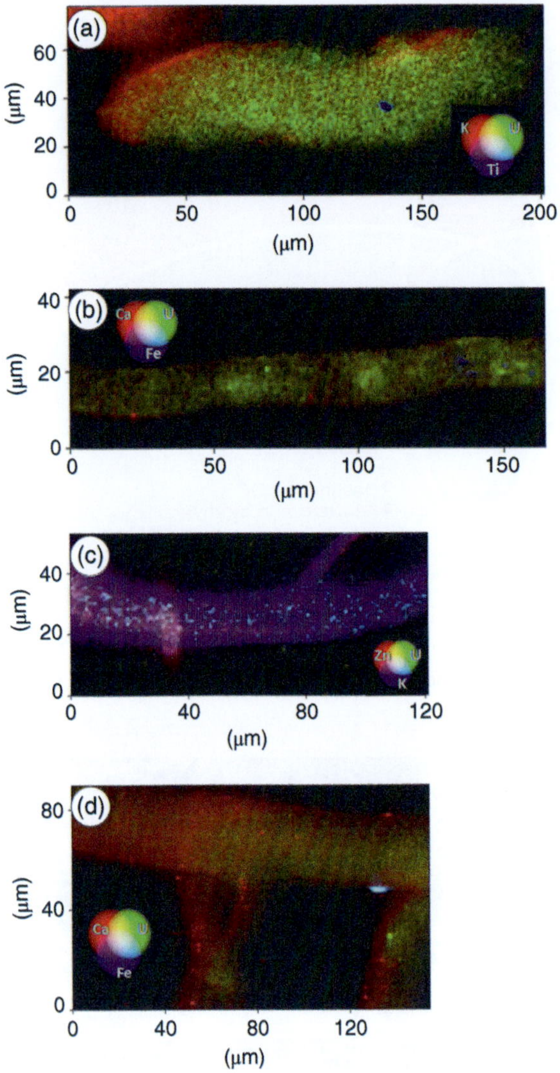

Figure 1.7 (a and b) Treatments with 100 and 200 ppm of U, respectively, without mycorrhiza. (c and d) Treatments with 100 and 200 ppm of U, respectively, with mycorrhiza. *Source:* Adapted from Alves et al. (2018).

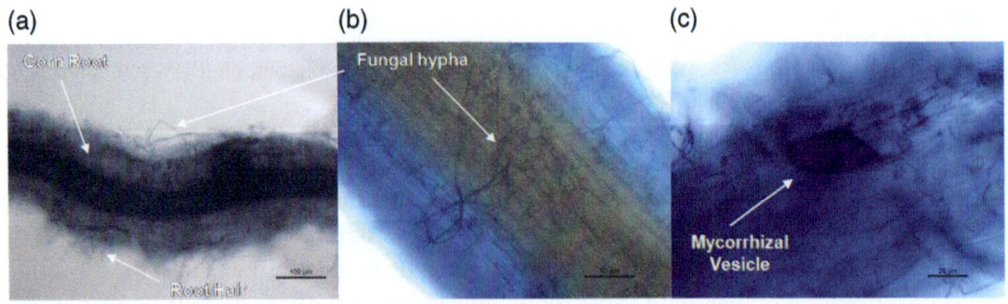

Figure 1.8 (a, b and c) Colonization of corn roots by arbuscular mycorrhizal fungi. *Source:* Adapted from Alves et al. (2018) UNIVERSIDADE ESTADUAL DE SANTA CRUZ.

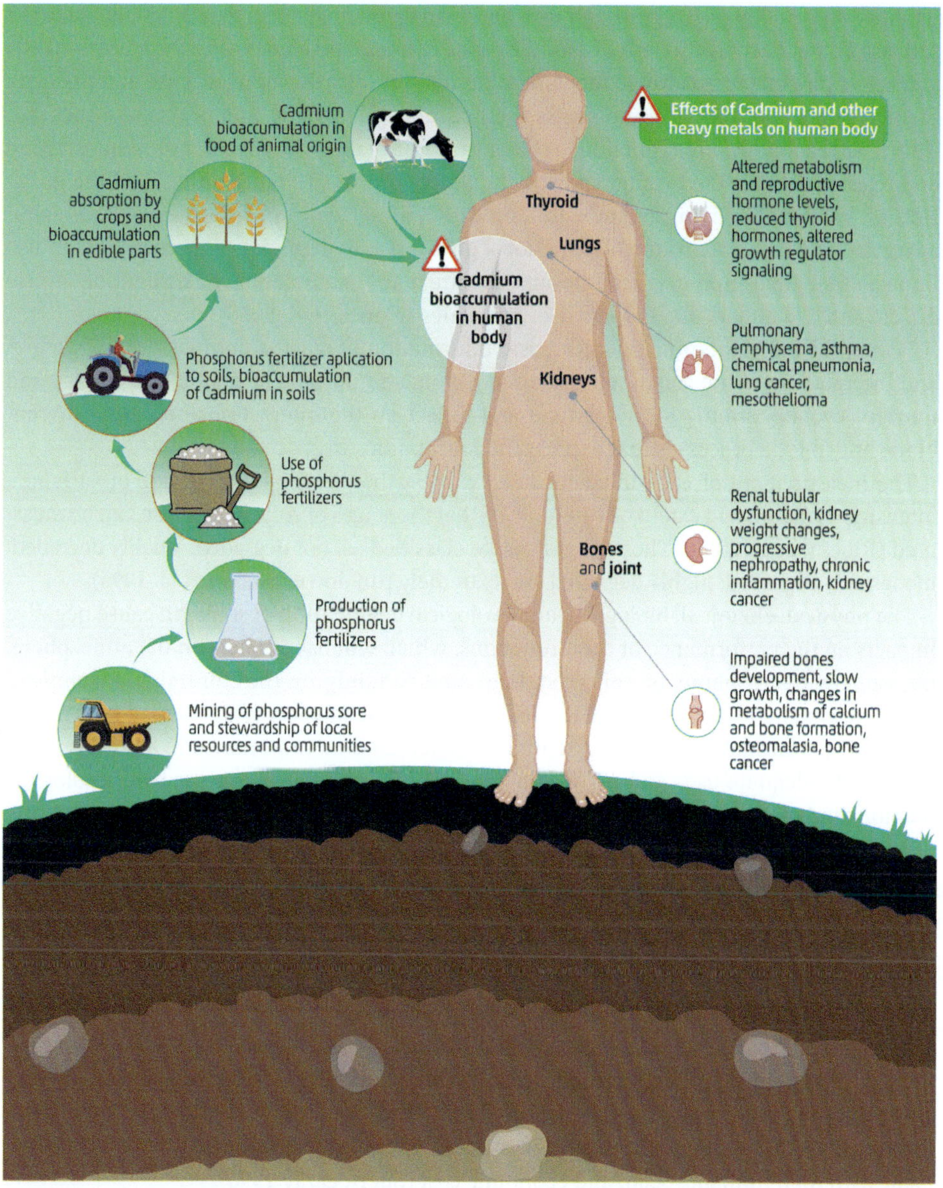

Figure 1.9 Life cycle of cadmium example in P fertilizers and soils. *Source:* Adapted from FAO and UNEP. 2021. Global assessment of soil pollution: Summary for policymakers. Rome, Italy, FAO. 84 pp. https://doi.org/10.4060/cb4827en.

substances. This phenomenon is called the "combined effect," as it can have infinite possibilities of effects with additive, antagonistic, and synergistic responses (Beyer et al. 2014).

Ascertaining the chemical and biological health of soils ensures nutritional and food security, as we must ensure that everyone has access to sufficient nutritious food free from organic and inorganic contaminants. It is necessary to guarantee the production of an adequate quantity and quality of food through uses and management systems that do not degrade the soil and do not harm the health of producers and consumers (Nunes et al. 2020; FAO 2018).

It is not possible to guarantee food and nutrition security through land use and management systems that mainly compromise their regulatory, support, and provision functions. This is especially true for functions related to primary productivity, carbon sequestration and regulation, nutrient cycling and supply, water control and purification, and the provision of habitats for living organisms. Conventional agriculture, for example, can cause adverse effects on important soil functions, giving rise to degradation processes that also impact other environments. In addition, there are many examples of contamination of producers and consumers due to the use of pesticides (Nunes et al. 2020).

Ecological indicators must gather basic information about the composition, structure, and interactions of the soil system, such as inputs and outputs of matter and energy and nutrient cycling, aiming to understand and reflect on the interactions between different biotic and abiotic processes that express transformations in space and in time.

The measurement of each indicator must express the direction, magnitude of variation, intensity, duration, and extent of variation, so that the stages of soil degradation can be measured (Ribeiro et al. 2009). These stages can be classified as: not degraded; weakly degraded; moderately degraded; highly degraded and extremely run down (Snakin et al. 1996).

The physical, chemical, biological, and ecological degradation of soils can cause negative impacts on the performance of their functions, which will have effects on the atmosphere, lithosphere, hydrosphere, biosphere systems, and certainly, on the cultural and technical-scientific systems (Nunes et al. 2021). Understanding the impacts of soil degradation on the functions performed by them is essential to understand its importance in environmental balance. Inadequate uses and management can progressively reduce the soil's ecological memory and make it more fragile. This, then, can contribute to increasing the susceptibility of the soil to degradation or reducing the capacity for regeneration, which would lead to environmental disasters, such as the effects caused by desertification (Nunes et al. 2020).

In arid, semi-arid and sub-humid regions, for example, the inappropriate use and exploitation of soils have been triggering changes over time in the various systems and leading to a strong environmental degradation called desertification. Desertification is a term that became widespread in the 1970s, especially from 1977 when the United Nations Environment Program held the Nairobi Conference in Kenya. In the 1990s, the discussion increased due to the tragic consequences that the phenomenon unleashed in some countries of the world, mainly in Africa, but it was from the United Nations Convention to Combat Desertification (1994) that desertification was recognized as a problem environment with high human, social, and economic cost (Suertegaray 1996).

Desertification is a notable example of the impacts of land degradation on climate and society. According to the United Nations Convention to Combat Desertification, desertification is the degradation of land in arid, semi-arid and sub-humid regions to drought resulting from various factors, ranging from natural causes such as climate change and variability, to activities such as overgrazing, deforestation, and agriculture (UE 2018).

Desertification results in socioeconomic deterioration through the progressive reduction of biomass, reduction of rainfall, increase in average temperature, soil infertility, intensification of erosion processes, reduction of the natural resilience of the land, reduction of water quality, decrease in food supply, increased malnutrition and hunger, economic stagnation, and rural exodus (EU 2018; Geist e Lambin 2004).

Figure 1.10 The Great Red Desert of Australia (Australian Outback). *Source:* Photo courtesy of Érica C. Nolasco.

It is estimated that each year around 20,000 km^2 of productive land will be destroyed by desertification (estimates from the United Nations Environmental Program–UNEP) and that the Sahara Desert has advanced, in some stretches, up to 100 km, which may also be happening in other deserts such as the Great Red Desert of Australia (Australian Outback), Atacama in Chile, and Uyuni in Bolivia (Figure 1.10). However, it is also important to emphasize the reduction of decertified areas in different areas, making the global assessment complex (Conti, 2002).

Soil degradation and soil sensitivity to desertification seem to have intensified in recent decades in many parts of the world. There are estimates suggesting a potential global increase in soil erosion driven by the expansion of agricultural land (Borrelli et al. 2017), which will produce heterogeneous spatial patterns determined by the interaction of factors such as climate, changes in land use, and human pressures. Increasing levels of soil degradation and soil sensitivity to desertification are reflected in increasingly complex (and non-linear) relationships between environmental and socioeconomic variables (Salvati et al. 2015).

1.3 Soil Protection, Conservation, and Recuperation Strategies

Due to the complexity and amplitude of soil degradation processes, before adopting protection, conservation, and recovery measures, it is necessary to identify, understand, and measure them (Salvati et al. 2015), using visual, physical, chemical, biological, and ecological variables, as presented in the previous item. Different authors have applied analytical strategies and statistical methodologies capable of approaching and quantifying the spatio-temporal evolution of degraded areas, among them Nicholson (2005), Salvati et al. (2015), and Nunes et al. (2022).

Nicholson (2005) used the vegetation index to study desertification in the Sahel between 1981 and 2005 and observed pulsating vegetation cover across seasons. The author emphasized that placing plant growth data side by side with rainfall data is a good method for assessing whether a productive area is becoming desert or not. It is important, however, to be cautious, as the use of vegetation can also represent shallow soils, rocky outcrops and fields that are no longer being cultivated. Fieldwork and using satellite images with higher spatial resolution are necessary for further clarification, but the vegetation index certainly helps to select places where studies should be conducted (Nicholson 2005; Nunes et al. 2022).

Salvati et al. (2015) used an approach in the study of soil degradation that considers integrated biophysical and socioeconomic aspects, called the complex adaptive systems approach (CAS). CAS simulate the non-linear relationships between its components, characterized by positive and negative feedback mechanisms. They are adaptive because system elements self-organize according to external and internal inputs that are both determinants and products of system function (Salvati and Zitti 2008). It is believed that due to these characteristics, CAS can better simulate the dynamics of complex systems such as the soil.

The study developed by Salvati et al. (2015) evaluated in detail changes in the level of soil degradation and land sensitivity to desertification for 773 agricultural districts in Italy during the period 1960–2010. The authors came to the conclusion that "fast" and "slow" factors were identified as underlying soil and land degradation according to the estimated speed of change for each of the indicators studied.

Different soil conservation techniques (SCT) have been used around the world to prevent soil degradation, which can be organized into three categories: (i) engineering techniques; (ii) management techniques; (iii) biological techniques. These techniques have been used especially for erosion control, including reforestation (Buendia et al. 2016), terraces (Chen et al. 2017; Wei et al. 2016), mulching (Liu et al. 2012; Prats et al. 2016; Prosdocimi et al. 2016a,b), and different soil management techniques, such as fallow and species consortium (Carpenter-Boggs et al. 2003; Keesstra et al. 2016); however, biological techniques seem to be the most effective.

Biological techniques (BTs) use living, dead, or sleeping beings. Organisms to reduce or avoid the "splash effect" (direct impact of rain on the soil) and surface runoff, incorporate organic matter and nutrients, improve soil structure, and thus reduce erosion. Xiong et al. (2018) compared BTs with soil management techniques (STs) and engineering techniques (ETs) and concluded that the former were more effective in reducing soil and water loss. We believe that BTs such as grass cover, reforestation, strips, green beans, hedges, hedge plants, exclusion, strip cutting, and mulching protect soil from erosion, maintain, recreate, and strengthen functional relationships between soil, vegetation, fungi, bacteria, and other living organisms, so they are more effective.

Xiong et al. (2018) and Chen et al. (2017) also emphasize the importance of SCTs and ETs as they reduce surface runoff (Buendia et al. 2016; Wang et al. 2017), improve soil fertility and productivity (Kagabo et al. 2013), increase crop yields (Borrelli et al. 2017), increase biodiversity (Buscardo et al. 2009) and improve water quality (Layman et al. 2014). However, it is important to note that they vary according to topography, climate and land use, whether biological, administration, or engineering. Therefore, predictive models of erosion, runoff, and productivity, as well as land use and conservation must take these variables into account. Soil conservation protection techniques mitigate or reduce the negative effects of soil degradation, whether of anthropic origin, and/or factors such as variability or climate change.

Conservation agriculture (CA) is an alternative that ensures sustainable food production and reduces adverse impacts on soil regulation, support, and provision functions, however, its large-scale use is still a challenge. Meeting the food and nutritional needs of the growing population of many countries and continents, such as Asia and Africa, will require, throughout this century, more investment in research to produce more food at low environmental cost.

Protection and conservation techniques are very important for maintaining the dynamic balance of soils. However, areas with highly degraded soils lack specific recovery techniques, among them we highlight bioremediation, as the technique can remedy the degraded area with inorganic and organic materials through natural biodegradation (Figure 1.11), unlike conventional methods that often introduce other elements and substances in the environment.

Faced with the need to use alternatives for the remediation of degraded soils and using greener technologies, especially those that present low cost, ease of execution and long-term effects, phytoremediation stands out (Alves, 2018; Alves et al. 2020; Nunes et al. 2022; Alves et al. 2022). Phytoremediation is an example of a low-cost, non-invasive approach to remediation of contaminated areas (Wang et al. 2017).

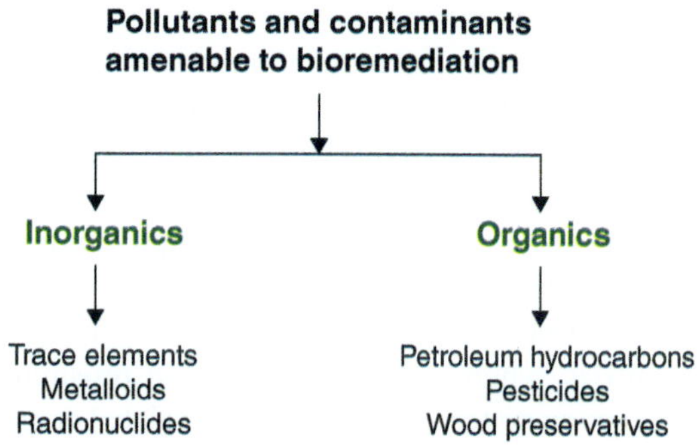

Figure 1.11 Pollutants and contaminants amenable to bioremediation.
Source: Adapted from Alves et al. (2020).

Table 1.3 Research in different fields of phytoremediation.

Phytoremediation technique	References
Plant growth-promoting bacteria (PGPB)	Wu et al. (2020)
Phytoaccumulation	He et al. (2020); Xu et al. (2020); Zhang et al. (2020)
Phytoextraction	Cao et al. (2020); Feng et al. (2019); Hu et al. (2019); Rabêlo et al. (2020); Zhou et al. (2020); Nunes et al. (2022)
Phytotoxicity alleviation	Kiran and Prasad (2019)
Phytofiltration	Favas et al. (2018); Nunes et al. (2022)
Phytostabilization	Alves et al. (2018); Favas et al. (2018); Prasad et al. (2016); Nunes et al. (2022)

Source: Prasad et al. (2021) / John Wiley & Sons.

Phytoremediation is the use of plants to extract, filter, degrade, stabilize, or reduce the mobility, bioavailability, or concentrations of pollutants. While positive publicity, habitat restoration, low secondary waste, and reduced cost are often advantages of phytoremediation over conventional methods, disadvantages can include a large timescale, limited root depth, plant vulnerability to toxins, variation in success with based on local conditions and transfer of contaminants into the food chain (Green and Hoffnagle 2004).

While phytoremediation techniques can have setbacks, many of them can be overcome with splicing applications or symbiotic microbes. A disadvantage may be the phytotoxicity of contaminants, which can arrest or reduce plant growth, however, this can be addressed by applying plant growth-promoting bacteria (Glick and Stearns 2011; Rajkumar et al. 2012; Rayu et al. 2012; Alves et al. 2020).

Bacterial-assisted phytoremediation is a rapidly growing field with applications for sodium, trace metals, organic contaminants, and phytopathogens. In addition, bacteria can be used directly to transform or degrade contaminants, while plants support these bacteria and help transfer them deeper into the soil (phytostimulation). In the table below we present the main processes related to phytoremediation (Table 1.3).

1.4 Challenges for the Twenty-First Century

One of the main challenges of the twenty-first century will be to produce safe food in adequate quantities to suppress international demand, degrading less the soil, and associated natural resources. To this end, we need to develop more research aimed at an even more productive green agriculture, in addition to investing in the recovery of degraded areas and their use. These purposes needed to apply engineering principles, which can be observed in the Table 1.4.

The effective implementation of these principles is a challenge, especially in developing countries, where resources devoted to research are scarce or insufficient. In addition, production systems designed to meet global demands for food go in the opposite direction

Table 1.4 Basic ecological principles of Ecological Engineering.

Principle	Description
Forcing functions determine the ecosystem	Ecosystem structure and functions are determined by the forcing functions of the system. This principle says that Ecological Engineering interventions should stimulate the system's forces to improve the ecosystem's performance, by strengthening its antifragility,[a] that is, improving its structure and functions
Relationships between biological function and chemical composition determine ecosystem homeostasis	Homeostasis of ecosystems requires accordance between biological function and chemical composition
Recycling management to reduce pollution	It is necessary, in environmental management to match recycling pathways and rates to ecosystems to reduce the effect of pollution.
Ecosystems are self-designing	The more one works with the self-designing ability of nature, the lower the costs of energy to maintain that system
Ecosystem-specific space-time scales	Processes of ecosystems have characteristic time and space scales that should be accounted for in environmental management
Chemical and biological diversity and self-designing ability of ecosystems	Chemical and biological diversity contribute to the spectrum of buffering capacities and the self-designing ability of ecosystems. A wide variety of chemical and biological components should be introduced or maintained for the ecosystem's self-designing ability to choose from. Thereby a wide spectrum of buffer capacities is available to meet the impacts from anthropogenic pollution.
Transition zones and agricultural management	The transition zones are as important for ecosystems as the membranes are for cells. Agricultural management should therefore consider the importance of the transition zones.
Coupling between ecosystems and management agricultural	The coupling between the ecosystems should be utilized to the benefit of the ecosystems in the application of ecotechnology and in environmental management agricultural systems.
Ecotechnology, environmental management, and components of the ecosystems	It is important that the application of ecotechnology and environmental management considers that the components of an ecosystem are interconnected, interrelated, and form a network, which implies that direct as well as indirect effects are of importance.
Ecosystem history	It is important to realize that an ecosystem has a history, in application of ecotechnology and environmental management in general.
Ecosystems and geographical edges	Ecosystems are most vulnerable at the geographical edges. Therefore, ecological management should take advantage of ecosystems and their biota in their optimal geographical range.
Ecosystem hierarchy	Ecosystems are hierarchical systems and all the components forming the various levels of the hierarchy make up a structure, that is important for the function of the ecosystem.

[a] Concept proposed and worked on by Nassin Nicholas Taleb in the book "Antifragile: Things That Gain from Disorder," Ed.: Random House, 2012. 544p. Antifragility is the opposite of fragility, a property of complex natural or social systems that benefits from stressors, volatility, and randomness. Intervening too much in natural systems that have survived over billions of years, depriving them of stressful, random, and volatile interactions, can weaken them and/or trigger unpredictable consequences, due to the multiplicative chains of unexpected effects.

Source: Nunes et al. (2021) / John Wiley & Sons.

of the principles of ecological engineering, which makes abrupt change difficult to carry out in a short time, unless one wants to condemn billions of people to hunger.

Agriculture is considered the most widely managed ecosystem on the planet, representing nearly a third of Earth's surface. Agricultural ecosystems provide food, fiber, fuel, and a host of other essential products to humans. However, for such services to be provided, agricultural ecosystems depend on a number of other supporting and regulatory services, such as adequate soil, pollinating agents, and water (MA 2005).

According to Nunes et al. (2021), considering that the services offered by natural or anthropic systems, such as agriculture, are indispensable for human survival, it can be considered, then, that there are so-called ecosystem disservices for agriculture, which are responsible for reducing productivity and increase in production costs (Zhang et al. 2007).

The idea that ecosystem processes must provide the products necessary for human survival is dangerous, since it provides an opportunity to understand that one can interfere unrestrictedly in processes that took thousands of years to exist and develop. We need to use natural resources, but trying to preserve the balance of natural processes, seeking to reach a middle ground, a way of producing that considers ecological processes as ecosystem services, excluding the idea of disservices to the agrosystem.

A large part of the ecosystem processes of agrosystems has the soil as a critical and dynamic regulatory center (Barrios 2007), being more than a substrate for the development of plants, contributing to all four dimensions of ecosystem services. The agrosystem is largely self-regulated by the high biodiversity of the soil, so soil conservation is the basic requirement for sustainable agriculture.

Regardless of the type of agriculture (conventional, ecological, conservationist), it is important to reflect that we will always be having some kind of impact on ecosystems. Our role is to reduce adverse environmental impacts as much as possible, developing techniques that respect the proper functioning of natural processes, in this sense, the principles advocated by agroecology can be great allies. Organic, conservation, biodynamic, ecological, syntropic agriculture, permaculture, and integrated crop-livestock-forest systems are examples of agricultural systems that use agroecological principles.

1.5 Final Considerations

Facing the challenge of preserving, conserving, recovering degraded soils and, in addition, producing healthy food in adequate amounts to meet the needs of the world population can be accomplished through the application of the principles of ecological engineering and agroecology. To do so, we need not only to invest in research on the subject, but also to engage in dialogue with professionals from different areas of knowledge and significantly transform our view of the world.

In addition, we need the knowledge produced to be correctly applied by the most direct users and by the professionals who support these users. The impact of the inadequate adoption of technologies by the producer in the degradation processes is still little discussed. The issue involves understanding their participation as subjects who, by implementing or intensifying practices, whose purpose in principle would be to conserve or increase soil productivity, end up promoting degradation due to the inadequate application of techniques.

This aspect is fundamental in the implementation of recovery programs, as it requires changes in the attitudes of individuals. These changes cross geographic and cultural barriers, the difficulty of decoding verbal language, the participation of technicians and producers in a management process that requires commitment or understanding of what is happening (Rifkin and Kangere 2002), as well as the low qualification of the professional directly related to the diffusion of technologies.

Modern science and technology in developing and underdeveloped countries was implemented in a way that provided training in diverse fields, but without creating an appreciation of science as an instrument for social and intellectual transformation, that is, many more elements considered useful for domination than for scientific culture (Schwartzman 2008).

We are facing a great challenge, but it is possible to overcome with dedication and humility!

References

Alves, L.A. (2018). Micorriza arbuscular como ferramenta biotecnológica para controle da transferência de urânio. *Tese de doutorado. UESC* 193.

Alves, A.G.C., Marques, J.G.W., Queiroz, S.B. et al. (2005). Caracterizac¸a˜o etnopedolo'gica de Planossolos utilizados em cera^mica artesanal no Agreste Paraibano. *Rev. Bras. Ciênc Solo* 29: 379–388. https://doi.org/10.1590/S0100-06832005000300008.

Alves, L.J., Nunes, F.C., Prasad, M.N.V. et al. (2018). Uranium mine waste phytostabilization with native plants – a case study from Brazil. In: *Bio-Geotechnologies for Mine Site Rehabilitation*, 1e, vol. 1 (ed. M.N.V. Prasad, P.J. de Campos Favase, and S.K. Maiti). (Org.), 299–322. Amsterdã: Elsevier.

Alves, L.J., Nunes, F.C., Santos, I.S. et al. (2020). Microbial approach for alleviation of potentially toxic elements in agricultural soils. In: *Climate Change and Soil Interactions: Managing Soil System for Environmental Sustainability*, 1e, vol. 1 (ed. M.N.V. Prasad and M. Pietrzykowski). (Org.), 271–303. Amsterdã: Elsevier.

Alves, L.J., Gross, E., Mangabeira, P.A.O. et al. (2022). Uranium transfer in grasses grown on mining waste and natural soil. *J. Environ. Radioact.* 251–252: 106973.

Amorim, R.S.S., Silva, D.D., da Pruski, F.F. et al. (2000). Influência da declividade do solo e da energia cinética de chuvas simuladas no processo de erosão entre sulcos. *Rev. bras. eng. agríc. ambient.* 5 (1): 124–130. https://doi.org/10.1590/S1415-43662001000100023.

Barrera-Bassols, N. and Zinck, J.A. (2003). Ethnopedology: a worldwide view on the soil knowledge of local people. *Geoderma* 111: 171–195. https://doi.org/10.1016/S0016-7061(02)00263-X.

Barrios, E. (2007). Soil biota, ecosystem services and land productivity. *Ecological Economics* 64: 269–285.

Beyer, J., Petersen, K., Song, Y. et al. (2014). Environmental risk assessment of combined effects in aquatic ecotoxicology: a discussion paper. *Mar. Environ. Res.* 96: 81–91.

Bordenave, J.E.D. (1994). *O que é participação*, 8e. São Paulo: Brasiliense. (Col. primeiros passos; 95).

Borrelli, P., Robinson, D.A., Fleischer, L.R. et al. (2017). An assessment of the global impact of 21st century land use change on soil erosion. *Nat. Commun.* 8. https://doi.org/10.1038/s41467-017-02142-7.

Buendia, C., Bussi, G., Tuset, J., et al. (2016). Effects of afforestation on runoff and sediment load in an upland Mediterranean catchment. *Sci. Total Environ.* 540: 144–157. https://doi.org/10.1016/j.scitotenv.2015.07.005.

Buscardo, E., Smith, G.F., Kelly, D.L. et al. (2009). The early effects of afforestation on biodiversity of grasslands in Ireland. *Plantation Forests and Biodiversity: Oxymoron or Opportunity? Topics in Biodiversity and Conservation*. Springer.

Cao, X., Luo, J., Wang, X. et al. (2020). Responses of soil bacterial community and Cd phytoextraction to a Sedum alfredii-oilseed rape (Brassica napus L. and Brassica juncea L.) intercropping system. *Sci. Total Environ.* 723: 138–152. https://doi.org/10.1016/j.scitotenv.2020.138152.

Carpenter-Boggs, L., Stahl, P.D., Lindstrom, M.J., and Schumacher, T.E. (2003). Soil microbial properties under permanent grass, conventional tillage, and no-till management in South Dakota. *Soil Tillage Res.* 71: 15–23. https://doi.org/10.1016/S0167-1987(02)00158-7.

Capra, G.F., Ganga, A., and Moore, A.F. (2017). Songs for our soils. How soil themes have been represented pop. Song. *Soil Sci. Plant Nutr.* 63: 517–525. https://doi.org/10.1080/00380768.2017.1369860.

Chapman, D. (1996). Water quality assessements. In: *A Guide to Use of Biota, Sediments and Water in Enviromental Monitoring*, 2e. London: E&FN Spon – Chapman & Hall.

Chen, D., Wei, W., and Chen, L. (2017). Effects of terracing practices on water erosion control in China: a meta-analysis. *Earth-Sci. Rev.* 173: 109–121. https://doi.org/10.1016/j.earscirev.2017.08.007.

Cherlet, M., Hutchinson, C., Reynolds, J. et al. (ed.) (2018). *World Atlas of Desertification*, 3e. Luxembourg: Publication Office of the European Union. Version online ISBN 978-92-79-75349-7. https://doi.org/10.2760/06292 https://wad.jrc.ec.europa.eu.

Comparato, F.K. (1998). *Fundamentos dos direitos humanos*, 27. São Paulo: LTr.

Conti, J.B. (2002). *As relações Sociedade/Natureza e os Impactos da Desertificação nos Trópicos*, Cadernos Geográficos. Universidade Federal de Santa Catarina, Florianópolis.

Dal Soglio, F.K. and Kubo, R.R. (Org.) (2016). *Desenvolvimento, agricultura e sustentabilidade*. Porto Alegre: Editora da UFRGS.

EU (2018). *Desertification in the EU*. European Court of Auditors, Luxembourg.

FAO (2015). Revised World Soil Charter. Presented at the Global Soil Partnership – World Soil Charter, Rome.

FAO (2018). *Building Climate Resilience for Food Security and Nutrition. The State of Food Security and Nutrition in the World*. Rome: FAO.

FAO (2021). *The State of the World's Land and Water Resources for Food and Agriculture – Systems at Breaking Point*. Synthesis Report 2021. Rome. https://doi.org/10.4060/cb7654en.

Favas, P.J.C., Martino, L.E., and Prasad, M.N.V. (2018). Abandoned mine land reclamation—challenges and opportunities (Holistic approach). In: *Bio-Geotechnologies for Mine Site Rehabilitation* (ed. M.N.V. Prasad, P.J. Favas, C. de, S.K. Maiti), 331. Elsevier. https://doi.org/10.1016/B978-0-12-812986-9.00001-4.

Feng, Y., Wang, Q., Meng, Q. et al. (2019). Chromosome doubling of Sedum alfredii Hance: a novel approach for improving phytoremediation efficiency. *J. Environ. Sci.* 86: 87–96. https://doi.org/10.1016/j.jes.2019.05.016.

Geist, H.J. and Lambin, E.F. (2004). Dynamic causal patterns of desertification. *BioScience* 54: 817. https://doi.org/10.1641/0006-3568(2004)054[0817:DCPOD]2.0.CO;2.

Ghaley, B., Rusu, T., Sande'n, T. et al. (2018). Assessment of benefits of conservation agriculture on soil functions in arable production systems in Europe. *Sustainability* 10: 794. https://doi.org/10.3390/su10030794.

Green, C. and Hoffnagle, A. (2004). *Phytoremediation Field Studies Database for Chlorinated Solvents, Pesticides, Explosives, and Metals*. Washington, DC: U.S. Environmental Protection Agency.

Glick, B.R. and Stearns, J.C. (2011). Making phytoremediation work better: maximizing a Plant's growth potential in the midst of adversity. *Int. J. Phytoremediation* 13: 4–16. https://doi.org/10.1080/15226514.2011.568533.

Hartemink, A.E., Balks, M.R., Chen, Z.-S. et al. (2014). The joy of teaching soil science. *Geoderma* 217–218: 1–9. https://doi.org/10.1016/j.geoderma.2013.10.016.

Hayhoe, D., MacIntyre, J., and Bullock, S. (2016). Soil science education for grades K_12 in Canada. *Can. J. Soil Sci.* 96: 336–345. https://doi.org/10.1139/cjss-2015-0079.

He, S., Wang, X., Wu, X. et al. (2020). Using rice as a remediating plant to deplete bioavailable arsenic from paddy soils. *Environ. Int.* 141: 105799. https://doi.org/10.1016/j.envint.2020.105799.

Hu, P., Zhang, Y., Dong, B. et al. (2019). Assessment of phytoextraction using Sedum plumbizincicola and rice production in Cd-polluted acid paddy soils of South China: a field study. *Agric. Ecosyst. Environ.* 286: 106651. https://doi.org/10.1016/j.agee.2019.106651.

Inda, A.V., Fink, J.R., and dos Santos, T.F. (2018). Pedogenic iron oxides in soils of the acre state, Brazil. *Ciênc Rural.* 48. https://doi.org/10.1590/0103-8478cr20180209.

Kagabo, D.M., Stroosnijder, L., Visser, S.M., and Moore, D. (2013). Soil erosion, soil fertility and crop yield on slow-forming terraces in the highlands of Buberuka, Rwanda. *Soil Tillage Res.* 128: 23–29. https://doi.org/10.1016/j.still.2012.11.002.

Keesstra, S., Pereira, P., Novara, A. et al. (2016). Effects of soil management techniques on soil water erosion in apricot orchards. *Sci. Total Environ.* 551–552: 357–366. https://doi.org/10.1016/j.scitotenv.2016.01.182.

Kiran, B.R. and Prasad, M.N.V. (2019). Biochar and rice husk ash assisted phytoremediation potentials of Ricinus communis L. for lead-spiked soils. *Ecotoxicol. Environ. Saf.* 183: 109574. https://doi.org/10.1016/j.ecoenv.2019.109574.

Ladeira, F.S.B. (2010). Solos do passado: origem e identificac¸a˜o. *Rev. Bras. Ciênc. Solo* 34: 1773–1786. https://doi.org/10.1590/S0100-06832010000600001.

Lal, R. (2018). Sustainable intensification of China's agroecosystems by conservation agriculture. *Int. Soil Water Conserv. Res.* 6: 1–12. https://doi.org/10.1016/j.iswcr.2017.11.001.

Layman, C.A., Jud, Z.R., Archer, S.K., and Riera, D. (2014). Provision of ecosystem services by human-made structures in a highly impacted estuary. *Environ. Res. Lett.* 9: 044009. https://doi.org/10.1088/1748-9326/9/4/044009.

van Leeuwen, J.P., Saby, N.P.A., Jones, A. et al. (2017). Gap assessment in current soil monitoring networks across Europe for measuring soil functions. *Environ. Res. Lett.* 12: 124007. https://doi.org/10.1088/1748-9326/aa9c5c.

Ling, L.L., Schneider, T., Peoples, A.J. et al. (2015). A new antibiotic kills pathogens without detectable resistance. *Nature* 517: 455–459. https://doi.org/10.1038/nature14098.

Liu, Y., Tao, Y., Wan, K.Y. et al. (2012). Runoff and nutrient losses in citrus orchards on sloping land subjected to different surface mulching practices in the Danjiangkou reservoir area of China. *Agric. Water Manag.* 110: 34–40. https://doi.org/10.1016/j.agwat.2012.03.011.

Minami, K. (2009). Soil and humanity: culture, civilization, livelihood, and health. *Soil Sci. Plant Nutr.* 55: 603–615. https://doi.org/10.1111/j.1747-0765.2009.00401.x.

Montanarella, L. (2007). Trends in land degradation in Europe. In: *Climate and Land Degradation* (ed. M.V.K. Sivakumar and N. Ndiang'ui), 83104. Springer, Berlin, Heidelberg. https://doi.org/10.1007/978-3-540-72438-4_5.

Nicholson, S. (2005). On the question of the "recovery" of the rains in the West African Sahel. *J. Arid. Environ.* 63: 615–641. https://doi.org/10.1016/j.jaridenv.2005.03.004.

Nunes, F.C., Alves, L.J., Carvalho, C.C.N. et al. (2020). Soil as a complex ecological system for meeting food and nutritional security. In: *Climate Change and Soil Interaction*, 1e, vol. 1 (ed. M.N.V. Prasad and M. Pietrzykowski). (Org.), 229–269. Amsterdã: Elsevier.

Nunes, F.C., Soares, T.M., Alves, L.J. et al. (2021). Ecological engineering and ecosystem services – theory and practice. In: *Handbook of Ecological and Ecosytem EngIneering*, 1e, vol. 1 (ed. M.N.V. Prasad). (Org.), 24. Wiley.

Nunes, F.C., Alves, L.J., Prasad, M.N.V. et al. (2022). Phytoremediation strategies for rehabilitation of soils affected by red mud: the Mariana tailing dam collapse (Minas Gerais, Brazil). *Eurasian Soil Sci.* 1: 1–13.

Oldeman, L.R., Hakkeling, R.T.A., and Sombroek, W.G. (1991). *World Map of the Status of Human-Induced Soil Degradation: An Explanatory Note*, 34. Wageningen, The Netherlands: ISRIC/UNEP.

Prasad, M.V.N., Alves, L.J., and Nunes, F.C. (2021). Global remediation industry and trends. In: *Handbook of Assisted and Amendment-Enhanced Sustainable Remediation Technology*, 1e, vol. 1 (ed. M.N.V. Prasad). (Org.), 3–28. Wiley.

Rabêlo, F.H.S., Borgo, L., Merloti, L.F. et al. (2020). Effects of winter and summer conditions on Cd fractionation and bioavailability, bacterial communities and Cd phytoextraction potential of Brachiariadecumbens and Panicum maximum grown in a tropical soil. *Sci. Total Environ.* 728: 138885. https://doi.org/10.1016/j.scitotenv.2020.138885.

Rajkumar, M., Sandhya, S., Prasad, M.N.V., and Freitas, H. (2012). Perspectives of plantas sociated microbes in heavy metal phytoremediation. *Biotechnol. Adv.* 30: 1562–1574. https://doi.org/10.1016/j.biotechadv.2012.04.011.

Rayu, S., Karpouzas, D.G., and Singh, B.K. (2012). Emerging technologies in bioremediation: constraints and opportunities. *Biodegradation* 23: 917–926. https://doi.org/10.1007/s10532-012-9576-3.

Ribeiro, M.R., Sampaio, E., and Galindo, I.C.L. (2009). Os solos e o processo de desertificação no semiárido brasileiro. Tópicos em Ciência do Solo. *Sociedade Brasileira de Ciência do Solo* 6: 497.

Rifkin, S.B. and Kangere, M. (2002). *What Is Participation? Chapter 3. CBR a Participatory Strategy in Africa: Gladnet Collection Year Community-Based Rehabilitation (CBR)*, 37–49. Cornell University. https://asksource.info/cbr-book/cbr03.pdf.

Sales, M.C.L. (2002). Evolução dos estudos de desertificação no nordeste brasileiro. *GEOUSP Espaço e Tempo* (11): 115–126.

Salvati, L. and Zitti, M. (2008). Regional convergence of environmental variables: empirical evidences from land degradation. *Ecol. Econ.* 68: 162–168. https://doi.org/10.1016/j.ecolecon.2008.02.018.

Salvati, L., Mavrakis, A., Colantoni, A. et al. (2015). Complex adaptive systems, soil degradation and land sensitivity to desertification: amultivariate assessment of Italian agro-forest landscape. *Sci. Total Environ.* 521–522: 235–245. https://doi.org/10.1016/j.scitotenv.2015.03.094.

Schaefer, V.H. (2011). Remembering our roots: a possible connection between loss of ecological memory, alien invasions, and ecological restoration. *Urban Ecosyst.* 14: 35–44. https://doi.org/10.1007/s11252-010-0138-3.

Schulte, R.P.O., Bampa, F., Bardy, M. et al. (2015). Making the most of our land: managing soil functions from local to continental scale. *Front. Environ. Sci.* 3. https://doi.org/10.3389/fenvs.2015.00081.

Schwartzman, S. (2008). *Ciência, universidade e ideologia: a política do conhecimento [online]*, 58–78. Rio de Janeiro: Centro Edelstein. Universidade, ciência e subdesenvolvimento. ISBN 978- 85-99662-50-2. Available from SciELO Books.

Sindelar, M. (2015). *Soils Support Recreation*. Soil Science Society of America: Soils SustainLife. https://www.soils.org/files/sssa/iys/june-soils-overview.pdf.

Snakin, V.V., Krechetov, P.P., Kuzovnikova, T.A. et al. (1996). The system of assessment of soil degradation. *Soil Technol.* 8: 331–343. https://doi.org/10.1016/0933-3630(95)00028-3.

Suertegaray, D.M.A. (1996, 1996). Desertificação: recuperação e desenvolvimento sustentável. In: *Geomorfologia e meio ambiente* (ed. A.T.E. Guerra and S.B. Cunha). (Org.), 249–297. Rio de Janeiro. Ed Bertrand Brasil, Capitulo V il: Desertificação: recuperação e desenvolvimento sustentável.

Volchko, Y., Norrman, J., Rose'n, L., and Norberg, T. (2014). SF box—a tool for evaluating the effects on soil functions in remediation projects: the SF box tool for soil function assessment. *Integr. Environ. Assess. Manag.* 10: 566–575. https://doi.org/10.1002/ieam.1552.

Wang, J., Peng, J., Zhao, M. et al. (2017). Significant trade-off for the impact of grain-for-green programme on ecosystem services in North-western Yunnan. *China. Sci. Total Environ.* 574: 57–64. https://doi.org/10.1016/j.scitotenv.2016.09.026.

Wei, W., Chen, D., Wang, L. et al. (2016). Global synthesis of the classifications, distributions, benefits and issues of terracing. *Earth-Sci. Rev.* 159: 388–403. https://doi.org/10.1016/j.earscirev.2016.06.010.

Weil, R.R. and Brady, N.C. (2017). *The Nature and Properties of Soils*, (global edition). Harlow: Pearson 1.104.

Wu, Y., Ma, L., Liu, Q. et al. (2020). Pseudomonas fluorescens accelerates a reverse and long-distance transport of cadmium and sucrose in the hyper accumulator plant sedum *alfredii. Chemosphere* 256: 127156. https://doi.org/10.1016/j.chemosphere.2020.127156.

WWF (2020). *Living Planet Report 2020 – Bending the Curve of Biodiversity Loss* (ed. R.E.A. Almond, M. Grooten, and T. Petersen), 57. Gland: WWF.

Xiong, M., Sun, R., and Chen, L. (2018). Effects of soil conservation techniques on water erosion control: a global analysis. *Sci. Total Environ.* 645: 753–760. https://doi.org/10.1016/j.scitotenv.2018.07.124.

Xu, W., Xiang, P., Liu, X., and Ma, L.Q. (2020). Closely-related species of hyperaccumulating plants and their ability in accumulation of As, Cd, Cu, Mn, Ni, *Pb and Zn. Chemosphere* 251: 126334. https://doi.org/10.1016/j.chemosphere.2020.126334

Zhang, W., Ricketts, T.H., Kremen, C. et al. (2007). Ecosystem services and dis-services to agriculture. *Ecological Economics* 64: 253–260.

Zhang, J., Zhang, M., Song, H. et al. (2020). A novel plasma membrane-based NRAMP transporter contributes to Cd and Zn hyper accumulation in Sedum alfredii Hance. *Environ. Exp. Bot.* 176: 104121. https://doi.org/10.1016/j.envexpbot.2020.104121.

Zhou, J., Li, Z., Zhou, T. et al. (2020). Aluminum toxicity decreases the phytoextraction capability by cadmium/zinc hyperaccumulator Sedum plumbizincicola in acid soils. *Sci. Total Environ.* 711: 134591. https://doi.org/10.1016/j.scitotenv.2019.134591.

2

Degradation of Agriculture Systems by Invasive Alien Plants and Agroecological Approaches for Sustainable Restoration

Prabhat Kumar Rai

Department of Environmental Science, School of Earth Sciences and Natural Resources Management, Mizoram University, Aizawl, India

2.1 Introduction

Biotic invasions have long been recognized as a significant threat to agrobiodiversity, socio-economy/livelihood, and agroecosystem resilience (Mack et al. 2000). In agriculture/agroforestry systems, invasive alien plants (IAPs) deploy multiple ecological mechanisms (e.g., novel weapon (NW)/ resource hypothesis (RH)) to result in their degradation (Rai 2015). Further, IAPs adversely influence the abiotic soil physical-chemical attributes and biotic (e.g., crop-pollinators niche overlap and microbial diversity) components of agroecosystems (Sollins et al. 2007; Raizada et al. 2008; Stefanowicz et al. 2018) (Figure 2.1). Allelopathic effects of several IAPs (e.g., *Parthenium hysterophorus, Ageratum conyzoildes*, and *Chromolaena odorata*) reduced seed germination and growth of food crops, thereby result in the decline of agricultural productivity. The detrimental effects of IAPs on agro/-forestry ecosystems received wide attention and still in need of affordable agroecological solutions (Tobin 2018). Therefore, IAPs are also considered to have the potential to cause significant impact on global agriculture, thereby hampering the global food security (Cook et al. 2011; Fleming et al. 2018). Increasing frequency of emerging IAPs introductions in novel landscapes adversely influence the crops yield, pasture output, and agroecosystem productivity (Seebens et al. 2017). Multiple effects of IAPs can adversely influence soil-crop interface in agriculture and agroforestry systems to result in enormous loss to agri-buisness and livelihood (Fried et al. 2017).

IAPs are a potential threat to sustainable agriculture with an estimated yearly global loss of over 40 billion dollars (Monaco et al. 2002). Due to IAPs competition with food crops for nutrient resources the annual crop loss can be up to 34% in global agroecosystems (Oerke 2006). Despite the efficiency of chemical pesticides to reduce the overall potential crop production losses of 50% to actual losses of about 30%, crop loss induced by pest still vary from 14% to 35% (Oerke 2006). In particular, heavy reliance of agriculture sector on synthetic chemical pesticides for crop protection against potential pests has been

Agroecological Approaches for Sustainable Soil Management, First Edition.
Edited by Majeti Narasimha Vara Prasad and Chitranjan Kumar.

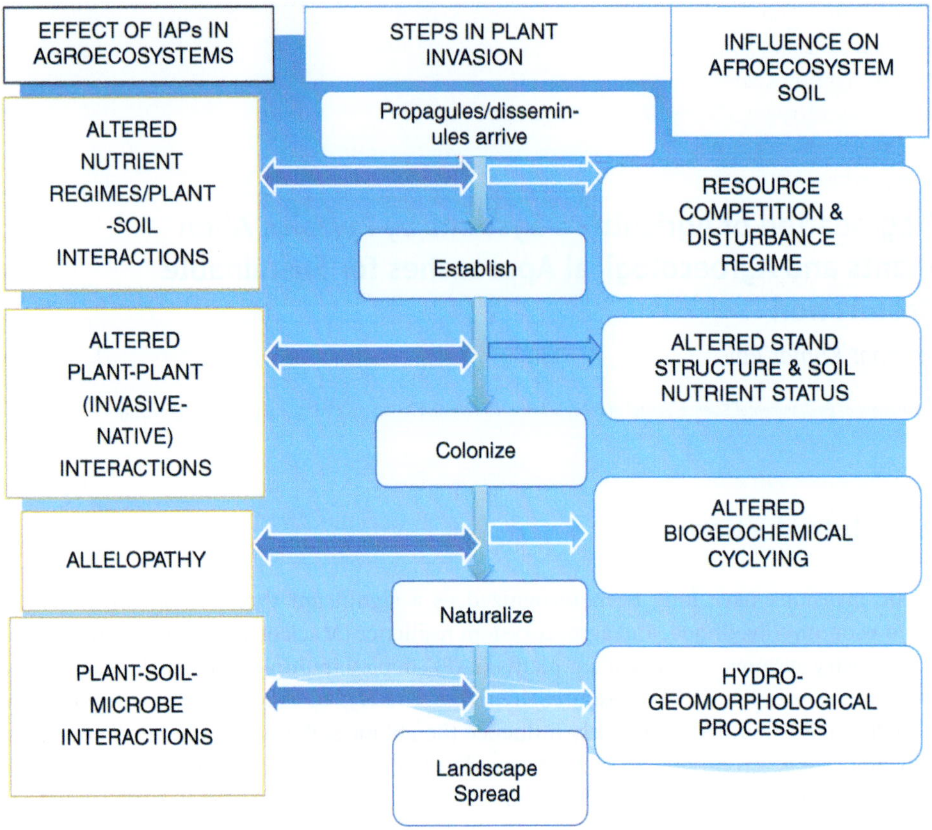

Figure 2.1 Steps associated with the spread of IAPs in agroecosystems and their effects on soil and crops.

recognized as potential source for water pollution (IFEN 2007) and biodiversity loss (Stoate et al. 2001). Thus, contemporary agriculture now faces conflicting challenges due to the urgent need of increasing crop production while simultaneously minimizing the adverse environmental impacts of modern intensive agriculture (Gaba et al. 2014).

Climate change, as a major driver of IAP propagules/disseminules dispersal, has a significant impact on their spread thereby influencing their infestation in agroecosystems (Taylor and Kumar 2013; Fandohan et al. 2015). Therefore, infestation of IAPs is exacerbated by their interactions with climate change which can have an imminent impact on productivity of agricultural and agroforestry systems (Arora 2019). Accordingly, several factors of climate change, such as increased temperature, atmospheric carbon dioxide (CO_2) levels, nitrogen (N) deposition, extreme weather conditions, and changes in precipitation, are expected to accelerate the spread, diversity, and distribution of IAPs (Jia et al. 2016; Ziska et al. 2011). In comparison to native plants at the same sites, several plant invaders have inherent traits like larger specific leaf areas, faster growth rates, and higher leaf nutrient levels, all of which have the ability to accelerate decomposition and bears tight linkage with biogeochemical cycles (Allison and Vitousek 2004) and tends to increase nitrification and mineralization of nitrogen which drastically alters ecosystem properties (Liao et al. 2008; Ehrenfeld 2004; Ehrenfeld et al. 2001). In this context, Rout and Callaway (2009) hypothesized differential IAPs effect which on one hand boost net primary productivity and nitrogen (N) cycling in invaded regions but not in their native regions.

Present chapter aims to document the multiple effects of IAPs on agroecosystems which can further influence the food security, livelihood, and human well-being. Also, the present chapter envisages the role of nature-based solutions through the quest of agroecological principles which can facilitate the sustainable restoration of degraded ecosystems. According to agroecology principles, by explicit understanding of ecological processes, agroecosystems can be manipulated to improve production efficiency and agricultural sustainability, with fewer negative environmental or social impacts and lower use of external inputs for weed management (Gliessman 2006; Altieri 1989; Altieri and Rosset 1995; Wezel et al. 2009; Wezel and Jauneau 2011). The major goal of agroecology principles in IAPs or weed control is to manage vegetation on land and water bodies in such a way that it encourages the growth of native edible plants that are beneficial to humans while suppressing the growth of undesired plants (Rana and Rana 2016). Therefore, the wise application of agroecological principles need to be prioritized for sustainable agriculture.

2.1.1 Effects of IAPs on Soil Attributes and Microbial Diversity of Agroecosystems

The infestation of IAPs can influence agroecosystems, crop growth, and metabolism. Several IAPs like *Amaranthus spinosus* competes for nutrients with neighboring crops in agroecosystems (Tiwari et al. 2005). Invasive plants alter the ecology of a habitat by altering the nitrogen cycle and soil pH (Drenovsky and Batten 2007). In relation to agroecosystems, the IAPs can influence several soil attributes such as soil organic carbon (SOC) and the soil organic matter (SOM), water holding capacity (WHC), nutrient status, organic matter, intensity of fine particles, and microbial diversity (Ravit et al. 2003; Edrisi et al. 2020; Shiferaw et al. 2021; El-Barougy et al. 2021; Kumar et al. 2021). Table 2.1 shows the effects of IAPs on agroecosystems in terms soil nutrients and microbial diversity which can exert profound effects on crop yield and food security.

For 11 germinations and cultivated seedlings of crops, *Argemone maxicana* has been found to exhibit harmful allelopathic effects. Soil fertility is reduced by *Cassia occidentalis* and *Bidens pilosa*, while soil productivity is reduced by *Xanthium strumarium*, which is widespread in wheat fields (Tiwari et al. 2005). The *X. strumarium* pigment xanthinosin has a considerable impact on the seedling development of *Lactuca sativa* and *Amaranthus mangostanus* (Shao et al. 2012). Guillerme et al. (2020) found that the establishment and development of a number of IAPs in the valley (particularly *Rickettsia japonica*, *Impatiens glandulifera*, *Buddleja davidii*, and *S. japonica*) has contributed to landscape alterations and is considered as an environmental impact of agriculture's demise. Figure 2.2 shows the effect of IAPs in modulating the soil-plant-microbe interactions which exert significant effects on agroecosystems.

Plants' enormous diversity of secondary metabolites and close interactions of agroecosystems with bio-synthesized allelo-chemicals possess the ability to alter soil substrates (Weidenhamer and Callaway 2010). In soils with inadequate nutrient accessibility and toxic metals, roots can produce and release secondary biochemicals that enhance availability of nutrients and reduce the bioavailability of toxic elements (Li et al. 2007). Application of

Table 2.1 Effects of invasive alien plants (IAPs) on agroecosystems in terms soil nutrients and microbial diversity.

S.No.	Invasive alien plant	Invaded region or landscape	Effects (structural and functional)	References
			Soil nutrients	
1.	*Prosopis juliflora*	Lucknow, India	Influence soil physical-chemical attributes and elevated soil nutrient levels compared to un-invaded sites; Enforced Groundwater depletion	Edrisi et al. (2020); Shiferaw et al. (2021)
2.	*Leucaena leucocephala, Lantana camara, Mikania mikrantha, and Chromolaena odorata*	Hailakandi District, Assam, North East India	Decline in soil organic carbon (SOC) and the soil organic matter (SOM) content at invaded or disturbed site	Sakachep and Rai (2021a,b)
3.	*Eucalyptus sp.*	Central Himalaya	IAP induced decline in proportion of finer particles, Water holding capacity, Organic matter, and nutrient concentration	El-Barougy et al. (2021)
			Microbial diversity	
4.	*Bromus tectorum* (cheatgrass)	National Park in South-eastern Utah	Increase in species richness of fungi and abundance of active bacteria; decrease in specialist pathogenic fungi; shifts in composition of AMF and bacteria	Kumar et al. (2021)
5.	*Phragmites australis* (common reed)	Brackish marshes, New Jersey	Significant shift in microbial communities at disturbed or IAP invaded sites	Ravit et al. (2003); Rai (2022)

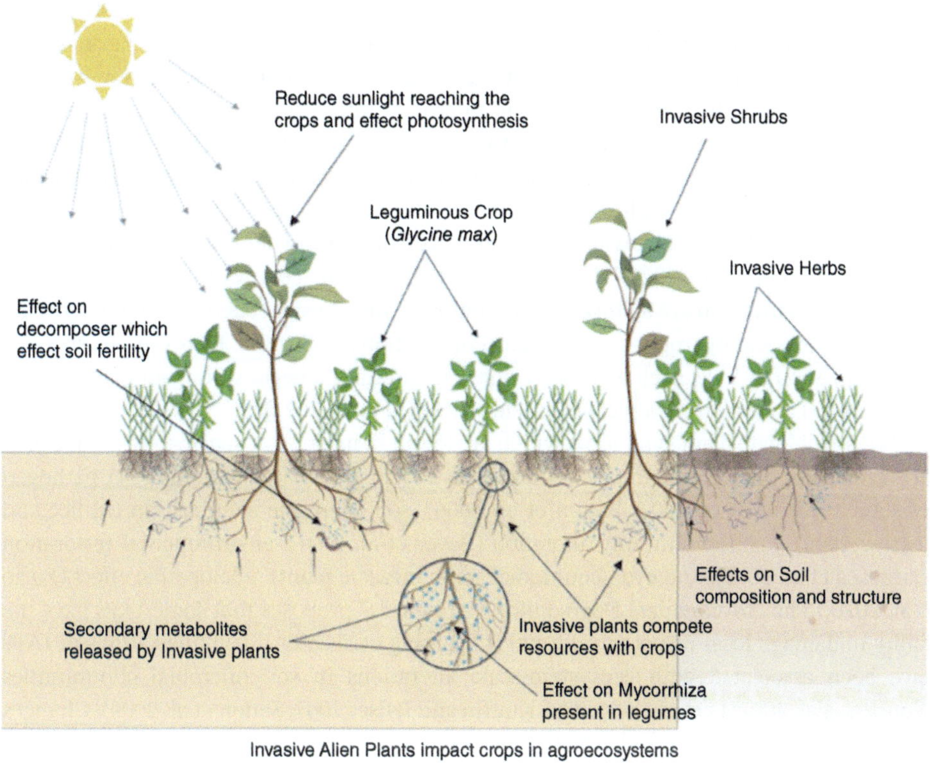

Figure 2.2 Invasive alien plants influence the above ground crop biomass and below ground microbial interactions in agroecosystems.

synthetic herbicides such as glyphosate to control IAPs potentially modify soil chemistry indirectly through the used them, (Baylis 2000; Borrgaard and Gimsing 2008; Duke and Powles 2008). Herbicides can have an indirect impact on plant ecosystems by modifying soil microbial communities, as well as plant chemistry and microorganism interactions, especially with mycorrhizal fungi and nitrogen-fixing bacteria (Weidenhamer and Callaway 2010). Therefore, the side effects of chemical methods in managing the effects of IAPs paved the way for nature based sustainable solutions.

Microbial communities are considered to operate as soil health indicators, which can be a crucial component in restoring a degraded agro-ecosystem (Li et al. 2015). Plant (IAPs)-soil-microbe interactions can be influenced by microbial diversity by enhancing vegetation establishment, organic matter, nutrient mobilization efficiency, and soil aggregation (Eviner and Hawkes 2008). IAPs induced alterations in soil attributes owing to invasive plant growth have been documented in past research (Callaway et al. 2004a,b; Elgersma and Ehrenfeld 2011; Kulmatiski and Beard 2011; Bozzolo and Lipson 2013; Lekberg et al. 2013; Lankau and Lankau 2014). The research on elucidating the role of plant (IAPs)-microbe interactions in success of plant invaders has been of immense interest in agroecosystems perspective (Chapman and Newman 2010; Dybzinski et al. 2008). IAPs tend to alter soil biological communities, which could lead to a positive feedback loop that favors exotic plants over the natives (Callaway et al. 2004a,b). For example, Hahn (2003) noted that *Zostera japonica* transformed the microbial community composition and hastened the rate of decomposition, resulting in greater carbon and nutrient exchange. Several IAPs like *I. glandulifera* also impacts plant-soil microbial interactions that have a significant influence on soil biogeochemistry and native plant community dynamics (Qu et al. 2021). Also, another IAP *Alliaria petiolata*, can inhibit mutual interaction between Arbuscular mycorrhizal fungi (AMF) with native plants and the emitted secondary metabolites which can remarkably influence belowground microbial diversity (Van der Putten and Klironomos Wardle 2007; Qu et al. 2021). The arrival, establishment, colonization, and landscape spread of IAPs invasion are all linked to alteration in soil physico-chemical characteristics, nutrient cycles, and hydrogeomorphological interactions (Rai 2022). For example, successful colonization of IAPs such as *Rhus typhina*, might have a negative impact on soil physico-chemical and microbiological characteristics (Qu et al. 2021).

AMF are obligate plant symbionts which are found in most plant species and are particularly crucial for phosphorus uptake (Batten et al. 2006; Lankau and Lankau 2014), pathogen resistance (Klironomos 2002; Batten et al. 2006), soil aggregate stabilization (Miller and Jastrow 2000), plant community alteration (Bever et al. 2001), environmental restoration (Caravaca et al. 2003), and even counteract some invasive plants' allelopathic effect (Barto et al. 2010). The "Mycorrhizal Degradation Hypothesis" proposes that some IAPs have the ability to damage local mycorrhizal fungi (Vogelsang et al. 2004). The infestations of IAPs have been associated with ecosystem-scale alterations in soil microbial communities (Kourtev et al. 2002; Li et al. 2006; Kao-Kniffin and Balser 2007; Sanon et al. 2009). Changes in morphology, phenology, and growth rate, as well as IAPs impact on the soil environment by altering the floristic composition and functional groups present, influenced the net primary productivity, nutrient cycling, and ecosystem potential to acquire carbon (C) (Ehrenfeld 2003; Reinhart and Callaway 2006; Vila et al. 2011). Alteration of ecosystem C-acquisition can change the decomposition rates, accumulation, and nutrient from organic matter, as well as rhizosphere interactions, which can affect soil microbial communities, nutrient cycling, and plant community interactions (Bohlen 2006). The phenomena of "fungal shifts" in response to IAPs is less understood which could have significant ramifications for sustaining biodiversity and ecological function in invaded habitats (Pickett et al. 2018).

Invasive pathogens or pests have been demonstrated to devastate food crops which can exert severe threat to sustainable agriculture (Desprez-Loustau et al. 2007). Invading pathogens like the *Phytophthora infestans* that inflict a destructive agricultural epidemic like the well-known Irish potato famine are those that coevolved with crop plants and reconnected with their host in a complex and intricate way (Desprez-Loustau et al. 2007). Paini et al. (2016) defined "invasion threat as the product of arrival likelihood (i.e., the chances of a particular pest or pathogen arriving in a new location) and establishment likelihood (i.e., the chances of a particular pest or pathogen establishing in a new location once it has arrived)". Local AMF degradation can alter the soil in ways that benefit invasive plants while hindering native plants (Pickett et al. 2018). Changes in the physical qualities of the soil due to increased distribution and diversity of IAPs can potentially influence the nutrient dynamics (Ehrenfeld 2001).

Several agroforestry systems depend significantly on alien plant taxa especially those that rely on planting trees or near treeless regions, which can cause issues when they spread from their points of introduction and culture to regions where they are not wanted

(Richardson et al. 2004). Exotic trees have been planted in agroforestry systems globally for a variety of reasons, including lumber, fodder, erosion control, shading, and esthetics, and some of them are now classified as invasive (Richardson and Rejmánek 2011). The relation between agroforestry and IAPs is rarely acknowledged, despite the fact that invasive plant expansion is a major concern, not just ecologically, but also economically and socially (Guillerme et al. 2020; Rai and Ruati 2021; Ruati and Rai 2022). IAPs can potentially influence vegetation dynamics, biogeochemical cycles, and hydrology and can result in loss of agro-biodiversity (Sankaran 2017). Rapid growth, efficient spreading abilities, enormous reproductive outputs, and adaptability to a wide range of environmental circumstances facilitate the spread of IAPs in agricultural and agroforestry systems (Campbell 2005).

2.2 Agroecological Solutions

The extent of IAPs induced agroecosystem degradation and secondary pollution cause by chemical fertilizers or pesticides pave the way to quest nature based agroecological solutions (Khadija et al. 2019). In the present section the agroecological methods to attain the food security and agricultural sustainability are discussed as listed below.

2.2.1 Physical Weed Control Methods

The oldest practice for controlling weeds is manual weeding, which is still popular in India and has been proven effective by numerous studies (Rao et al. 2014; Shrestha et al. 2004). In this physical method, initially IAPs or weeds were first pulled out by hand and get eliminated, but later, hand tools were created to gain some mechanical advantage in performing this process (Sridhar 2013). With the harnessing of animal power for farming, a number of implements were designed and used for carrying out weeding operations. Currently, many machine operated weeders are available and are being used by farmers of agrarian countries like India (GOI 2012). In this sense, Raffaelli et al. (2005) revealed that all the hoeing systems effectively controlled weeds and increased maize yield as compared to harrowing method.

2.2.2 Cultural Control Method

Cultural weed control refers to any technique that involves maintaining the ideal field conditions to minimize IAPs establishment and/or infestation in agroecosystems. Planting a crop is a sure way to reduce growth because the crop interferes with the weeds (Rana and Rana 2016). In an agroecosystem, weed prevention is preferable over eradication because the former is more expensive, labor intensive, and calls for more resources (Kawakami et al. 2012). Using clean seeds, properly composting waste, and maintaining the cleanliness of farm equipment helps stop weed seeds from spreading in the field and it is best to prevent animals from moving from weed-infested areas to weed-free fields, especially when the weeds are established (Panetta and Roger 2005). Wise utilization of cultural techniques can improve a crop's ability to compete with weeds may also help minimize the load of herbicides on the soil (Matloob et al. 2020). Examples of cultural weed control would be crop rotation, optimum date of sowing, plant density, planting pattern/crop architecture, selection of quick growing varieties, avoiding overgrazing of pastures or rangeland, using well-adapted competitive forage species, stale seed bed or dab system, proper water management, method of fertilizer application, and maintaining good soil fertility (Rana and Rana 2016).

2.2.3 Stale Seed Bed

This technique involves irrigating the area 10–15 days prior to seeding to promote the germination of weed seeds lying on the soil surface which are then subsequently eliminated with the use of non-selective herbicides like glyphosate and ammonium glufosinate. The primary advantage of employing a stale seedbed is that the crop emerges weed-free, providing it a competitive advantage against late-emerging weed seedlings. This strategy minimizes subsequent weed emergence by emptying the seed bank from the soil surface (Singh et al. 2016). The practice of false seedbed technique may decrease weed infestation in agroecosystems by 80% or more when compared to standard seedbed preparation (Van Der Weide et al. 2002). The stale seedbed technique is widely used in many countries to manage weedy rice and awnless barnyard grass in rainfed rice (Fischer 1996). Stale seedbeds have been reported to reduce weed populations in direct-seeded rice (Rao et al. 2007) and may be especially effective when combined with sustainable agronomy practices (Chauhan et al. 2006). Pittelkow et al. (2012) reported that stale seedbed practice was effective at reducing the population of sedges and grasses, but not for controlling redstem weeds. This practice is very effective in case of specific wheat variant in the north-western Indo-Gangetic Plains (Mahajan et al. 1999).

2.2.4 Cover Cropping

Cover crops are non–cash crops can be grown before or together with a cash crop with the purpose of keeping the soil covered with vegetation as long as possible throughout the year (Melander et al. 2005). Cover crops offer a plethora of beneficial effects to the agroecosystem such as conservation of soil fertility and moisture, activation of soil nutrient dynamics, regulation of pest, pathogen, and weed populations. Therefore, cover crops represent an important agro-ecological technique for agroecosystems. Effects of cover crop are weed specific (Ba'rberi and Mazzoncini 2001). The introduction of fibrous organic debris from *Polygonum hydropiper*, *Oxalis corymbosa*, and *O. repensa* has been suggested as a way to slow the growth of dreaded IAP i.e. *Mikania micrantha* (Gogoi 2001).

2.2.5 Intercropping

Intercropping involves growing a smother crop between rows of the main crop such that the competition for water or nutrients does not occur. Intercropping of short-duration, quick-growing and early-maturing legume crops with long-duration and wide-spaced crops leads to quick ground cover, with higher total weed suppressing ability than sole cropping. This technique enhances weed control by increasing shade and crop competition. Like cover crops, intercrops increase the ecological diversity in a field. In addition, they often compete better with weeds for light, water, and nutrients. When compared to leek and celery sole crops, Baumann et al. (2000) found that a leek–celery intercrop sown in a row-by-row replacement design reduced relative soil cover of weeds by 41% and reduced common ragwort (*Senecio vulgaris* L.) density and biomass by 58% and 98%, respectively, owing to reduced light transmittance through the canopy. Success of intercropping relies on the best match between the requirements of the component species for light, water, and nutrients, which increases resource use. Fast-growing legumes like cowpeas and beans cover the soil early in the growing season, reducing the impact of rainfall before cotton or maize build a canopy to protect the soil. This method dramatically minimizes agricultural land degradation agents, particularly soil erosion, and aids in the regeneration of the land (Saturday 2021).

2.2.6 Crop Rotation

Crop rotation is the process of cultivating a variety of unrelated crops in the same area over the course of several seasons. It has several advantages, including preventing the disease and pest buildup that results from consistently producing one kind of crop. Crop rotation traditionally includes replenishing nitrogen through the use of green manure and legumes planted after cereals and other crops. Crop rotation can also help the soil's structure and fertility by alternating shallow- and deep-rooted plants. Crop rotation ensures soil fertility restoration, a decrease in diseases and pests, the addition of humus, and erosion prevention (Saturday 2021). A study conducted by Bationo and Ntare (2000) in Niger found out that legume–millet rotation with 30 kg Nha maintained a high level in SOC pool and sustained crop yields.

Crop rotation is an essential tool for ecological weed management (Liebman and Gallandt 1997). It challenges weeds by providing a broad range of stress and mortality factors associated with different crops grown in the rotation (Liebman and Staver 2001). Under various tillage regimes, including zero, conventional, and deep ploughing, Shahzad et al. (2016) found that IAPs and weed growth was inhibited in the sorghum-wheat rotation. Due to the fact that parasitic weeds are host-specific plants, crop rotation is also essential for managing them (Rubiales and Fernandez-Aparicio 2012).

2.2.7 Crop Selection

Additional cover crops may be needed if there is a too-long delay between harvesting one crop and planting the following one. Cover crops strengthen the conservation agriculture system's stability, improved soil qualities lessen the effects of erosion, and the agro-biodiversity ecosystems is encouraged for its potential benefits. Perennial plants outperform annual ones in terms of combating soil erosion. Sugar cane, fodder grasses, sweet potatoes, and tea are the most productive (Saturday 2021).

2.2.8 Cover Cropping

Cover crops are "close-growing crops that provide soil protection, seeding protection, and soil improvement between periods of normal crop production or between trees in orchards and vines in vineyards" (Saxton and Rawls 2006). These are also referred to as green manure crops. The use of cover crops is an ancient practice and dates back to the ancient civilizations in Greece, Rome, China, and others (Lamb et al. 2005). The practice of growing crops to cover the soil surface to reduce wind and water erosion is called cover cropping. This practice creates a favorable habitat for microorganisms by regulating the soil heat and temperature. These also sources of organic matter in soil as the fallen are decomposed (Saturday 2018). Leguminous plants called "cover crops" are typically grown to ensure permanent soil cover, add organic matter to the soil, and fix atmospheric nitrogen. Even in heavily populated places, where long-term fallows are invariably impossible, these aid in reversing soil deterioration. In some parts of Africa, the use of Mucuna spp. as a cover crop has improved soil organic matter, increased nitrogen availability in soils, and boosted yields (FAO 2015). Figure 2.3 represents restoration strategy of IAPs infested agriculture and forestry systems with the agroecological approaches and role of agroforestry plants in such nature-based solutions. Henceforth, the plantation of screened native or agroforestry species can be beneficial in restoration of degraded agroecosystem, associated services, and food security (Chazdon 2008; Rai 2022).

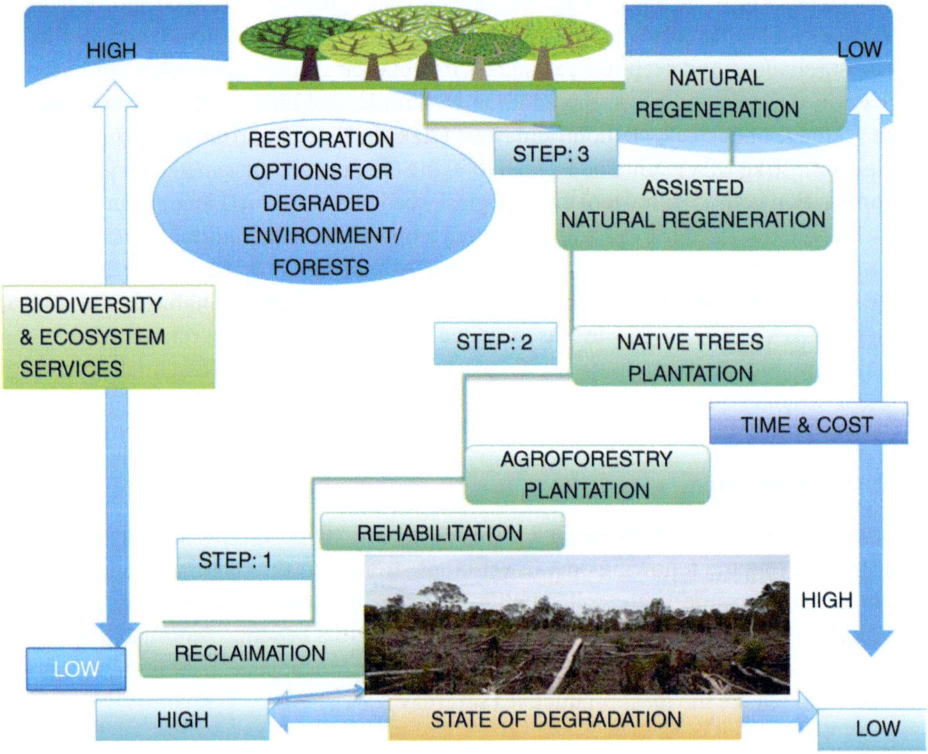

Figure 2.3 Restoration of IAPs infested agriculture and forestry systems with the agroecological approaches and role of agroforestry plants in nature-based solutions. *Source:* Modified and redrawn after Chazdon 2008.

2.3 Biological Control Methods

Biological weed management is a method of reducing weed population, growth, reproduction, and impact on crops by using live organisms or their products (Charudattan 2005; Gage and Schwartz-Lazaro 2019). It is a strategy that uses weed plants' natural enemies (e.g., insects, disease organisms, herbivorous fish, other animals, and competitive plants) to prevent weed seeds from germinating or spreading of established plants (Rana and Rana 2016). Biocontrol agents are environmentally friendly, cost-effective, energy-efficient, biodegradable, and leave no hazardous residues (Goeden and Andres 1999). Depending on whether the control agent is bacterial, fungal, insect, or other, there are two approaches to biological weed control i.e., classical biological control and inundative biocontrol (Harding 2015). Classical control relies upon the introduction of a pathogen, herbivore, or predator of the target pest. For classical biocontrol to be effective, the environment must be conducive to the biocontrol agent's survival, population growth, and metabolism (Harding and Raizada 2015; Hokkanen and Sailer 1985). Inundative biocontrol might be more suited for agricultural systems, such as bioherbicides, which can be used in a manner similar to chemical herbicides, where insects, fungal spores, bacterial preparations, or inoculum may be used (Harding and Raizada 2015).

2.4 Classical or Inoculative Biological Control

The classical or inoculative strategy involves introducing and establishing a naturally hostile organism in a new region to suppress the plant invaders in agroecosystems (Scheepens et al. 2001). The biocontrol agent is applied to the newly invaded area after being harvested from the dynamically balanced population of the weed in its native zone (Charudattan and Dinoor 2000). Classical control agents maintain their population balance with the host population once they are discharged into the environment (Abbas 2017; Abbas et al. 2018). Classical biocontrol is most effective when agents are introduced to control a plant species that is usually unrelated to significant indigenous plant species since the host range of these agents is often large. Conservation biocontrol techniques which involve promoting the environmental and growth conditions of the biocontrol agent can help protect and improve populations of these biocontrol agents in nature or field conditions (Ghorbani et al. 2005). Classical biological weed control using natural enemies has sparked interest as an economical and environmentally responsible weed control method (Rao et al. 2020). In this sense, *Zygogramma bicolorata* (a leaf-feeding beetle), *Epiblema strenuana* (a stem-boring moth), and *Puccinia abrupta* var. partheniicola and *P. melampodii* f. sp. *Parthenium hysterophorae* (rust fungi) were introduced in India as the potential biocontrol agents of IAPs in agroecosystems (Sankaran 2017). Different microorganisms including *Puccinia spegazzinii* de Toni., *Exserohilum monoceras*, *P. abrupta* Diet, and fungal pathogen ITCC-4896 of *P. abrupta* have been reported to possess herbicidal potential against several IAPs such as *M. micrantha* Kunth., *Echinochloa crusgalli*, *P. hysterophorus*, and *Lantana camara* (Thomas et al. 1999; Saxena and Kumar 2007; Kadir et al. 2008).

2.4.1 Inundative or Augmentative Biological Control

The underlying principle of this approach is the periodic application of high populations of natural antagonists of weeds that are already present in the area. These biocontrol agents are also called bioherbicides. Inundative agents are expected to immediately suppress target weeds and are faster in mode of action and efficacy when compared with the classical biocontrol agents (Dagno et al. 2012). The important characteristics of these biocontrol agents include high virulence against target plants, establishment of high populations in host plants, easy to produce and store, suitable formulation, and safety of non-target species (Auld et al. 2003).

Microbes may be used as bioherbicides in an inundative approach to biocontrol. To date, there is no microbial herbicide registered for use in agricultural crops, although there are products available for invasive weeds in natural systems (Watson 2018). Bioherbicides may be limited in their efficacy, providing only short-term residual effects, but may provide an additional selection pressure in an Integrated Weed Management system (Wolfe et al. 2016). For example, bacterial species *Xanthomonas campestris* (strain LVA-987), may provide a new management option for horseweed (*Conyza canadensis* (L.) Cronquist) (Boyette and Hoagland 2015). Most bioherbicides tested and marketed in the United States are species of fungi from three genera: *Colleto trichum* (e.g. BioMal and Collego), *Phoma,* and *Sclerotinia* (Harding and Raizada 2015). Bioherbicides are less hazardous to the environment than chemical herbicides because the plant pathogens used in them usually occur naturally in the locations where they are applied (Rana and Rana 2016). In addition to biological control using bacteria and fungi, viruses may also be bioherbicidal agents. In this respect, the Tobacco Mild Green Mosaic Tobamovirus has been approved by the Environmental Protection Agency (EPA) for control of tropical soda apple (*Solanum viarum* Dunal) in pastures (Harding and Raizada 2015).

2.5 Allelopathy in Agroecosystems

Allelopathy is the process by which other invasive plants and microbes produce phytotoxic compounds into the environment, thereby inhibiting the growth of native plants or crops in agroecosystems (Nongtri et al. 2022; Torres et al. 1996; Syngkli et al. 2022 a,b). The production and release of phytotoxic chemicals by higher plants and microorganisms plays a part in their defense against other coexisting species as well as their competitiveness with them (Mattner 2006). Allelopathic effects of several IAPs (e.g., *P. hysterophorus*, *Ageratum conyzoildes,* and *C. odorata*) reduced seed germination and growth of food crops, thereby causing decline in agricultural productivity (Rai and Singh 2020). Nevertheless, allelopathy has grown as tremendous field of interest for use in biological weed management. Allelochemicals produced by these allelopathic plants have been demonstrated to interfere with other plants' physiological functions such as respiration, photosynthesis, enzyme activity, balance of water and hormones, growth, and germination (Farooq and Cheema 2013). Allelochemicals have various chemical structures and operate on numerous crops which limits their specificity as noted in case of herbicides (Nimbal et al. 1996). Allelochemic compounds emitting plants can be cultivated as intercrops, allowing for simultaneous production and weed-suppression benefits (Makoi and Ndakidemi 2012). To exploit wise use of both weed control and cover crop benefits, some plants with allelopathic potential like *Brassica* sp. may be preferred over others (Altieri et al. 2011). Rye waste contains considerable amounts of the allelopathic substances such as isothiocyanate benzyl and isothiocyanate allyl which leach away and subsequently prevent the germination of small-seeded weeds in the agroecosystems (Singh et al. 2016). The common weed management strategies therefore include use of allelopathic mulches in agroecosystems (Abbas et al. 2017a), use of allelopathic cover crops, use of aqueous crop extracts, and residues from allelopathic plants (Ali et al. 2013; Safdar et al. 2016; Abbas et al. 2017b), crop rotation, use of allelopathic cultivars (developed through conventional breeding or biotechnology), intercropping of allelopathic crops with non-allelopathic crops (Farooq et al. 2011; Jabran et al. 2015).

2.6 Restoration and Carbon Sequestration Approaches in Agro/Ecosystem/Forestry Systems

Modern intensive agriculture and traditional unregulated agricultural practices such as shifting cultivation facilitated the IAPs infestation, causing multiple effects, eventually impeding the SDGs (Figure 2.4). Degradation of soil can be driven by biological, chemical, or physical factors which can be remarkably modulated through infestation of IAPs in

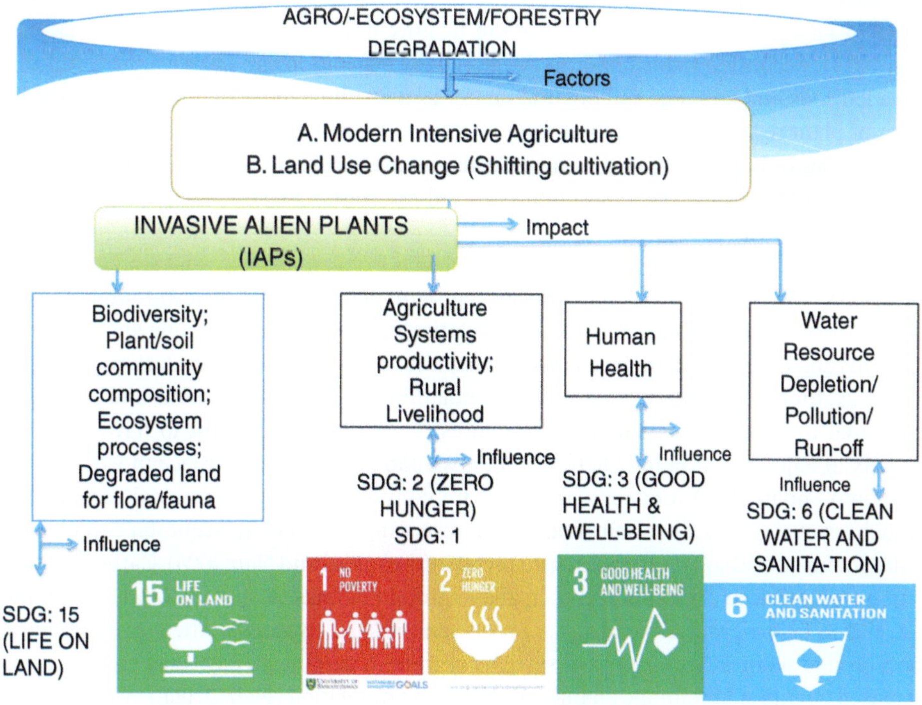

Figure 2.4 Effects of modern intensive agriculture on spread of IAPs, agroecosystem resilience, human well-being, and UN-Sustainable Development Goals (SDGs).

agroecosystems (Rai 2009, 2012, 2013, 2015) (Figure 2.1). Approximately 33% of the world's soils are currently moderately to severely degraded and in this perspective the depletion of soil quality in agroecosystems due to infestation of IAPs are of wide concern (Rai and Singh 2021). The global food insecurity and poverty with 40% of them being found in Africa. The close link between soil health and food security necessitates swift, targeted action, particularly at the local level, to reverse soil degradation, enhance food production, and reduce food insecurity in areas where it is most needed, all while taking climate action mechanisms (FAO 2015). The process of restoration is a component of the environmental strategy that can occasionally help to minimize the soil degradation caused by mining operations. Revegetation is therefore viewed as a long-term, economical, and environmentally acceptable plant-based method of restoring deteriorated soils (Rai and Kim 2020). To initiate restoration measures in degraded ecosystems, the selection of plant species is crucial (Rai 2022; Ait Elallem et al. 2021).

The agroecological method is initiated to revive soil life in order to revive and improve the many biological processes rooted in the soil. Since soil organic matter is thought to be the most prevalent shortfall in degraded soils and is the primary indicator for soil quality, this urges for raising and encouraging frequent soil organic matter monitoring and management efforts to enrich soil quality and sustain a plethora of ecosystem processes (FAO 2015). The installation of plant cover generally has a significant role in enhancing the physicochemical characteristics of the soil. Through the infestation of broad root systems, the potential bioagents helps in maintaining the integrity of soil structure (Daynes et al. 2013). The cohesiveness of soil particles is enhanced by the plant root systems of screened bioagents, which also enables the healthy development of aggregates (Daynes et al. 2013). According to Muñoz-Rojas et al. (2016), vegetation can improve the fertility of degraded soils and enhance soil organic matter as a result of root exudates and the breakdown of plant residues. In totality the application of biogents decrease in the bulk density of the soil and result in increase in the porosity, permeability, and water retention capacity of the soil. Additionally, vegetation improves soil restoration by enhancing microbial variety and activity (Rai 2022).

Carbon (C)sequestration is the net removal of CO_2 from atmosphere and storage in long-lived pools of C. Such pools include the aboveground plant biomass, belowground biomass such as roots, soil microorganisms, and the relatively stable forms of organic and inorganic C in soils and deeper subsurface environments, and the durable products derived from biomass (e.g., timber) (Takimoto et al. 1997; Montagnini and Nair 2004; Lorenz and Lal 2014). Compared to pastures or field crops, agroforestry systems are thought to have a greater capacity to sequester carbon (Sanchez 2000; Roshetko et al. 2002; Sharrow and Ismail 2004; Kirby and Potvin 2007). Conceptually trees are considered to be a terrestrial carbon sink (Houghton et al. 1998). Therefore, managed forests can, theoretically, sequester carbon both in situ (biomass and soil) and ex-situ (products) (Montagnini and Nair 2004). Carbon sequestration in agroforestry systems occurs in above-ground biomass, i.e., stem, branch, and foliage, and in below-ground biomass, i.e., roots, and in soil. Especially, the large volume of aboveground biomass and deep root systems of trees in agroforestry systems have received increased attention for climate change adaption and mitigation (Nair 2012).

The Soil Science Society of America (SSSA) recognizes that C is sequestered in soils in two ways i.e., direct and indirect (Soil Science Society of America 2001) – "Direct soil C sequestration occurs by inorganic chemical reactions that convert CO_2 into soil inorganic C compounds such as calcium and magnesium carbonates." Indirect plant C sequestration occurs as plants photosynthesize atmospheric CO_2 into plant biomass. Some of this plant biomass is indirectly sequestered as SOC during decomposition processes (Montagnini and Nair 2004).

In relation to forests/crops, grazing-land management and revegetation with adequate realization of the role of trees as an important means to capture and store atmospheric CO_2 in vegetation, soils, and biomass products (Malhi et al. 2008; Sakachep and Rai 2021a,b). In this sense, agroforestry plantation has been widely recognized as a C-sequestration activity especially under the afforestation and reforestation that have been approved as GHG-mitigating strategies under the Kyoto Protocol (Albrecht and Kandji 2003; Nair and Nair 2003; Makundi and Sathaye 2004; Takimoto et al. 2008). Thus, agroforestry systems attracted attention as a C-sequestration strategy from both industrialized and developing countries (Sharrow and Ismail 2004; Haile et al. 2008; Takimoto et al. 2008).

Crop residues show high C : N ratios and yield high amounts of litter, which increases the SOM content in the soil. Leguminous crops also produce considerable litter, but their residues have low C : N ratios (Sakachep and Rai 2021b). For example, crops of Brassicaceae produce small amounts of litter and the C : N ratio of their residues is low (Komatsuzaki 1999). These low C : N residue-producing crops result in quick decomposition of residues in the soil. Other temperate-zone agroforestry systems include the ancient tree-based agriculture involving a large number of multipurpose trees such as chestnuts (*Castanea* spp.), oaks (*Quercus* spp.), carob (*Ceratonia siliqua*), olive (*Olea europa*), and figs (*Ficus* spp.) in the Mediterranean region (Nair et al. 2008; Rigueiro-Rodriguez et al. 2008).

2.7 Conclusions

Increased infestation of IAPs remarkably influenced the agriculture/agroforestry systems, food security, ecosystem services, and human well-being. The synergy of IAPs with climate change exacerbated the agroecosystem and agroforestry systems' resilience which pave the way to quest agroecological solutions. Though the effects of IAPs on forests are widely studied however, their impacts on agroecosystems (soil-crop subsystem) are scantily documented. In agriculture and agroforestry systems, IAPs owing to multiple ecological mechanisms like NWH or allelopathy influence ecosystem functioning. Further, IAPs adversely influence the abiotic (e.g., soil physical-chemical attributes) and biotic (crop-pollinators niche overlap and microbial diversity) components of agroecosystems. Allelopathic effects of several IAPs reduce seed germination and growth of food crops, thereby cause decline in agricultural productivity. Therefore, by adversely impacting food security and livelihood, IAPs can concern the UN-sustainable Development Goals (SDGs) and UN-Decade on Ecosystem Restoration targets. To this end, the management prospects especially nature based sustainable solutions with incorporation of underlying agroecological principles such as cover cropping/inter-cropping with native plants can accelerate the C-sequestration and sustainable restoration of degraded agroecosystems. Last, the wise application of agroecological principles need to be prioritized for sustainable agriculture to help achieve global food security and human well-being.

2.7.1 Declaration of Competing Interest

The author/s declare that they have no known competing financial interests or personal relationships that could have appeared to influence the work reported in this paper.

Acknowledgment

PKR thanks the financial assistance received from Department of Biotechnology (DBT) vide research project no.BT/PR24917/NER/95/907/2017 and Department of Science and Technology (DST) vide research project no. DST/TMD/EWO/WTI/2 K19/EWFH/2019 C. PKR also thank Ms. Emaracee and Mr. Roger Bruce Syngkli for their cooperation.

References

Abbas, T. (2017). Effect of allelopathic bacteria on the growth and yield of wheat (*Triticum aestivum* L.) and its associated weeds. PhD dissertation. University of Agriculture, Faisalabad, Pakistan.

Abbas, T., Nadeem, M.A., Tanveer, A. et al. (2017a). Role of allelopathic crop mulches and reduced doses of tank-mixed herbicides in managing herbicide resistant Phalaris minor in wheat. *Crop Prot.* https://doi.org/10.1016/j.cropro.2017.06.012.

Abbas, T., Nadeem, M.A., Tanveer, A. et al. (2017b). Allelopathic influence of aquatic weeds on agro-ecosystems: a review. *Planta Daninha* 35: 017163146.

Abbas, T., Zahir, Z.A., Naveed, M., and Kremer, R.J. (2018). Limitations of existing weed control practices necessitate development of alternative techniques based on biological approaches. *Adv. Agron.* 147: 239–280.

Ait Elallem, K., Sobeh, M., Boularbah, A., and Yasri, A. (2021). Chemically degraded soil rehabilitation process using medicinal and aromatic plants. *Environ. Sci. Pollut. Res.* 28 (1): 73–93.

Albrecht, A. and Kandji, S.T. (2003). Carbon sequestration in tropical agroforestry systems. *Agric. Ecosyst. Environ.* 99: 15–27.

Ali, H., Abid, S.A., Ahmad, S. et al. (2013). Impact of integrated weed management on flat-sown cotton (*Gossypium hirsutum* L.). *J. Anim. Plant Sci.* 23: 1185–1192.

Allison, S.D. and Vitousek, P.M. (2004). Rapid nutrient cycling in leaf letter from invasive plants in Hawai'i. *Oecologia* 141: 612–619.

Altieri, M.A. (1989). Agroecology: a new research and development paradigm for world agriculture. *Agric. Ecosyst. Environ.* 27: 37–46.

Altieri, M.A. and Rosset, P. (1995). Agroecology and the conversion of large scale conventional systems to sustainable management. *Int. J. Environ. Stud.* 50: 165–185.

Altieri, M.A., Lana, M.A., Bittencourt, H.V. et al. (2011). Enhancing crop productivity via weed suppression in organic no-till cropping systems in Santa Catarina. *Brazil. J. Sustain. Agric.* 35: 855–869.

Arora, N.K. (2019). *Impact of Climate Change on Agriculture Production and its Sustainable Solutions.* Springer.

Auld, B.A., Hetherington, S.D., and Smith, H.E. (2003). Advances in bioherbicide formulation. *Weed Biol. Manag.* 3: 61–67.

Ba'rberi, P. and Mazzoncini, M. (2001). Changes in weed community composition as influenced by cover crop and management system in continuous corn. *Weed Sci.* 49: 491–499.

Barto, K., Friese, C., and Cipollini, D. (2010). Arbuscular mycorrhizal fungi protect a native plant from allelopathic effects of an invader. *J. Chem. Ecol.* 36: 351–360.

Bationo, A. and Ntare, B.R. (2000). Rotation and nitrogen fertilizer effects on pearl millet, cowpea and groundnut yield and soil chemical properties in a sandy soil in the semi-arid tropics, West Africa. *J. Agric. Sci.* 134 (3): 277–284.

Batten, K.M., Scow, K.M., Davies, K.F., and Harrison, S.P. (2006). Two invasive plants alter soil microbial community composition in serpentine grasslands. *Biol. Invasions* 8 (2): 217–230.

Baumann, D.T., Kropff, M.J., and Bastiaans, L. (2000). *Intercropping Reference Manual*. Oxford, England: Oxford University Press.

Baylis, A.D. (2000). Why glyphosate is a global herbicide: strengths, weaknesses and prospects. *Pest Manag. Sci.* 56: 299–308.

Bever, J.D., Schultz, P.A., Pringle, A., and Morton, J.B. (2001). Arbuscular mycorrhizal fungi: more diverse than meets the eye, and the ecological tale of why. *Bioscience* 51 (11): 923–931.

Bohlen, P.J. (2006). Biological invasions: linking the aboveground and belowground consequences. *Appl. Soil Ecol.* 32: 1e5.

Borrgaard, O.K. and Gimsing, A.L. (2008). Fate of glyphosate in soil and the possibility of leaching to ground and surface waters: a review. *Pest Manag. Sci.* 64: 441–456.

Boyette, C.D. and Hoagland, R.E. (2015). Bioherbicidal potential of *Xanthomonas campestris* for controlling *Conyza canadensis*. *Biocontrol Sci. Tech.* 25: 229–237.

Bozzolo, F.H. and Lipson, D.A. (2013). Differential responses of native and exotic coastal sage scrub plant species to N additions and the soil microbial community. *Plant Soil* 371 (1–2): 37–51.

Callaway, R.M., Thelen, G.C., Barth, S. et al. (2004a). Soil fungi alter interactions between the invader *Centaurea maculosa* and north American natives. *Ecology* 85 (4): 1062–1071.

Callaway, R.M., Thelen, G.C., Rodriguez, A., and Holben, W.E. (2004b). Soil biota and exotic plant invasion. *Nature* 427: 731–733.

Campbell, S. (2005). A global perspective on forest invasive species: the problem, causes and consequences. In: *Proceedings of the Asia-Pacific Forest Invasive Species Conference*, Kunming, China (17–23 August 2003) (ed. P. McKenzie, C. Brown, J. Su, and J. Wu). The Unwelcome Guests, 9–10. Bangkok, Thailand: FAO Regional Office for Asia and the Pacific.

Caravaca, F.J. et al. (2003). Establishment of shrub species in a degraded semiarid site after inoculations with native allocthonous arbuscular mycorrhizal fungi. *Appl Ecol.* 22: 103–111.

Chapman, S.K. and Newman, G.S. (2010). Biodiversity at the plant-soil interface: microbial abundance and community structure respond to litter mixing. *Oecologia* 162: 763–769.

Charudattan, R. (2005). Ecological, practical and political inputs into selection of weed targets: what makes a good biological control target? *Biol. Control* 35: 183–196.

Charudattan, R. and Dinoor, A. (2000). Biological control of weeds using plant pathogens: accomplishments and limitations. *Crop Prot.* 19: 691–695.

Chauhan, B.S., Gill, G., and Preston, C. (2006). Tillage system effects on weed ecology, herbicide activity and persistence: a review. *Aust. J. Exp. Agric.* 46: 1557–1570.

Chazdon, R.L. (2008). Beyond deforestation: restoring forests and ecosystem services on degraded lands. *Science* 320: 1458–1460.

Cook, D.C., Fraser, R.W., Paini, D.R. et al. (2011). Biosecurity and yield improvement technologies are strategic complements in the fight against food insecurity. *PLoS One* 6 (10): e26084.

Dagno, K., Lahlali, R., Diourte, M., and Jijakli, M.H. (2012). Present status of the development of mycoherbicides against water hyacinth: successes and challenges – a review. *Biotechnol. Agron. Soc. Environ.* 16 (3): 360–368.

Daynes, C.N., Field, D.J., Saleeba, J.A. et al. (2013). Development and stabilisation of soil structure via interactions between organic matter, arbuscular mycorrhizal fungi and plant roots. *Soil Biol. Biochem.* 57: 683–694.

Desprez-Loustau, M.L., Robin, C., Buee, M. et al. (2007). The fungal dimension of biological invasions. *Trends Ecol. Evol.* 22 (9): 472–480.

Drenovsky, R.E. and Batten, K.M. (2007). Invasion by *Aegilops triuncialis* (barb Goatgrass) slows carbon and nutrient cycling in a serpentine grassland. *Biol. Invasions* 9: 107–116.

Duke, S.O. and Powles, S.B. (2008). Glyphosate: a once-in-a-century herbicide. *Pest Manag. Sci.* 64: 319–325.

Dybzinski, R., Fargione, J.E., Zak, D.R. et al. (2008). Soil fertility increases with plant species diversity in a long-term biodiversity experiment. *Oecologia* 158: 85–93.

Edrisi, S.A., El-Keblawy, A., and Abhilash, P.C. (2020). Sustainability analysis of *Prosopis juliflora* (Sw.) DC based restoration of degraded land in North India. *Land* 9: 59.

Ehrenfeld, J.G. (2001). Plant-soil interactions. In: *Encyclopedia of Biodiversity* (ed. S. Levin), 689–709. San Diego, CA: Academic Press.

Ehrenfeld, J.G. (2003). Effects of exotic plant invasions on soil nutrient cycling processes. *Ecosystems* 6: 503e523.

Ehrenfeld, J.G. (2004). Implications of invasive species for belowground community and nutrient processes. *Weed Technol.* 18: 1232–1235.

Ehrenfeld, J.G., Kourtev, P., and Huang, W. (2001). Changes in soil functions following invasions of exotic understory plants in deciduous forests. *Ecol. Appl.* 11: 1287–1300.

El-Barougy, F. et al. (2021). Invasion risk assessment using trait-environment and species distribution modelling techniques in an arid protected area: towards conservation prioritization. *Ecol. Indic.* 129: 107951.

Elgersma, K.J. and Ehrenfeld, J.G. (2011). Legacy effects overwhelm the short-term effects of exotic plant invasion and restoration on soil microbial community structure, enzyme activities, and nitrogen cycling. *Oecologia* 167: 733–745.

Eviner, V.T. and Hawkes, C.V. (2008). Embracing variability in the application of plant–soil interactions to the restoration of communities and ecosystems. *Restor. Ecol.* 16: 713–729.

Fandohan, A.B., Oduor, A.M., Sod'e, A.I. et al. (2015). Modeling vulnerability of protected areas to invasion by *Chromolaena odorata* under current and future climates. *Ecosyst. Health Sustain.* 1 (6): 1–12.

FAO (2015). Agroecology to reverse soil degradation and achieve food security.

Farooq, O. and Cheema, Z.A. (2013). Impact of planting time and seedbed conditions on little seed canary grass and lambsquarters dynamics in wheat. *Int. J. Agric. Biol.* 15: 1103–1007.

Farooq, M., Jabran, K., Cheema, Z.A. et al. (2011). The role of allelopathy in agricultural pest management. *Pest Manag. Sci.* 67: 493–506.

Fischer, A.J. (1996). Integrated red rice management in Latin American rice fields. In: *Proceedings of the Second International Weed Control Congress*, Copenhagen, Denmark, vol. 2, 53–66.

Fleming, P.J., Ballard, G., Reid, N.C., and Tracey, J.P. (2018). Invasive species and their impacts on Agri-ecosystems: issues and solutions for restoring ecosystem processes. *Rangel. J.* 39 (6): 523–535.

Fried, G., Chauvel, B., Reynaud, P., and Sache, I. (2017). Decreases in crop production by nonnative weeds, pests, and pathogens. In: *Impact of Biological Invasions on Ecosystem Services*, 83–101. Springer.

Gaba, S., Fried. G., Kazakou. E., et al. (2014). Agroecological weed control using a functional approach: a review of cropping systems diversity. *Agron. Sustain. Dev.* 34: 103–119.

Gage, K.L. and Schwartz-Lazaro, L.M. (2019). Shifting the paradigm: an ecological systems approach to weed management. *Agriculture* 9 (8): 179.

Ghorbani, R., Leifert, C., and Seel, W. (2005). Biological control of weeds with antagonistic plant pathogens. *Adv. Agron.* 86: 191–225.

Gliessman, S.R. (2006). *Agroecology: The Ecology of Sustainable Food Systems*, 2e. Boca Raton: CRC Press.

Goeden, R.D. and Andres, L.A. (1999). Biological control of weeds in terrestrial and aquatic environments. In: *Handbook of Biological Control* (ed. T.S. Bellows and T.W. Fisher), 871–890. San Diego, CA: Academic Press.

Gogoi, A.K. (2001). Status of Mikania infestation in North-Eastern India: management options and future research thrust. In: *Alien Weeds in Moist Tropical Zones: Banes and Benefits. Proceedings of a Workshop, Kerala Forest Research Institute*, Peechi, India (2–4 November 1999) (ed. K.V. Sankaran, S.T. Murphy, and H.C. Evans), 77–79. Peechi, India and CABI Bioscience, UK Centre (Ascot), Ascot, UK: Kerala Forest Research Institute.

GOI (2012). A users compendium on small agricultural machinery and implements (state/region/agroclimatic zones specific low cost, light weight, compact, multi utility farm machinery and implements affordable to small and marginal farmers). In: *Government of India, Ministry of Agriculture*. Krishi Bhawan, New Delhi, India: Department of Agriculture & Cooperation.

Guillerme, S., Barcet, H., de Munnik, N. et al. (2020). Evolution of traditional agroforestry landscapes and development of invasive species: lessons from the Pyrenees (France). *Sustain. Sci.* 15 (5): 1285–1299.

Hahn, D.R. (2003), Alteration of microbial community composition and changes in decomposition associated with an invasive intertidal macrophyte. *Biol. Invasions* 5: 45–51.

Haile, S.G., Nair, P.K.R., and Nair, V.D. (2008). Carbon storage of different soil-size fractions in Florida silvopastoral systems. *J. Environ. Qual.* 37: 1789–1797.

Harding, D.P. and Raizada, M.N. (2015). Controlling weeds with fungi, bacteria and viruses: a review. *Front. Plant Sci.* 6: 659.

Hokkanen, H.M. and Sailer, R.I. (1985). Success in classical biological control. *Crit. Rev. Plant Sci.* 3: 35–72.

Houghton, R.A., Davidson, E.A., and Woodwell, G.M. (1998). Missing sinks, feedbacks, and understanding the role of terrestrial eco systems in the global carbon balance. *Global Biogeochem. Cy.* 12: 25–34.

IFEN (2007). *Les Pesticides dans les Eaux-Données 2005*. Orléans.: Institut Français de l'Environnement.

Jabran, K., Mahajan, G., Sardana, V., and Chauhan, B.S. (2015). Allelopathy for weed control in agricultural systems. *Crop Prot.* 72: 57–65.

Jia, J., Dai, Z., Li, F., and Liu, Y. (2016). How will global environmental changes affect the growth of alien plants? *Front Plant Sci.* 7: 1623.

Kadir, J., Sajili, M.H., Juraimi, A.S., and Napis, S. (2008). Effect of *Exserohilum monoceras* (Drechslera) Leonard & suggs on the competitiveness of *Echinocloa crussgalli* (L.) P. Beauv. *Pertanika J. Trop. Agric. Sci.* 31: 19–26.

Kao-Kniffin, J. and Balser, T.C. (2007). Elevated CO_2 differentially alters belowground plant and soilmicrobial community structure in reed canary grass-invaded experimental wetlands. *Soil Biol. Biochem.* 39: 517–525.

Kawakami, K., Kai, W., and Paul, R.D. (2012). Invasive or hyper abundant species. In: *Understanding world heritage in Asia and the Pacific – N 35 – the second cycle of periodic reporting 2010–2012*, 62–63. UNESCO.

Khadija, A.E., Mansour, S., Ali, B., and Abdelaziz, Y. (2019). Chemically degraded soil rehabilitation process using medicinal and aromatic plants. *Environ. Sci. Pollut. Res. Int.* 28 (1): 73–93.

Kirby, K.R. and Potvin, C. (2007). Variation in carbon storage among tree species: implications for the management of a small-scale carbon sink project. *For. Ecol. Manag.* 246: 208–221.

Klironomos, J. (2002). Feedback with soil biota contributes to plant rarity and invasiveness in communities. *Nature* 417: 67–70.

Komatsuzaki, M. (1999). Cropping system and rotation. In: *Sustainable Agriculture System Management (in Japanese)* (ed. O. Endo). Tokyo, Japan: Nourin Toukei Kyoukai.

Kourtev, P.S., Ehrenfeld, J.G., and Häggblom, M. (2002). Exotic plant species alter the microbial community structure and function in the soil. *Ecology* 83: 3152–3166.

Kulmatiski, A. and Beard, K.H. (2011). Longterm plant growth legacies overwhelm short-term plant growth effects on soil microbial community structure. *Soil Biol. Biochem.* 43 (4): 823–830.

Kumar, M. et al. (2021). Invasion of *Lantana camara* and *Ageratina adenophora* alters the soil physico-chemical characteristics and microbial biomass of chir pine forests in the central Himalaya, India. *Catena* 207: 105624.

Lamb, D., Erskine, P.D., and Parrotta, J.A. (2005). Restoration of degraded tropical forest landscapes. *Science* 310 (5754): 1628–1632.

Lankau, E.W. and Lankau, R.A. (2014). Plant species capacity to drive soil fungal communities contributes to differential impacts of plant-soil legacies. *Ecology* 95: 3221–3228.

Lekberg, Y., Gibbons, S.M., Rosendahl, S., and Ramsey, P.W. (2013). Severe plant invasions can increase mycorrhizal fungal abundance and diversity. *ISME J.* 7 (7): 1424–1433.

Li, W.H., Zhang, C.B., Jiang, H.B. et al. (2006). Changes in soil microbial community associated with invasion of the exotic weed, *Mikania micrantha* H.B.K. *Plant Soil* 281: 309–324.

Li, L., Li, S., Sun, J. et al. (2007). Diversity enhances agricultural productivity via rhizosphere phosphorus facilitation on phosphorus-deficient soils. *Proc. Natl. Acad. Sci. U. S. A.* 104: 11192–11196.

Li, C. et al. (2015). Indigenous trees restore soil microbial biomass at faster rates than exotic species. *Plant Soil* 396: 151–161.

Liao, C., Peng, R., Luo, Y. et al. (2008). Altered ecosystem carbon and nitrogen cycles by plant invasion: a meta-analysis. *New Phytol.* 177: 706–714.

Liebman, M. and Gallandt, E.R. (1997). Many little hammers: ecological management of crop weed interactions. In: *Ecology in Agriculture* (ed. L. Jackson), 291–341. New York, NY: Academic Press.

Liebman, M. and Staver, C.P. (2001). Crop diversification for weed management. In: *Ecological Management of Agricultural Weeds* (ed. M. Liebman, C.L. Mohler, and C.P. Staver), 322–374. UK: Cambridge University Press.

Lorenz, K. and Lal, R. (2014). Soil organic carbon sequestration in agroforestry systems. A review. *Agron. Sustain. Dev.* 34 (2): 443–454.

Mack, R.N., Simberloff, D., Mark Lonsdale, W. et al. (2000). Biotic invasions: causes, epidemiology, global consequences, and control. *Ecol. Appl.* 10 (3): 689–710.

Mahajan, G., Brar, L.S., and Sardana, V. (1999). Effect of tillage and time of sowing on the efficacy of herbicides against *Phalaris minor* in wheat. *Proceeding of 17th Asian Pacific Weed Science Society Conference*, Bangkok, Thailand, 193–198.

Makoi, J.H. and Ndakidemi, P.A. (2012). Allelopathy as protectant, defense and growth stimulants in legume cereal mixed culture systems. *N.Z.J. Crop. Hortic. Sci.* 40: 161–186.

Makundi, W.R. and Sathaye, J.A. (2004). GHG mitigation potential and cost in tropical forestry – relative role for agroforestry. *Environ. Dev. Sustain.* 6: 235–260.

Malhi, Y., Roberts, J.T., Betts, R.A. et al. (2008). Climate change, deforestation, and the fate of the Amazon. *Science* 319: 169–172.

Matloob, A., Safdar, M.E., Abbas, T. et al. (2020). Challenges and prospects for weed management in Pakistan: a review. *Crop Prot.* 134: 104724.

Mattner, S.W. (2006). The impact of pathogens on plant interference and allelopathy. In: *Allelochemicals: Biological Control of Plant Pathogens and Diseases* (ed. M. Inderjit and K.G.). Dordrecht, The Netherlands: Springer.

Melander B., Rasmussen I.A., and Ba`rberi P. (2005). Integrating physical and cultural methods of weed control – examples from European research. *Weed Sci.* 53: 369–381.

Miller, R.M. and Jastrow, J.D. (2000). Mycorrhizal fungi influence soil structure. In: *Arbuscular Mycorrhizas: Physiology and Function*, 3–18. Dordrecht: Springer.

Monaco, T.J., Weller, S.C., and Ashton, F.M. (2002). *Weed Science Principles and Practices*, 671. Wiley.

Montagnini, F. and Nair, P.K.R. (2004). Carbon sequestration: an under exploited environmental benefit of agroforestry systems. *Agric. Syst.* 61: 281–295.

Muñoz-Rojas, M., Erickson, T.E., Dixon, K.W., and Merritt, D.J. (2016). Soil quality indicators to assess functionality of restored soils in degraded semiarid ecosystems. *Soc. Ecol. Restor.* 24: 1–10. https://doi.org/10.1111/rec.12368.

Nair, P.K.R. (2012). Carbon sequestration studies in agroforestry systems: a reality-check. *Agrofor. Syst.* 86: 243–253. https://doi.org/10.1007/s10457-011-9434-z.

Nair, P.K.R. and Nair, V.D. (2003). Carbon storage in north American agroforestry systems. In: *The Potential of U.S. Forest Soils to Sequester Carbon and Mitigate the Greenhouse Effect* (ed. J. Kimble, L.S. Heath, R.A. Birdsey, and R. Lal), 333–346. Boca Raton, FL, USA: CRC Press.

Nair, P.K.R., Gordon, A.M., and Mosquera-Losada, M.-R. (2008). Agro forestry. In: *Ecological Engineering. Encyclopedia of Ecology*, vol. 1 (ed. S.E. Jorgensen and B.D. Fath), 101–110. Oxford, U.K: Elsevier.

Nimbal, C.I., Yerkes, C.N., Weston, L.A., and Weller, S.C. (1996). Herbicidal activity and site of action of the natural product sorgoleone. *Pestic. Biochem. Physiol.* 54 (2): 73–83.

Nongtri, E.S., Lallianpuii, S., Rai, P.K. (2022). Microcosm Study to Assess Allelochemical Potential of Lantana camara. *Indian J. Ecol.* 49(6): 2163–2166.

Oerke, C.E. (2006). Centinary review on crop losses to pests. *J. Agric. Sci.* 144: 31–43.

Paini, D.R., Sheppard, A.W., Cook, D.C. et al. (2016). Global threat to agriculture from invasive species. *Proc. Natl. Acad. Sci.* 113 (27): 7575–7579.

Panetta, F.D. and Roger, L. (2005). Evaluation of weed eradication programs: the delimitation of extent. *Divers. Distrib.* 11: 435–442.

Pickett, B., Maltz, M., and Aronson, E. (2018). Impacts of invasive plants on soil fungi and implications for restoration. In: *Diversity and Ecology of Invasive Plants*, 45. IntechOpen.

Pittelkow, C.M., Fischer, A.J., and Moechnig, M.J. (2012). Agronomic productivity and nitrogen requirements of alternative tillage and crop establishment systems for improved weed control in direct-seeded rice. *Field Crop Res.* 130: 128–137.

Qu, T. et al. (2021). Invasive species allelopathy decreases plant growth and soil microbial activity. *PLoS One* 16: e0246685.

Raffaelli, M., Barberi, P., Peruzzi, A., and Ginanni, M. (2005). Mechanical weed control in maize: evaluation of weed harrowing and hoeing systems. *Agric. Mediterr.* 135: 33–43.

Rai, P.K. (2009). Comparative assessment of soil properties after bamboo flowering and death in a tropical forest of Indo-Burma hotspot. *Ambio* 38 (2): 118–120.

Rai, P.K. (2012). Assessment of multifaceted environmental issues and model development of an Indo-Burma hot spot region. *Environ. Monit. Assess.* 184: 113–131.

Rai, P.K. (2013). *Plant Invasion Ecology: Impacts and Sustainable Management*, 196. New York: Nova Science Publisher.

Rai, P.K. (2015). Paradigm of plant invasion: multifaceted review on sustainable management. *Environ. Monit. Assess.* 187: 759.

Rai, P.K. (2022). Environmental degradation by invasive alien plants in the Anthropocene: challenges and prospects for sustainable restoration. *Anthropocene Sci. ANPS* 1: 5–28.

Rai, P.K. and Kim, K.H. (2020). Invasive alien plants and environmental remediation: a new paradigm for sustainable restoration ecology. *Restor. Ecol.* 28 (1): 3–7.

Rai, P.K. and Singh, J.S. (2020). Invasive alien plant species: their impact on environment, ecosystem services and human health. *Ecol. Indic.* 111: 1–20.

Ruati, V.L. and Rai, P.K. (2022). Societal perception on environmental and socio-economic implications of Tithonia diversifolia (Hemsl.) A. Gray invasion in an Indo Burma biodiversity hotspot. *Environ. Socioecon. studies* 10: 59–66

Rai, P.K. and Singh, J.S. (2021). Plant invasion in protected areas, the Indian Himalayan region, and the north East India: progress and prospects. *Proc. Natl. Acad. Sci. India A* 87: 19–35.

Raizada, P., Raghubanshi. A.S., and Singh, J.S. (2008). Impacts of invasive alien plant species on soil processes: a review. *Proc. Natl. Acad. Sci. Ind. Biol. Sci.* 78: 288–298.

Rana, S.S. and Rana, M.C. (2016). *Principles and Practices of Weed Management*, 138. CSK Himachal Pradesh Krishi Vishvavidyalaya, Palampur: Department of Agronomy, College of Agriculture.

Rao, A.N., Johnson, D.E., Sivaprasad, B. et al. (2007). Weed management in direct-seeded rice. *Adv. Agron.* 93: 153–255.

Rao, A.N., Wani, S.P., and Ladha, J.K. (2014). Weed management research in India – an analysis of the past and outlook for future. In: *Souvenir (1989–2014). DWR Publication No. 18*, 1–26. Jabalpur, India: Directorateof Weed Research.

Rao, A.N., Singh, R.G., Mahajan, G., and Wani, S.P. (2020). Weed research issues, challenges, and opportunities in India. *Crop Prot.* 134: 104451.

Ravit, B., Ehrenfeld, J.G., and Haggblom, M.M. (2003). A comparison of sediment microbial communities associated with *Phragmites australis* and *Spartina alterniflora* in two brackish wetlands of New Jersey. *Estuaries* 26: 465–474.

Reinhart, K.O. and Callaway, R.M. (2006). Soil biota and invasive plants. *New Phytol.* 170: 445–457.

Richardson, D.M. and Rejmánek, M. (2011). Trees and shrubs as invasive alien species—a global review. *Divers. Distrib.* 17 (5): 788–809.

Richardson, D.M., Binggeli, P., and Schroth, G. (2004). Invasive agroforestry trees: problems and solutions. In: *Agroforestry and Biodiversity Conservation in Tropical Landscapes*, 371–396.

Rigueiro-Rodriguez, A., McAdam, J.H., and Mosquera-Losada, M.R. (ed.) (2008). *Agroforestry in Europe*, Advances in Agroforestry, 5. Dordrecht, the Netherlands: Springer.

Roshetko, M., Delaney, M., Hairiah, K., and Purnomosidhi, P. (2002). Carbon stocks in Indonesian homegarden systems: can small holder systems be targeted for increased carbon storage? *Am. J. Altern. Agric.* 17: 125–137.

Rout, M. and Callaway, R.M. (2009). An invasive plant paradox. *Science* 324: 734–735.

Ruati, V.L. and Rai, P.K. (2021). The impact of *Tithonia diversifolia* (Hemsl.) a. gray on phytosociology in Aizawl, Mizoram. *Ecol. Environ. Conserv.* 27: S209–S214.

Rubiales, D. and Fernandez-Aparicio, M. (2012). Innovations in parasitic weeds management in legume crops. *Agron. Sustain. Dev.* 32: 433–449.

Safdar, M.E., Tanveer, A., Khaliq, A., and Maqbool, R. (2016). Critical competition period of parthenium weed (*Parthenium hysterophorus* L.) in maize. *Crop Prot.* 80: 101–107.

Shahzad, M., Farooq, M., and Hussain, M. (2016). Weed spectrum in different wheat-based cropping systems under conservation and conventional tillage practices in Punjab, Pakistan. *Soil Tillage Res.* 163: 71–79.

Sakachep, Z.K. and Rai, P.K. (2021a). Influence of invasive alien plants on vegetation of Hailakandi District, Assam, north-east, India. *Indian J. Ecol.* 48 (1): 261–266.

Sakachep, Z.K. and Rai, P.K. (2021b). Impact assessment of invasive alien plants on soil organic carbon (SOC) status in disturbed patches of Hailakandi District in an Indo Burma hotspot region. *Indian J. Ecol.* 48 (6): 1698–1704.

Sanchez, P.A. (2000). Linking climate change research with food security and poverty reduction in the tropics. *Agric. Ecosyst. Environ.* 82: 371–383.

Sankaran, K.V. (2017). Control options for invasive alien plants in agroforestry in the Asia-Pacific region. In: *Invasive Alien Plants: Impacts on Development and Options for Management*, 138–161. UK: CABI.

Sanon, A., Béguiristain, T., Cébron, A. et al. (2009). Changes in soil diversity and global activities following invasions of the exotic invasive plant, *Amaranthus viridis* L., decrease the growth of native Sahelian Acacia species. *FEMS Microbiol. Ecol.* 70: 118–131.

Saturday, A. (2018). Restoration of degraded agricultural land: a review. *J. Environ. Health Sci.* 4 (2): 44–51.

Saxena, S. and Kumar, M. (2007). Mycoherbicidal potential of *Alternaria alternata* ITCC4896 for the control of *Parthenium hysterophorus*. *J. Plant Protect. Res.* 47: 213–218.

Saxton, K.E. and Rawls, W.J. (2006). Soil water characteristic estimates by texture and organic matter for hydrologic solutions. *Soil Sci. Soc. Am. J.* 70 (5): 1569–1578.

Scheepens, P.C., Muller-Scharer, S., and Kempenaar, C. (2001). Opportunities for biological weed control in Europe. *Biol. Control* 46: 127–138.

Seebens, H., Blackburn, T.M., Dyer, E.E. et al. (2017). No saturation in the accumulation of alien species worldwide. *Nat. Commun.* 8 (1): 1–9.

Shao, H., Huang, X., Wei, X., and Zhang, C. (2012). Phytotoxic effects and a phytotoxin from the invasive plant *Xanthium italicum* Moretti. *Molecules* 17 (4): 4037–4046.

Sharrow, S.H. and Ismail, S. (2004). Carbon and nitrogen storage in agroforests, tree plantations, and pastures in western Oregon, USA. *Agrofor. Syst.* 60: 123–130.

Shiferaw, W. et al. (2021). Invasion of *Prosopis Juliflora* and its effects on soil physicochemical properties in Afar Region, Northeast Ethiopia. *Int. Soil Water Conserv. Res.* https://doi.org/10.1016/j.iswcr.2021.04.003.

Shrestha, A., Clements, D.R., and Upadhyaya, M.K. (2004). Weed management in agroecosystems: towards a holistic approach.

Singh, R.P., Verma, S.K., and Singh, R.K. (2016). Effects of herbicides on growth and yield of *Cicer arietinum* L. under rainfed condition. *Bangladesh J. Bot.* 45: 305–311.

Soil Science Society of America (2001). *Carbon Sequestration: Position of the Soil Science Society of America (SSSA)*. Madison, WI: SSSA. http://www.soils.org/pdf/pos_paper_carb_seq.pdf (accessed 20 January 2008).

Sollins, P., Swanston, C., and Kramer, M. (2007). Stabilization and destabilization of soil organic matter—a new focus. *Biogeochemistry* 85: 1–7. https://doi.org/10.1007/s10533-007-9099-x.

Sridhar, H.S. (2013). Development of single wheel multi use manually operated weed remover. *Int. J. Mod. Eng. Res. (IJMER)* 3: 3836–3840.

Stefanowicz, A.M. et al. (2018). Differential influence of four invasive plant species on soil physicochemical properties in a pot experiment. *J. Soils Sediments* 18: 1409–1423.

Stoate, C., Boatman, N.D., Borralho, R.J. et al. (2001). Ecological impacts of arable intensification in Europe. *J. Environ. Manag.* 63: 337–365.

Syngkli, R.B., Lalnunhlui, Lallianpuii, S., and Rai, P.K. (2022a). Microcosm investigation on the allelochemical potential of Mikania micrantha to the selected food crop. *Int. J. Plant Environ.* 8(2): 137–142.

Syngkli, R.B., Lalnunhlui, Lallianpuii, S., and Rai, P.K. (2022b). Microcosm investigation on the allelochemic potential of Ageratum conyzoides on selected food crop. *Eco. Environ. Cons.* 28 (November Suppl. Issue): S298–S304

Takimoto, A., Nair, P.K.R., and Alavalapati, J.R.R. (2008). Socio economic potential of carbon sequestration through agroforestry in the west African Sahel. Mit. Adapt. Strat. *Global Change* 13: 745–761. https://doi.org/10.1007/s11027-007-9140-3.

Taylor, S. and Kumar, L. (2013). Potential distribution of an invasive species under climate change scenarios using CLIMEX and soil drainage: a case study of *Lantana camara* L. in Queensland, Australia. *J. Environ. Manag.* 114: 414–422.

Thomas, H.J., Sauerborn, D., and Muller-Stover, K.J. (1999). Fungi of *Orobanche aegyptiaca* in Nepal with potential as biological control agents. *Biocontrol Sci. Tech.* 9: 379–381.

Tiwari, S., Adhikari, B., Siwakoti, M., and Subedi, K. (2005). *An Inventory and Assessment of Invasive Alien Plant Species of Nepal*. Nepal: IUCN-The World Conservation Union.

Tobin, P.C. (2018). Managing invasive species. *F1000Res*. 7: F1000 Faculty Rev-1686.

Torres, A., Oliva, R.M., Castellano, D., Cross, P. (1996).

Van der Putten, W.H. and Klironomos Wardle, D.A. (2007). Microbial ecology of biological invasions. *ISME J.* 1: 28–37.

Van Der Weide, R.Y., Bleeker, P.O., and Lotz, L.A.P. (2002). Simple innovations to improve the effect of the false seed bed technique. *Proceedings of the 5th Workshop of the EWRS Working Group on Physical and Cultural Weed Control*, Pisa, Italy, 3–4.

Vila, M., Espinar, J.L., Hejda, M. et al. (2011). Ecological impacts of invasive alien plants: a meta-analysis of their effects on species, communities and ecosystems. *Ecol. Lett.* 14: 702e708.

Vogelsang, K.M., Bever, J.D., Griswold, M. and Schultz, P.A. (2004). The use of mycorrhizal fungi in erosion control applications. Final Report. Caltrans Contract No. 65A0070, no. 65.

Watson, A.K. (2018). Microbial herbicides. In: *Weed Control: Sustainability, Hazards, and Risks in Cropping Systems Worldwide*, 133. Boca Raton, FL, USA: CRC Press.

Weidenhamer, J.D. and Callaway, R.M. (2010). Direct and indirect effects of invasive plants on soil chemistry and ecosystem function. *J. Chem. Ecol.* 36 (1): 59–69.

Wezel, A. and Jauneau, J.C. (2011). Agroecology – interpretations, approaches and their links to nature conservation, rural development and ecotourism. In: *Integrating Agriculture, Conservation 1 and Ecotourism: Examples from the Field, Issues in Agroecology – Present Status and Future Prospectus*, vol. 1 (ed. W.B. Campbell and S.L. Ortíz). https://doi.org/10.1007/978-94-007-1309-3_1.

Wezel, A., Bellon, S., Dore, T. et al. (2009). Agroecology as a science, a movement and a practice. *A Rev. Agron. Sustain. Dev.* 29 (4): 503–515. https://doi.org/10.1051/agro/2009004.

Wolfe, J.C., Neal, J.C., and Harlow, C.D. (2016). Selective broadleaf weed control in turfgrass with the bioherbicides Phoma macrostoma and thaxtomin a. *Weed Technol.* 30: 688–700.

Ziska, L.H., Blumenthal, D.M., Runion, G.B. et al. (2011). Invasive species and climate change: an agronomic perspective. *Climate Change* 105 (1–2): 13–42.

3

Soil Management for Carbon Sequestration

Taoufik El Rasafi[1,2], *Ahmed El Moukhtari*[3,4], *Ayoub Haouas*[5], *Anas Tallou*[5], *Wassila Bouta*[6,7], *Yassine Aallam*[6], *Soumia Amir*[5], *Hanane Hamdali*[6], *Mohamed Farissi*[4], *Abdelmajid Haddioui*[6], and *Abdallah Oukarroum*[2]

[1] *Laboratory of Health and Environment, Faculty of Sciences Ain Chock, Hassan II University, PO 5366 Maarif, Casablanca, 20100, Morocco*
[2] *High Throughput Multidisciplinary Research Laboratory, University Mohammed VI Polytechnic (UM6P), Ben Guerir, 43150, Morocco*
[3] *Laboratory of Ecology and Environment, Faculty of Sciences Ben Msik, Hassan II University, PO 7955 Sidi Othmane, Casablanca, Morocco*
[4] *Laboratory of Biotechnology and Sustainable Development of Natural Resources, Polydisciplinary Faculty, USMS, Beni Mellal, Morocco*
[5] *Centre d'Agrobiotechnologie et Bioingénierie, Unité de Recherche Labellisée CNRST (Centre AgroBiotech-URL-CNRST-05), Cadi Ayyad University, Marrakech*
[6] *Laboratory of Agro-Industrial and Medical Biotechnologies, Faculty of Science and Techniques, University of Sultan Moulay Slimane, Beni Mellal, Morocco*
[7] *Office de la formation professionnelle et de la promotion du travail (Ofppt), Pole agriculture, La Cité des Métiers et des Compétences de la région Béni Mellal-Khénifra, Béni Mellal, Morocco*

3.1 Introduction

Global warming and climate change are mostly due to the uncontrolled release of greenhouse gases emissions (Zafar et al. 2018), such as methane (CH_4), carbon dioxide (CO_2), and nitrous oxide (N_2O) in atmosphere. The gradual development of industrial processes and insensitivity of agriculture activities are the most, among others, reasons behind the excessive emission of these gases. According to the *World Meteorological Organization* (WMO), concentrations of N_2O, CH_4, and CO_2 were higher by 122% (330 ppb), 257% (1859 ppb), and 146% (405 ppm), respectively, compared to pre-industrial (before 1750) levels (WMO 2018). Moreover, emission of other gases, such as ammonia (NH_3) is reported to reduce air quality and cause climate change (Chaopu et al. 2019). As a result of climate change effects, soil organic carbon (SOC) decreased dramatically with time in agricultural soils lowering as consequence its normal capacity (Qambrani et al. 2017) leading to soil degradation. SOC has been reported to be crucial for water and nutrients Nitrogen (N), Phosphorus (P), Potassium (K) retention and serve as carbon source and habitat for soil microorganisms (Qambrani et al. 2017). Increasing SOC in soil has been reported to be an effective strategy in minimizing the global greenhouse gases emissions (Bolinder et al. 2020)

Agroecological Approaches for Sustainable Soil Management, First Edition.
Edited by Majeti Narasimha Vara Prasad and Chitranjan Kumar.

and therefore mitigation of climate change. The SOC is mainly affected by the agricultural management practices and climate conditions interactions (Zhang et al. 2017).

To overcome this problem and reduce the loss of carbon, soil carbon sequestration (SCS) has a key role not only in minimizing the emission of gases (e.g., CO_2) but also in the increase soil fertility. SCS mainly occurs through abiotic and biotic strategies (Mandal et al. 2020). Abiotic C-sequestration is ensured by the natural injection of CO_2 into geologic strata or ocean (Chowdhury et al. 2021). However, biotic sequestration of carbon involves the capture and transformation of CO_2 into soil mainly by plants and microorganisms. SCS is the capture of atmospheric CO_2 by plants and its incorporation into soil organic matter (SOM). There are two factors affecting SCS: SOC decomposition and biomass input to the soil (Chowdhury et al. 2021). Various sustainable approaches have been adopted to reduce the soil C loss. Soil management, such as agricultural practices is the major processes that can be applied to reduce C loss and enhance its sequestration. For example, organic amendments, such as biochar have been reported to play a vital role in carbon sequestration (Cornelissen et al. 2013; Hussain et al. 2017) and reduction of greenhouse gas emission (Dissanayake et al. 2020), which mitigate climate change (Zhao et al. 2016; Majumder et al. 2019). The present chapter is focused on discussing the recent advances in effective agricultural management practices to sequester C in soil, such as (i) soil tillage system, (ii) amendment of soil with different organic substances (biochar, animal manure, compost, etc.), (iii) crop rotation (iv) agroforestry, (v) nutrient management, (vi) soil microorganisms utilization, and (vii) water irrigation.

3.2 Agronomic Management Practices

3.2.1 Tillage

The crucial role of the tillage system was found in SCS. Conventional and conservation tillage were age-old agricultural practice and have large and strong impacts on SCS. Conventional tillage refers to the physical disturbance of the upper and deeper soil layer by reversing and mixing both of them with a crop residue burring (de Medeiros Barbosa 2015; Tanveer et al. 2020). However, conservation tillage (no till [NT] or minimum till) is considered very effective in SCS compared to conventional tillage. Conservation tillage has been evidenced to show a strong positive effect on carbon sequestration in cropped lands (Chowdhury et al. 2021) due to its reduced effect on destruction of soil macroaggregates, which is very important in storage of labile carbon in NT soils (Helgason et al. 2010). These soil macroaggregates are very important for the soil microbial community habitats and protection. Higher percentage of Arbuscular mycorrhizal fungi and microbial biomass carbon were found in NT as compared to tilled soil (Helgason et al. 2010). A meta-analysis was conducted by Li et al. (2020) where the authors concluded that conservation tillage practices (minimum tillage, no-tillage) and NT plus residue retention increased soil microbial population size, diversity, and microbial community structure compared to conventional tillage. Generally, conservation tillage may help the sequestration and cycling of soil carbon by the improvement of soil microbial community, reduction of SOC decomposition, reduction of soil disturbance, decrease of soil erosion, and soil water conservation (Chowdhury et al. 2021; Aduhene-chinbuah et al. 2022).

3.2.2 Nutrient Management

Nutrient application at appropriate and frequent intervals has been shown to improve plant growth and yield (Oppong et al. 2019) as well as influence soil characteristics such as pH, microbial community, and soil C sequestration (Chowdhury et al. 2021; Bhattacharyya et al. 2022). In addition, supply of organic or inorganic nutrients as amendments has been reported to enhance soil fertility by enhancing SOC concentration and plays a key role in carbon sequestration (Bolinder et al. 2020; Tanveer et al. 2020). Recently, the effect of organic amendment (crop residue), inorganic fertilizers (N: urea; P: single superphosphate; K: muriate of potash), and combination of both types of amendments on SCS under an intensive maize-sunflower cropping system grown in Indian semi-arid tropical soil (Anandakumar et al. 2022) was tested. The authors found that annual mean value of carbon sequestration was higher after addition of organic amendments (0.18 mg Carbon ha^{-1}) followed by the combination of both amendments (0.14 mg Carbon ha^{-1}). Moreover, they concluded that active carbon pools viz., microbial biomass carbon, labile carbon, water-soluble carbon, particulate carbon fraction, and passive pools, like humic acid and fulvic acid were enhanced after continuous application of farmyard manure in presence or absence of inorganic fertilizers. Nitrogen fertilization using adequate dose at an adequate moment has improved crop yield and stabilize carbon in soil (Snyder et al. 2009). In a meta-analysis conducted by Ngaba et al. (2022), the addition of low N doses (<30 kg N ha^{-1} yr^{-1}) in a period less than one year has notably increased SCS in Chinese soil. Moreover, N addition has promoted the decomposition of SOC and increased its concentration in soil (Li et al. 2018). However, application of nutrients may lead to a decrease of soil carbon due to the stimulation of soil microbial community that increases SOM decomposition (Chowdhury et al. 2021; Schipper et al. 2007) that indicated that soil carbon decreased after nutrient application, which may be due to the changes in SOMSOM, its leaching or losses through erosion.

3.2.3 Organic Amendments

Organic amendment addition to soil is very effective to improve SCS and soil aggregate. Various amendments, such as biochar, animal manure, and compost are promising strategies for soil carbon storage. Table 3.1 showed the effect of some organic amendments on SCS.

3.2.3.1 Biochar

Biochar is the result of thermal decomposition of different biomass in presence of low or no oxygen supply (Sasongko et al. 2021). It has been used as soil amendments for waste management, soil remediation, wastewater treatment, GHG mitigation, carbon sequestration, and soil fertility improvement. Large terrestrial amount of C is present in soil with 72% of wetland, 47% of grassland and 9% of forest (Yin et al. 2022). C is highly increased upon biochar addition to soil since biochar is rich carbon that can remain in soil for a longtime period due to its ability to resist chemical and biological degradation. Moreover, biochar application can improve SOC quality and reduce CO_2 emission (Geng et al. 2022). In a study conducted by Cui et al. (2013), the application of biochar increased SOC by up to 51% under wheat straw biochar treatments. Similar results have been demonstrated using mixed crop residues of wheat straw and peanut shell (El-Naggar et al. 2018) where SOC

Table 3.1 Effect of some organic amendments on soil carbon sequestration.

Feedstock	Dose used	Effects	References
Wheat straw biochar	$40\,t\,ha^{-1}$	SOC increased by 51%	Cui et al. (2013)
Municipal biowaste		SOC increased by 20%	Butnan et al. (2015)
Mixed crop residues of wheat straw and peanut shell	$8\,t\,ha^{-1}$	SOC increased by 56%	El-Naggar et al. (2018)
Woodchips of Japanese Cedar (*Cryptomeria japonica*)	5%	C sequestration ranged between 41.3 and 41.5 g C/kg soil	Lai et al. (2013)
Farmyard manure	$30\,Mg\,ha^{-1}$ farmyard	SOC stock increased by 37.1%	Li et al. (2021b)
Biosolid compost	$0.20\,kg$ of $C\,m^{-2}$	Net C sequestration up to 50% in the topsoil	Panettieri et al. (2022)

increased by 56%, respectively (Table 3.1). Carbon sequestration using biochar refers to the quantity of C restored biochar itself, which is equal biochar residues in soil after several years' abiotic and biotic aging (Yin et al. 2022). Between 21% and 33% of the carbon of the original biomass feedstock would be captured in soils for long term (Windeatt et al. 2014). This process is mainly due to the transformation of active C to stable or inert passive pool (Qambrani et al. 2017). The long-term stability of biochar is a fundamental criterion in carbon sequestration since the rate of carbon stability is estimated to reach 80% during the first 100 years (Papageorgiou et al. 2021). Once in soil, biochar may sequestrate C due to its recalcitrant nature, reduction of mineralization of native SOC, and the increase of the atmospheric C capture by plant (Yin et al. 2022).

3.2.3.2 Organic Residues

Organic fertilizers, such as animal manure and compost have proven to have a crucial role not only in plant growth and soil improvement but also in SCS. Various kinds of animal manure (chicken, cow, horse, sheep, and goat) and compost (animal manure, crop residues) have been used by farmers to increase soil fertility by increasing nutrient content which helps in ameliorating crop production. It has been reported that manure application effect is due to their alteration of soil characteristics, such as water holding capacity (WHC), soil pH, soil reaction, plant nutrient availability, cation exchange capacity (CEC), and organic matter content (Thangarajan et al. 2013). Animal manure has shown great influence on SOC level. According to Rui and Zhang (2010), animal manure application enhanced SOC level by $0.34\,Mg\,ha^{-1}\,yr^{-1}$. In a short-term (4 years) and long-term (12 years) field experiment conducted by Antonio et al. (2021), the application of animal manure increased SOC content in the topsoil layers (up to 10 cm) as compared to the control. Furthermore, five years of composted poultry manure application has significantly increased SOC level and favored SOC sequestration due to increased plant derived organic carbon input or direct poultry manure stabilization (Poblete-grant et al. 2020). Carbon content in soil and its sequestration is highly enhanced after animal manure application.

1.77 t C ha^{-1} was found sequestrated after application of a mixture of NPK and 5 t ha^{-1} of farmyard manure. Additionally, 4.9 and 1.04 t ha^{-1} was sequestrated after application of poultry manure and cow dung, respectively (Rahman et al. 2016). Beside its positive role in carbon sequestration, several researches have reported that the application of organic residues may cause significant increase of greenhouse gases (GHGs) (CO_2, CH_4, and N_2O) emissions (Mandal et al. 2020; Jat et al. 2022) due to its fast rate of degradation (Bolan et al. 2012).

3.2.4 Crop Rotation

Crop rotation consists of growing different crops sequentially on the same land parcel. The success of this agronomic practice is quite complex and governed by climate, weather, field size, and characteristics of its soil and crop proprieties (Schöning et al. 2022). This process has been reported to increase crop production, ameliorate soil characteristics by increasing its nutrient content and bacterial community, and reduce pesticide use as well as diseases and pests (Chowdhury et al. 2021; Li et al. 2021a; Young et al. 2021). Crop rotation has also been suggested as one of the soil management practices that have a positive impact on SCS. It has been reported that crop rotation can ameliorate soil quality by enhancing C pool in soil and change SOC concentration since it can reduce SOM decomposition (Chowdhury et al. 2021).

The effect of crop rotation on carbon sequestration is highly related to crop type and rotation length. These two factors may change SOC turnover due to the alteration of both quantity and quality of crop residue-derived carbon input into the soil (Liu et al. 2022). Crop rotation regime using rice–chickpea–mungbean has been found to increase SOC, particulate organic carbon, and carbon management index as compared to rice–chickpea and rice–wheat (Kumar et al. 2019). The authors suggested that carbon sequestration was favored in the order rice–chickpea–mungbean > rice–chickpea > rice–wheat. In fact, several cropping regimes, such as rice-based cropping system (Ghimire et al. 2017), sweet sorghum-based cropping system (Malobane et al. 2021) as well as soybean-based cropping system (Gura and Mnkeni 2019) have positive effects on carbon sequestration and soil aggregation. It was indicated that soybean-based cropping regime performed better with no-till, organic fertilizers, straw retention, more rotation cycles, longer rotation length, and a range of 100–200 kg N ha^{-1} y^{-1} of N fertilizer input rate (Liu et al. 2022).

3.2.5 Carbon Sequestration Potential of Agroforestry Systems

Agroforestry is an intermediate land-use practice in which trees are grown in combination with crops or pasture in the same field and thus combine the ecosystem characteristics of forests and cropland or pastures (Shi et al. 2018). Because of this combination, agroforestry is considered an integrated approach to sustainable land use, as it has several benefits on soil quality and health. Indeed, agroforestry has been demonstrated to be a promising approach for C sequestration in several countries, such as Canada (Stinson and Freedman 2001) and India (Singh et al. 2000). In fact, regarding its beneficial roles, recent projections indicated that the area of the world under agroforestry will increase substantially in the near future (Albrecht and Kandji 2003), which will have a great impact on C sequestration and greenhouse gas emission.

According to the Soil Science Society of America (SSSA), in the soil, C is sequestered in two ways: direct and/or indirectly (Dhyani et al. 2020). Direct soil C sequestration consists of the conversion of CO_2 into soil inorganic C compounds such as calcium and magnesium carbonates due to the inorganic chemical reactions. Indirect C sequestration concerns the atmospheric CO_2 captured by plants during photosynthesis and stored as plant biomass. This later represents the long living C pools. The long-living C pools in which C stored include the aboveground plant standing biomass such as timber, and belowground plant biomass such as roots, in wood products and in the soil, mostly through increased SOM content (Atangana et al. 2014). C pools sequestered by agroforestry could also include fruits, soil microorganisms, and any form of organic and inorganic C in soils (Atangana et al. 2014).

According to estimates, agroforestry could sequester C at a rate of 0.2–3.1 $t\,C\,ha^{-1}$ per year (Pandey 2002). Moreover, between 1995 and 2050, C sequestration in agroforestry is estimated of 7 Gt C, which cost a total of US$ 30 × 109 (Sathaye and Ravindranath 1998), indicating the promising role of agroforestry in climate change attenuation and greenhouse gas emission reduction. However, C sequestration potential of agroforestry systems varies depending on several factors, including environment and specific management systems. For example, in the same country (India), C sequestration in tree-bearing arid agroecosystems of Rajasthan was estimated of 23.46–47.36 $t\,C\,ha^{-1}$ per year, while the amount sequestered in the agroforestry of Uttar Pradesh is only 19.56 $t\,C\,ha^{-1}$ per year (Singh et al. 2000). Moreover, in Mexico, several farm forestry systems are viable, including live fences, coffee with shade trees, plantations, tree enrichment of fallows, and taungya, with a C sequestration potential ranging from 17.6 to 176.3 $Mg\,C\,ha^{-1}$ (Mg = 106 g) (De Jong et al. 1997). Likewise, in west and central Africa, up to 62% of C stocks found in primary forests are held by the cacao agroforests in the humid regions (Duguma et al. 2001). These results clearly indicated the important of the environment on C sequestration in agroforestry. Moreover, agricultural practices could also play an important role in C sequestration. Among others, integrating N-fixing trees into agroforestry has demonstrated to improve C sequestration (Lorenz and Lal 2014). For example, the introduction of *Acacia auriculiformis* and *Acacia mangium* in the agroforestry of Democratic Republic of Congo and the Republic of the Congo significantly improves the soil health through C sequestration (Koutika et al. 2021). In a similar study, using stable carbon isotope technique, Resh et al. (2002) compared the soil C pools under N-fixer trees with the non-N-fixer *Eucalyptus* at four tropical sites (two sites on Andisol soils in Hawaii and two sites on Vertisol and Entisol soils in Puert). They found that agroforestry with N-fixing trees (N-fixers) accumulate about 55% of C in soils than similar agroforestry but without N-fixing trees. It has been reported that the fact that mixing plantings with N-fixing trees could improve biomass production, and eventually soil organic C sequestration and pools (Nair et al. 2009; Gärdenäs et al. 2011). In addition, reports evidenced that the use of N-fixing trees in agroforestry induces a change in microbial C composer community (Resh et al. 2002). This result in a greater retention of relatively stable soil organic C.

3.2.6 Effect of Water Quality and Irrigation Practices on Soil Sequestration

In irrigated soils, inorganic C reactions are among the most significant chemical processes that take place and may contribute to the overall amount of C sequestered in soil

(Entry et al. 2004). In fact, water quality and management and irrigation practices have been considered among the most sustainable approaches that could help in C sequestration and climate change mitigation (Trost et al. 2013). Several previous studies indicated that an improved water supply led to an increased biomass generation, which contributed to a higher input of C into the soil mainly in the form of roots, leaves, etc. (Entry et al. 2008; Kochsiek et al. 2009). In the line with these findings, Li et al. (2009) and Su et al. (2010) investigated the changes in soil organic C after converting uncultivated dry land to irrigated cropland. Ten years after conversion, a strong increase in plant growth and biomass was observed and consequently a higher input of organic matter into the soil (C sequestration). Water irrigation management could also protect soil against decomposition by the induction of micro- and macro-aggregates formation, leading to higher C sequestration. In fact, some authors consider the content of micro-aggregates an ideal indicator for the C sequestration (Kong et al. 2005). In the line with this, Gillabel et al. (2007) reported that C inputs and soil organic C stocks were higher and larger in the irrigated fields than the dryland cultivated fields. The authors explained the difference by the fact that irrigation helped increase micro-aggregates leading to a sequestration of C inside micro-aggregates. Likewise, the intensity of irrigation may affect aggregate building and therefore the SOC content. By assessing the effect of irrigated farming on soil C sequestration and dynamics in two California soils, Wu et al. (2008) indicated that long-term irrigated farming significantly increased soil organic C. This was explained by an increase in the soil organic C below the 10-cm soil depth, along with a significant increase in the soil inorganic C because of the transformation of CO_2 to carbonates. Similarly, in their investigations on deficit irrigation, Blanco-Canqui et al. (2010) indicated that soil aggregates with high diameter (over 0.5 mm) increased as the amount of applied irrigation water was increased, while aggregates under 0.5 mm decreased.

Consequently, the soil organic C content increased. Water quality is another factor that could have an impact on C sequestration. To confirm this hypothesis, Eshel et al. (2007) conducted a field study to assess if irrigation water quality could have an impact on C sequestration in the upper 4 m of soil compared to a non-irrigated field. Based on stable isotope analyses (^{13}C and ^{18}O) and radiocarbon dating, authors demonstrated that the irrigation with fresh water, for more than 75 years, could sequester about 7.15 $kg\,m^{-2}$ $(4\,m)^{-1}$ of C, while the sequestration of C under treated effluent was estimated of 0.9–2.4 $kg\,m^{-2}$ $(4\,m)^{-1}$. Besides the rate, duration, and the quality, the irrigation system may also affect C sequestration by influencing soil structural stability.

3.2.7 Contribution of Microorganisms to Soil Carbon Sequestration

Soil microorganisms could increase the level of C inputs and decrease the levels of C outputs in the soils (Figure 3.1). In fact, soil microbial communities have an important role in C sequestration (Fang et al. 2014). Previous reports evidenced that plant growth promoting rhizobacteria (PGPR) have the ability to reduce CO_2 and could contribute to C sequestration and mitigate atmospheric CO_2 levels (Ramakrishna et al. 2020). In a study of Nie et al. (2015), plant inoculation with *Pseudomonas fluorescens* increased plant productivity and showed high ability to mitigate high atmospheric CO_2 levels by increasing terrestrial C sequestration, especially in high-CO_2 ecosystems. These findings clearly indicate the important role of PGPR in atmospheric CO_2 sequestration. Other PGPR can also

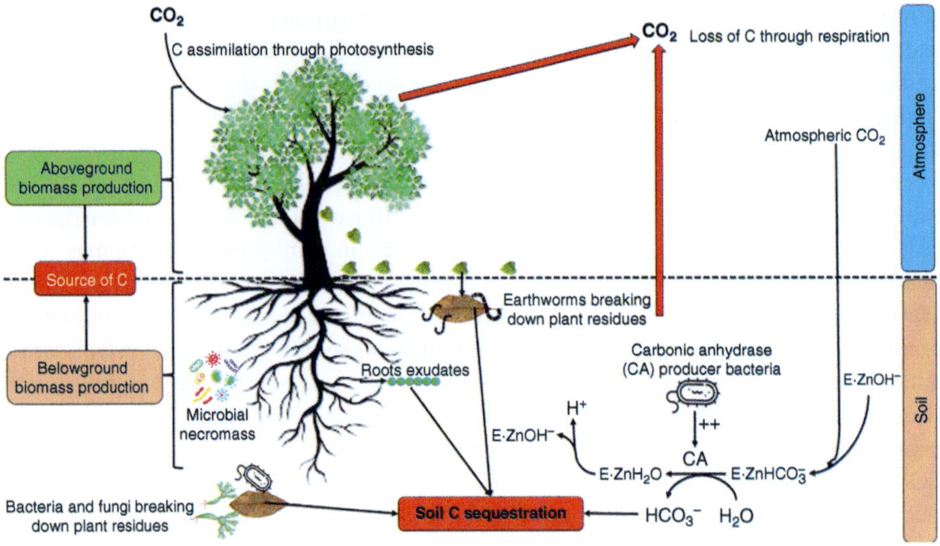

Figure 3.1 Schematic summarizing the effects of trees and soil microorganisms on carbon sequestration in agroforestry.

contribute to C sequestration through synthesis of carbonate, which has the potential to capture atmospheric CO_2 and sequester C (Peng et al. 2010). For example, *Bacillus mucilaginosus* has been reported to produce carbonic anhydrase, which firstly captures the atmospheric CO_2 and then fixes it through bacterial metabolism (Zhang et al. 2011). Thus, by fixing atmospheric CO_2, this mechanism stimulates carbonate production, resulting in the synthesis of organic compounds or carbonated minerals, lowering atmospheric CO_2 levels, and increasing soil C sequestration. Beside carbonate synthesis, other PGPR such as *Bacillus pumilus* (Komala and Khun 2014) and *Bacillus cereus* (Han et al. 2013), which have the ability to sediment carbonates, were reported effective to reduce CO_2 and contribute to C sequestration. Bacteria can also contribute to C sequestration through their interaction with plants. Indeed, atmospheric CO_2 is captured by plants during their interactions with microorganisms (plant-microbe interactions), deposited as plant root material, and eventually incorporated into the soil microorganisms and SOM (Wu et al. 2009). Moreover, in the case of legumes-rhizobia symbiosis, plant secret flavonoids compound in rhizospheric soil to attract bacteria (Cooper 2007). These compounds are particularly C-based compound, which represent an excellent option for C sequestration.

Besides bacteria, fungi have been known for their contribution in C sequestration, directly or indirectly (Treseder and Holden 2013). Thus, since C builds up fungal mycelium, C sequestration is more important as fungal mycelia more produced (Treseder and Holden 2013; Ahmed et al. 2020). Therefore, fungal mycelia could be considered as a vital C sink to the soils. Likewise, in soil, glomalin is an important reservoir of C and N (Sarfraz et al. 2019). Indeed, to protect the hyphae from drying out, mycorrhizal fungi and *Scutellospora*, *Acaulospora*, *Entrophospora*, *Lomus*, and *Gigaspora* released glomalin from spore and hyphae, which remarkably contribute to C sequestration (Singh et al. 2000; Pal and Pandey 2014). Moreover, fungi could improve C sequestration indirectly. Indeed, during their interaction with plants, mycorrhizal fungi improve plant growth through enhancing water and nutrient use efficiency (Malyan et al. 2019). Thus, the better growth of plants results in faster removal of atmospheric CO_2 through its conversion into plant biomass and as a results C sequestration in soil.

3.3 Conclusion

SCS is a key means of facing the increase in CO_2 emission and C loss. Capture of carbon in soil may ameliorate soil characteristics, increase SOM, which affect positively soil fertility, thus increase crop production, and ensure food security. Sequestrating carbon in soil mainly occurs through abiotic and biotic approaches. The present chapter is focused on biotic strategies used for C sequestration. Numerous agricultural management practices have been used, such as amendment of soil with different organic substances (biochar, animal manure, compost. . .) and chemical nutrients that affects positively soil proprieties and C storage. Moreover, no, or less tillage, crop rotation, water management, and agroforestry microorganisms are widely applied to increase C sequestration in soil.

References

Aduhene-chinbuah, J., Sugihara, S., Komatsuzaki, M. et al. (2022). No tillage increasesti SOM in labile fractionlf but not stable fractionnsf of andosolsa from a long-term experimentlte in Japan. *Agronomy* 12 (479): 1–15. https://doi.org/10.3390/agronomy12020479.

Ahmed, A.A.Q., Odelade, K.A., and Babalola, O.O. (2020). Microbial inoculantsi for improving carbon sequestrationics in agroecosystemsa to mitigate climate change. In: *Handbook of Climate Change Resilience* (ed. W.L. Filho), 381–401. https://doi.org/10.1007/978-3-319-93336-8_119.

Albrecht, A. and Kandji, S.T. (2003). Carbon sequestration in tropical agroforestry systems. *Agric. Ecosyst. Environ.* 99 (1–3): 15–27. https://doi.org/10.1016/S0167-8809(03)00138-5.

Anandakumar, S., Bakhoum, N., Chinnadurai, C. et al. (2022). Impact of long-term nutrient management on sequestration and dynamics of soil organic carbon in a semi-arid tropical Alfisol of India. *Appl. Soil Ecol.* 177 (June): 104549. https://doi.org/10.1016/j.apsoil.2022.104549.

Antonio, L., Rodrigues, T., Jos, S. et al. (2021). Short- and long-term effects of animal manures and mineral fertilizer on carbon stocks in subtropical soil under no-tillage. *GeodermaGeoderma* 386: (December 2019). https://doi.org/10.1016/j.geoderma.2020.114913.

Atangana, A., Khasa, D., Chang, S., and Degrande, A. (2014). Carbon sequestrations in agroforestry systems. In: *Tropical Agroforestry*, 1e (ed. A. Atangana et al.), 1–380. Springer Science+Business Media Dordrecht. https://doi.org/10.1007/978-94-007-7723-1_10.

Bhattacharyya, S.S., Ros, G.H., Furtak, K. et al. (2022). Soil carbon sequestration – ana interplay between soil microbial community and soil organic matter dynamics. *Sci. Total Environ.* 815. https://doi.org/10.1016/j.scitotenv.2022.152928.

Blanco-Canqui, H., Klocke, N.L., Schlegel, A.J. et al. (2010). Impacts of deficit irrigationdi on carbon sequestrationcs and soil physical propertiesspp under no-till. *Soil Sci. Soc. Am. J.* 74 (4): 1301–1309. https://doi.org/10.2136/sssaj2009.0364.

Bolan, N.S., Kunhikrishnan, A., Choppala, G.K. et al. (2012). Stabilization of carbon in composts and biochars in relation to carbon sequestration and soil fertility. *Sci. Total Environ.* 424: 264–270. https://doi.org/10.1016/j.scitotenv.2012.02.061.

Bolinder, M.A., Crotty, F., Elsen, A. et al. (2020). The effect of crop residues, cover crops, manures and nitrogen fertilization on soil organic carbon changes in agroecosystems: a synthesis of reviews. *Mitig. Adapt. Strateg. Glob. Chang.* 25: 929–952.

Butnan, S., Deenik, J.L., Toomsan, B. et al. (2015). Biochar characteristics and application rates affecting corn growth and properties of soils contrasting in texture and mineralogy. *Geoderma* 237: 105–116. https://doi.org/10.1016/j.geoderma.2014.08.010.

Chaopu, T., Xia, L., Chang, S.X., and Yan, X. (2019). Potential for mitigating global agricultural ammonia emission: a meta-analysis. *Environ. Pollut.* 245: 141–148. https://doi.org/10.1016/j.envpol.2018.10.124.

Chowdhury, S., Bolan, N., Farrell, M. et al. (2021). Role of cultural and nutrient management practices in carbon sequestration in agricultural soil. In: *Advances in Agronomy*, 1e, 131–196. Elsevier Inc. https://doi.org/10.1016/bs.agron.2020.10.001.

Cooper, J.E. (2007). Early interactions between legumes and rhizobia: disclosing complexity in a molecular dialogue. *J. Appl. Microbiol.* 103 (5): 1355–1365. https://doi.org/10.1111/j.1365-2672.2007.03366.x.

Cornelissen, G., Martinsen, V., Shitumbanuma, V. et al. (2013). Biochar effecte on maize yield and soil characteristics in five conservation farming sites in Zambia. *Agronomy* 3 (2): 256–274. https://doi.org/10.3390/agronomy3020256.

Cui, L., Yan, J., Yang, Y. et al. (2013). Influence of biochar on microbial activities of heavy metals contaminated paddy fields. *Bioresources* 8 (4): 5536–5548. https://doi.org/10.15376/biores.8.4.5536-5548.

De Jong, B.H.J., Tipper, R., and Taylor, J. (1997). A framework for monitoring and evaluating carbon mitigation by farm forestry projects: example of a demonstration project in Chiapas, Mexico. *Mitig. Adapt. Strateg. Glob. Chang.* 231–246. https://doi.org/10.1023/b:miti.0000004479.17784.f8.

Dhyani, S.K., Ram, A., Newaj, R. et al. (2020). Agroforestry for carbon sequestrationcs in tropicalt India. In: *Carbon Management in Tropical and subs-Tropical Terrestrial Systems* (ed. P.K. Ghosh et al.), 312–331. Springer: Singapore. https://doi.org/10.1007/978-981-13-9628-1_19.

Dissanayake, P.D., You, S., Igalavithana, A.D. et al. (2020). Biochar-based adsorbents for carbon dioxide capture: a critical review. *Renew. Sust. Energ. Rev.* 119 (November 2019): 109582. https://doi.org/10.1016/j.rser.2019.109582.

Duguma, B., Gockowski, J., and Bakala, J. (2001). Smallholder cacao (Theobroma cacao Linn.) cultivation in agroforestry systems of West and Central Africa: challenges and opportunities. *Agrofor. Syst.* 51 (3): 177–188. https://doi.org/10.1023/A:1010747224249.

El-Naggar, A., Awad, Y.M., Tang, X.Y. et al. (2018). Biochar influences soil carbon pools and facilitates interactions with soil: a field investigation. *Land Degrad. Dev.* 29 (7): 2162–2171. https://doi.org/10.1002/ldr.2896.

Entry, J.A., Sojka, R.E., and Shewmaker, G.E. (2004). Irrigation increases inorganic carbon in agricultural soils. *Environ. Manag.* 33 (Suppl. 1): 309–317. https://doi.org/10.1007/s00267-003-9140-3.

Entry, J.A., Mills, D.E., Mathee, K. et al. (2008). Influence of irrigated agriculture on soil microbial diversity. *Appl. Soil Ecol.* 40 (1): 146–154. https://doi.org/10.1016/j.apsoil.2008.03.012.

Eshel, G., Fine, P., and Singer, M.J. (2007). Total soil carbons and water quality: an implication for carbon sequestration. *Soil Sci. Soc. Am. J.* 71 (2): 397–405. https://doi.org/10.2136/sssaj2006.0061.

Fang, H., Cheng, S., Wang, Y. et al. (2014). Changes in soil heterotrophic respiration, carbon availability, and microbial function in seven forests along a climate gradient. *Ecol. Res.* 29 (6): 1077–1086. https://doi.org/10.1007/s11284-014-1194-6.

Gärdenäs, A.I., Ågren, G.I., Bird, J.A. et al. (2011). Knowledge gaps in soil carbon and nitrogen interactions – from molecular to global scale. *Soil Biol. Biochem.* 43 (4): 702–717. https://doi.org/10.1016/j.soilbio.2010.04.006.

Geng, N., Kang, X., Yan, X. et al. (2022). Biochar mitigation of soil acidification and carbon sequestration is influenced by materials and temperature. *Ecotoxicol. Environ. Saf.* 232: 113241. https://doi.org/10.1016/j.ecoenv.2022.113241.

Ghimire, R., Lamichhane, S., Acharya, B.S. et al. (2017). Tillage, crop residue, and nutrient management effects on soil organic carbon in rice-based cropping systems: a review. *J. Integr. Agric.* 16 (1): 1–15. https://doi.org/10.1016/S2095-3119(16)61337-0.

Gillabel, J., Denef, K., Brenner, J. et al. (2007). Carbon sequestrations and soil aggregationsa in center-pivot irrigatedcpi and dryland cultivated farming systems. *Soil Sci. Soc. Am. J.* 71 (3): 1020–1028. https://doi.org/10.2136/sssaj2006.0215.

Gura, I. and Mnkeni, P.N.S. (2019). Crop rotation and residue management effects under no till on the soil quality of a haplich Cambisol in Alice, eastern capeec, South Africa. *Geoderma* 337 (November 2018): 927–934. https://doi.org/10.1016/j.geoderma.2018.10.042.

Han, J., Lian, B., and Ling, H. (2013). Induction of calcium carbonate by *Bacillus cereus*. *Geomicrobiol J.* 30 (8): 682–689. https://doi.org/10.1080/01490451.2012.758194.

Helgason, B.L., Walley, F.L., and Germida, J.J. (2010). No-till soil management increases microbial biomass and alters community profiles in soil aggregates. *Appl. Soil Ecol.* 46 (3): 390–397. https://doi.org/10.1016/j.apsoil.2010.10.002.

Hussain, M., Farooq, M., Nawaz, A. et al. (2017). Biochar for crop production: potential benefits and risks. *J. Soils Sediments* 17 (3): 685–716. https://doi.org/10.1007/s11368-016-1360-2.

Jat, M.L., Chakraborty, D., Ladha, J.K. et al. (2022). Carbon sequestration potential, challenges, and strategies towards climate action in smallholder agricultural systems of South Asia. *Crop. Environ.* 1 (1): 86–101. https://doi.org/10.1016/j.crope.2022.03.005.

Kochsiek, A.E., Knops, J.M.H., Walters, D.T., and Arkebauer, T.J. (2009). Impacts of management on decomposition and the litter-carbon balance in irrigated and rainfed no-till agricultural systems. *Agric. For. Meteorol.* 149 (11): 1983–1993. https://doi.org/10.1016/j.agrformet.2009.07.004.

Komala, T. and Khun, T.C. (2014). Biological carbon dioxide sequestration potential of bacillus pumilus. *Sains Malays.* 43 (8): 1149–1156.

Kong, A.Y.Y., Six, J., Bryant, D.C. et al. (2005). The relationship between carbon input, aggregation, and soil organic carbon stabilization in sustainable cropping systems. *Soil Sci. Soc. Am. J.* 69 (4): 1078–1085. https://doi.org/10.2136/sssaj2004.0215.

Koutika, L.S., Taba, K., Ndongo, M., and Kaonga, M. (2021). Nitrogen-fixing trees increase organic carbon sequestration in forest and agroforestry ecosystems in the Congo basin. *Reg. Environ. Chang.* 21 (4): https://doi.org/10.1007/s10113-021-01816-9.

Kumar, N., Nath, C.P., Hazra, K.K. et al. (2019). Impact of zero-till residue management and crop diversification with legumes on soil aggregation and carbon sequestration. *Soil Tillage Res.* 189 (December 2018): 158–167. https://doi.org/10.1016/j.still.2019.02.001.

Lai, W., Lai, C., Ke, G. et al. (2013). The effects of woodchip biochar application on crop yield, carbon sequestration and greenhouse gas emissions from soils planted with rice or leaf beet. *J. Taiwan Inst. Chem. Eng.* 44 (6): 1039–1044. https://doi.org/10.1016/j.jtice.2013.06.028.

Li, X.G., Li, Y.K., Li, F.M. et al. (2009). Changes in soil organic carbon, nutrients and aggregation after conversion of native desert soil into irrigated arable land. *Soil Tillage Res.* 104 (2): 263–269. https://doi.org/10.1016/j.still.2009.03.002.

Li, J.H., Li, F., Li, W.J., and Chen, S. (2018). Nitrogen additions promote decomposition of soil organic carbon in a Tibetan Alpine Meadow. *Soil Sci. Soc. Am. J.* 82: 614–621. https://doi.org/10.2136/sssaj2017.12.0417.

Li, Y., Zhang, Q., Cai, Y. et al. (2020). Minimum tillage and residue retention increase soil microbial population size and diversity: implicationsi for conservation tillage. *Sci. Total Environ.* 716: 137164. https://doi.org/10.1016/j.scitotenv.2020.137164.

Li, M., Guo, J., Ren, T. et al. (2021a). Crop rotation history constrains soil biodiversity and multifunctionality relationships. *Agric. Ecosyst. Environ.* 319 (September 2020): 107550. https://doi.org/10.1016/j.agee.2021.107550.

Li, X. et al. (2021b). Organic amendments affect soil organic carbon sequestration and fractions in fields with long-term contrasting nitrogen applications. *Agric. Ecosyst. Environ.* 322 (October 2020): 107643. https://doi.org/10.1016/j.agee.2021.107643.

Liu, X., Tan, S., Song, X. et al. (2022). Response of soil organic carbon content to crop rotation and its controls: a global synthesis. *Agric. Ecosyst. Environ.* 335 (May): 108017. https://doi.org/10.1016/j.agee.2022.108017.

Lorenz, K. and Lal, R. (2014). Soil organic carbon sequestration in agroforestry systems. A review. *Agron. Sustain. Dev.* 34 (2): 443–454. https://doi.org/10.1007/s13593-014-0212-y.

Majumder, S., Neogi, S., Dutta, T. et al. (2019). The impact of biochar on soil carbon sequestration: Meta-analytical approach to evaluating environmental and economic advantages. *J. Environ. Manag.* 250 (November 2018): 109466. https://doi.org/10.1016/j.jenvman.2019.109466.

Malobane, M.E., Nciizah, A.D., Bam, L.C. et al. (2021). Soil microstructure as affected by tillage, rotation and residue management in a sweet sorghum-based cropping system in soils with low organic carbon content in South Africa. *Soil Tillage Res.* 209 (December 2019): 104972. https://doi.org/10.1016/j.still.2021.104972.

Malyan, S.K., Kumar, A., Baram, S. et al. (2019). Role of Fungi in climate change abatement through carbon sequestration. In: *Recent Advancement in White Biotechnology throught Fungi, Fungal Biology* (ed. A.N. Yadav, S.S.S. Mishra, and A. Gupta), 1–13. Springer https://doi.org/10.1007/978-3-030-25506-0_11.

Mandal, A., Majumder, A., Dhaliwal, S.S. et al. (2020). Impact of agricultural management practices on soil carbon sequestration and its monitoring through simulation models and remote sensing techniques: a review. *Crit. Rev. Environ. Sci. Technol.* 0 (0): 1–49. https://doi.org/10.1080/10643389.2020.1811590.

de Medeiros Barbosa, V.F.A. (2015). Planning. In: *Sugarcane: Agricultural Production, Bioenergy and Ethanol* (ed. F. Santos, A. Borém, and C. Caldas), 35–51. Elsevier Inc. https://doi.org/10.1016/B978-0-12-802239-9.00003-7.

Nair, P.K.R., Kumar, B.M., and Nair, V.D. (2009). Agroforestry as a strategy for carbon sequestration. *J. Plant Nutr. Soil Sci.* 172 (1): 10–23. https://doi.org/10.1002/jpln.200800030.

Ngaba, M.Y.J., Uwiragiye, Y., Bol, R. et al. (2022). Low-level nitrogen and short-term addition increase soil carbon sequestration in Chinese forest ecosystems. *Catena* 215 (April): 106333. https://doi.org/10.1016/j.catena.2022.106333.

Nie, M., Bell, C., Wallenstein, M.D., and Pendall, E. (2015). Increased plant productivity and decreased microbial respiratory C loss by plant growth-promoting rhizobacteria under elevated CO_2. *Sci. Rep.* 5: 1–6. https://doi.org/10.1038/srep09212.

Oppong, E., Yakubu, A., Kugblenu, Y.O. et al. (2019). Field Crops Research Impact of rice straw biochar and irrigation on maize yield, intercepted radiation and water productivity in a tropical sandy clay loam. *Field Crops Res.* 243 (February): 107628. https://doi.org/10.1016/j.fcr.2019.107628.

Pal, A. and Pandey, S. (2014). Role of glomalin in improving soil fertility: a review. *Int. J. Plant Soil Sci.* 3 (9): 1112–1129. https://doi.org/10.9734/ijpss/2014/11281.

Pandey, D.N. (2002). Carbon sequestration in agroforestry systems. *Clim. Pol.* 2 (4): 367–377. https://doi.org/10.3763/cpol.2002.0240.

Panettieri, M., Moreno, B., De Sosa, L.L. et al. (2022). Soil management and compost amendment are the main drivers of carbon sequestration in rainfed olive trees agroecosystems: an evaluation of chemical and biological markers. *Catena* 214 (April): https://doi.org/10.1016/j.catena.2022.106258.

Papageorgiou, A., Azzi, E.S., Enell, A., and Sundberg, C. (2021). Biochar produced from wood waste for soil remediation in Sweden: carbon sequestration and other environmental impacts. *Sci. Total Environ.* 145953. https://doi.org/10.1016/j.scitotenv.2021.145953.

Peng, Y.L., Zhao, H.Z., Yang, Q.Z., and Song, Z. (2010). Progress in microbial and enzyme immobilization of carbon dioxide. *ChemBioEngChemBioEng* 27 (7): 10–13.

Poblete-grant, P., Suazo-hernández, J., Condron, L. et al. (2020). Regional soils available P, soil organic carbon and aggregation as affected by long-term poultry manure application to Andisols under pastures in southerns Chile. *Geoderma Regional.Reg.* 21: 1–10. https://doi.org/10.1016/j.geodrs.2020.e00271.

Qambrani, N.A., Rahman, M.M., Won, S. et al. (2017). Biochar properties and eco-friendly applications for climate change mitigation, waste management, and wastewater treatment: a review. *Renew. Sust. Energ. Rev.* 79 (May): 255–273. https://doi.org/10.1016/j.rser.2017.05.057.

Rahman, F., Rahman, M., Rahman, G.K.M.M. et al. (2016). Effect of organic and inorganic fertilizers and rice straw on carbon sequestration and soil fertility under a rice – rice cropping pattern. *Carbon Management* 7 (1–2): 41–43. https://doi.org/10.1080/17583004.2016.1166425.

Ramakrishna, W., Rathore, P., Kumari, R., and Yadav, R. (2020). Brown gold of marginal soil: plantp growth promoting bacteria to overcome plant abiotic stress for agriculture, biofuels and carbon sequestration. *Sci. Total Environ.* 711: 135062. https://doi.org/10.1016/j.scitotenv.2019.135062.

Resh, S.C., Binkley, D., and Parrotta, J.A. (2002). Greater soil carbon sequestration under nitrogen-fixing trees compared with Eucalyptus species. *Ecosystems* 5 (3): 217–231. https://doi.org/10.1007/s10021-001-0067-3.

Rui, W. and Zhang, W. (2010). Effect size and duration of recommended management practices on carbon sequestration in paddy field in Yangtze Delta plainp of China: a meta-analysis. *Agric. Ecosyst. Environ.* 135: 199–205. https://doi.org/10.1016/j.agee.2009.09.010.

Sarfraz, R., Hussain, A., Sabir, A. et al. (2019). Role of biochar and plant growth promoting rhizobacteria to enhance soil carbon sequestration—a review. *Environ. Monit. Assess.* 191 (4). https://doi.org/10.1007/s10661-019-7400-9.

Sasongko, D., Gunawan, D., and Indarto, A. (2021). Biochar-based water treatment systems for clean water provision synthesis of biochar. In: *Handbook ofAssisted and Amendment-Enhanced Sustainable Remediation Technology*, First Edit (ed. M.N.V. Prasad), 77–101. Wiley.

Sathaye, J.A. and Ravindranath, N.H. (1998). Climate change mitigation in the energy and forestry sectors of developing countries. *Annu. Rev. Energy Environ.* 23 (1): 387–437. https://doi.org/10.1146/annurev.energy.23.1.387.

Schipper, L.A., Baisden, W.T., Parfitt, R.L. et al. (2007). Large losses of soil C and N from soil profiles under pasture in New Zealand during the past 20 years. *Glob. Chang. Biol.* 13: 1138–1144. https://doi.org/10.1111/j.1365-2486.2007.01366.x.

Schöning, J., Wachter, P., and Trautz, D. (2022). Crop rotation and management tools for every farmer? The current status on crop rotation and management tools for enabling sustainable agriculture worldwide. *SSRN Electron. J.* 3: (March 2022). https://doi.org/10.2139/ssrn.4072691.

Shi, L., Feng, W., Xu, J., and Kuzyakov, Y. (2018). Agroforestry systems: Meta-analysis of soil carbon stocks, sequestration processes, and future potentials. *Land Degrad. Dev.* 29 (11): 3886–3897. https://doi.org/10.1002/ldr.3136.

Singh, T.P., Varalakshmi, V., and Ahluwalia, S.K. (2000). Carbon sequestration through farm forestry: case from India. *Indian Forester* 126: 1257–1264.

Snyder, C.S., Bruulsema, T.W., Jensen, T.L., and Fixen, P.E. (2009). Review of greenhouse gas emissions from crop production systems and fertilizer management effects. *Agric. Ecosyst. Environ.* 133: 247–266. https://doi.org/10.1016/j.agee.2009.04.021.

Stinson, G. and Freedman, B. (2001). Forests and agroecosystems. *Mitig. Adapt. Strateg. Glob. Chang.* 6: 1–23.

Su, Y.Z., Yang, R., Liu, W.J., and Wang, X.F. (2010). Evolution of soil structure and fertility after conversion of native sandy desert soil to irrigated cropland in arid region, China. *Soil Sci.* 175 (5): 246–254. https://doi.org/10.1097/SS.0b013e3181e04a2d.

Tanveer, S.K., Lu, X., Shah, S.-U.-S. et al. (2020). Soil carbon sequestrationcs through agronomic management practices. In: *CO2 Sequestration* (ed. L.A. Frazão, A.M. Silva-Olaya, and J.C. Silva), 1–17. IntechOpen. https://doi.org/10.5772/intechopen.87107.

Thangarajan, R., Bolan, N.S., Tian, G. et al. (2013). Organic amendment application on greenhouse gas emission from soil. *Sci. Total Environ.* 465: 72–96. https://doi.org/10.1016/j.scitotenv.2013.01.031.

Treseder, K.K. and Holden, S.R. (2013). Fungal carbon sequestration. *Science* 340 (6127): 1528–1529. https://doi.org/10.1126/science.1236338.

Trost, B., Prochnow, A., Drastig, K. et al. (2013). Irrigation, soil organic carbon and N2O emissions. A review. *Agron. Sustain. Dev.* 33 (4): 733–749. https://doi.org/10.1007/s13593-013-0134-0.

Windeatt, J.H., Ross, A.B., Williams, P.T. et al. (2014). Characteristics of biochars from crop residues: potentialp for carbon sequestration and soil amendment. *J. Environ. Manag.* 146: 189–197. https://doi.org/10.1016/j.jenvman.2014.08.003.

WMO (World Meteorological Organization) (2018). *The State of Greenhouse Gases in the Atmosphere Based on Global Observations through 2016*. World Meteorological Organization Bulletin. ISSN 2078-0796.

Wu, L., Wood, Y., Jiang, P. et al. (2008). Carbon sequestrations and dynamicsd of two irrigated agricultural soilstias in California. *Soil Sci. Soc. Am. J.* 72 (3): 808–814. https://doi.org/10.2136/sssaj2007.0074.

Wu, C.H., Bernard, S.M., Andersen, G.L., and Chen, W. (2009). Developing microbe-plant interactions for applications in plant-growth promotion and disease control, production of useful compounds, remediation and carbon sequestration. *Microb. Biotechnol.* 2 (4): 428–440. https://doi.org/10.1111/j.1751-7915.2009.00109.x.

Yin, J., Zhao, L., Xu, X. et al. (2022). Evaluation of long-term carbon sequestration of biochar in soil with biogeochemical fi eld model. *Sci. Total Environ.* 822: 153576. https://doi.org/10.1016/j.scitotenv.2022.153576.

Young, M.D., Ros, G.H., and de Vries, W. (2021). Impacts of agronomic measures on crop, soil, and environmental indicators: a review and synthesis of meta-analysis. *Agric. Ecosyst. Environ.* 319 (July): 107551. https://doi.org/10.1016/j.agee.2021.107551.

Zafar, S.A., Hameed, A., Nawaz, M.A. et al. (2018). Mechanisms and molecular approaches for heat tolerance in rice (Oryza sativa L.) under climate change scenario. *J. Integr. Agric.* 17 (4): 726–738. https://doi.org/10.1016/S2095-3119(17)61718-0.

Zhang, Z., Lian, B., Hou, W. et al. (2011). Bacillus mucilaginosus can capture atmospheric CO_2 by carbonic anhydrase. *Afr. J. Microbiol. Res.* 5 (2): 106–112.

Zhang, L., Wang, G., Zheng, Q. et al. (2017). Quantifying the impacts of agricultural management and climate change on soil organic carbon changes in the uplands of easterne China. *Soil Tillage Res.* 174 (June): 81–91. https://doi.org/10.1016/j.still.2017.06.005.

Zhao, B., Xu, R., Ma, F. et al. (2016). Effects of biochars derived from chicken manure and rape straw on speciation and phytoavailability of cdc to maize in artificially contaminated loess soil. *J. Environ. Manag.* 184: 569–574. https://doi.org/10.1016/j.jenvman.2016.10.020.

4

Soil Degradation, Resilience, Restoration, and Sustainable Use

Diana Cota-Ungson[1], *Yolanda González-García*[2], *and Antonio Juárez-Maldonado*[2]

[1] *Doctorado en Agricultura Protegida, Universidad Autónoma Agraria Antonio Narro, Saltillo, Coahuila, 25315, México*
[2] *Departamento de Botánica, Universidad Autónoma Agraria Antonio Narro, Saltillo, Coahuila, 25315, México*

4.1 Introduction

The soil is considered one of the most diverse habitats on the planet, 1 g of soil can host millions of microorganisms, as well as hundreds or thousands of invertebrate species per square meter; furthermore, most insects spend at least one stage of their life in the soil, while for terrestrial plants it is the most important substrate (Rodrigo-Comino et al. 2020).

The soil is a fundamental resource for the development of the human population, since it is the basis of several processes such as agriculture and livestock from which food is obtained, so the health of the soil is essential to ensure the production of food in the world (Rinot et al. 2019). But the soil also participates in other important processes that have to do with water quality, ecosystem services, among other important functions (Brevik et al. 2019; Lehmann et al. 2020). It is considered that the soil is related up to 98.8% in the production of food for humanity (Kopittke et al. 2019), being perhaps the most important function, in addition to the ecosystem services that are essential for human health and well-being (Brevik et al. 2020, 2019). Hence, the health of the soil must be a priority for humanity, with the objective of maintaining the capacity of the soil to function, sustaining the productivity of crops and animals, maintaining, or improving environmental sustainability and improving human health (Yang et al. 2020). Currently, not only is the capacity of the soil for crop production considered as part of its health, but also the influence it has on water quality, environmental impact such as climate change, and in human health (Lehmann et al. 2020). Ultimately, land, water, energy, and food, are resources that are intimately intertwined due to the complex interactions that exist between them, and the growing demand of the population that is increasing (Rodrigo-Comino et al. 2020).

Soil is a limited resource on our planet, which is why it is imperative to develop strategies to conserve it and, if possible, regenerate it. Since due to human activities such as

Agroecological Approaches for Sustainable Soil Management, First Edition.
Edited by Majeti Narasimha Vara Prasad and Chitranjan Kumar.

agriculture, the soil is rapidly degrading, the organic matter of the soil is being lost, which in turn releases greenhouse gases into the environment, the overapplication of fertilizers causes pollution problems, coupled with erosion, acidification, salinization, and loss of genetic diversity (Kopittke et al. 2019).

For this reason, it is essential to determine the causes that affect the health of the soil with the primary objective of finding alternatives to conserve it. Fortunately, today various strategies have been sought that have made it possible to achieve this goal. In agricultural production systems, a conservation agriculture approach has been sought to protect soil biological diversity, and influence other areas such as climate change, ecosystem services, food security, and water security (Mugandani et al. 2021). In addition to this, what is known as regenerative agriculture is being developed, which considers various practices such as the incorporation of crop residues, cover crops, reduced tillage, and other good agricultural practices that consider the agricultural system with a holistic vision (Giller et al. 2021).

In this chapter, the impacts of human activity on soil degradation and loss will be described, in such a way that the causes and origins that affect soil health will be adequately located. In addition, the different strategies that can be used to achieve soil restoration and good health will be presented. And finally, a soil management approach will be proposed that allows achieving a sustainable use, and therefore, that allows maintaining the soil health over time.

4.2 Impacts of Human Activity on Soil Degradation

Human activity definitely affects the health of the soil through degradation. There is a diversity of human activities that negatively impact the soil, among which are mainly agriculture (Li et al. 2022), ranching through overgrazing (Bayat et al. 2022), and mining (Silveira et al. 2021) (Figure 4.1).

4.2.1 Agriculture

Agriculture in general impacts on the soil when it converts natural ecosystems into farmland (Li et al. 2022). Particularly intensive farming practices, such as monoculture and simplified two-crop rotation systems that endanger land quality and degradation, also threaten essential ecosystem services for crop production, reducing agricultural productivity (Wang et al. 2021).

Tillage also has a direct impact on soil erosion by depleting soil organic matter and reducing plant cover (Zhang et al. 2016). From a mechanical point of view, tillage presupposes several mechanisms of mechanical disturbance of the soil, which drastically modify its physical structure (Mileusnić et al. 2022).

The continuous cultivation of crops has resulted in soil degradation, affecting both the physicochemical characteristics of the soil as well as the biological ones (composition, abundance, diversity, and functioning of soil micro and macro organisms, microbial networks, enzyme activities, and soil food web interactions), which has led in turn to affect soil health, fertility, and ultimately the productivity of soils (Pervaiz et al. 2020).

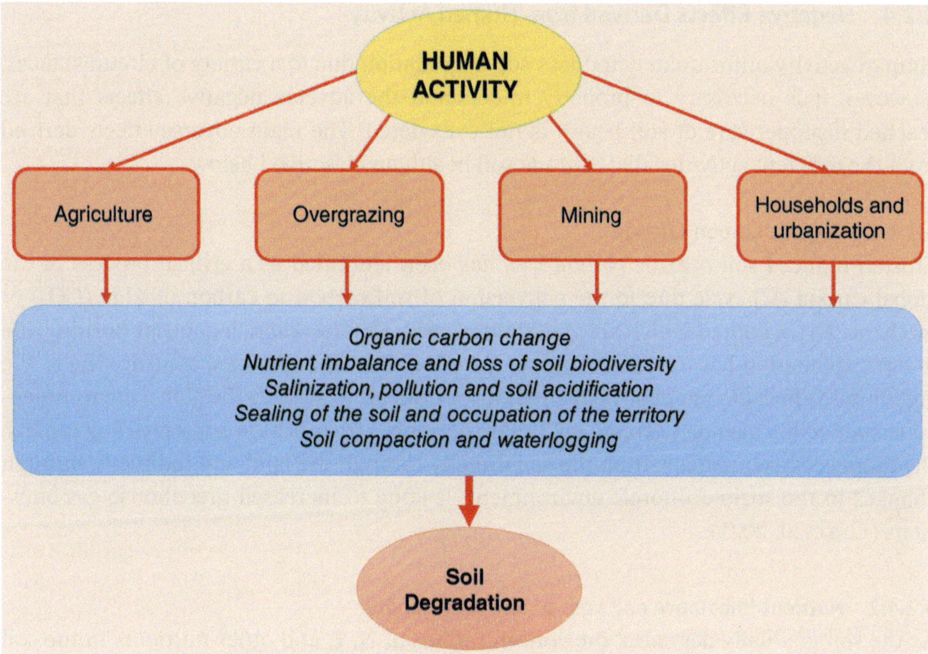

Figure 4.1 Impacts of different human activities on soil degradation.

4.2.2 Overgrazing

Arid and semi-arid environments are sensitive to poor soil management practices, such as overgrazing. Trampling and overgrazing by animals can lead to land degradation through soil compaction (Bayat et al. 2022).

Increased grazing intensity affects plant functional group and soil chemistry through soil pH and C/N ratio, which in turn affects soil bacterial and fungal communities, suppresses dominant bacterial phyla such as Proteobacteria and Bacteriodetes, and increases the abundance of oligotrophic bacteria and promotes the dominant fungal phylum of Ascomycetes due to the nutrient-poor soil integer (Z. Wang et al. 2022).

4.2.3 Mining

Mining, although important for the economic liberation of the country, is considered one of the most destructive human activities, inducing soil degradation as the main negative impact, which can lead to food insecurity, loss of biodiversity and climate change (Silveira et al. 2021).

Some of the negative impacts of mining are the loss of vegetation cover, the massive destruction of bodies of water, the loss of biodiversity, changes in land use and food insecurity, the increase in vices and social conflicts, high cost of living and air pollution (Worlanyo and Jiangfeng 2021).

4.2.4 Negative Effects Derived from Human Activity

Human activity unfortunately induces soil degradation due to a variety of circumstances. However, it is necessary to properly understand the adverse negative effects that are reached if proper care of soil health is not considered. The main adverse effects derived from the different activities that degrade soil health are described below.

4.2.4.1 Organic Carbon Change

Erosion-induced soil organic carbon loss has been identified as a critical process in the global carbon (C) cycle due to the conversion of soil carbon to carbon dioxide (CO_2) or methane (CH_4), both of which are greenhouse gases. Deforestation, logging or burning due to overexploitation has become a source of carbon dioxide emissions, contributing to the greenhouse effect (Azhar et al. 2022), making it difficult to achieve the global environmental sustainability agenda (Zaman 2022). Furthermore, urbanization is progressing rapidly, shedding ecosystem effects from the soil directly through the land and indirectly through changes in the socio-economic environment, leading to increased greenhouse gas emissions (Luo et al. 2021).

4.2.4.2 Nutrient Imbalance and Loss of Soil Biodiversity

As the soil gradually degrades, the concentrations of N, P, and other nutrients in the soil decrease significantly, reducing the loss of carbon in the soil, being related to the decoupling of the nitrogen and carbon cycles of the soil, affecting other physical and chemical properties, such as pH, organic matter, porosity, and density (Jin et al. 2022).

Rapid changes in soil chemistry caused by long-term acid deposition directly or indirectly alter natural habitats and ecosystem functions (Beloica et al. 2022). For example, one of the heavy metals such as Thallium (TI) is extremely toxic and can influence the composition of the soil microbial community, geochemical factors reshape the microbial distribution under the stress of TI contamination (She et al. 2022).

4.2.4.3 Salinization, Pollution, and Soil Acidification

Salinization and sodification are types of degradation associated with the accumulation of salts in the soil and are considered an environmental threat. Gradually they advance unnoticed until a severe degradation if there is no proper control. Many countries do not routinely maintain their relevant soil information, which contributes to its advancement, specifically in agricultural areas where its impact has food security consequences (Omuto et al. 2022), since the high content of salt hinders the growth of plants (Celleri et al. 2022). Soil salinity is commonly associated with regional climate aridity, while subsurface hydrology, topography, and land use join with climate to generate landscape-scale salinization patterns (Celleri et al. 2022).

Environmental contamination with heavy metals continues to affect soil quality and crop yields (Gavrilescu 2022). The presence of heavy metals in the environment is mainly related to industrial development, agricultural practices, and other excessive anthropogenic activities, they are considered significant sources of heavy metals, causing an accumulation of heavy metals in the soil that can enter the food chain and cause health risks to humans (Manoj et al. 2020; Yan et al. 2020; Yang et al. 2022).

Acid rain, as one of the global environmental problems, not only alters the biogeochemistry of ecosystems, but also significantly threatens human health and survival (Ziqiang Liu et al. 2022; Ma et al. 2021).

4.2.4.4 Sealing of the Soil and Occupation of the Territory

Soil sealing is one of the hidden manifestations of desertification and is the implicit explanation for the loss of land for the agricultural production process (AbdelRahman et al. 2021).

In recent decades, the occupation of land for urbanization has increased significantly, which has caused a serious loss and degradation of land capital, causing an important environmental and urban planning problem (Hassoun et al. 2022). Urbanization increases the rate of greenhouse gas emissions and atmospheric compositions. Nitrous oxide (N_2O) is an important greenhouse gas that contributes significantly to global warming, and N_2O emissions are sensitive to changes in substrate availability such as litter and N, and to microenvironmental factors caused by changes in land use after urbanization (Xu et al. 2022).

4.2.4.5 Soil Compaction and Waterlogging

The use of heavy agricultural machinery has resulted in widespread soil compaction in many regions of the world (She et al. 2022). Compacted soil limits crop access to soil water and nutrients and is expected to reduce crop productivity by imposing physical–chemical constraints on plant growth and development (Correa et al. 2022; Ziqiang Liu et al. 2022).

Soil erosion and torrential flooding are worsening due to overexploitation of agricultural and forest land, coupled with urbanization (Dwarapureddi et al. 2022). On the other hand, intermittent flooding in rain-rich areas can affect the phytoremediation of heavy metals in the soil, for example, cadmium (Cd) and pyrene (PYR) contamination, which have gained worldwide attention due to their negative effects about the environment (Zang et al. 2017).

4.3 Methods to Restore the Soil

Once the origins of soil degradation are known, as well as the adverse effects we observe on soil health, it is then possible to propose strategies to reverse the negative effects observed. Fortunately, more information is currently available to carry out a more adequate management of the soil and to be able to preserve its health, and even restore soils affected by different human activities (Figure 4.2).

4.3.1 Conservation Agriculture

Conservation agriculture (CA) involves the use of three interrelated methods: minimal or no mechanical soil disturbance, biomass mulch cover, crop diversity, and other complementary agricultural practices of integrated crop and production management (Kassam et al. 2019).

Conservation tillage regulates the assembly, network structure, and ecological function of the soil bacterial community and contributes to soil carbon sequestration, while stimulating nitrogen fixation and nitrification (Zhuxiu Liu et al. 2022). Cover crops can improve internal

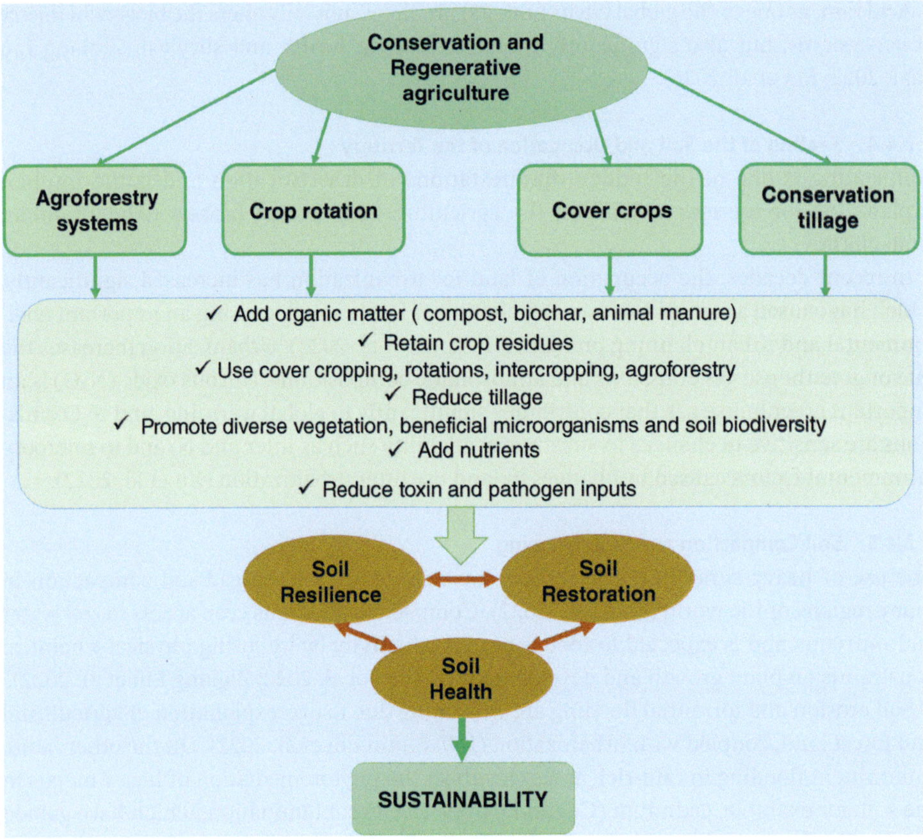

Figure 4.2 Good soil management practices that allow increasing its health and achieving sustainability.

nitrogen (N) cycling in agricultural ecosystems and can be particularly useful for organic vegetable growers who have limited opportunities to meet crop nitrogen demand with certified organic products (Allar and Maltais-Landry 2022). Biomass addition also significantly increases soil carbon stocks and decreases net global warming potential (GWP) (J. Lee et al. 2022).

One of the main challenges of the twenty-first century is to meet the growing demand for food and protect the environment. To meet this challenge, conservation agriculture can rely on the centuries-old practice of crop rotation (Mugandani et al. 2021). Crop rotation is an important tool to improve the climate resilience of agricultural production systems and reduce greenhouse gas emissions from the soil without sacrificing yields, effectively resolving the shortcomings of current continuous cultivation methods (Xiao et al. 2022; Yu et al. 2022).

4.3.2 Soil Amendments

Soil erosion is one of the main drivers of land degradation worldwide, which is why the soil conservation service has received increasing attention (An et al. 2022). Organic materials as soil amendments have been widely used to improve soil properties, plant growth, and soil microflora and microfauna (Chaudhari et al. 2021).

Biochar, weathered charcoal, and peat grass have been widely used around the world for a long time. Biochar increases plant growth and cover, effectively reducing soil losses (M.-H. Lee et al. 2022). In addition, it increases the water retention capacity of the soil mainly in areas prone to drought with little rainfall and strong evaporation and with the application of weathered carbon it improves and stabilizes soil aggregates to reduce the risk of soil erosion (Xiaoyuan Zhang et al. 2022). Peat grass reduces soil particle size, soil acidification, and viscosity, while increasing soil organic matter, soil microbial biomass carbon, available phosphorus, available potassium, alkaline hydrolysable nitrogen, and decreases heavy metal content (S. Zhang et al. 2022).

The application of substrates and manure and biological waste called compost are also effective in mitigating soil degradation by significantly increasing the activity of soil enzymes, increasing soil organic matter, nutrients, and soil microbial diversity (Kacprzak et al. 2022; C. Wang et al. 2022).

Humic amendments improve crop growth and yield, efficiency in water use, improve soil microenvironmental conditions and these, in turn, promote greater activity of soil enzymes such as urease, invertase, catalase, and turnover and availability of nutrients in the soil (Ma et al. 2022). And the combination of humic substances with calcareous substances help to significantly decrease the mobility of metals in soils due to their large surface area and abundant organic functional groups and increase the pH in acid soils (Hao Liu et al. 2022).

4.3.3 Plant Growth Promoting Rhizobacteria (PGPR)

Plant growth promoting rhizobacteria (PGPR) are root-colonizing bacteria that promote plant growth and development through direct and indirect mechanisms, such as atmospheric nitrogen fixation, phosphorus solubilization, the production of siderophores that chelate iron and make it available to plant roots, synthesize plant hormones such as auxins, gibberellins, and cytokinins, or synthesize enzymes that relieve stress, such as 1-aminocyclopropane-1-carboxylate deaminase, or enzymes that degrade cell wall (Rani et al. 2022).

Rhizobacteria not only play a role in promoting plant growth and biological control of diseases but can also help decrease the toxicity of various contaminants in the soil (Kaur 2021). Therefore, inoculation of suitable and specific heavy metal tolerant PGPR strains associated with plants can take full advantage for phytoremediation (Roy Chowdhury 2022). In addition, PGPRs in agriculture can be an excellent support to counteract the damaging effects of abiotic stress, such as excessive salinity and drought, replacing costly and environmentally harmful inorganic fertilizers (Vocciante et al. 2022).

4.3.4 Grazing Management

Intense grazing reduces the content of nitrogen, potassium, and urease in the soil, and causes an immediate response in the chemical environment of the soil, the bacterial and fungal communities, and consequently soil erosion. For the development of grazing management, it is clear, heavy grazing should be avoided and soil structure should be continuously monitored to explore possible long-term consequences of grazing (Van Syoc et al. 2022; Xuran Zhang et al. 2022). Currently the best performing practice is controlled grazing with high grazing density since it increases soil organic carbon by an average of 21% (Phukubye et al. 2022).

4.3.5 Phytoremediation

Biotechnology offers a number of environmentally friendly options, one of which is phytoremediation. Phytoremediation cleans metal-contaminated environments based on factors such as substrate conditions, plant characteristics, and metal availability (Rana et al. 2022).

The polyaromatic hydrocarbons (PAH) that are present in the effluents of various industries reach the environment and end up contaminating the biota. To eliminate the different PAH, it can be done through rhizofiltration, phytoextraction, and other actions of the plants that hyperaccumulate heavy metal ions, it is also beneficial in terms of feasibility and low costs (Bose et al. 2022; Gavrilescu 2022).

4.4 Sustainable Use of the Soil

Currently, different views have emerged about the proper management of agricultural production systems that ultimately lead to the conservation of soil health (Figure 4.2). In accordance with Lehmann et al. (2020) soil health is the continued capacity of soil to function as a vital living ecosystem that sustains plants, animals, and humans, and connects agricultural and soil science to policy, stakeholder, needs and sustainable supply chain management. Hence, approaches such as regenerative agriculture or conservation agriculture are becoming more common when it comes to soil conservation (Giller et al. 2021; Mugandani et al. 2021).

Regenerative agriculture considers different practices in agricultural production systems such as retention and incorporation of crop residues, cover crops, and reduced tillage, among others (Giller et al. 2021). While conservation agriculture considers that it is based on three interconnected elements to achieve multiple benefits that can lead to sustainability, these are minimum soil tillage, permanent soil cover, and crop rotation or intercropping (Mugandani et al. 2021).

Conservation agriculture provides a variety of soil health benefits including (Mugandani et al. 2021):

1) Protection of soil biological diversity
2) Soil carbon sequestration resulting in climate adaptation and mitigation
3) Provision of ecosystem services
4) Food security through soil health
5) Water safety

Furthermore, Giller et al. (2021) mention that there is still a debate about what regenerative agriculture represents, however, they propose that there are two main challenges associated with regenerative agriculture from an agronomic perspective:

1) Restoration of soil health, including, the capture of carbon to mitigate climate change
2) Reversal of biodiversity loss

This demonstrates that both conservation agriculture and regenerative agriculture are two approaches that enable sustainable soil use, and therefore promote soil health. The health of the soil is fundamental, since different ecosystem services of the soil are derived from it, such as plant production, water quality, human health, and climate change (Lehmann et al. 2020). For this reason, soil health impacts human health either directly or indirectly through crop production, quality, storage and transportation, food quality and taste, soil contamination, or through climate change, recreation, and culture (Brevik et al. 2019; Lehmann et al. 2020; Rodrigo-Comino et al. 2020).

Therefore, to achieve sustainable land use and maintain soil health, it is necessary to use production systems that promote the increase of key elements such as soil organic carbon, soil biodiversity, biomass, and microbial activity. Lehmann et al. (2020) adequately list a series of management strategies that allow meeting these objectives, and therefore maintain or increase soil health:

1) Add organic matter (compost, biochar, animal manure)
2) Retain crop residues
3) Use cover cropping, rotations, intercropping, agroforestry
4) Reduce tillage
5) Promote diverse vegetation, beneficial microorganisms, and soil biodiversity
6) Add nutrients
7) Reduce toxin and pathogen inputs

In this regard, there is a diversity of agroecological production systems that allow meeting the objectives of maintaining soil health and achieving sustainability, among which are polycultures, crop rotation, agroforestry systems, cover crops, and green manures, as well as conservation tillage. The proper use of these production systems provides a series of benefits, so they can definitely be an alternative that protects the soil and takes advantage of all the benefits it brings to society.

4.4.1 Production Systems Based on Polycultures

Possibly one of the main approaches that should be considered to improve soil health is biological diversity, specifically in agricultural production systems the diversity of crops is essential. Through adequate planning in the diversity of the crops that will be established, the health of the soil can be increased through the balance of biodiversity, greater efficiency in the use of nutrients, changes in the structure of the soil, among other benefits (Yang et al. 2020).

Sprunger et al. (2020) observed that crop diversity increased soil biological health in both annual and perennial systems. They evaluated 10 production systems in their study and found several positive effects on soil health. Annual rotations with a cover crop increased oxidizable carbon permanganate and soil organic matter, perennial polycultures increased mineralizable carbon relative to annual perennial and monoculture systems. The authors conclude that incorporating perennial polycultures in rotations is the most effective way to increase soil biological health and improve carbon stabilization.

4.4.2 Agroforestry Systems

Agroforestry systems present a greater diversity of crops compared to monoculture production systems, since they include the presence of trees and mixtures of crops that can be perennial or annual. In particular, woody perennial cropping systems have the potential to produce food while maintaining or improving soil health relative to annual cropping systems (Eddy and Yang 2022). Marsden et al. (2020) present an excellent review of agroforestry systems, where they show that most of the studies carried out focus on perennial crops, silvoarable systems, and to a lesser extent silvopastoral systems. In their review, they found that agroforestry systems generally have a positive impact on the abundance and diversity of soil fauna. Dollinger and Jose (2018) present a review in which solid evidence of the positive effect of agroforestry systems on soil quality and health is observed, since they have the ability to increase soil organic carbon, improve the availability of nutrients in the soil and therefore improve its fertility, as well as improve the microbial dynamics of the soil. Muchane et al. (2020) carried out a meta-analysis that demonstrates several positive effects of agriculture on soil health: agroforestry can reduce soil erosion rates by 50% compared to crop monocultures, soil organic carbon is increased, as well as the availability of N and P. Derived from this, they conclude that agroforestry can contribute significantly to the provision of soil ecosystem services in the humid and sub-humid tropics.

Eddy and Yang (2022) compared various agroforestry production systems and found that converting land to agroforestry improved soil health and increased soil carbon sequestration. Specifically, they found that soil organic carbon and nitrogen, as well as carbon and nitrogen content of microbial biomass, were higher in the agroforestry system than in an adjacent no-till corn-soybean rotation system.

4.4.3 Crop Rotation

Crop rotation is an agroecological production system that increases soil biodiversity, and therefore soil health, in addition to favoring the control of pathogens and pests (Yang et al. 2020).

There is evidence that demonstrates the positive effects of crop rotation, and even more so if it is combined with other agroecological production systems such as conservation tillage. Alhameid et al. (2019) showed that after four years of applying crop rotation in combination with no-till, they observed an increase in microbial biomass and microbial biomass N, hot-water-extractable C, urease, and β-glucosidase enzyme activity. These results positively impact soil health from both a biological and chemical point of view, which is essential to achieve sustainable agricultural production.

Behnke et al. (2018) observed that corn-soybean (*Zea mays* L. and *Glycine max* [L.] Merr.) crop rotation increased corn yield while decreasing greenhouse gas (N_2O) emissions, compared to continuous corn or continuous soybean. These results demonstrate the positive impact of using this practice, even when crop rotation is simple.

Silva et al. (2020) recommend crop rotation that includes legumes and grasses but should be combined with a no-tillage system and high biomass-yielding cover crops intercropped with maize to see the best results. Similar results have been reported by other authors. Haruna and Nkongolo (2019) observed a significant increase in carbon in corn/soybean rotation and no-till compared to continuous corn and soybean. In addition, the authors mention that the chemical properties of the soil are positively affected due to no-till, cover crop, and crop rotation.

Laamrani et al. (2020) observed that the crop rotation corn–corn–oats–barley with red clover cover crop achieved the highest total carbon content, compared to other systems and tillage methods. Furthermore, the authors conclude that conservation tillage and the incorporation of a cover crop in the crop rotation had a positive effect on total carbon accumulation in the upper soil layer and could be suitable management practices to promote fertility and soil sustainability.

It has also been observed that crop rotation can positively influence crop pest control. Neher et al. (2019) observed that crop rotation had a positive impact on the control of nematodes (*Heterodera glycines*) due to the trophic diversity generated.

4.4.4 Cover Crops

Maximum ground cover with crops is an approach used in conservation agriculture management. Cover cropping is considered as a promising sustainable agricultural method that has the potential to improve soil health and mitigate the consequences of soil degradation, as it has been documented that it can form an agroecosystem distinct from bare fallow, which significantly increases parameters of soil microbial abundance, activity, and diversity (Kim et al. 2020).

Farmaha et al. (2022) present a review that covers a variety of soil types in the United States, where they observed that cover crops and conservation tillage (no-till or reduced tillage) improve soil health, through the increase of C, N, and nutrients in the soil. The authors conclude that the agroecological approach to food and fiber production has great potential to improve soil health.

Ghimire et al. (2019) in the southern region of the Ogallala Aquifer found a positive influence of cover crops on soil organic carbon and total nitrogen in a winter wheat-based cropping system, while the impact on soil water content was minimal. They evaluated various cover crops such as pea (*Pisum sativum* L.), oat (*Avena sativa* L.), canola (*Brassica napus* L.), hairy vetch (*Vicia villosa* L.), forage radish (*Raphanus sativus* L.), barley (*Hordeum vulgare* L.), and various combinations between them. The authors conclude that cover crops with oats and their mixture with other species have the potential to improve soil health and fertility, increase C and N sequestration, and maintain winter wheat yields.

In addition, positive effects of cover crops on pest and disease control have also been documented. Acharya et al. (2020) observed in the maize crop (*Z. mays* L.) established after a cover crop of winter camelina (*Camelina sativa* L.) that the root disease decreased since a lower population of *Pythium* was observed in the roots of the plants, which can contribute to the sustainability of corn production systems.

4.4.5 Conservation Tillage

Reduced tillage has been shown to be an effective tool for improving soil health. In semiarid vertisols of India, reduced tillage coupled with retention of crop residues increased nutrient availability as well as their availability (Jayaraman et al. 2021). In China it has been shown that conservation tillage (no tillage and subsoiling tillage) increased the soil organic matter content, as well as the availability of phosphorous and nitrogen, and the abundance of *Acidobacteria* (Wang et al. 2020). These authors showed that after eight years of applying conservation tillage, the diversity of bacteria increased and the availability of nutrients in the soil was improved compared to conventional tillage. In Bangladesh, after two years of experimentation with conservation tillage (zero tillage and minimum tillage) in wheat crop, a significant increase in soil organic carbon was observed compared to conventional tillage; in addition, the water footprint and the emission of greenhouse gases were also reduced (Rahman et al. 2021).

However, it has also been shown that conservation tillage alone may be insufficient to significantly improve soil health. Cooper et al. (2020) showed that after five years of conservation tillage (direct drill and shallow non-inversion), no significant changes were observed in the physical, chemical, or biological characteristics of the soil compared to conventional tillage, in addition to not reducing nutrient leaching losses. The authors conclude that conservation tillage alone is not enough to improve environmental sustainability in the short term, and that other aspects such as cover cropping, crop rotations, and precision farming techniques must be incorporated. This aspect has also been considered by other authors, who recommend improving conservation tillage with the support of other conservation practices such as cover crops or crop rotation (Nunes et al. 2018).

Mondal et al. (2020) observed that the rice–maize cropping systems under reduced tillage and incorporation of crop residues increased the total organic carbon of the soil, in addition to presenting positive effects on the rate of soil infiltration, which improved the physical health of the soil (bulk density, soil aggregation, infiltration rate) and organic carbon status. Crop residues, which are basically the post-harvest remains of agriculture, are organic matter and nutrients that can be reincorporated, generating positive effects on soil health, especially in physicochemical characteristics such as soil water content, total porosity, aggregate stability, cation exchange capability, organic carbon, phosphorus, potassium, and other nutrients; in addition to this, they can provide nutrients for microorganisms (Fu et al. 2021). Therefore, the incorporation of crop residues can be a good alternative to improve soil health, especially if it is combined with other agroecological approaches such as conservation tillage.

Nunes et al. (2018) observed that the soil managed under long-term no-till showed the most favorable biological, physical, and chemical conditions of the soil for the development of maize (*Z. mays* L.) plants, since higher levels of organic matter were present, respiration, and protein, water stable aggregation, water infiltration rate, total N, P, and Zn. In addition, more benefits were seen when introducing a mixture of grass and legume cover crops with no-till.

4.5 Conclusions

Soil is essential for the production of crops, fibers, and other products of importance to humanity. In addition, not only does it have that function, but it also provides a series of ecosystem services that make it even more important. Hence, conserving the soil, and above all maintaining or increasing the soil health, is imperative for humanity. Fortunately, today there are various tools and sufficient knowledge to achieve proper soil management, and to promote its health, and even restore soils with various problems such as pollution by heavy metals or toxins. Agriculture is one of the human activities that most degrade the soil, however, different techniques or production systems can be applied that will allow the conservation or restoration of soil health, and therefore achieve a sustainable use of this resource. The use of various agroecological systems for agricultural production such as polycultures, agroforestry, cover crops, and conservation tillage, are essential for proper soil management, and can prevent its degradation, as well as increase its resilience and achieve restoration.

References

AbdelRahman, M.A.E., Afifi, A.A., and Scopa, A. (2021). A time series investigation to assess climate change and anthropogenic impacts on quantitative land degradation in the North Delta, Egypt. *ISPRS Int. J. Geo-Information* 11: 30.

Acharya, J., Moorman, T.B., Kaspar, T.C. et al. (2020). Cover crop rotation effects on growth and development, seedling disease, and yield of corn and soybean. *Plant Dis.* 104: 677–687.

Alhameid, A., Singh, J., Sekaran, U. et al. (2019). Soil biological health: influence of crop rotational diversity and tillage on soil microbial properties. *Soil Sci. Soc. Am. J.* 83: 1431–1442.

Allar, J. and Maltais-Landry, G. (2022). Limited benefits of summer cover crops on nitrogen cycling in organic vegetable production. *Nutr. Cycl. Agroecosyst.* 122: 119–138.

An, Y., Zhao, W., Li, C., and Sofia Santos Ferreira, C. (2022). Temporal changes on soil conservation services in large basins across the world. *Catena* 209: 105793. https://doi.org/10.1016/j.catena.2021.105793.

Azhar, M.F., Qadir, I., Shehzad, M.M., and Jamil, A. (2022). Changing climate impacts on forest resources. In: *Building Climate Resilience in Agriculture* (ed. W.N. Jatoi, M. Mubeen, A. Ahmad, et al.), 111–130. Cham: Springer International Publishing.

Bayat, H., Sheklabadi, M., Moradhaseli, M. et al. (2022). Grazing management, slope aspect and canopy effects on the compression characteristic of soils of the Gonbad experimental watershed in Hamedan. *Iran. Geoderma* 409: 115641.

Behnke, G.D., Zuber, S.M., Pittelkow, C.M. et al. (2018). Long-term crop rotation and tillage effects on soil greenhouse gas emissions and crop production in Illinois, USA. *Agric. Ecosyst. Environ.* 261: 62–70.

Beloica, J., Belanović Simić, S., Čavlović, D. et al. (2022). Soil acidification patterns due to long-term Sulphur and nitrogen deposition and how they affect changes in vegetation composition in eastern Serbia. In: *Advances in Understanding Soil Degradation*, 737–754. Springer, Cham.

Bose, V.G., Shreenidhi, K.S., and Malik, J.A. (2022). Phytoremediation of PAH-contaminated areas. In: *Advances in Bioremediation and Phytoremediation for Sustainable Soil Management* (ed. J.A. Malik), 141–156. Cham: Springer.

Brevik, E.C., Steffan, J.J., Rodrigo-Comino, J. et al. (2019). Connecting the public with soil to improve human health. *Eur. J. Soil Sci.* 70: 898–910.

Brevik, E.C., Slaughter, L., Singh, B.R. et al. (2020). Soil and human health: current status and future needs. *Air, Soil Water Res.* 13: 1–23.

Celleri, C., Pratolongo, P., and Arena, M. (2022). Spatial and temporal patterns of soil salinization in shallow groundwater environments of the Bahía Blanca estuary: influence of topography and land use. *L. Degrad. Dev.* 33: 470–483.

Chaudhari, S., Upadhyay, A., and Kulshreshtha, S. (2021). Influence of organic amendments on soil properties, microflora and plant growth. In: *Sustainable Agriculture Reviews* (ed. E. Lichtfouse), 147–191. Cham: Springer.

Cooper, R.J., Hama-Aziz, Z.Q., Hiscock, K.M. et al. (2020). Conservation tillage and soil health: lessons from a 5-year UK farm trial (2013–2018). *Soil Tillage Res.* 202: 104648.

Correa, J., Postma, J.A., and Wojciechowski, T. (2022). Phenotypic response to soil compaction varies among genotypes and correlates with plant size in sorghum. *Plant Soil.* 472: 59–76.

Dollinger, J. and Jose, S. (2018). Agroforestry for soil health. *Agrofor. Syst.* 92: 213–219.

Dwarapureddi, B.K., Dash, S., Raj, A. et al. (2022). Prevention of soil erosion and torrential floods. In: *Prevention and Management of Soil Erosion and Torrential Floods*, 92–111. IGI Global.

Eddy, W.C. and Yang, W.H. (2022). Improvements in soil health and soil carbon sequestration by an agroforestry for food production system. *Agric. Ecosyst. Environ.* 333: 107945.

Farmaha, B.S., Sekaran, U., and Franzluebbers, A.J. (2022). Cover cropping and conservation tillage improve soil health in the southeastern United States. *Agron. J.* 114: 296–316.

Fu, B., Chen, L., Huang, H. et al. (2021). Impacts of crop residues on soil health: a review. *Environ. Pollut. Bioavailab.* 33: 164–173.

Gavrilescu, M. (2022). Enhancing phytoremediation of soils polluted with heavy metals. *Curr. Opin. Biotechnol.* 74: 21–31.

Ghimire, R., Ghimire, B., Mesbah, A.O. et al. (2019). Soil health response of cover crops in winter wheat–fallow system. *Agron. J.* 111: 2108–2115.

Giller, K.E., Hijbeek, R., Andersson, J.A., and Sumberg, J. (2021). Regenerative agriculture: an agronomic perspective. *Outlook Agric.* 50: 13–25.

Haruna, S.I. and Nkongolo, N.V. (2019). Tillage, cover crop and crop rotation effects on selected soil chemical properties. *Sustainability* 11: 1–11.

Hassoun, G., Madrau, S., El Arab, M. et al. (2022). Soil sealing along the Beirut coast during half a century marked by population growth and refugee inflows. *Arab. J. Geosci.* 15: 266.

Jayaraman, S., Sinha, N.K., Mohanty, M. et al. (2021). Conservation tillage, residue management, and crop rotation effects on soil major and micro-nutrients in semi-arid vertisols of India. *J. Soil Sci. Plant Nutr.* 21: 523–535.

Jin, Q., Wang, C., Sardans, J. et al. (2022). Effect of soil degradation on the carbon concentration and retention of nitrogen and phosphorus across Chinese rice paddy fields. *Catena* 209: 105810. https://doi.org/10.1016/j.catena.2021.105810.

Kacprzak, M., Kupich, I., Jasinska, A., and Fijalkowski, K. (2022). Bio-based waste' substrates for degraded soil improvement – advantages and challenges in european context. *Energies* 15: 385.

Kassam, A., Friedrich, T., and Derpsch, R. (2019). Global spread of conservation agriculture. *Int. J. Environ. Stud.* 76: 29–51.

Kaur, J. (2021). PGPR in management of soil toxicity. In: *Rhizobiont in Bioremediation of Hazardous Waste* (ed. V. Kumar, R. Prasad, and M. Kumar), 317–344. Singapore: Springer Singapore.

Kim, N., Zabaloy, M.C., Guan, K., and Villamil, M.B. (2020). Do cover crops benefit soil microbiome? A meta-analysis of current research. *Soil Biol. Biochem.* 142: 107701.

Kopittke, P.M., Menzies, N.W., Wang, P. et al. (2019). Soil and the intensification of agriculture for global food security. *Environ. Int.* 132: 105078.

Laamrani, A., Voroney, P.R., Berg, A.A. et al. (2020). Temporal change of soil carbon on a long-term experimental site with variable crop rotations and tillage systems. *Agronomy* 10: 840.

Lee, J.G., Chae, H.G., Kim, G.W. et al. (2022). Cover cropping and its biomass incorporation: not enough to compensate the negative impact of plastic film mulching on global warming. *Sci. Total Environ.* 807: 151015.

Lee, M.-H., Lin, H.-H., and Jien, S.-H. (2022). Effects of biochar application on vegetation growth, cover, and erosion potential in sloped cultivated soil derived from mudstone. *Processes* 10: 306.

Lehmann, J., Bossio, D.A., Kögel-Knabner, I., and Rillig, M.C. (2020). The concept and future prospects of soil health. *Nat. Rev. Earth Environ.* 1: 544–553.

Li, H., Zhu, H., Liang, C. et al. (2022). Soil erosion significantly decreases aggregate-associated OC and N in agricultural soils of Northeast China. *Agric. Ecosyst. Environ.* 323: 107677.

Liu, H., Zhang, T., Tong, Y. et al. (2022). Effect of humic and calcareous substance amendments on the availability of cadmium in paddy soil and its accumulation in rice. *Ecotoxicol. Environ. Saf.* 231: 113186. https://doi.org/10.1016/j.ecoenv.2022.113186.

Liu, Z., Gu, H., Liang, A. et al. (2022). Conservation tillage regulates the assembly, network structure and ecological function of the soil bacterial community in black soils. *Plant Soil.* 472, 207–223.

Liu, Z., Shi, Z., Wei, H., and Zhang, J. (2022). Acid rain reduces soil CO2 emission and promotes soil organic carbon accumulation in association with decreasing the biomass and biological activity of ecosystems: a meta-analysis. *Catena* 208: 105714.

Luo, Y., Shen, J., Chen, A. et al. (2021). Loss of organic carbon in suburban soil upon urbanization of Chengdu megacity, China. *Sci. Total Environ.* 785: 147209.

Ma, S., Liu, X., Jia, Z. et al. (2021). Response of Quercus acutissima foliage to different types of simulated acid rain. *Atmos. Pollut. Res.* 12: 10112.

Ma, B., Ma, B.-L., McLaughlin, N.B. et al. (2022). Improvement in dryland crop performance and soil properties with multiple annual applications of lignite-derived humic amendment. *Soil Tillage Res.* 218: 105306.

Manoj, S.R., Karthik, C., Kadirvelu, K. et al. (2020). Understanding the molecular mechanisms for the enhanced phytoremediation of heavy metals through plant growth promoting rhizobacteria: a review. *J. Environ. Manag.* 254: 109779.

Marsden, C., Martin-Chave, A., Cortet, J. et al. (2020). How agroforestry systems influence soil fauna and their functions – a review. *Plant Soil* 453: 29–44.

Mileusnić, Z.I., Saljnikov, E., Radojević, R.L., and Petrović, D.V. (2022). Soil compaction due to agricultural machinery impact. *J. Terramechanics* 100: 51–60.

Mondal, S., Naik, S.K., Haris, A.A. et al. (2020). Effect of conservation tillage and rice-based cropping systems on soil aggregation characteristics and carbon dynamics in eastern Indo-Gangetic plain. *Paddy Water Environ.* 18: 573–586.

Muchane, M.N., Sileshi, G.W., Gripenberg, S. et al. (2020). Agroforestry boosts soil health in the humid and sub-humid tropics: a meta-analysis. *Agric. Ecosyst. Environ.* 295: 106899.

Mugandani, R., Mwadzingeni, L., and Mafongoya, P. (2021). Contribution of conservation agriculture to soil security. *Sustainability* 13: 1–11.

Neher, D.A., Nishanthan, T., Grabau, Z.J., and Chen, S.Y. (2019). Crop rotation and tillage affect nematode communities more than biocides in monoculture soybean. *Appl. Soil Ecol.* 140: 89–97.

Nunes, M.R., van Es, H.M., Schindelbeck, R. et al. (2018). No-till and cropping system diversification improve soil health and crop yield. *Geoderma* 328: 30–43.

Omuto, C.T., Vargas, R.R., Elmobarak, A.A. et al. (2022). Digital soil assessment in support of a soil information system for monitoring salinization and sodification in agricultural areas. *L. Degrad. Dev.* 33: 1204–1218.

Pervaiz, Z.H., Iqbal, J., Zhang, Q. et al. (2020). Continuous cropping alters multiple biotic and abiotic indicators of soil health. *Soil Syst.* 4: 59.

Phukubye, K., Mutema, M., Buthelezi, N. et al. (2022). On the impact of grassland management on soil carbon stocks: a worldwide meta-analysis. *Geoderma Reg.* 28: e00479.

Rahman, M.M., Aravindakshan, S., Hoque, M.A. et al. (2021). Conservation tillage (CT) for climate-smart sustainable intensification: assessing the impact of CT on soil organic carbon accumulation, greenhouse gas emission and water footprint of wheat cultivation in Bangladesh. *Environ. Sustain. Indic.* 10: 100106.

Rana, V., Bandyopadhyay, S., and Maiti, S.K. (2022). Potential and prospects of weed plants in phytoremediation and eco-restoration of heavy metals polluted sites. In: *Phytoremediation Technology for the Removal of Heavy Metals and Other Contaminants from Soil and Water* (ed. V. Kumar, M.P. Shah, and S.K. Shahi), 187–205. Elsevier.

Rani, R., Kumar, V., and Gupta, P. (2022). Plant growth-promoting rhizobacteria-assisted bioremediation of toxic contaminant: recent advancements and applications. In: *Microbial Biodegradation and Bioremediation* (ed. S. Das and H.R. Dash), 327–341. Elsevier.

Rinot, O., Levy, G.J., Steinberger, Y. et al. (2019). Soil health assessment: a critical review of current methodologies and a proposed new approach. *Sci. Total Environ.* 648: 1484–1491.

Rodrigo-Comino, J., López-Vicente, M., Kumar, V. et al. (2020). Soil science challenges in a new era: a transdisciplinary overview of relevant topics. *Air Soil Water Res.* 13: 1–17.

Roy Chowdhury, A. (2022). Plant growth-promoting rhizobacteria (PGPR): strategies to improve heavy metal stress under sustainable agriculture. In: *Sustainable Agriculture* (ed. S.A. Bandh), 189–208. Cham: Springer.

She, J., Liu, J., He, H. et al. (2022). Microbial response and adaption to thallium contamination in soil profiles. Mater: J. Hazard. Mater. 423.

da Silva, P.C.G., Tiritan, C.S., Echer, F.R. et al. (2020). No-tillage and crop rotation increase crop yields and nitrogen stocks in sandy soils under agroclimatic risk. *F. Crop. Res.* 258: 107947.

Silveira, P.G., Venâncio, M.D., and Leite, J.R.M. (2021). Brumadinho's disaster, mine tailing dams and the environmental licencing in Brazil: preventing risks to human wellbeing, soils, and the environment. In: *International Yearbook of Soil Law and Policy* (ed. H. Ginzyky, E. Dooley, I.L. Heuser, et al.), 241–259. Cham: Springer.

Sprunger, C.D., Martin, T., and Mann, M. (2020). Systems with greater perenniality and crop diversity enhance soil biological health. *Agric. Environ. Lett.* 5: 1–6.

Van Syoc, E., Albeke, S.E., Scasta, J.D., and van Diepen, L.T.A. (2022). Quantifying the immediate response of the soil microbial community to different grazing intensities on irrigated pastures. *Agric. Ecosyst. Environ.* 326: 107805.

Vocciante, M., Grifoni, M., Fusini, D. et al. (2022). The role of plant growth-promoting rhizobacteria (PGPR) in mitigating plant's environmental stresses. *Appl. Sci.* 12: 1231.

Wang, H., Wang, S., Wang, R. et al. (2020). Conservation tillage increased soil bacterial diversity and improved soil nutrient status on the loess plateau in China. *Arch. Agron. Soil Sci.* 66: 1509–1519.

Wang, T., Jin, H., Fan, Y. et al. (2021). Farmers' adoption and perceived benefits of diversified crop rotations in the margins of U.S. Corn Belt. *J. Environ. Manag.* 293: 112903. https://doi.org/10.1016/j.jenvman.2021.112903.

Wang, C., Ning, P., Li, J. et al. (2022). Responses of soil microbial community composition and enzyme activities to long-term organic amendments in a continuous tobacco cropping system. *Appl. Soil Ecol.* 169: 104210. https://doi.org/10.1016/j.apsoil.2021.104210.

Wang, Z., Ding, Y., Jin, K. et al. (2022). Soil bacterial and fungal communities are linked with plant functional types and soil properties under different grazing intensities. *Eur. J. Soil Sci.* 73: e13195. https://doi.org/10.1111/ejss.13195.

Worlanyo, A.S. and Jiangfeng, L. (2021). Evaluating the environmental and economic impact of mining for post-mined land restoration and land-use: a review. *J. Environ. Manag.* 279: 111623.

Xiao, H., van Es, H.M., Amsili, J.P. et al. (2022). Lowering soil greenhouse gas emissions without sacrificing yields by increasing crop rotation diversity in the North China plain. *F. Crop. Res.* 276: 108366.

Xu, X., He, C., Zhong, C. et al. (2022). Soil N2O emission in Cinnamomum camphora plantations along an urbanization gradient altered by changes in litter input and microbial community composition. *Environ. Pollut.* 299: 118876. https://doi.org/10.1016/j.envpol.2022.118876.

Yan, A., Wang, Y., Tan, S.N. et al. (2020). Phytoremediation: a promising approach for revegetation of heavy metal-polluted land. *Front. Plant Sci.* 11: 1–15.

Yang, T., Siddique, K.H.M., and Liu, K. (2020). Cropping systems in agriculture and their impact on soil health-a review. *Glob. Ecol. Conserv.* 23: e01118.

Yang, L., Ren, Q., Zheng, K. et al. (2022). Migration of heavy metals in the soil-grape system and potential health risk assessment. *Sci. Total Environ.* 806: 150646.

Yu, T., Mahe, L., Li, Y. et al. (2022). Benefits of crop rotation on climate resilience and its prospects in China. *Agronomy* 12: 436. https://doi.org/10.3390/agronomy12020436.

Zaman, K. (2022). Environmental cost of deforestation in Brazil's Amazon rainforest: controlling biocapacity deficit and renewable wastes for conserving forest resources. *For. Ecol. Manag.* 504: 119854.

Zang, F., Wang, S., Nan, Z. et al. (2017). Accumulation, spatio-temporal distribution, and risk assessment of heavy metals in the soil-corn system around a polymetallic mining area from the loess plateau, Northwest China. *Geoderma* 305: 188–196.

Zhang, X., Sun, N., Wu, L. et al. (2016). Effects of enhancing soil organic carbon sequestration in the topsoil by fertilization on crop productivity and stability: evidence from long-term experiments with wheat-maize cropping systems in China. *Sci. Total Environ.* 562: 247–259.

Zhang, S., Zhao, R., Wu, K. et al. (2022). Effects of the rapid construction of a high-quality plough layer based on woody peat in a newly reclaimed cultivated land area. *Agriculture* 12.

Zhang, X., Wang, K., Sun, C. et al. (2022). Differences in soil physical properties caused by applying three organic amendments to loamy clay soil under field conditions. *J. Soils Sediments* 22: 43–55.

Zhang, X., Zhang, W., Sai, X. et al. (2022). Grazing altered soil aggregates, nutrients and enzyme activities in a Stipa kirschnii steppe of Inner Mongolia. *Soil Tillage Res.* 219: 105327.

5

Organic Farming – a Sustainable Option to Reduce Soil Degradation

Ana Paula Pinto[1,2], *Jorge M.S. Faria*[1,3], *A. V. Dordio*[2,4], *and A. J. Palace Carvalho*[2,5]

[1] *Mediterranean Institute for Agriculture, Environment and Development, MED & Global Change and Sustainability Institute, CHANGE, Institute for Advanced Studies and Research, University of Évora, Pólo da Mitra, Ap. 94, 7006-554, Évora, Portugal*
[2] *Chemistry and Biochemistry Department, School of Sciences and Technology, University of Évora, Rua Romão Ramalho n°59, 7000-671, Évora, Portugal*
[3] *National Institute for Agrarian and Veterinarian Research, INIAV, I. P., Quinta do Marquês, 2780-159, Oeiras, Portugal*
[4] *MARE – Marine and Environmental Sciences Centre, School of Sciences and Technology, University of Évora, Évora, Portugal*
[5] *LAQV-REQUIMTE, School of Sciences and Technology, University of Évora, Évora, Portugal*

Life on earth depends on soil – humanity needs to become aware of this reality

5.1 Introduction

The planet faces urgent sustainability challenges that require effective solutions. This century's global challenge is posed by the way of achieving development across the world that keeps economic, environmental, and societal aspects all balanced (European Environment Agency 2019).

Citizens' expectations of living in a healthy environment must be met. In this respect, we not only have to do more, but we also must do things in a different way. In the next decade, we will need more innovative responses to the world's environmental and climate challenges than the ones we have provided in the last 50 years (FAO 2017b; European Environment Agency 2019; IPCC 2019).

Future policies must be based on responses to our current environmental and climate challenges–the *acquis*–and must also reflect the state-of-the-art knowledge, which requires essentially different approaches–both in terms of *what* we will have to do, as well as *how* we should do it (FAO 2017b; European Environment Agency 2019; IPCC 2019).

In the last year, major global scientific reports from the Intergovernmental Panel on Climate Change (IPCC), Intergovernmental Science-Policy Platform on Biodiversity and Ecosystem Services (IPBES), International Resource Panel (IRP), and UN Environment have been published, all with similar messages: current trajectories are unsustainable;

Agroecological Approaches for Sustainable Soil Management, First Edition.
Edited by Majeti Narasimha Vara Prasad and Chitranjan Kumar.

these trajectories are interconnected and linked to our main production and consumption systems; and time is running out to find safe answers that change the trend.

Since the mid of past century, the world's population has tripled to 7.7 billion, people living in cities has quadrupled its number to more than 4 billion, primary energy use has increased fivefold and economic output has expanded 12-fold, which was matched by a similar increase in the use of nitrogen, phosphate, and potassium fertilizers. These global developments impose growing pressure on the environment. It is predicted that by 2050 the world's population grows by almost one third to 9.8 billion, and by 2100 this number is expected to surpass 11 billion. Globally, the use of resources may double by 2060, with a growth of 55% for water requirements by 2050, and an increase in energy demand of 30% by 2040 (FAO 2017a; European Environment Agency 2019; United Nations 2022).

This accelerated development surely enabled great benefits. However, it was also the cause of widespread damages to ecosystems, with around 75% of the terrestrial environment and 40% of the marine environment being now seriously altered. Earth is undergoing an exceptionally rapid loss of biodiversity and more species now face the threat of extinction than at any other period in the history of Humanity. Actually, there is evidence that points to the possibility that a sixth mass extinction is currently in progress (FAO 2017b; European Environment Agency 2019; IPCC 2019).

Many of the climate changes observed globally since the 1950s are similarly unprecedented over decades to millennia. In large part, they result from greenhouse gas emissions produced by human activities, such as burning fossil fuels, agricultural activities, and deforestation (FAO 2017b; European Environment Agency 2019; IPCC 2019).

These pressure are causing vast damages to human health and well-being, both directly and indirectly, and may lead to even more far-reaching threats if the collapse of ecosystems such as the Arctic, coral reefs, and the Amazon rainforest, is triggered. This kind of abrupt and irreversible changes can severely disrupt Nature's capacity of providing essential services, such as supply of food and resources, maintenance of clean water and fertile soils, and protection against natural disasters (FAO 2017a; European Environment Agency 2019).

The pathways that social and economic development is currently taking have been destroying the ecosystems that ultimately sustain Humanity. A shift toward sustainable routes hopefully may allow rapid and large-scale reductions in environmental pressures.

The Seventh Environmental Action Program (7th EAP) that defines the comprehensive vision for Europe's environment and society, predicts that by 2050:

> We live well, within the planet's ecological limits. Our prosperity and healthy environment stem from an innovative, circular economy where nothing is wasted and where natural resources are managed sustainably, and biodiversity is protected, valued, and restored in ways that enhance our society's resilience. Our low-carbon growth has long been decoupled from resource use, setting the pace for a safe and sustainable global society.

Three thematic policy priorities in the 7th EAP provide guidance for EU environmental policies: (i) to protect, conserve, and enhance the EU's natural capital; (ii) to turn the EU into a resource-efficient, green, and competitive low-carbon economy; and (iii) to safeguard the EU's citizens from environment-related pressure and risks to their health and well-being.

Europe's sustainability vision of "living well, within the limits of our planet" will not be reached simply by promotion of economic growth and placing the expectations on environmental and social policy tools to manage its harmful side effects. Instead, ambitious and coherent policies and actions must be undertaken across society having sustainability as the guiding principle. The time for action is now.

5.2 Land Degradation–What Are we Doing to our Soil?

It is not widely recognized how much our future depends on the thin layer beneath our feet. Land is the solid surface of Earth that is not permanently covered by water, while soil is the ecosystem in the immediate surface of the earth that serves as a natural medium for plant growth. Currently, land and soil degradation is a global concern.

Soil is finite in extent; it is non-renewable over the human timescale, and it is prone to degradation by natural and anthropogenic causes (Lal 2015). It serves as a large storehouse of carbon, which is critical for the control of climate change by helping regulated emissions of carbon dioxide and other greenhouse gases (FAO 2017b). Indeed, soil is the world's largest terrestrial pool of carbon (IPCC 2000; FAO 2004; FAO 2017b). Furthermore, the soil is where most of the food consumed worldwide is produced (FAO 2015b; FAO 2017b). Notwithstanding, the Status of the World's Soil Resources (SWSR) report as well as other studies have been presenting evidence of moderate or severe degradation for over 30% of global soils (FAO 2011; FAO 2015c; FAO 2017b; IPCC 2019).

Approximately 70–75% of the ice-free land area shows human influence, with about 50% of this area being highly influenced by human action (FAO 2022b). The global forest cover today is 32–43 Mkm^2, whereas the pre-human forest cover has been estimated as approximately 50–55 Mkm^2. Approximately two-thirds of remaining forests are currently under some anthropogenic form of management (Arneth et al. 2019; Arneth et al. 2021).

Degradation negatively impacts people's livelihoods in more than a quarter of Earth's ice-free land area, with possibly up to 3.2 (1.3–3.2) billion people affected globally (Mirzabaev et al. 2019; Olsson et al. 2019). However, the impacts of degradation extend beyond the local land surface and societies, affecting marine and freshwater systems, as well as people and ecosystems far away from the site of degradation.

Soils are very diverse in terms of their chemical, physical, and biological properties. Therefore, they differ in their response to management practices, their inherent ability to deliver ecosystem services, as well as their resilience to disturbance or vulnerability to degradation.

Land degradation is a global issue, frequently due to the combined effect of several different factors. The degradation of soil may occur as a consequence of different processes such as loss of soil biodiversity, erosion by water and wind, acidification, salinization, nutrient imbalance, organic carbon loss, contamination, soil sealing, compaction, and waterlogging. These different threats differ in terms of intensity and trend, depending on soil inherent properties (i.e., physical, chemical, biological), climate (i.e., precipitation, temperature), the characteristics of the terrain (i.e., slope, drainage) and the vegetation (i.e., biomass) (Lal 1997; Lal 2015; FAO 2017b). An important element in land degradation is protecting soil as its degradation constitutes a serious threat for human health, food security, climate change mitigation, and the preservation of healthy ecosystems.

Soil and the multitude of organisms that live there provide us with raw materials, food, energy, regulate carbon and nutrient cycles as well as water availability, and support terrestrial biodiversity. Soil forms a fundamental part of the landscape. It preserves the remains that are testimony of our past and is itself an important part of our cultural heritage (European Environment Agency 2000). However, to produce just a few millimeters of this precious carpet it takes some thousands of years (European Commission 2021a).

Soil hosts more than 25% of all biodiversity on the planet and contains around twice the amount of carbon that is found in the atmosphere and three times the amount found in vegetation (FAO et al. 2020; European Commission 2021a). This fragile layer is expected to feed and filter water for consumption by a global population of nearly 10 billion people by 2050 (European Commission 2021a).

The Special IPPC report on Climate Change and Land, Europe's 2020 State of the Environment Report as well as the IPBES Assessment Report on Land Degradation and Restoration, they all alert to the fact that land and soil continue to be subject to severe degradation processes at European and global level. Land and soil degradation have transboundary effects and come with high costs and negative impacts on human health, natural ecosystems and climate, as well as on our economy (Gomiero 2016; European Environment Agency 2019).

The main causes are poor land management, such as unsustainable farming and forestry practices, land use change (e.g., deforestation, drainage of wetlands, and conversion of permanent grasslands into arable land), construction activities and soil sealing, as well as pollution from industrial emissions to the atmosphere, poor waste management or contaminants present in fertilizers or sewage sludge that are applied to soil (Panagos et al. 2015; European Commission 2021a).

Soil is a major partner in a resource-efficient and circular economy, since it is inarguably the planet's biggest recycling machine: it recycles water, carbon, and nutrients, and can break down and filter pollutants (European Commission 2021a).

Soil, sediment, and water are intimately connected. Soil absorbs and buffers water, and it can provide the functions of a filter. However, it can also get eroded and become polluted. Furthermore, when soil is sealed, the transport of water is altered, flowing across surfaces. Methods that allow floodwaters to infiltrate soil, in order to reduce disasters and pollution to water bodies, are important for water management (European Commission 2021a).

Soil's resilience is perhaps soil's worst enemy. Soil's buffering capacity, resilience, and its ability to filter and absorb contaminants mean that damage is not perceived until it is very advanced. This may be the major cause for soil protection usually not being promoted to the same extent as air or water protection. It is also an important reason why a demanding monitoring process is required. After many years of misuse, the evidence and the effects are now more clearly manifesting. There is a need for effective responses, both preventive and corrective (which, in this case, are usually more costly and not always fully feasible), so that the problem does not continue to be transferred to future generations (European Environment Agency 2000).

An important consequence of soil multifunctionality and the limited availability of the resource–and one of the main causes of soil degradation–is an increasing competition between concurrent uses of soil and land (feed people, increase space for living and industrial production, enhance sources of energies, improve mobility). The question is how these uses can be developed in a sustainable way without damage and depletion of the soil resource (European Environment Agency 2000; Blum et al. 2009).

In fact, the current increase in the pressure on soil resources in Europe and globally is linked to the growing demand for soil and land resources to satisfy diverse needs (European Environment Agency 2000).

Soil is being put under pressure by the demographic growth and urbanization, as agricultural intensification is leading soil to get more prone to erosion. At the global level, the need to guarantee food security to a growing population will require more agricultural land or, more probably, an intensification of agriculture in currently cultivated land. Sealing of soil surfaces due to changes in urban patterns and behavior (sub-urbanization and urban sprawling), together with the need for increased mobility, are the main causes of soil degradation in the most industrialized and populated countries of western and northern Europe (European Environment Agency 2000; European Environment Agency 2022).

Really, soil is a limited resource, and while limited remediation of some functions can be made, soil is not renewable within the timespan needed for its regeneration (European Environment Agency 2000).

Soil is crucial for climate change mitigation and adaptation, agricultural production and food security, nature and biodiversity preservation, and are the foundation of our health and our wealth. Soil functions are meritorious of protection because of their socio-economic as well as environmental importance.

Healthy soil is the largest terrestrial carbon pool on the planet, which associated with its sponge-like ability to absorb water (and reduce the risks of flooding and drought), makes soil an invaluable aid for climate change mitigation and adaptation (European Commission 2021a; European Commission 2021b).

Soil is healthy when it has the capacity to function, within natural or managed ecosystem boundaries, to sustain plant and animal productivity, maintain or enhance water and air quality, and promote plant and animal health (European Commission 2021a).

Therefore a healthy soil: (i) provides food and biomass production, including in agriculture and forestry; (ii) absorbs, stores, and filters water and transforms nutrients and substances, thus protecting water bodies; (iii) provides the basis for life and biodiversity, including habitats, species and genes; (iv) acts as a carbon reservoir; (v) provides a physical platform and cultural services for humans and their activities; (vi) acts as a source of raw materials; (vii) constitutes an archive of geological, geomorphological, and archeological heritage (European Commission 2021a). In conclusion, optimal soil health claims a balance between soil functions for productivity, environmental quality, and plant and animal health all of which are greatly affected by management and land-use decisions. Beneficial management practices that consider soil health must include all functions, rather than focus on single functions, such as crop productivity.

One handful of healthy soil can contain up to a billion bacteria, more than 1 km of fungi, essential to plant and animal life. Yet we only know a little percentage of them (European Commission 2021a). Furthermore, soil is the host for the first stages of life of a variety of pollinator insects. In fact, soil biodiversity greatly contributes to human health.

In unmanaged and natural ecosystems, soils are, in general, healthy. In managed ecosystems, soils can also be maintained healthy if a sustainable soil management is employed. This includes a set of practices that can maintain the soil in, or restore it to, a healthy condition yielding multiple benefits, including for water and air. These practices increase soil biodiversity, fertility, and resilience, which are needed for the vitality of rural areas (European Commission 2021a).

These practices are also part of broader agroecological approaches that attempt to reduce nutrient losses and pesticides use, increase the proportion of agricultural land under organic farming and increment the level of soil organic matter (SOM). There is evidence that soil carbon levels are probable to improve if organic farming is applied to agricultural production systems (Gattinger et al. 2012; European Commission 2021a).

From an economic viewpoint it makes sense to invest in the prevention and restoration of degraded soil. As the EU's largest terrestrial ecosystem, healthy soils sustain many sectors of the economy while soil degradation is costing the EU several tens of billions of euros per year (European Commission 2020; European Commission 2021a). Management practices that sustain and enhance soil health and biodiversity improve cost efficiency and limit the inputs (e.g., pesticides, fertilizers), needed to maintain yields.

Soil degradation, environmental quality, sustainable development, and climate change are clearly linked. Climate change is expected to exacerbate the degradation processes in many regions, while degradation-related greenhouse gas emissions and the loss of carbon sink capacity, in turn, contribute to climate change. Soil plays an important role in the sequestration of atmospheric carbon, through the dynamic process affecting the content of SOM (European Environment Agency 2000; Toor et al. 2020; Arneth et al. 2021). A decline in soil quality leads to a reduction in the soil organic carbon pool, and an increase in the emission of CO_2 to the atmosphere. The decline in soil quality and structure leads to a loss in the capacity to retain water, and therefore, in plant productivity (Abrudan et al. 2014).

Although climate change has until now had relatively minor impacts on degradation, it is expected to become more important for land degradation in the future, exacerbating the effects of land-use change (Mirzabaev et al. 2019; Olsson et al. 2019).

Avoiding, reducing, and reversing land degradation needs to be part of climate change mitigation and adaptation strategies, with many possible co-benefits for ecosystems and human societies (Arneth et al. 2021).

Globally, tackling soil degradation will help offset greenhouse gas emissions, will provide a better environment, will guarantee more food to an increasing population, and will contribute to the economic progress of future generations (European Environment Agency 2000).

Agricultural systems contribute to carbon emissions using fossil fuels in agricultural operations and through practices that result in loss of organic matter in soils. On the other hand, agricultural systems can compensate for carbon losses by accumulating organic matter in the soil, or by increasing above-ground woody biomass, which acts as a permanent sink or can be used as an energy source that replaces fossil fuel. For soil restoration, it is important to properly manage all factors (European Environment Agency 2000).

Innovation in agricultural production management is a prerequisite for developing the rural economy and promoting sustainable agricultural systems.

Comparing a conventional crop management with conservation agricultural practices, the former results in a typically more degraded soil, depleted in organic matter and with increased erosion and greenhouse gas emissions (De Meyer et al. 2011; Haider et al. 2022).

Future research approaches should focus on sustainable and simple methods that can effectively remediate degraded soils (Haider et al. 2022). In this context, organic farming is an environmentally, economically, and socially accepted way to produce food, and it has attracted substantial attention in recent years.

On a per unit area basis, organic agriculture has been shown to have several environmental benefits compared to conventional agriculture. It reduces pesticide use, it can increase biodiversity (Bengtsson et al. 2005; Hole et al. 2005), reduce soil erosion (Siegrist et al. 1998), increase soil fertility (Leifeld and Fuhrer 2010), use less energy, and reduce agricultural greenhouse gas emissions (Gomiero 2016), and may reduce nitrogen losses from the system (Drinkwater et al. 1998).

Organic farming focuses on the natural equilibrium of the soil–plant system, producing high quality and healthy food by banning harmful residues and toxic substances for humans and animals (Mondelaers et al. 2009; Durrer et al. 2021).

5.3 Organic Farming–An Environmentally Sustainable Trend Expanding Worldwide

One of the most important challenges of human society is meeting the food demands of the world's growing population, which is predicted to increase to 9.8 billion by 2050 from a current 7.7 billion (United Nations. Population Division 1999; Davis et al. 2016; Aghasafari et al. 2020; United Nations 2022).

Organic agriculture is an ecologically safer production system, expanding globally as the demand for sustainability increases, being known as a viable solution for the modern farming crisis (Eyhorn et al. 2019; Smith et al. 2019). Organic farming has had a significant worldwide development in recent years and is defined as a production system that promotes human, plant, animal, and soil health, sustains ecological systems and biodiversity, ensures equity in relation to the environment and life opportunities, and preserves the health of future generations and the environment, which refer to IFOAM's four principles of Health, Ecology, Fairness, and Care (IFOAM 2016; Aghasafari et al. 2020).

The use of synthetic or chemical fertilizers and pesticides in organic farming is interdicted, relying instead on bio-fertilizers, natural pathogen, and pest control (Fess and Benedito 2018; Aghasafari et al. 2020). Thus, organic farming pursues sustainable agricultural development in the long run (Smith and Lampkin 2019; Aghasafari et al. 2020) and contributes to improving the sustainability of food systems (Muller et al. 2017; Aghasafari et al. 2020).

The latest data on organic farming worldwide presented by the Research Institute of Organic Agriculture FiBL and IFOAM indicated that a total of 72.3 million hectares were organically managed at the end of 2019, representing a growth of 1.6% or 1.1 million hectares compared to 2018 (Willer et al. 2019; Eurostat 2021). The largest areas used for organic agriculture are located in Australia (35.7 million hectares), Argentina (3.7 million hectares), and Spain (2.4 million hectares). Half of the world's organic agricultural land is in Oceania (36.0 million hectares), owing to the extensive organic farmland area in Australia. Europe (16.5 million hectares) and Latin America (8.3 million hectares) follow with next largest areas. Organic farmland area has increased in all continents in comparison with the situation in 2018, except in Asia (mainly because of the decline in organic farmland in China) and Oceania. Globally, as much as 1.5% of farmland is organic nowadays.

In 2019, organic food achieved a global market of 106 billion euros. The leading market is that of the United States (44.7 billion euros), with Germany (12.0 billion euros) and France (11.3 billion euros) following in the top three markets. Many major markets continued in 2019 to show solid rates of growth. For example, the French market showed an increase of over 13%. Consumers in Denmark and Switzerland did the highest spending on organic food (344 and 338 € per capita, respectively). Denmark also attained the highest organic market share, corresponding to 12.1% of its global food market (Willer et al. 2019).

In 2019, 3.1 million organic producers were reported. India remains the country with most producers (over 1.3 million). Next in this list are Uganda (210,000) and Ethiopia (204,000) (Willer et al. 2019).

By the end of 2019, in Europe, as much as 16.5 million hectares of agricultural land (14.6 million hectares within the European Union) were managed organically by over 430, 000 producers (European Union: almost 344, 000). In Europe, 3.3% of the agricultural area was organic (European Union: 8.1%). Organic farmland has increased by over 0.97 million hectares compared to 2018. The countries with the largest organic agricultural areas were Spain (2.4 million hectares), France (2.2 million hectares), and Italy (2.0 million hectares). In 12 countries, at least 10% of the farmland was organic: Liechtenstein has the lead (41.0%), followed by Austria (26.1%) and Estonia (22.3%) (Willer et al. 2019; Eurostat 2021).

In Europe, agricultural land use is well known, and the main crop categories are well documented. Permanent grassland accounts for almost 40% of the organic agricultural land. Arable land (48%) is mainly used for the cultivation of cereals and by green fodder (almost 3 million hectares and almost 2.6 million hectares, respectively). Permanent crops account for almost 11 % of the organic agricultural land. More than one-third of this land was used for olives, followed by grapes, nuts, and temperate fruits (Willer et al. 2019).

The coronavirus pandemic is described as a multi-layered crisis with health, social, and economic impacts. Since it began in spring 2020, it caused a deep impact on our daily lives, as well as on the organic food industry with organic food sales increasing extensively during the coronavirus pandemic. Indeed, consumer demand for organic foods increased substantially during the pandemic because consumers are turning to organic products as they look to boost their personal immunity.

Since organic foods avoid synthetic pesticides, growth promoters, fertilizers, and agrochemicals, they are considered healthier and safer than conventional foods.

Evidence shows that retail sales have risen by up to 30% in some countries (Willer et al. 2019). Organic food retailers have benefited from the emergency measures (lockdowns or circuit-breakers) introduced by national governments. Organic food and health food shops remained open during the crisis, attracting new shoppers. The pandemic has highlighted the importance of healthy eating, and consumers are shifting their lifestyles toward healthy and accountable consumption (Willer et al. 2019). COVID-19 is undoubtedly having a huge impact on us as consumers, changing the way we shop and eat.

In conclusion, organic farming, as the most sustainable form of agriculture, will certainly benefit in the post-COVID world. In accordance with the International Federation of

Organic Agriculture Movements (IFOAM), organic farming is defined as a "production system that sustains the health of soils, ecosystems, and people. It relies on ecological processes, biodiversity and cycles adapted to local conditions, rather than the use of inputs with adverse effects. Organic farming combines tradition, innovation, and science to benefit the shared environment and promote fair relationships and a good quality of life for all involved" (IFOAM 2014).

According to the EU Council Regulation 848/2018, organic farming is "an overall system of farm management and food production that combines best environmental practices, a high level of biodiversity, the preservation of natural resources, the application of high animal welfare standards and a production method in line with the preference of certain consumers for products produced using natural substances and processes. The organic production method thus plays a dual societal role, where it on the one hand provides for a specific market responding to a consumer demand for organic products, and on the other hand delivers public goods contributing to the protection of the environment and animal welfare, as well as to rural development."

Furthermore, according to this directive, organic production must pursue the following general objectives:

a) contribute to the protection of the environment and climate;
b) maintain the long-term soil fertility;
c) contribute to a high level of biodiversity;
d) contribute substantially to a non-toxic environment;
e) contribute to high standards of animal welfare and, in particular, to meet the specific behavioral needs of animals;
f) encourage short distribution channels and local production in the different areas of the Union;
g) encourage the preservation of rare and native breeds in danger of extinction;
h) contribute to the development of the supply of plant genetic material adapted to the specific needs and objectives of organic farming;
i) contribute to a high level of biodiversity, namely through the use of diversified plant genetic material, such as heterogeneous organic material and organic varieties suitable for organic production; 2019
j) foster and develop organic plant breeding activities in order to contribute to the favorable economic prospects of the organic sector.

In accordance with Sivaranjani and Rakshit (2019), organic production must be a sustainable management system based on respect for nature's systems and cycles and on maintaining and improving the state of the soil, water and the air, the health of plants and animals, and the balance between them, through of the combination of the following:

- *Increase soil organic matter* by returning organic materials to the soil and choosing practices that support a biologically active humus complex;
- *Composting* organic residues, such as animal manure, so as to obtain a more uniform and chemically stable fertilizing material;
- *Timing* the release of nutrients resulting from organic matter mineralization to coincide with times when plants are actively growing and taking up nutrients;

- *Using the rotation of crops for nitrogen fixation* and for recycling nutrients from the soil profile, originate different types of crop residues and increase soil tilth by roots' growth;
- *Using intercropping practices* that introduce a diversify crops in the field, which may reduce pest pressures and improve the fertility of the soil and the efficiency of nutrient use;
- *Planting catch crops or cover crops* to recover nutrients that may otherwise be leached along the soil profile;
- *Using conservation practices* that minimize wind and water erosion and diminish the potential for water runoff phenomena;
- *Providing buffers or filter* areas between crop areas and water bodies to protect against nutrient and sediment movement into lakes and streams;
- *Managing and monitoring irrigation practices* to increase nutrient uptake, decrease nutrient leaching, and minimize root and stem diseases; and
- *Controlling pest populations* through cultural practices, greater balance between pests and predators, and use of biodegradable pesticides with low toxicity to beneficial insects, fish, birds, and mammals.

A positive effect of organic farming on biodiversity is reported by several studies (Maeder et al. 2002; Verbruggen et al. 2010; Seitz et al. 2019) soil carbon sequestration (Gattinger et al. 2012), and environmental sustainability (Reganold and Wachter 2016). Additionally, earlier studies based on erosion modeling and comparison of different field sites with variable farming systems indicated that organic farming practices can reduce soil erosion when compared to conventional farming systems, as a consequence of higher soil surface cover and higher SOM content under organic farming. A further attenuation of soil erosion can be attained from the reduction of tillage intensity in organic farming in comparison to intensive tillage practices (Lockeretz et al. 1981; Auerswald et al. 2003; Pacini et al. 2003; Arnhold et al. 2014; Seitz et al. 2019).

The increased SOM content in the topsoil layer in organic farming might be attributed to the combined effect of organic fertilization and reduced tillage intensity. SOM is one of the key factors that affect aggregate stability and size by binding soil particles (and small aggregates), thus helping to define soil structure (Six et al. 2000; Blanco-Canqui and Lal 2010; Seitz et al. 2019; Ferreira et al. 2020).

In organic farming, tillage and cultivation practices shall be used that maintain or increase soil organic matter, enhance soil stability, and soil biodiversity, and prevent soil compaction and erosion (European Union 2018).

Conservation agriculture (CA) is a resource-saving strategy encompassing three broad principles: (i) minimal soil disturbance or absence of deep plowing; (ii) diversified crop rotation; and (iv) permanent soil cover with green manure during the non-cultivation period. The main objective of CA is to protect the soil from mechanical and weather-caused disturbances (Peigné et al. 2016; Mandal et al. 2021; FAO 2022a). Currently, organic farmers are encouraged to adopt conservation practices to assure soil quality and sustainability, along with food security (Mandal et al. 2021). The possible potential benefits of adopting CA in organic farming systems are to reduce soil erosion and degradation, due to improved soil structure, increased soil carbon stock, augmented biomass, and diversity of macro- as well as micro-organisms, and reduced nutrient loss through runoff and/or leaching. As CA and OF rely on principles of contrasting strategies of practices, introduction of CA in organic farming also poses challenges like difficulties in weed control, soil compaction, limited availability of nutrients resulting from the low rates of mineralization and restricted crop choices. Major challenges for the adoption of conservation tillage in OF are: (i) the preservation of soil fertility and of the main soil functions such as the mineralization of organic matter and (ii) weed control and maintenance of crop yields (Peigné et al. 2016).

Other constraints apart from these (for example, the low incentives and funding from governments, unavailability of suitable equipment, and lack of strategies for the development of the organic market) are also holding up a spontaneous adoption of CA by the organic farmers. Therefore, the practice of conservation techniques in OF resolutely needs a change of traditional management toward well optimized practices but having some range of flexibility depending on the availability of equipment, the state of pedoclimatic conditions, and the farmers' targeted objectives (Mandal et al. 2021).

In conclusion, OF practices have presented beneficial effects by enhancing biological activity and biodiversity, nutrient availability, soil moisture and organic matter content, soil porosity, structural stability, and the reduction of soil erosion (Reganold et al. 1987; Marinari et al. 2006; Fließbach et al. 2007; Wang et al. 2017b; Durrer et al. 2021).

5.4 Organic Farming and Soil Fertility

The term *soil fertility* has ancient origins and has been used over the centuries to refer to the ability of soil to support plant production in agricultural contexts (Abbott and Johnson 2017). Soil fertility is the ability of a soil to support plant growth and optimize crop yields, by providing essential plant nutrients and ensuring propitious chemical, physical, and biological characteristics as a habitat for plant growth.

Soil fertility is fundamental in determining the productivity of all farming systems and results from interactions between physical, chemical, and biological processes.

Soil fertility maintenance is one of the pivotal benefits of the organic farming systems (Barnwal et al. 2021). Organic farming increases soil fertility by ameliorating physical, biological, and chemical properties of soil. This farming method mainly aims to keep nutrients in their natural reservoirs and in bioavailable forms that can be easily absorbed by plants, making the addition of chemical fertilizers unnecessary (JLI Social Media 2020).

Soil organic carbon plays the key role in maintaining soil fertility. Soil organic carbon is both a source and a sink of nutrients and plays a vital role in maintaining soil fertility (Barnwal et al. 2021).

Soil is the central component of organic farming system and organic matter is critical for the efficient management of soil fertility. Organic farming depends mainly on the mineralization of nutrients from organic residues, occurring the whole process with the help of soil organisms such as bacteria, fungi, nematodes, and the full range of soil flora and fauna (Stockdale et al. 2006; Barnwal et al. 2021).

Continuous losses of organic matter due to rapid turnover regarding climatic, anthropogenic, as well as management practices create serious concerns in the modern-day agriculture. Cultivation of land without the addition of organic inputs can result in the depletion of total carbon content. On the other hand, the addition of soil improvers is critical to maintain optimal soil physical conditions (e.g., soil structure, aeration, and water holding capacity) and a way to promote soil microbiome, providing opportunities to improve crop production, enhance feed efficiency, and increase resilience to stress and disease (Barnwal et al. 2021; American Society for Microbiology 2022).

The physical degradation of soil results, in general, in a deterioration of structural attributes such as pore geometry and continuity, which, in turn, aggravates the soil's susceptibility to crusting, compaction, increased obstacle to water infiltration and potential for surface runoff, wind and water erosion, higher soil temperature fluctuations, and an increased propensity for desertification (Lal 2015).

The next sections of this chapter describe the main parameters and processes, that should be assessed and their contribution, when it is intended to improve soil fertility in organic farming.

5.4.1 Organic Matter

Organic matter is essential for soil fertility. It contains most of the soil's reserve of nitrogen (N) and large proportions of other nutrients such as phosphorus (P) and sulfur (S) (Figure 5.1) (Department for Environment Food and Rural Affairs 2006), and in most soil produce the majority of CEC (Hatten and Liles 2019). A very important role is played by the soil's organic matter in ecosystems sustainability by the regulation of nutrient cycling and impact on physical properties, as was described above. Increased levels of SOM, and

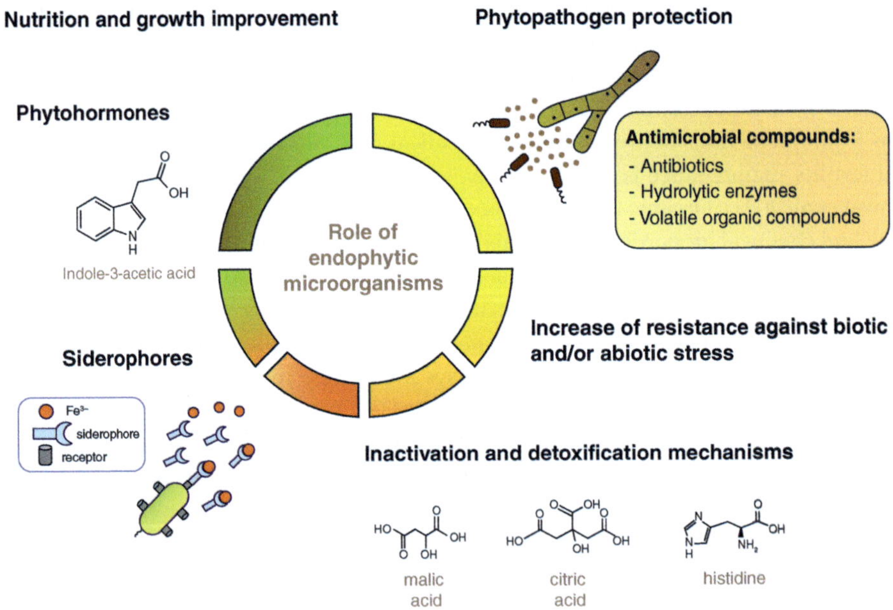

Figure 5.1 **The role of endophytic microorganisms.** Endophytic microorganisms are capable of promoting plant growth through: (1) enhancement of nutrients absorption, fixation of atmospheric nitrogen and solubilization of organic and inorganic P; (2) production of phytohormones; (3) production of siderophores; (4) increase of the resistance to biotic and abiotic stress; and (5) protection against phytopathogens.

consequently increased organic reserves of nutrients, are widely reported for organic systems (Stockdale et al. 2001; Stockdale et al. 2006).

A meta-analysis carried out by Bai et al. (Bai et al. 2018) on various long-term experiments provides a confirmation that those soils managed according to the principles of OF have a content in SOM which is larger than that in soils managed by conventional farming (CF).

SOM is a vital part of soil as its presence provides the main reason for the several benefits induced by organic farming on soil resilience and on the sustainability of soil biological fertility (Vujinović et al. 2020).

The amount of soil organic matter depends on the balance between the decomposition rate and the primary productivity. SOM typically comprises just 1–5% of the soil mass but plays a critical role in soil health because it exerts a powerful effect on soil properties and function. SOM is the largest terrestrial reservoir of organic C, a large depository for all essential nutrients, and a crucial factor contributing to aggregate formation and stability. As a result, SOM also regulates soil trafficability and workability, as well as hydrologic functions, with increases in SOM reducing infiltration and runoff rates, and enhancing flood control (Hatten and Liles 2019, and references therein). Furthermore, SOM can bind toxic compounds (e.g., heavy metals), decreasing soil toxicity with benefits to plants. Thereby, high SOM soils are generally more fertile than low SOM soils and promote an extensive range of positive benefits to plants including adequate soil moisture and a better rooting environment (Hatten and Liles 2019). A central role is played by SOM in preventing the degradation of soil and maintaining its functions.

Soils constitute the largest organic carbon pool on Earth and play a critical role in the climate regulation and mitigating climate change through trade-offs between greenhouse gas emission and carbon sequestration. For this reason, SOM is strategic for climate change adaptation and mitigation, and global reserves of SOM should be stabilized or increased. A loss of soil organic carbon due to inappropriate land use or the use of unsuitable soil management or cropping practices can promote a decline in soil quality and soil structure, and increase soil erosion, potentially leading to carbon emissions into the atmosphere. Otherwise, appropriate land use and soil management can lead to increased soil organic carbon and improved soil quality that can partially mitigate the rise of atmospheric CO_2 (Stockdale et al. 2006; Seufert 2012).

One of the key goals of organic agriculture is to improve soil fertility by returning organic matter to the soil (Seufert 2012). Soil managed with organic methods have therefore typically a higher organic matter content, which results in soil that can hold more water and that are less likely to suffer from erosion. Organic management methods can therefore provide practicable ways to restore degraded soil or prevent further soil degradation in regions prone to land degradation (Colla et al. 2000; Lotter et al. 2003; Seufert 2012).

Organic matter plays a key role in binding mineral particles together in most soil and making their architecture resistant (Rabot et al. 2018; Baveye et al. 2020). Surface horizons depleted in organic matter tend to disperse more easily during rainfall events, and in some cases tend to form impermeable crusts, both situations accelerating the occurrence of ponding and surface runoff (Bresson and Boiffin 1990; Bielders and Baveye 1995; Bielders et al. 1996; Obale-ebanga 2001; Janeau et al. 2003; Baveye et al. 2020).

As a corollary of a less stable architecture resulting from the loss of organic matter, research has demonstrated that it also makes soil more vulnerable to compaction, which is increasingly considered a major cause of soil degradation. This effect has become particularly acute in situations where mechanized agriculture is practiced and where, over the years, increasingly heavier equipment is used (Baveye et al. 2020).

Organic farming systems rely on the management of soil organic matter to ameliorate the soil's chemical, biological, and physical properties, in order to preserve and improve soil fertility and optimize crop production (Watson et al. 2006).

Soil fertility maintenance is ensured through the use of certified organic fertilizers and slow acting minerals (Fortune et al. 2006; Pampuro et al. 2018; Ferreira et al. 2020), which provide sustainable nutrient cycles (Ritchie 2017; Kolbe 2019; Ferreira et al. 2020), being particularly valued those derived from composting residues or by-products from agricultural, livestock, industrial or forestry activities.

Using animal manure to nutrient addition is the most common and traditional approach in agriculture. Many studies reveal that soil supplied with animal manure has a larger population of microorganisms than soil supplied with inorganic fertilizers (Edmeades 2003; Herencia et al. 2007; Barnwal et al. 2021). Distinct sources of organic matter may be adopted as fertilizers to substitute the mineral fertilization and to establish an organic system. Composted animal manure (e.g., poultry, pig, or cow) and green manure or mixes with straw are the most advised fertilizers to be used in organic practices (Govasmark et al. 2011; Zhen et al. 2014; Sacco et al. 2015; Chen et al. 2020; Durrer et al. 2021). Organic fertilizers should have a C/N ratio of less than 30 parts of carbon to each part of nitrogen by weight, particularly when the objective is to increase mineralization rate (Durrer et al. 2021).

The quality and quantity of added crop residue in the field define the rate of nutrient release. Wide C : N ratio crop residue like cereals take time to decompose and promote net immobilization. This can be advantageous in reducing leaching losses of nutrients, especially N. Green manure and leguminous crop with low C : N ratio leads to rapid mineralization and nutrient release. The incorporation of crop residues with different C : N ratios can provide essential nutrients synchronized with crop demands (Watson et al. 2006). Inclusion of leguminous in crop rotation increases soil N, by symbiotic N fixation, and suppresses pest and diseases (Ritchie 2017; Krishnaprabu 2019; Ferreira et al. 2020; Barnwal et al. 2021).

In resume, SOM has a vital role in enhancing soil fertility and plant nutrition since it provides food for microbial activity, improves the soil condition by influencing physicochemical properties of soil and affects the nutrient cycle (Barnwal et al. 2021).

5.4.2 Nutrient Cycling

Nutrient cycling is the flux of nutrients within and between the different biotic or abiotic pools in which nutrients occur in the soil environment (Barnwal et al. 2021). These nutrients can either be obtained from their mineral or atmospheric sources or recycled from their organic forms by conversion into the ionic forms, promoting uptake by plants (Barnwal et al. 2021).

The sustainability of any plant community is regulated by the cycling of nutrients. Without cycling, nutrients would be lost or immobilized, and the plant community would not be able to develop.

One of the major challenges the organic farmer is faced with is the management of soil fertility. Soil fertility needs to consider soil nutrient status, soil physical condition, and biological "health" of the soil (Department for Environment Food and Rural Affairs 2006).

Organic farming aims to build, or at least maintain, soil nutrient reserves while maximizing nutrient recycling and reducing external inputs. Only balanced management can solve this apparent conflict (Department for Environment Food and Rural Affairs 2006). Nutrient management in organic production systems focuses on agricultural productivity maintenance with inputs of on-farm or minimally processed materials.

Nutrient cycling is initiated and carried out by a wide number of soil microorganisms; whose microbial activity requires an organic substrate. Nutrient inputs in organic production are typically focused on carbon-based nutrient sources (e.g., crop residues, compost, manure) and unprocessed mineral sources (e.g., rock phosphate, lime, and gypsum) (Nelson and Mikkelsen 2008).

It is known that the availability of nitrogen in the soil is one of the most important yield-determining factors (Oehl et al. 2005; Tsvetkov et al. 2018). In organic farming, nitrogen is acquired from the leguminous plants, crop residues, manure, and compost, which may not provide enough nitrogen, causing lower crop yields compared to conventional farming.

Conversion from conventional to organic management systems demands a transition period, which is needed for adaptation of the soil ecosystems to the new conditions.

For long-term sustainability, a gradual transition from synthetic nitrogen fertilizers to their replacement with legume species and crop rotations, typical of organic farming, is necessary (Tsvetkov et al. 2018). Overall, during the periods of rapid plant growth, the insufficient nitrogen content in organic fields usually causes stress, which negatively influences the crop yield (Tsvetkov et al. 2018). In comparison to the conventional systems, with high inputs of synthetic nitrogen fertilizers, the productivity in organic farms is generally lower, more frequently during the first years (Drinkwater et al. 1995; Tsvetkov et al. 2018). Some evidence is revealing that the lower yield of organic systems in comparison to conventional ones is not so significant after the first five years (Tsvetkov et al. 2018). The yield could be even similar, according to other reports (Pimentel et al. 2005; De Ponti et al. 2012; Tsvetkov et al. 2018). Graham et al. (2017) discussed the possibility of improving soil nutrient content through integrated nutrient management (combined application of organic, inorganic, and biological nutrient sources).

The availability of nutrients in soils and respective cycling can be defined by the following soil fertility management practices: (i) use of organic residues as soil amendments or sparingly soluble minerals; (ii) biological nitrogen fixation as the main nitrogen source; (iii) crop rotation including a mixture of fertility-enhancing leguminous crops and cash crops; (iv) planting diverse crop species to fulfill different functions.

In summary, farmers need to maximize the productivity of their crops through specific actions, such as crop rotation and application of non-synthetic fertilizers, looking for the integration and synergy of different actions like crop diversification and techniques of cultivation (Benoit et al. 2016; Viaene et al. 2016; Mu et al. 2017; Flores-Félix et al. 2019).

Crop residues can be an important source of nutrients and organic matter, whose amounts and nature vary among crop types. For example, cereal straw contains around 35 kg N/ha and has a high C : N ratio, compared with more than 150 kg N/ha for some vegetable residues, with a low C : N ratio. The low C : N ratio of green leafy residues means that

N will be released much faster than from cereal straw. As a result, different types of residues will require different management for maximum benefit.

After incorporation of green leafy residues, there is likely to be a rapid rise in soil N. Rapid establishment of the next crop will allow this N to be used. If that is not possible, a cover crop could be used to retain N. Incorporating a low N residue, such as cereal straw can have the opposite effect, immobilizing N from the soil (Department for Environment Food and Rural Affairs 2006).

As described earlier the process of converting atmospheric nitrogen by soil microbes into ammonia that can be rapidly assimilated by plants is known as biological nitrogen fixation. In legumes, a large proportion of the crop's N demand can be met through symbiotic N fixation, with subsequent crops, however, being highly dependent on the mineralization of this fixed N (Berry et al. 2002; Barnwal et al. 2021).

It is important to plan and manage crop rotations rigorously considering nitrogen supply from legumes. Achieving an ideal balance between fertility building and fertility depleting rotation phases assures there is enough nitrogen available in the rotation. Furthermore, nitrogen leaching can be restricted by cover and catch crops. Crop rotation is also very important for the balance of soil microflora in organically managed fields because different plants have a specific impact on it (Baggs et al. 2000; Tsvetkov et al. 2018).

From the beginning of the twentieth century, a concept began to be considered, which was the use of different crops not intended for production but for the maintenance of the covered land surface.

Traditional agriculture has always banned other plants to maximize the use of resources by the crop of interest, leading to a massive use of herbicides. Therefore, we must first define the concept of cover crops to differentiate it from crop rotation systems (Flores-Félix et al. 2019). This is a short-term rotation that happens when the principal crop is not sowed (Robačer et al. 2016; Flores-Félix et al. 2019).

The main objective of cover crops is to prevent the loss of edaphic resources by erosion. These crops reduce the kinetic energy of water or wind, decreasing the aggression they cause on agricultural systems (Reeves and Stallard 1994; Flores-Félix et al. 2019). In this sense, it is possible to prevent erosion, and reduce losses of soluble nutrients through run-off (Klodd et al. 2016).

In terms of soil fertility, the most notable advantages of using cover crops are in addition to reducing nitrogen losses (Novara et al. 2013; Flores-Félix et al. 2019), also the stabilization and recycling of other nutrients such as phosphorus, potassium, and magnesium, ensuring that they remain in the upper soil horizon for longer (Rosolem and Calonego 2013). Moreover, crops with pivotal or highly developed root systems are often used to reach deeper horizons, being beneficial for transporting nutrients from deep to superficial horizons and providing available nutrients to crops with superficial root systems such as wheat (Flores-Félix et al. 2019).

Each action produces different effects on the organic farming system; for example, the application of intercropping, which is the mixed cultivation of different crops, is one of the agricultural practices that perfectly match the philosophy of organic farming (Mao et al. 2015; Flores-Félix et al. 2019). In many cases, intercropping practices involve growing cereal rows with leguminous plants, due to the biological nitrogen fixation occurring in their root nodules (Hauggaard-Nielsen et al. 2009; Ngwira et al. 2012; Scalise et al. 2015; Flores-Félix et al. 2019).

Legumes act as suppliers of nitrogen and make available other essential nutrients, due to the microbiota associated with their nodules and roots, which can solubilize phosphates and potassium and to mobilize other nutrients as well as produce phytohormones enhancing plant development (Chapagain and Riseman 2014; Latati et al. 2016; Flores-Félix et al. 2019). Moreover, through indirect mechanisms, the microbiota associated with leguminous plants might play a role as biocontrollers to prevent or avoid plant diseases (Bhardwaj et al. 2014; Flores-Félix et al. 2019).

Organic farming does not allow the use of inorganic fertilizers, and therefore introduces a further incentive to recycle P from residues or permitted supplements, such as rock phosphate (Foereid 2017). Ideally, on-farm recycling would be ideal, but as P is exported with produce, calculated P balances indicate that more P is removed with the products than applied as fertilizer (Watson C.A. et al. 2002; Berry et al. 2003; Bhat et al. 2017; Foereid 2017). Chemical transformations and the action of soil microbes make soil phosphorus available to the plants.

Phosphorous (P) is one of the major essential plant nutrients as it plays a crucial role in the overall growth of plants by influencing various key metabolic processes such as cell division and development, energy transport, signal transduction, macromolecular biosynthesis, photosynthesis, root development, and respiration of plants (Ahemad 2015; Kumar 2016; Ma et al. 2016; Pinto et al. 2018). P in soil is immobilized or becomes less soluble by chemical precipitation, so that the amount of readily bioavailable P is very low, compared with the total amount of P. Hence, in agricultural soils P is often a limiting nutrient (Guiñazú et al. 2010; Kumar 2016; Pinto et al. 2018).

Management of P can be difficult in organic production since approved sources are limited and the consequences of under- or over-fertilization can be significant. Since P is an essential macronutrient for plant growth, and is involved in many critical plant metabolic functions, sustainable agricultural production depends on an adequate P supply (Nelson and Mikkelsen 2008). Therefore, it is imperative to try to replace the removed P in harvested crops in order to avoid declines in yield and quality. Although organic farming looks for to minimize non-agricultural inputs, it is essential that farmers replace P removed in harvested crops (Nelson and Mikkelsen 2008).

Solubilization and mineralization of applied and native P to plant available form is mediated through a different biological and biochemical processes that are strongly influenced by organic matter, besides other biotic and abiotic factors. Organic matter rich soil usually has higher availability of phosphorus, which potentially is due to higher biological activity. In conventional farming systems, mineral fertilizers are used to provide P for plant growth, while organic systems largely rely on inputs of organic origin. Soil under organic management are assumed to be biologically more active, and therefore, have a greater capacity to mobilize native or applied P (Bhat et al. 2017).

Low availability of native soil P for plant growth acts as a limiting factor for crop productivity (Lynch and Brown 2008; Khan and Joergensen 2009; Malik et al. 2012; Johnston et al. 2014; Bhat et al. 2017). As is well known, there is an appreciable amount of total phosphorous in most soil. However, usual P concentrations in soil solutions are peculiarly low, which constitutes an impediment for sufficient P assimilation by plants (Hinsinger 2001; Bhat et al. 2017). As P is subjected to precipitation reactions and sorption reactions on soil colloids, substantial proportions of applied and native soil P remain unavailable (Alam and Ladha 2004; Brady and Weil 2008; Khan and Joergensen 2009; Bhat et al. 2017).

Inorganic P fertilizers are, however, costly and are either out of the reach of resource poor farmers in most of the developing countries or need to be heavily subsidized by taxpayers' money. Moreover, to rely only on inorganic P fertilizers is not a sustainable strategy as accessible natural P resources are rapidly depleting (Cordell et al. 2009; Bhat et al. 2017). Therefore, it is of great importance that the alternative agricultural management strategies, applied are profitable, P efficient and sustainable (Harvey et al. 2009; Sánchez 2010). Besides the input of mineral P fertilizers, organic matter management is among the agricultural strategies that can be adopted for mobilizing soil P for plant assimilation (Damodar Reddy et al. 1999; Aulakh et al. 2003; Singh et al. 2007; Bhat et al. 2017), tillage interventions (Basamba et al. 2006; Shi et al. 2013; Bhat et al. 2017), microbial inoculation (Ramesh et al. 2011; Kumar et al. 2014; Ramesh et al. 2014; Bhat et al. 2017), and crop rotation (Aulakh et al. 2003; Ciampitti et al. 2011; Bhat et al. 2017).

In nature, P is known to occur in several discrete chemical forms, varying in solubility and availability. P is found in both inorganic and organic forms in agricultural soils, the latter being the predominant form (Turner et al. 2002; Kong et al. 2009; Richardson et al. 2011; Condron et al. 2015; Bhat et al. 2017).

Most organic P exists as phytate-P and to a lesser extent as other phosphate esters such as phospholipids (Turner et al. 2007; Richardson et al. 2011; Bhat et al. 2017). The presence of high phytate-P in soils could be attributed to its low solubility and close affinity with the solid phase (soil colloids) due to its higher stability (George et al. 2005; Tang et al. 2006; Bhat et al. 2017). This has been a major impediment to the availability of P for plant uptake. P availability for crop assimilation results from several simultaneous occurring processes, predominantly inorganic P mobilization, organic P mineralization, applied P immobilization and P diffusion rate. These processes are influenced and mediated by diverse physical, chemical, and biological processes (Bhat et al. 2017).

Organic matter contributes 20–80% to the organic phosphorous in soil (Richardson 1994; Nelson and Mikkelsen 2008), which in turn, is hydrolyzed by phosphatases – enzymes of plant or microbial origin – which make P available to plants (Tarafdar and Claassen 1988).

Soil organic matter contains a diversity of organic P compounds, e.g., inositol phosphate, phospholipids, and nucleic acids. These compounds must be first converted to inorganic phosphate by soil enzymes before being absorbed for plant growth. These phosphatase enzymes are produced by soil microorganisms, mycorrhizal fungi, or excreted by plant roots. Some organic P compounds are stable for many years in the soil, while others are converted to inorganic P within a few days or weeks (Nelson and Mikkelsen 2008).

Organic manure mineralization not only provides available P for plant uptake, but also plays a significant role in mobilizing native P forms through different mechanisms. For example, organic anions evolved during manure decomposition, metal complexation or dissolution reactions mediate the release of P from exchange sites (Bolan et al. 1994; Iyamuremye and Dick 1996; Bhat et al. 2017).

Also, the addition of organic matter acts as a substrate for microbial proliferation that helps in changing the P dynamics (both organic and inorganic forms) in the rhizosphere, positively affecting root architecture and biological properties, such as phosphatases released by roots and/or microorganisms (Gichangi et al. 2009; Richardson et al. 2011; Guan et al. 2013; Malik et al. 2013; Bhat et al. 2017).

Organic P is mineralized by the activity of extracellular phosphatases, but soil microorganisms can also immobilize significant amounts of P to meet their own needs (Barnwal et al. 2021). However, this immobilization is short-term and may protect available forms of P from longer-term sorption in soil. The activity of acid phosphatases has been found to be higher in organic and biodynamically managed soil than under conventional management (Barnwal et al. 2021).

As described earlier, cover crops are frequently grown in rotation with cash crops for a variety of beneficial purposes. Many studies reveal that a P nutritional benefit may be provided by some cover crops for the next crop in comparison to crops grown without a preceding cover crop. Some cover crops can be excellent hosts for mycorrhizal fungi, which may allow a higher exploitation of the soil P reserves. There are significant differences in the ability of different cover crops to provide additional P for the subsequent crop (Brito et al. 2014; Faria et al. 2020; Faria et al. 2021b). Studies have generally shown that legume cover crops confer a greater P benefit than grass cover crops. However, it should be noted that the effects on P nutrition provided by cover crops can be highly variable (Nelson and Mikkelsen 2008). However, we recall that cover crops do not provide any new P to the soil, but only allow the existing P reserve in the soil to be used more efficiently.

The most important soil parameter for guaranteeing nutrient supply to the crop is soil pH. The maintenance of a neutral to slightly acidic medium is required for converting poorly soluble P compounds into crop available P. Even when all plant nutrients are present in sufficient amounts in the soil, if the pH is not maintained at the proper level (6.0–7.0), the crop can display nutrient deficiency/toxicity symptoms and will not reach its yield potential. This deficiency is due, in part, to the fact that in acidic conditions (low pH, below 5.5), soil biological activity is reduced, delaying the release of nutrients. Furthermore, at either end of the pH scale, some macro and micronutrients become unavailable to the crop. Other effects of low pH or acidity include deterioration of soil structure, reduced crop quality, reduced fertilizer efficiency, and increased nutrient losses.

Some soils are naturally rich in lime (calcareous) and will never need liming. In general, acidic soils should be limed to raise the pH to 6.0 for grassland and 6.5 for arable soils, but it is important not to add too much lime in short periods (≤1 year). Over-liming can result in nutrient deficiency and yield loss due to antagonistic and imbalanced effects. As a guideline, it is inadvisable to raise the pH by more than half a point in any one application.

Not all liming materials are easily at the disposal of organic farmers. Fastest acting materials, slaked lime and quicklime are not permitted. Ground chalk, limestone, and calcified seaweed are the main sources. The thinner and softer the product, the greater the solubility and the faster its action. Therefore, ground chalk with a diameter of 1 mm diameter will have approximately the same solubility as hard limestone ground to a powder of 0.07 mm in diameter (Department for Environment Food and Rural Affairs 2006).

It is difficult to provide universal recommendations for rock P application because many factors affect its dissolution and plant availability. However, the key factors to consider include:

- Soil pH is important in dissolving P rock. P Rock is much more soluble in acidic soils (soil pH < 5.5). In neutral pH to alkaline soils, rock P normally offers little benefit for plant nutrition, except under special conditions (Nelson and Mikkelsen 2008);
- Particle size influences P rock dissolution by controlling the surface area available for reaction (Nelson and Mikkelsen 2008).

Finally, in organic farming, managing soil fertility requires a better understanding of the effects of different practices.

Organic farming does not only consider the addition of organic input but focuses on those practices that tend to increase soil organic matter and minimize their loss (Leifeld and Fuhrer 2010; Barnwal et al. 2021).

No-till practices with or without residue application enhances soil aggregation and organic fertility. Higher soil carbon sequestration and macro aggregate associated C can be achieved by adopting zero tillage along with residue retention (Gupta Choudhury et al. 2014; Barnwal et al. 2021). The use of tillage in agriculture was one of the first implemented techniques by humans and generated many benefits in terms of weed control and soil aeration, making nutrients assimilable or mixing depths of soil, which were not explored by plant roots (Drakopoulos et al. 2016; Flores-Félix et al. 2019). However, tillage is one of the practices that most actively contributes to the loss of edaphic material, due to surface exposure of unconfined material that can be washed away by runoff or wind action (Crittenden et al. 2015; Flores-Félix et al. 2019).

Conservational tillage is based on the minimal disturbance of the soil composition, structure, and biodiversity, reducing soil degradation and water contamination (Holland 2004). Climatic conditions, high temperatures and precipitation rates, tend to accelerate the mineralization process of organic matter in agricultural soil, which is also benefited by tillage, aiding in the development of aerobic microorganisms, and causing the pathway of organic matter to CO_2 at the upper soil horizon (Chivenge et al. 2007; Lienhard et al. 2013; Raphael et al. 2016; Flores-Félix et al. 2019).

So-called conservation agriculture is being widely promoted in many areas mainly for the recovery of degraded soils (Tsvetkov et al. 2018). This practice aims to improve agricultural productivity, profits and food security based on three principles: minimal soil mechanical disturbance, permanent soil cover, and crop rotation (FAO 2015a; Tsvetkov et al. 2018). Positive effects of conservation agriculture on agroecosystems have been widely reported (e.g., preventing or minimizing soil erosion and organic carbon loss, improving water use efficiency, nutrient cycling, and mitigation of greenhouse gas emissions) (Kassam et al. 2009; Palm et al. 2014; Tsvetkov et al. 2018).

Many works have employed soil metagenomics to explore the complexity of soil microbiology in many different soil types and in diverse agricultural conditions (Tsvetkov et al. 2018). There are significant differences in the microbiota structures and taxonomic composition between cultivated soils under conventional and no-tillage systems (Carbonetto et al. 2014; Brígido et al. 2017; Tsvetkov et al. 2018; Campos et al. 2019). It seems, therefore, that agronomical use and the type of tillage system could induce shifts in the soil microbiota. Usually, the microbiota of conventionally cultivated soils, under higher nutrient amendment, tends toward copiotrophy, whereas the microbiota of non-cultivated soils appears to be more oligotrophic. It seems that conventional tillage might trigger copiotrophy more than no tillage, with the consequence of decreasing the soil organic matter stability and increasing the nutrient availability (Souza et al. 2013; Carbonetto et al. 2014; Tsvetkov et al. 2018).

In organically managed systems where no synthetic fertilizers are allowed, soil microbial biomass (e.g., bacteria, fungi, earthworms, and a whole range of organisms – the soil microflora, macro- and micro-fauna), are an essential part of the system helping to supply and recycle nutrients by mineralization processes and to avoid nutrient leaching (Friedel et al. 2001). Most of the activities of soil organisms are beneficial and it should be one of the central aims of the organic farmer to stimulate soil biological activity and ameliorate fertility.

5.4.3 Microbial Biomass

As mentioned earlier, soil microbiome plays a key role in cycling the nutrients present in the soil and supplying them to the plant. The most important source of nutrients for crops is previous crop residues that are rich in nitrogen, phosphorus, sulfur, potassium, and others. These crop residues are degraded by soil microbes that break them down into forms that are assimilated by plants. A rich and active soil microbiota plays an important role for different soil-based ecosystem services such as SOM mineralization, nutrient cycling, erosion control, soil structure stability, and pest and disease regulation. These important functions in the soil may be impacted by any changes in soil microbial biomass.

According to a large meta-analysis study done by Peltoniemi et al. (2021), long-term effects in organic and conventional cropping systems, were detected regarding the bacterial and fungal community composition. In comparison to the conventional system, biomass C and N as well as microbial activity were higher in the organic system for cereal crop rotation.

Overall, the results seem to show that changes in the soil microbiome were induced by agricultural practices typically employed in organic farming, such as the use of green manure and plant cover, which is evidence that those organic systems comprise a higher microbial richness.

Lori et al. (2017) conducted a meta-analysis on current literature to quantify possible differences in key indicators for soil microbial abundance and activity in organic and conventional cropping systems. The study integrated data from 56 mainly peer-reviewed papers, including 149 pairwise comparisons originating from different climatic zones and experimental duration ranging from 3 to more than 100 years. Overall, this analysis found that organic systems had 32% to 84% higher microbial biomass carbon, microbial biomass nitrogen, total phospholipid fatty-acids, and dehydrogenase, urease, and protease activities than conventional systems. Solely the metabolic quotient as an indicator for stresses on microbial communities remained unaffected by the farming systems. Crop rotation, the inclusion of legumes in the crop rotation and organic inputs are important farming practices that categorical subgroup analysis revealed to have an effect in soil microbial community activity and size. Furthermore, differences in microbial size and activity between organic and conventional farming systems vary as a function of land use (arable, orchards, and grassland), plant life cycle (annual and perennial), and climatic zone. To summarize, it is shown in this study that total microbial abundance and activity is enhanced in agricultural soil on a global scale by organic farming.

Physical damage to soil structure and leaching of essential nutrients may be promoted by intensive soil cultivation, eventually negatively changing the ecosystem services that are provided by the soil microbiome. Long-term reduced tillage and no-till practices have been identified as acting positively on both, microbial indicators such as fungal hyphae length and fungal abundance (Peltoniemi et al. 2021), but also on soil aggregate structure and soil chemical properties (Peltoniemi et al. 2021).

Soil living organisms are therefore at the heart of organic farming. Although earthworms are the most visible soil organisms and perform an important role in organic matter breakdown, it is the microscopic organisms that are the key players. In agricultural soil, the largest group of these, both in number and in biomass, are bacteria, followed by fungi, followed by protozoa and nematodes. This results in a highly complex "food web" within the soil, with different organisms feeding on soil organic matter and other soil organisms (Singh et al. 2016).

Bacteria play an important role in the degradation of organic materials, especially in the early stages of degradation when moisture levels are high. In the later stages of decomposition fungi tend to dominate (Singh et al. 2016).

Plant growth promoting bacteria (PGP, a term referring to bacteria with the potential for enhancing the development of plants) may establish symbiotic/mutualistic associations with plants and can be found as endophytic or free living rhizospheric bacteria. Genera that include *Bacillus*, *Burkholderia*, *Beijerinkia*, *Enterobacter*, *Erwinia*, *Flavobacterium*, *Gluconacetobacterium*, *Klebsiella*, *Pseudomonas*, and *Serratia* are among the beneficial PGP bacteria that are capable of controlling plant growth (Dardanelli et al. 2010; Nadeem et al. 2010; Ashraf et al. 2017; Pinto et al. 2018).

PGP bacteria can also enhance plant development by acting as biofertilizers (increasing the availability of essential nutrients through, e.g., N_2 fixation and phosphate and iron solubilization); as phytostimulants (by producing plant growth regulators and hormones like the indoleacetic acid, cytokinins, or other auxins); as biodegraders of organic contaminants; as stress controllers (by decreasing the production of ethylene through the synthesis of 1-aminocyclopropane-1-carboxylic acid deaminase-ACCD); and as plant defense inducers against phytopathogens (by producing siderophores, antibiotics, or fungicidal compounds) (Lugtenberg and Kamilova 2009; Weyens et al. 2009a; Weyens et al. 2009b; Compant et al. 2010; Zafar et al. 2012; Becerra-Castro et al. 2013a; Becerra-Castro et al. 2013b; Balseiro-Romero et al. 2017; Pinto et al. 2018; Pinto et al. 2021).

Nitrogen-fixing bacteria, i.e., bacteria which convert stable atmospheric nitrogen into a form that is biologically useful, are known as diazotrophs. These organisms reduce di-nitrogen to ammonia with the help of the enzyme nitrogenase (Ullah et al. 2015; Pinto et al. 2018; Sarkar et al. 2021).

Endophytic bacteria that have a powerful associative nitrogen-fixing ability, help plants to survive in soil environments that are poor in nitrogen. In comparison to other rhizospheric microbes, these bacteria play a very important role in promoting plant health and growth (Hurek and Reinhold-Hurek 2003; Gupta et al. 2013a; Ma et al. 2016; Pinto et al. 2018).

Some endophytic bacteria can produce substances that may effectively limit phytopathogens, such as antibiotics, siderophores, a variety of hydrolytic enzymes (such as chitinases, proteases and glucanases) and antimicrobial volatile organic compounds (Figure 5.1) (Sheoran et al. 2015; Ma et al. 2016; Pinto et al. 2018; Sarkar et al. 2021).

Furthermore, endophytic bacteria may be effective as competitors of pathogens for bioavailable nutrients and colonization niches, thereby promoting plant productivity indirectly (Alvin et al. 2014; Ma et al. 2016; Pinto et al. 2018).

It is well known that endophytic bacteria assimilate soluble P and prevent its subsequent sorption (Khan and Joergensen 2009; Ma et al. 2016; Pinto et al. 2018). Hence, these endophytic bacteria may serve as a sink for phosphorus by rapid microbial phosphorus mobilization even under phosphate limiting conditions. Subsequently, these endophytes become a source of phosphorus to the plant hosts upon its release from bacterial cells.

Phosphate solubilizing bacteria are one of the main contributors of plant nutrition in agriculture. They convert the nonavailable inorganic phosphate to its soluble form, thus increasing the availability of phosphorus by secreting organic acids like citric, oxalic, succinic, tartaric, and malic acids in the soil rhizosphere, playing a pivotal role in making soluble phosphorus available to plants (Saritha and Prasad Tollamadugu 2019; Pinto et al. 2021). The principal mechanism in soil for solubilization of P is lowering of soil pH by microbial production of organic acids or the release of protons and mineralization by producing acid phosphatases (Tarafdar and Claassen 1988; Dwivedi 2020) ultimately resulting in P availability in soil (Khosro and Yousef 2012). The 2-keto-gluconic acid produced from direct oxidation of glucose by these bacteria play an important role in weathering and solubilization of phosphates in soil (Dwivedi 2020).

Iron (Fe) is one of the important elements for life and almost all organisms (apart from certain lactobacilli) need iron-containing proteins involved in physiological processes and a series of enzymatic reactions (Ma et al. 2016; Pinto et al. 2018). However, the main form in which iron exists in soil's aerobic environment is Fe (III), which tends to form insoluble hydroxides and oxyhydroxides, thereby making it mostly unavailable to microorganisms. To acquire enough iron, bacteria had to develop strategies to solubilize this metal prior uptake.

Under such conditions, siderophore producing bacteria (e.g., rhizospheric; endophytic) might offer a biological rescue system that is able of scavenging Fe^{3+} and making it available to iron-deficient plants (Pinto et al. 2018).

Many organic farmers encourage the associations of mycorrhizal fungi with crop roots using cover crops and rotations. However, frequent tillage commonly used for weed control causes a disruption of the soil fungal network and may reduce its effectiveness for providing nutrients to the plant (Nelson and Mikkelsen 2008).

Many authors pay particular attention to mycorrhizal fungi as a source of innovation in organic farming practices for improvement of nutrient uptake, biocontrol, and microbial ecology (Briar et al. 2016; Tsvetkov et al. 2018). Their benefits to agriculture are well-known (Figure 5.2).

Arbuscular mycorrhizal fungi (AMF) are ubiquitous and form symbiotic associations with the majority of terrestrial plants (Leung et al. 2010b; Leung et al. 2010a; Deng and Cao 2017; Pinto et al. 2018; Pinto et al. 2021). Consequently, they represent an important part of the soil microbiome and are considered the major component of living organisms in the root zone, interacting with the roots of more than 80% of terrestrial plants and providing their hosts with benefits including increased access to nutrients through the enlargement of soil volume that can be accessed by roots/mycelia (He et al. 2013; Sarwar et al. 2017), especially phosphorus; they also improve water acquisition; sequester potentially deleterious metal(loid)s and reinforce pathogen resistance (Qiao et al. 2015; Deng and Cao 2017; Sarwar et al. 2017; Chaturvedi et al. 2018; Li et al. 2018; Pinto et al. 2018; Faria et al. 2021a; Faria et al. 2021b; Pinto et al. 2021). Furthermore, AMF can explore more efficiently the soil microsites that are not accessible to plant roots. Therefore, mycorrhizal plants invest less energy into the extension of the root system than non-mycorrhizal plants and are more resistant to drought stress and pathogen attacks (Coninx et al. 2017).

Mycorrhizal fungi keep the number of pathogens low through the acidification of the rhizosphere and/or increased production of antibiotics (Fortin et al. 2009; Coninx et al. 2017).

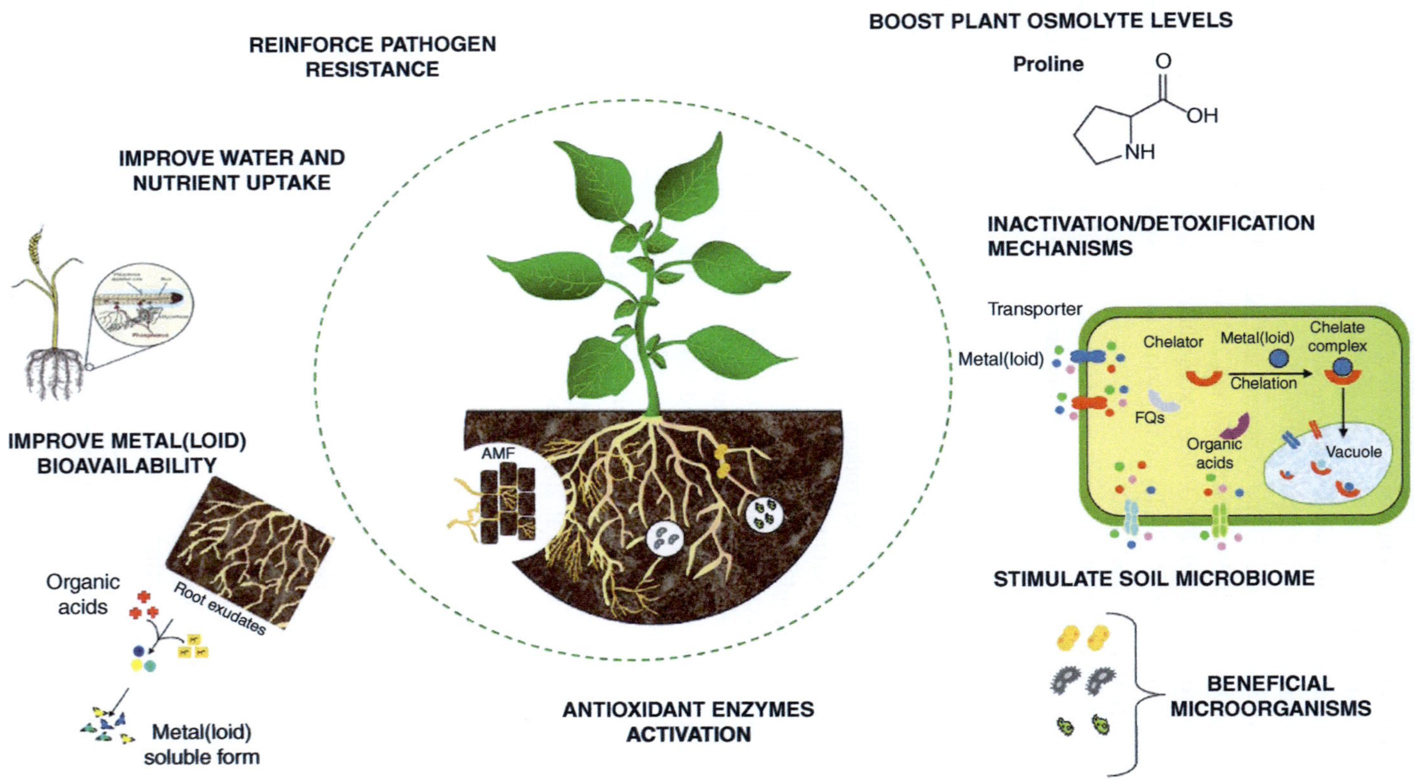

Figure 5.2 AMF **benefits in organic farming.** AMF improve access to nutrients, especially phosphorous, through an increase of the soil's volume that can be explored by plant roots; they reinforce the resistance to pathogens; they stimulate osmolyte levels of the plant; they produce chelating agents with high affinity for metals and/or metalloids; they improve the response of the antioxidant system of the plant.

Mycorrhizal fungi can also increase the resistance of the plant at a physiological level. It has been shown that AMF can lead to greater resistance to herbivores through the action of jasmonic acid (Fortin et al. 2009; Coninx et al. 2017).

AMF of the Glomeromycota are the most common soil microorganisms in natural and agricultural soils (He et al. 2013; Mohammad and Mittra 2013; Pinto et al. 2018). So far approximately 160 AM fungal taxa of the order Glomales (Glomeromycota) have been described. Based on morphological and molecular identification methods, the AM fungi have been divided by 12 genera (*Acaulospora*, *Ambispora*, *Archaeospora*, *Diversispora*, *Entrophospora*, *Kuklospora*, *Geosiphon*, *Gigaspora*, *Glomus*, *Intraspora*, *Paraglomus*, *Scutellospora*) with the *Glomus* species as dominant (Morton and Redecker 2001; Sieverding and Oehl 2006; Walker et al. 2007; Leung et al. 2013; Pinto et al. 2018).

Moreover, mycorrhizal associations could promote a faster closure of the leaf stomata, preventing plant wilting (Fortin et al. 2009; Coninx et al. 2017) and boost plant osmolyte levels such as proline (Garg and Chandel 2012; Coninx et al. 2017).

Plants benefit in many ways from the association between fungi and their roots, namely through the establishment of a mycorrhizal network which offers a number of basic advantages to the host plant for the acquisition of mineral nutrients: (i) fungal hyphae extend beyond the zone of nutrient depletion that surrounds plant roots; (ii) fungal hyphae largely increase the available surface for absorption of nutrients; (iii) the hyphae are able to spread into soil pores that are too small for plant roots to enter; and (iv) some mycorrhizal fungi can access forms of N and P that are not accessible to non-mycorrhizal plants, particularly the organic forms of these nutrients (Leung et al. 2013; Pinto et al. 2018).

The extensive hyphal networks in the soil may also provide a significant contribution to the stabilization of soil aggregates, change the chemical composition of root exudates and control soil pH and the metal(loid)s bioavailability in the soil.

AMF brings significant agroecosystem sustainability as it leads to the maintenance and improvement of the soil structure and microbial ecology (Smith and Read 2008a; Tsvetkov et al. 2018).

For example, glomalin and other glycoproteins released by AMF, may improve soil structure by increasing particle aggregation and aggregates stabilization against wind and water erosion (Wang et al. 2017a; Pinto et al. 2018).

Verbruggen et al. (2010) reported that organic management enhances the diversity of AMF assemblages, when compared with conventionally managed agricultural fields. AMF communities are richer and more diverse across organically managed fields and are more like those of natural, undisturbed systems (Tsvetkov et al. 2018). In addition, the mycorrhiza richness increases significantly with the time since conversion from conventional to organic management (Verbruggen et al. 2010; Tsvetkov et al. 2018). The positive effect of organic management on AMF diversity could be explained by higher frequency of crop rotation with a grass–clover mixture as a forage crop (Tsvetkov et al. 2018). Organic farming may sustain the mycorrhizal component of soil fertility and agroecosystem functioning. Destruction of the mycorrhizal fungal network in the soil system is a critical event of soil disturbance, and their resettlement is an essential approach of habitat restoration.

In resume, soil microorganisms play a crucial role in sustaining favorable soil conditions (e.g., soil aggregation, biological control, nutrient cycling, organic matter degradation, nitrogen fixation) (Bender et al. 2016; Durrer et al. 2021). Rich soil microbial activity is indispensable for efficient crop production, the ability to maintain healthy plants, and ensuring a good yield under different environmental conditions. Therefore, the use soil management practices that improve microbial community, and its functions can be a promissory approach for better crop yield (Panke-Buisse et al. 2015; Castiglione et al. 2021).

As mentioned above from earthworms to bacteria, microorganisms play vital roles in nutrient cycling, soil structural development and pest and disease control. The options available to stimulate the activity of soil microorganisms, which will in turn enhance the benefits they provide, are simple: (i) provide adequate living conditions–good soil structure; (ii) provide a food (energy) source–fresh organic matter; (iii) avoid the application of materials that are antagonistic to biological activity.

Global environmental changes can compromise plant and soil biodiversity, suggesting a complex feedback loop between plants and microorganisms under stressful environmental conditions (De Vries et al. 2020; Castiglione et al. 2021; Yang et al. 2021). Several studies have shown that climate change has decreased plant diversity and yield, and a more negative effect has been observed under reduced soil biodiversity (Castiglione et al. 2021; Yang et al. 2021). This is regardless of plant genotypes, indicating that the negative effect of soil biodiversity loss could generally come from soil microbes (Prudent et al. 2020; Castiglione et al. 2021; Yang et al. 2021). In support of this evidence, plant growth under high microbial diversity showed higher productivity and greater recovery under stress conditions. Furthermore, evidence highlights the importance of the microbial component and its diversity not only for crop management, but also as a promising biological tool to recover degraded soils and implement revegetation activities (Caravaca et al. 2003; Diagne et al. 2018).

If the soil is rich in organic matter and all biological and biochemical processes are working well, the soil is considered to be healthy, containing an efficient microbiome and presenting all ecosystem services for maximum productivity (Tuomisto et al. 2012; Durrer et al. 2021). Thus, healthy soils also maintain a harmonic functioning of microorganisms, enzyme activity, and soil fertility.

Previously to implanting an organic system at a site that had been cultivated using a conventional system, a transitional period should be necessary to restore and improve soil microbial biomass content. An understanding of the microbial dynamics during this transitional period is important for providing information about the suitable conversion strategies and for minimizing yield losses (Tu et al. 2006; Santos et al. 2012).

Moreover, the yield losses can be mitigated in the presence of elevated soil microbial communities, suggesting their potential and crucial role as yield stabilizers after global change disturbances (Prudent et al. 2020; Castiglione et al. 2021).

The status of the microbial ecosystem can be assessed through the genetic characteristics of the soil microbial community, and it also gives the idea about quality of the soil and the progress of restoration after degradation (Harris 2003; Singh 2015; Singh et al. 2016).

5.4.4 Biostimulants

Considering the continuous growth of the global population, the reduction of arable land area, and the depletion of the genetic potential of crops, there is an urgent need for the implementation of novel agricultural technologies. A more sustainable agricultural production system, aimed at improving plant resilience to adverse soil conditions, preserving soil fertility, and reducing soil biodiversity loss, has become indispensable in guaranteeing the high demand for food with high nutritional value (Szparaga et al. 2019; Castiglione et al. 2021; Mannino et al. 2021).

A promising tool and sustainable approach would be the use of bioeffectors, formally known as plant biostimulants (PBs), which is gaining interest globally (Colla and Rouphael 2015; Rouphael et al. 2017; de Pascale et al. 2018). PBs market is constantly growing with an expected compound annual rate of over 11% in the next two years (Sessitsch et al. 2018; Vasseur-Coronado et al. 2021). The main reasons of this growth are: (i) increasing importance of the organic farming, (ii) more PBs use in developed countries, and (iii) good acceptance of PBs by consumers (Vasseur-Coronado et al. 2021).

PBs are an innovative technology able to ensure agricultural yield with high nutritional values, overcoming yield-limiting factor in many organic production systems (Castiglione et al. 2021). They can not only act directly on plants, but also sustain productivity through the selection and stimulation of beneficial soil microorganisms (Hellequin et al. 2020; Rouphael and Colla 2020b; Castiglione et al. 2021). Indeed, PBs are a new generation of products, which may be useful for achieving agricultural sustainability policies. In Europe they are defined by the EU Regulation 2019/1009 as "products stimulating plant nutrition processes independently of the product's nutrient content with the sole aim of improving one or more of the following characteristics of the plant or the plant rhizosphere: nutrient use efficiency, tolerance to abiotic stress, quality traits, availability of confined nutrients in soil or rhizosphere." They are by nature more like fertilizing products than to most categories of plant protection products. PBs can be composed of certain substances, mixtures, and microorganisms, and therefore they be classified as microbial or non-microbial plant biostimulants. They act in addition to fertilizers, with the aim of optimizing the effectiveness of those fertilizers and reducing the nutrient application rate (European Comission 2019; Campobenedetto et al. 2020; Campobenedetto et al. 2021; Castiglione et al. 2021; Joshi et al. 2021). Thereby, PBs application can overcome nutrient limitation typical of organic systems by promoting nutrient availability, uptake and assimilation, and consequently reducing the gap between organic and conventional yields (Joshi et al. 2021).

PBs stimulate natural processes in plants to increase nutrient uptake and use efficiency, crop quality, and tolerance to abiotic stress, benefitting both plant yield and vigor (EBIC 2021). This ensures that plants have good resistance and health, making them less vulnerable to stress, pests, and other threats, including weather shocks. Biostimulant products can help farmers adapt their agricultural systems to an increasingly volatile climate, while enhancing food production sustainability, thereby contributing to a climate-smart farm model for the future, which is both resilient and flexible (EBIC 2021).

Here are some of the ways by PBs have been shown to contribute to climate-smart agriculture, ameliorating resilience and adaptation:

1) Biostimulants help improve agricultural resilience to climate shocks. Plant biostimulants help farmers adapt to climate change by making crops more resilient to abiotic stresses, including climate shocks such as floods, drought, and extreme temperatures, for example: (i) Some biostimulant products increase root growth, improving access to water even in drought conditions; (ii) plants that are heat tolerant will survive heat waves better and start normal growth again more quickly when temperatures cool

down; (iii) protecting against abiotic stresses also reduces the chances of plants falling victim to opportunistic pests or diseases, and researchers expect pest and disease challenges to increase with climate change;

2) Plant biostimulants improve soil health: (i) by enriching soil micro biodiversity, microbial plant biostimulants help improve soil fertility, health, and structure to benefit plant growth and climate resilience, improving capacity of agricultural soils to store carbon; and
3) The circular nature of plant biostimulants increases resource effectiveness and reduces nutrient losses (EBIC 2021).

Microbes of the phytomicrobiome are associated with all plant tissues, and in combination with the plant, form the holobiont. Plants carefully regulate the composition and activity of their associated bacterial community. These microbes provide a large range of services and benefits to the plant; in return, the plant provides the microbial community with reduced carbon and other metabolites. The rhizomicrobiome is of high importance to agriculture due to the rich diversity of root exudates and plant cell remains, that can promote diverse and unique patterns of microbial colonization. Rhizomicrobiome microbes play key roles in nutrient uptake and assimilation, improving soil texture, secreting, and modulating extracellular molecules such as hormones, secondary metabolites, antibiotics, and different signal compounds, all leading to improved plant growth. The microbes and compounds they secrete constitute valuable biostimulants and play pivotal roles in modulating plant stress responses (Backer et al. 2018).

A microbial plant biostimulant consists of a microorganism or a consortium of microorganisms (European Comission 2019; Campobenedetto et al. 2020; Campobenedetto et al. 2021; Castiglione et al. 2021; Joshi et al. 2021). PBs can be composed by microbial inoculants including plant growth promoting rhizobacteria (PGPRs) belonging to the genus *Azotobacter*, *Azospirillum*, and *Rizhobium* spp. arbuscular mycorrhizal fungi (AMF) and *Trichoderma* spp. and/or by natural substances, as vegetal- or animal-based protein hydrolysates (signaling peptides and free amino acids), macro- and micro-algal extracts; and humic substances (Battacharyya et al. 2015; Canellas et al. 2015; Colla et al. 2015a; López-Bucio et al. 2015; Rouphael et al. 2015; Ruzzi and Aroca 2015; Savvas and Ntatsi 2015; Colla et al. 2017; Chiaiese et al. 2018; de Pascale et al. 2018; Rouphael and Colla 2020b; Rouphael and Colla 2020a; Joshi et al. 2021; Kapoore et al. 2021).

AMFs or PGPRs, may influence plant physiological mechanisms by exerting a biostimulant action (Copetta et al. 2006; Boyer et al. 2015; Chitarra et al. 2016; Lally et al. 2017; Zhang et al. 2018; Santander et al. 2019; Castiglione et al. 2021).

This ability of the Glomeromycota phylum is related to the colonization of plant roots through endomycorrhizal symbioses. Even though a large number of AMF species have been identified, only a select few are used in agricultural practices (Giovannini et al. 2020; Castiglione et al. 2021). Accordingly, the available commercial inocula contain species almost exclusively belonging to *Rhizophagus* and *Funneliformis* genera that are generalist symbionts, present in almost all soils and under a wide range of climate zones (Smith and Read 2008b; Giovannini et al. 2020; Castiglione et al. 2021).

On the other hand, PGPRs are a very heterogeneous group of bacteria found in the rhizosphere and able to promote plant growth, with recent scientific evidence showing the importance of their role in enhancing soil productivity and tolerance to abiotic stress in plants (Mannino et al. 2020; Castiglione et al. 2021).

PGPRs can promote/stimulate plant growth through several mechanisms that include (i) biological nitrogen fixation; (ii) solubilization of insoluble minerals (P, K, Zn); (iii) release of siderophores, organic acids and enzymes; (iv) rhizosphere engineering; (v) modulation of the hormone balance through the release of indole-3-acetic acid (IAA) and synthesis of 1-amino-cyclopropane-1-carboxylate (ACC) deaminase; (vi) quorum sensing (QS) signal interference and inhibition of biofilm formation; (vii) stress-responsive/induced phytohormones; (viii) exhibiting antifungal activity; (ix) production of volatile organic compounds (VOCs); (x) induction of systemic resistance; (xi) promoting beneficial plant-microbe symbioses; (xii) interference with pathogen toxin production, etc. (Bhattacharyya and Jha 2012; Glick 2014; Hamid et al. 2021; Vasseur-Coronado et al. 2021). Moreover, these beneficial soil microbes may play an important role in recycling organic nutrients, since soil-nitrifying bacteria are able to mineralize organic nitrogen into nitrite and then into nitrate, which can be readily absorbed by crops (de Pascale et al. 2018).

Many PGPRs are also capable of producing plant hormones such as auxins, cytokinins, gibberellins, ethylene, and abscisic acid (Hayat et al. 2010; de Pascale et al. 2018). It is well known that auxins can stimulate root growth improving plant uptake of nutrients and water, while cytokinins promote mitotic cell division in shoots and roots, and delay leaf senescence. Gibberellins affect several physiological processes, such as fruit and flower formation, dormancy of vegetative organs, as well as seed germination (Métraux 1987; de Pascale et al. 2018). Furthermore, abscisic acid plays a crucial role in adaptive responses to environmental stresses such as drought and high salinity (controlling stomata closure), as well as in plant development such as seed maturation and dormancy (Miyakawa et al. 2013; de Pascale et al. 2018).

PGPRs can enhance solubility of phosphates, resulting in an increased availability of phosphate in the soil that can be assimilated by the plant (Canbolat et al. 2006; Liao et al. 2008). Different bacterial strains belonging to the genera *Aspergillus*, *Bacillus*, *Flavobacterium*, *Micrococcus*, and *Pseudomonas* have been reported to be active in the solubilization of inorganic phosphate compounds (i.e., dicalcium and tricalcium phosphate and rock phosphate) (Rodŕiguez and Fraga 1999; Rodríguez et al. 2007; Pindi 2012; de Pascale et al. 2018). Moreover, several *Bacillus* species (*Bacillus megaterium*, *Bacillus circulans, Bacillus subtilis*, *Bacillus polymyxa*, *Bacillus sircalmous*) have been described as effective phosphate solubilizers (Khan et al. 2009).

Through the production of siderophores, PGPRs as well as *Trichoderma atroviride* can increase iron solubility and therefore uptake and translocation by plant (Lugtenberg and Kamilova 2009; Colla et al. 2015b; de Pascale et al. 2018).

Research has demonstrated that inoculating plants with PGPRs offers an attractive way to replace the use of chemical fertilizers, and agrochemicals, and for this reason the use of multifunctional PGPRs based formulations for commercial agriculture is steadily increasing.

Protein hydrolysates (PHs) defined as "mixtures of polypeptides, oligopeptides, and amino acids that are manufactured from protein sources using partial hydrolysis" (Schaafsma 2009; Colla et al. 2015a; de Pascale et al. 2018), are basically available as soluble granular or powder as well as liquid extracts and can be applied as foliar sprays or dosed close to the root system (Colla et al. 2015a; de Pascale et al. 2018).

PHs are mainly produced by enzymatic and/or chemical hydrolysis of proteins from animal or plant-derived raw materials (Colla et al. 2015a).

PHs have gained prominence due to their potential to improve germination, seedling growth, plant growth, fruits, and vegetable quality, as well as crop productivity, especially under conditions of environmental stress (e.g., thermal stress, salinity, drought, alkalinity) (Colla et al. 2015a; Colla et al. 2017; Rouphael and Colla 2020b).

PHs can improve soil respiration, microbial biomass, and respective activity, since microorganisms can easily use amino acids and peptides as C and N source (Farrell et al. 2014; de Pascale et al. 2018). They also impact on plant nutrition by forming complexes and chelates between peptides/amino acids and soil micronutrients (i.e., Cu, Fe, Mn, and Zn), therefore may contribute to nutrients availability and uptake by the root system (du Jardin 2012; Colla et al. 2015a). Moreover, the ability of peptides/amino acids to form complexes and chelates with some macro (i.e., K, Ca, and Mg), and micronutrients (i.e., Cu, Fe, Mn, and Zn) are currently used by some industries, to develop fertilizers having high nutrient use efficiency (Colla et al. 2015a; du Jardin 2015; de Pascale et al. 2018).

PHs have been identified to improve the performance and productivity of several horticultural crops (Kunicki et al. 2010; Lisiecka et al. 2011; Paradiković et al. 2011; Colla et al. 2014; Ertani et al. 2014; Colla et al. 2015a; Halpern et al. 2015). The higher nutrient uptake in PHs treated plants has been attributed to (i) an increase in soil microbial activity and soil enzymatic activities, (ii) improvement of micronutrient availability and solubility, in particular Fe, Zn, Mn, and Cu, (iii) changes in the root architecture of plants, in particular root length, density and number of lateral roots, and (iv) an increase in nitrate reductase, glutamine synthetase and Fe(III)-chelate reductase activities (Cerdán et al. 2009; Ertani et al. 2009; García-Martínez et al. 2010; Colla et al. 2014; Colla et al. 2015a; Lucini et al. 2015). PHs can also interfere with the plant phytohormone balance, influencing plant development due to the presence of specific peptides and phytohormone biosynthesis precursors, such as tryptophan (Colla et al. 2014). Several bioactive peptides produced in a variety of plants have been identified as having hormone-like activities (Ito et al. 2006; Kondo et al. 2006; Colla et al. 2015a). In addition, many scientific papers reported that the application of plant-derived PHs elicited auxin- and gibberellin-like activities, thus promoting crop performance (Schiavon et al. 2008; Ertani et al. 2009; Matsumiya and Kubo 2011; Colla et al. 2014; Colla et al. 2015a). PHs have been shown not only to improve plant nutrition, but also fruit and vegetable quality in terms of phytochemicals (i.e., carotenoids, flavonoids, polyphenols) (Parrado et al. 2008; Paradiković et al. 2011; Gurav and Jadhav 2013; Ertani et al. 2014; Colla et al. 2015a), and they can reduce unwanted compounds, such as nitrates (Liu et al. 2007; Colla et al. 2015a).

Protein hydrolysate-based biostimulants are effective tools for making agriculture more sustainable. Many research studies have documented the benefits of PH applications on growth, yield, product quality, resource use efficiency and tolerance to environmental and chemical soil stresses of several horticultural crops (Colla et al. 2015a, and references therein).

Research has shown that algae (both microalgae and macroalgae) have biostimulant potential to promote overall plant development and growth along with improved tolerance to abiotic stresses (Crouch and Van Staden 1993; Craigie 2011; Sharma et al. 2014; Blanke 2016; Michalak et al. 2017; Kapoore et al. 2021).

A wide range of biostimulating compounds have been identified in macroalgae extracts, including mineral nutrients, amino acids, polysaccharides, vitamins, fatty acids, mineral elements, phenolics, and traces of phytohormones. Furthermore, seaweed extracts confer tolerance to abiotic stress and improve yield and shelf-life of products (Battacharyya et al. 2015).

Extracts of brown seaweeds are widely used in horticulture, largely for their plant growth promoting effects and for their ameliorating effect on crop tolerance to abiotic stresses such as salinity, extreme temperatures, nutrient deficiency, and drought (Battacharyya et al. 2015).

In the case of agricultural applications, microalgae have been less explored compared to macroalgae. Traditionally, cyanobacteria (blue-green algae) are known for their nitrogen fixing ability and are beneficial for a variety of crops (Chakdar et al. 2012; Kapoore et al. 2021). Soil application of harvested microalgal biomass is known to act as a slow-release biofertilizer and soil conditioner (Dasgan et al. 2012; Garcia-Gonzalez and Sommerfeld 2016; Kapoore et al. 2021). The application of living cyanobacteria is known to act as a potential biocontrol agent, via activation of plant defense enzymes and production of hydrolytic enzymes and antimicrobial compounds, against plant pathogens (Gupta et al. 2013b; Kapoore et al. 2021).

Chiaiese et al. (2018) proposed microalgae as a renewable source of PBs. In their review paper, the authors described the extraction techniques and the bioactive compounds (carbohydrates, proteins, and amino acids) as well as the biostimulatory action of microalgal extracts belonging to the species: *Chlorella vulgaris*, *Acutodesmus dimorphus*, *Spirulina platensis*, *Scenedesmus quadricauda*, *Dunaliella salina*, *C. ellipsoida*, *S. maxima*, and *Calothrix elenkinii*.

The ability of rhizospheric humic substances (HS) to improve plant growth has been well established by many studies performed using different plant species grown under many different conditions. These beneficial effects of HS on plant development are expressed both in the root and in the shoot ([Olaetxea et al. 2018] and references therein).

Soil HS are generally considered to be the result of partial degradation and resynthesis of organic material, especially plant residues. They originate from polymerization/polycondensation of phenolic compounds, mainly resulting from microbial lignin degradation. As a consequence, soil HS have a strong aromatic nature; however, during the condensation process, various organic molecules, including aliphatic chains, peptides, amino acids, fatty acids, and sugars, can be incorporated, forming substances from medium to high molecular weight (Stevenson 1994; Zanin et al. 2019).

Soil HS could also originate from associations of relatively small humic molecules linked together by hydrophobic interactions and hydrogen bonds (Piccolo 2002; Zanin et al. 2019). Humic molecules of different molecular weights can bind together forming a supramolecular humic network, depending the degree of aggregation on the pH, ionic strength and mineral composition of the solution (García-Mina 2007; Esfahani et al. 2015; Zanin et al. 2019).

HS can contribute to the regulation of many crucial ecological and environmental processes. For example, HS sustain plant growth and terrestrial life in general, regulate both soil carbon and nitrogen cycling, the growth of plants and microorganisms, the fate and transport of anthropogenic derived compounds and heavy metals, and the stabilization of soil structure (Piccolo 1996; Canellas et al. 2015). The HS include humins, humic acids, and fulvic acids, which are natural constituents of soil organic matter.

Current evidence suggests that the biostimulant effects of HS are characterized by both, structural and physiological changes in roots and shoots related to nutrient uptake, assimilation, and distribution (nutrient use efficiency traits). In addition, they can induce shifts in plant primary and secondary metabolism related to abiotic stress tolerance, which collectively modulate plant growth as well as promoting fitness (Canellas et al. 2015).

Increases on root growth were accompanied by a higher release of low molecular weight exudates from HA-treated maize plants (Canellas et al. 2019). On the other hand, it has been reported that organic acids, such as those released by the roots, can disaggregate supramolecular structure of HS releasing low molecular weight humic fractions (Piccolo et al. 2003), which in turn might exert their effects on roots. Regarding this point, it is worth mentioning that the accumulation of HS on the root surface and in the apoplast was also observed (García et al. 2012; Kulikova et al. 2014; Zanin et al. 2019).

Different research has shown that the treatment of plants with water extractable humic substances, from peat and vermicompost, were able to induce changes in root morphology and modulate plant membrane activities related to nutrient acquisition, pathways of primary and secondary metabolism, hormonal, and reactive oxygen balance (Varanini and Nannipieri 2000; Nardi et al. 2002; Canellas and Olivares 2014; Olaetxea et al. 2018; Zanin et al. 2019; Vujinović et al. 2020). These effects, which vary according to the origin, molecular size, and chemical characteristics of HS, suggest an action of these organic fractions in promoting growth and resistance to stress in plants (Zanin et al. 2019).

Numerous studies have been carried out to understand the molecular mechanisms activated by plant exposure to HS (Canellas et al. 2015) and references therein. Varanini and Pinton (2000) distinguished between indirect effects (such as increased nutrient availability through metal binding) and direct effects. Within the latter, improvement of root ion uptake ability, and rhizosphere acidification via stimulation of plasma membrane H^+-ATPase, and root proliferation involving hormone-like activity have been reported for HS (Canellas et al. 2002; Nardi et al. 2002; Canellas et al. 2015; Zanin et al. 2015; Zamboni et al. 2016). HS can induce H^+-ATPase activity, which, in turn, can energize secondary ion transporters and promote nutrient absorption (Canellas et al. 2015).

The mechanisms by which HS influence the rhizosphere seem to be the result of many chemical and biological processes acting together and depend on the type of plant, on the microorganisms present in the rhizosphere and on the soil physicochemical properties (Olaetxea et al. 2018; Barros-Rodríguez et al. 2020).

Transcriptomic studies indicated that root exposure to HS also induced changes in the expression profile of genes involved in the acquisition and assimilation of several nutrients (Trevisan et al. 2011; Jannin et al. 2012; Olaetxea et al. 2018; Zanin et al. 2018; Vujinović et al. 2020). It has also been shown that HS can promote the physiological mechanisms involved in Fe acquisition acting at the transcriptional and post-transcriptional level (Aguirre et al. 2009; Olaetxea et al. 2018; Zanin et al. 2018; Zanin et al. 2019; Vujinović et al. 2020).

A recognized target of HS action is the root plasma membrane (PM)-H^+-ATPase (Zandonadi et al. 2016; Zanin et al. 2019). Evidence of PM proton pump activation was observed both at transcriptional and post-transcriptional level and related to proton extrusion (Varanini et al. 1993; Canellas et al. 2002) and uptake of ions, such as nitrate (Pinton et al. 1999; Quaggiotti et al. 2004; Tavares et al. 2017), phosphate (Jindo et al. 2016), and sulfate (Jannin et al. 2012).

In several experiments performed with different plant species growing under diverse soil conditions, HS application to either the rhizosphere or the leaves, caused significant improvements in N root uptake and subsequent assimilation by the plant (Rose et al. 2014) and reference therein. Moreover, this increase in N root uptake, in this case in nitrate form, was linked to significant increases in the activity of the root PM-H^+-ATPase in plants growing in hydroponics (Pinton et al. 1997; Santi et al. 2013; Priyadarshini et al. 2021). On the other hand, numerous studies reported that these effects of HS on N root uptake were also associated to increased activity of enzymes involved in N metabolism,–such as nitrate reductase and other enzymes involved in ammonium assimilation–in several plant species, such as maize (Nardi et al. 2000; Vaccaro et al. 2009), cucumber (Mora et al. 2010), rapeseed (Jannin et al. 2012), or recently, rice (Tavares et al. 2017).

Moreover, sulphate root uptake is also linked to root PM-H^+-ATPase activity as in the case of nitrate (Marschner 1996; Marschner 2012). Thus, it is possible that HS are able to enhance sulphate root uptake and further assimilation within the plant (Olaetxea et al. 2018).

Different studies have reported that the root application of HS with diverse origin was able to improve overall P plant nutrition (Urrutia et al. 2014) and references therein. Typically, these results have been linked to the HS ability to mobilize and solubilize phosphate from insoluble forms in the rhizosphere (Weil and Magdoff 2004; Olaetxea et al. 2018).

It is well established that HS improves Fe availability for plants growing in soils prone to Fe deficiency, particularly alkaline and calcareous soils (Chen 1996; Chen et al. 2004; Chen and Aviad 2015). Such favorable action of HS is due to the ability to form complexes with Fe, thereby increasing its solubility in soil solution (Chen 1996; Chen et al. 2004; Olaetxea et al. 2018).

Effects of HS on root growth (signaling pathways), ion uptake (primary and secondary membrane transporters), primary metabolism (nitrogen and carbon), secondary metabolism (phenylpropanoids), and root exudation might be important for Fe acquisition and could improve the response of plants to Fe deprivation (Zanin et al. 2019).

In resume, HS bioactivity can help to reduce fertilizer application rates, increase nutrient assimilation (e.g., N, P, S, K, and Fe) and their use efficiency, promote carbon metabolism (glycolysis and Krebs cycle); replace synthetic plant regulators, improve fruit quality, increase water stress tolerance, decrease disease incidence, enhance early growth and flowering, and impact the soil microbial community in the rhizosphere (Mora et al. 2010; Trevisan et al. 2011; Jannin et al. 2012; Puglisi et al. 2013; Canellas et al. 2015; Vaccaro et al. 2015; Zanin et al. 2018; Zanin et al. 2019).

Since most HS used in agriculture are currently derived from non-renewable resources such as coal and peat, the promotion of this technology will require, in the near future, the development of new sustainable sources (e.g., organic wastes) (Canellas et al. 2015).

5.5 Conclusions

Land degradation is a ubiquitous challenge to human societies, driven mostly by socioeconomic factors. Climate change is expected to exacerbate the effects of land use and management on degradation processes in the future, while degradation-related greenhouse gas emissions and loss of carbon sink capacity, in turn, contribute to climate change.

Soil loss and deterioration will continue and probably accelerate if adequate and prompt actions are not taken to de-couple the progress of economic sectors and their pressures on the soil resource, through the integration of soil protection measures into sectoral policies. As soil is a cross-cutting issue, soil assessment approaches, and soil protection policies need to incorporate a broad perspective.

Strategies for soil protection, and soil monitoring systems, are not adequately developed, as compared with air and water, for which monitoring, assessment, and policy frameworks are already in place. A policy framework is needed to recognize the environmental importance of soil, consider the problems arising from competition between its concurrent uses (ecological and socioeconomical), and aims to maintain its multiple functions. Moreover, the intensification of agriculture has now reached a critical point whereby the negative impacts derived from this activity are now resulting in irreversible global climate change and loss in many ecosystem services. Maintaining food production for a growing world population without compromising the natural resources of future generations represents one of the greatest challenges for agriculture worldwide. Therefore, there is a requirement for new approaches that may help promote sustainable intensification. Organic agriculture has been recognized as an innovative and increasingly important approach to food production because of its potential to balance multiple sustainability goals. Indeed, OF has increased tremendously in importance over the past 20 years, including in developing countries, and the global market for organic products has grown. Nowadays, the pursuit of productivity is increasingly allied to a quest and even demand for sustainability.

Sustainable agriculture involves successful management of agricultural resources to satisfy human needs, maintaining environmental quality and protecting natural resources for future generations. Improvement in agricultural sustainability requires the optimal use and management of soil's fertility and biodiversity.

Organic systems are suitable for reducing land degradation and improving soil quality, but crop productivity and environmental efficiency can be further improved if these methods are complemented using PBs, which function as a biological tool capable of improving plant resilience to nutrient limitation typical of these systems. Thus, very attractive results may be obtained by using mixed inoculants of various bacterial strains or mixtures of fungi and bacteria.

Developed countries' consumers consider organic food to be healthier and organic agriculture more environmentally friendly than conventional farming methods, which together with the COVID-19 pandemic, has caused the organic sector to recently expand worldwide. However, there are several areas that warrant more research to further improve our understanding of the effects of organic agriculture. Such research should include different cropping and manure regimes on different substrates and within different climates. It should also cover the influences of microorganisms on fertility, structure, resilience, water holding capacity, nutrient cycling, and carbon sequestration. Further research will also be needed to elucidate the mode of action of PBs, either to re-establish soil biodiversity or to overcome nutrient limitation by improving availability and uptake, thereby reducing the gap between organic and conventional yields.

Finally, it is of great interest to carry out further research on the effects of reduced tillage techniques on different types of organic farming systems and their individual application in different environments. In this context, the consideration of reduced tillage within strategies to increase yields in organic farming becomes important, as these strategies will mainly contribute to the general acceptance of organic principles in agriculture. Soil management on OF systems can have considerable and positive effects on the control of greenhouse gas emissions and plays a key role in the conservation of fertility and biodiversity, in addition the productivity itself.

Improvement of organic matter is another consequence of management techniques, ensuring a healthy microbial population and improving crop yields.

The future challenges in the agricultural sector are immense. Protecting climate and biodiversity, while achieving high yields in resilient agricultural systems will require transformative changes in agriculture and food systems worldwide. More importantly, any future political initiative related to the agricultural sector must consider the long-term climate and biodiversity goals. This inevitably implies converging the two farming approaches (conventional and organic).

As a key solution, healthy soils contribute to meeting our main challenges of achieving climate neutrality and becoming resilient to climate change, developing a clean, and circular (bio)economy, reversing biodiversity loss, protecting human health, halting desertification and reverse land degradation. Agriculture and food systems have already changed importantly but will need to adjust further in this evolving global environment.

References

Abbott, L.K. and Johnson, N.C. (2017). Chapter 6 – Introduction: perspectives on mycorrhizas and soil fertility. In: *Mycorrhizal Mediation of Soil* (ed. N.C. Johnson, C. Gehring, and J. Jansa), 93–105. Elsevier. https://doi.org/10.1016/C2015-0-01928-1.

Abrudan, I.V., Popa, B., Codreanu, C. et al. (2014). Combating climate change by restoration of degraded land. In: *Climate Change and Restoration of Degraded Land* (ed. M. Bermúdez-Cañete, J. Cerezal, I. Florin, et al.). Madrid, Spain: Colegio de Ingenieros de Montes.

Aghasafari, H., Karbasi, A., Mohammadi, H., and Calisti, R. (2020). Determination of the best strategies for development of organic farming: a SWOT – fuzzy analytic network process approach. *Journal of Cleaner Production* 277: 124039. https://doi.org/10.1016/j.jclepro.2020.124039.

Aguirre, E., Leménager, D., Bacaicoa, E. et al. (2009). The root application of a purified leonardite humic acid modifies the transcriptional regulation of the main physiological root responses to Fe deficiency in Fe-sufficient cucumber plants. *Plant Physiol. Biochem.* 47: 215–223. https://doi.org/10.1016/j.plaphy.2008.11.013.

Ahemad, M. (2015). Phosphate-solubilizing bacteria-assisted phytoremediation of metalliferous soils: a review. *3 Biotech* 5 (2): 111–121.

Alam, M.M. and Ladha, J.K. (2004). Optimizing phosphorus fertilization in an intensive vegetable-rice cropping system. *Biology and Fertility of Soils* 40 (4): 277–283. https://doi.org/10.1007/s00374-004-0778-7.

Alvin, A., Miller, K.I., and Neilan, B.A. (2014). Exploring the potential of endophytes from medicinal plants as sources of antimycobacterial compounds. *Microbiological Research* 169 (7): 483–495. https://doi.org/10.1016/j.micres.2013.12.009.

American Society for Microbiology (2022). Microbes in Agriculture. https://asm.org/Articles/Policy/2019/November-19/Microbes-in-Agriculture (accessed 19 November 2019).

Arneth, A. et al. (2019). Framing and context. In: *Climate Change and Land: An IPCC Special Report on Climate Change, Desertification, Land Degradation, Sustainable Land Management, Food Security, and Greenhouse Gas Fluxes in Terrestrial Ecosystems* (ed. P.R. Shukla et al.).

Arneth, A. et al. (2021). Restoring degraded lands. *Annual Review of Environment and Resources* 46 (1): 569–599. https://doi.org/10.1146/annurev-environ-012320-054809.

Arnhold, S. et al. (2014). Conventional and organic farming: soil erosion and conservation potential for row crop cultivation. *Geoderma* 219–220: 89–105. https://doi.org/10.1016/j.geoderma.2013.12.023.

Ashraf, U., Hussain, S., Anjum, S.A. et al. (2017). Alterations in growth, oxidative damage, and metal uptake of five aromatic rice cultivars under lead toxicity. *Plant Physiology and Biochemistry* 115: 461–471. https://doi.org/10.1016/j.plaphy.2017.04.019.

Auerswald, K., Kainz, M., and Fiener, P. (2003). Soil erosion potential of organic versus conventional farming evaluated by USLE modelling of cropping statistics for agricultural districts in Bavaria. *Soil Use and Management* 19 (4): 305–311. https://doi.org/10.1079/SUM2003212.

Aulakh, M.S., Kabba, B.S., Baddesha, H.S. et al. (2003). Crop yields and phosphorus fertilizer transformations after 25 years of applications to a subtropical soil under groundnut-based cropping systems. *Field Crops Research* 83 (3): 283–296. https://doi.org/10.1016/S0378-4290(03)00078-9.

Backer, R. et al. (2018). Plant growth-promoting rhizobacteria: context, mechanisms of action, and roadmap to commercialization of biostimulants for sustainable agriculture. *Frontiers in Plant Science* 9. https://doi.org/10.3389/fpls.2018.01473.

Baggs, E.M., Watson, C.A., and Rees, R.M. (2000). The fate of nitrogen from incorporated cover crop and green manure residues. *Nutrient Cycling in Agroecosystems* 56 (2): 153–163. https://doi.org/10.1023/A:1009825606341.

Bai, Z.G., Caspari, T., Gonzalez, M.R. et al. (2018). Effects of agricultural management practices on soil quality: a review of long-term experiments for Europe and China. *Agriculture, Ecosystems & Environment* 265: 1–7. https://doi.org/10.1016/j.agee.2018.05.028.

Balseiro-Romero, M., Gkorezis, P., Kidd, P.S. et al. (2017). Use of plant growth promoting bacterial strains to improve Cytisus striatus and Lupinus luteus development for potential application in phytoremediation. *Science of The Total Environment* 581–582: 676–688. https://doi.org/10.1016/j.scitotenv.2016.12.180.

Barnwal, P., Devika, S., Singh, S. et al. (2021). Soil fertility management in organic farming. In: *Advances in Organic Farming* (ed. V. Meena, A. Rakshit, C. Srinivasarao, et al.), 39–46. Elsevier. https://doi.org/10.1016/B978-0-12-822358-1.00016-X.

Barros-Rodríguez, A., Rangseekaew, P., Lasudee, K. et al. (2020). Regulatory risks associated with bacteria as biostimulants and biofertilizers in the frame of the European regulation (EU) 2019/1009. *Science of The Total Environment* 740: 140239. https://doi.org/10.1016/j.scitotenv.2020.140239.

Basamba, T.A., Barrios, E., Amézquita, E. et al. (2006). Tillage effects on maize yield in a Colombian savanna oxisol: soil organic matter and P fractions. *Soil and Tillage Research* 91 (1): 131–142. https://doi.org/10.1016/j.still.2005.11.010.

Battacharyya, D., Babgohari, M.Z., Rathor, P., and Prithiviraj, B. (2015). Seaweed extracts as biostimulants in horticulture. *Scientia Horticulturae* 196: 39–48. https://doi.org/10.1016/j.scienta.2015.09.012.

Baveye, P.C., Schnee, L.S., Boivin, P. et al. (2020). Soil organic matter research and climate change: merely re-storing carbon versus restoring soil functions. *Frontiers in Environmental Science* 8. https://doi.org/10.3389/fenvs.2020.579904.

Becerra-Castro, C. et al. (2013a). Improving performance of Cytisus striatus on substrates contaminated with hexachlorocyclohexane (HCH) isomers using bacterial inoculants: developing a phytoremediation strategy. *Plant and Soil* 362 (1–2): 247–260. https://doi.org/10.1007/s11104-012-1276-6.

Becerra-Castro, C., Kidd, P.S., Rodríguez-Garrido, B. et al. (2013b). Phytoremediation of hexachlorocyclohexane (HCH)-contaminated soils using Cytisus striatus and bacterial inoculants in soils with distinct organic matter content. *Environmental Pollution* 178: 202–210. https://doi.org/10.1016/j.envpol.2013.03.027.

Bender, S.F., Wagg, C., and van der Heijden, M.G.A. (2016). An underground revolution: biodiversity and soil ecological engineering for agricultural sustainability. *Trends in Ecology & Evolution* 31 (6): 440–452. https://doi.org/10.1016/j.tree.2016.02.016.

Bengtsson, J., Ahnström, J., and Weibull, A.C. (2005). The effects of organic agriculture on biodiversity and abundance: a meta-analysis. *Journal of Applied Ecology* 42 (2): 261–269. https://doi.org/10.1111/j.1365-2664.2005.01005.x.

Benoit, M., Garnier, J., Beaudoin, N., and Billen, G. (2016). A participative network of organic and conventional crop farms in the Seine Basin (France) for evaluating nitrate leaching and yield performance. *Agricultural Systems* 148: 105–113. https://doi.org/10.1016/j.agsy.2016.07.005.

Berry, P.M., Sylvester-Bradley, R., Philipps, L. et al. (2002). Is the productivity of organic farms restricted by the supply of available nitrogen? *Soil Use and Management* 18 (s1): 248–255. https://doi.org/10.1111/j.1475-2743.2002.tb00266.x.

Berry, P.M., Stockdale, E.A., Sylvester-Bradley, R. et al. (2003). N, P and K budgets for crop rotations on nine organic farms in the UK. *Soil Use and Management* 19 (2): 112–118. https://doi.org/10.1079/SUM2003176.

Bhardwaj, D., Ansari, M.W., Sahoo, R.K., and Tuteja, N. (2014). Biofertilizers function as key player in sustainable agriculture by improving soil fertility, plant tolerance and crop productivity. *Microbial Cell Factories* 13 (1): 66. https://doi.org/10.1186/1475-2859-13-66.

Bhat, N.A., Riar, A., Ramesh, A. et al. (2017). Soil biological activity contributing to phosphorus availability in vertisols under long-term organic and conventional agricultural management. *Frontiers in Plant Science* 8. https://doi.org/10.3389/fpls.2017.01523.

Bhattacharyya, P.N. and Jha, D.K. (2012). Plant growth-promoting rhizobacteria (PGPR): emergence in agriculture. *World Journal of Microbiology and Biotechnology* 28 (4): 1327–1350. https://doi.org/10.1007/s11274-011-0979-9.

Bielders, C.L. and Baveye, P. (1995). Processes of structural crust formation on coarse-textured soils. *European Journal of Soil Science* 46 (2): 221–232. https://doi.org/10.1111/j.1365-2389.1995.tb01830.x.

Bielders, C.L., Baveye, P., Wilding, L.P. et al. (1996). Tillage-induced spatial distribution of surface crusts on a Sandy Paleustult from Togo. *Soil Science Society of America Journal* 60 (3): 843–855. https://doi.org/10.2136/sssaj1996.03615995006000030024x.

Blanco-Canqui, H. and Lal, R. (2010). *Principles of Soil Conservation and Management*. Dordrecht: Springer Netherlands. https://doi.org/10.1007/978-1-4020-8709-7.

Blanke, M. (2016). Biostimulantien – von Algenextrakten bis Shrimpsschalen – eine Branche (er-)findet sich. *Erwerbs-Obstbau* 58 (2): 81–87. https://doi.org/10.1007/s10341-016-0276-8.

Blum, E.H., Horak, O., and Mentler, A. (2009). Trace Elements. In: *Environmental and Ecological Chemistry*, vol. 2 (ed. A. Sabljic), 156–164. Oxford, united Kingdom: Eolss Publishers Co. Ltd.

Bolan, N.S., Naidu, R., Mahimairaja, S., and Baskaran, S. (1994). Influence of low-molecular-weight organic acids on the solubilization of phosphates. *Biology and Fertility of Soils* 18 (4): 311–319. https://doi.org/10.1007/BF00570634.

Boyer, L.R., Brain, P., Xu, X.-M., and Jeffries, P. (2015). Inoculation of drought-stressed strawberry with a mixed inoculum of two arbuscular mycorrhizal fungi: effects on population dynamics of fungal species in roots and consequential plant tolerance to water deficiency. *Mycorrhiza* 25 (3): 215–227. https://doi.org/10.1007/s00572-014-0603-6.

Brady, N.C. and Weil, R.R. (2008). *The Nature and Properties of Soils*, 4e (ed. N.C. Brady and R.R. Weil). Upper Saddle River, NJ: Prentice Hall.

Bresson, L.-M. and Boiffin, J. (1990). Morphological characterization of soil crust development stages on an experimental field. *Geoderma* 47 (3–4): 301–325. https://doi.org/10.1016/0016-7061(90)90035-8.

Briar, S.S., Wichman, D., and Reddy, G.V.P. (2016). Plant-parasitic nematode problems in organic agriculture. In: *Organic Farming for Sustainable Agriculture, Sustainable Development and Biodiversity* (ed. D. Nandwani), 107–122. Cham, Switzerland: Springer https://doi.org/10.1007/978-3-319-26803-3_5.

Brígido, C., van Tuinen, D., Brito, I. et al. (2017). Management of the biological diversity of AM fungi by combination of host plant succession and integrity of extraradical mycelium. *Soil Biology and Biochemistry* 112: 237–247. http://linkinghub.elsevier.com/retrieve/pii/S0038071717305011.

Brito, I., Carvalho, M., Alho, L., and Goss, M.J. (2014). Managing arbuscular mycorrhizal fungi for bioprotection: Mn toxicity. *Soil Biology and Biochemistry* 68: 78–84. http://dx.doi.org/10.1016/j.soilbio.2013.09.018.

Campobenedetto, C., Grange, E., Mannino, G. et al. (2020). A biostimulant seed treatment improved heat stress tolerance during cucumber seed germination by acting on the antioxidant system and glyoxylate cycle. *Frontiers in Plant Science* 11. https://doi.org/10.3389/fpls.2020.00836.

Campobenedetto, C., Mannino, G., Beekwilder, J. et al. (2021). The application of a biostimulant based on tannins affects root architecture and improves tolerance to salinity in tomato plants. *Scientific Reports* 11 (1): 354. https://doi.org/10.1038/s41598-020-79770-5.

Campos, C., Nobre, T., Goss, M.J. et al. (2019). Transcriptome analysis of wheat roots reveals a differential regulation of stress responses related to arbuscular mycorrhizal Fungi and soil disturbance. *Biology* 8 (4): 93. https://www.mdpi.com/2079-7737/8/4/93.

Canbolat, M.Y., Bilen, S., Çakmakçı, R. et al. (2006). Effect of plant growth-promoting bacteria and soil compaction on barley seedling growth, nutrient uptake, soil properties and rhizosphere microflora. *Biology and Fertility of Soils* 42 (4): 350–357. https://doi.org/10.1007/s00374-005-0034-9.

Canellas, L.P. and Olivares, F.L. (2014). Physiological responses to humic substances as plant growth promoter. *Chemical and Biological Technologies in Agriculture* 1 (1): 1–11. https://doi.org/10.1186/2196-5641-1-3.

Canellas, L.P., Olivares, F.L., Okorokova-Façanha, A.L., and Façanha, A.R. (2002). Humic acids isolated from earthworm compost enhance root elongation, lateral root emergence, and plasma membrane H+-ATPase activity in maize roots. *Plant Physiology* 130 (4): 1951–1957. https://doi.org/10.1104/pp.007088.

Canellas, L.P., Olivares, F.L., Aguiar, N.O. et al. (2015). Humic and fulvic acids as biostimulants in horticulture. *Scientia Horticulturae* 196: 15–27. https://doi.org/10.1016/j.scienta.2015.09.013.

Canellas, L.P., Olivares, F.L., Canellas, N.O.A. et al. (2019). Humic acids increase the maize seedlings exudation yield. *Chemical and Biological Technologies in Agriculture* 6 (1): 3. https://doi.org/10.1186/s40538-018-0139-7.

Caravaca, F., Barea, J.M., Palenzuela, J. et al. (2003). Establishment of shrub species in a degraded semiarid site after inoculation with native or allochthonous arbuscular mycorrhizal fungi. *Applied Soil Ecology* 22 (2): 103–111. https://doi.org/10.1016/S0929-1393(02)00136-1.

Carbonetto, B., Rascovan, N., Álvarez, R. et al. (2014). Structure, composition and metagenomic profile of soil microbiomes associated to agricultural land use and tillage systems in Argentine pampas. *PLoS ONE* 9 (6): e99949. https://doi.org/10.1371/journal.pone.0099949.

Castiglione, A.M., Mannino, G., Contartese, V. et al. (2021). Microbial biostimulants as response to modern agriculture needs: composition, role and application of these innovative products. *Plants* 10 (8): https://doi.org/10.3390/plants10081533.

Cerdán, M., Sánchez-Sánchez, A., Oliver, M. et al. (2009). Effect of foliar and root applications of amino acids on iron uptake by tomato plants. *Acta Horticulturae* 830 (830): 481–488. https://doi.org/10.17660/ActaHortic.2009.830.68.

Chakdar, H., Jadhav, S.D., Dhar, D.W., and Pabbi, S. (2012). Potential applications of blue green algae. *Journal of Scientific and Industrial Research* 71 (1): 13–20.

Chapagain, T. and Riseman, A. (2014). Barley–pea intercropping: effects on land productivity, carbon and nitrogen transformations. *Field Crops Research* 166: 18–25. https://doi.org/10.1016/j.fcr.2014.06.014.

Chaturvedi, R., Favas, P., Pratas, J. et al. (2018). Assessment of edibility and effect of arbuscular mycorrhizal fungi on *Solanum melongena* L. grown under heavy metal(loid) contaminated soil. *Ecotoxicology and Environmental Safety* 148 (November 2017): 318–326. https://doi.org/10.1016/j.ecoenv.2017.10.048.

Chen, Y. (1996). Organic matter reactions involving micronutrients in soils and their effect on plants. In: *Humic Substances in Terrestrial Ecosystems* (ed. A. Piccolo), 507–529. Elsevier. https://doi.org/10.1016/B978-044481516-3/50014-1.

Chen, Y. and Aviad, T. (2015). Effects of humic substances on plant growth. In: *Humic Substances in Soil and Crop Sciences: Selected Readings* (ed. P. MacCarthy, C. Clapp, R. Malcolm, and P.R. Bloom), 161–186. Madison, WI, USA: Soil Science Society of America https://doi.org/10.2136/1990.humicsubstances.c7.

Chen, Y., Clapp, C.E., and Magen, H. (2004). Mechanisms of plant growth stimulation by humic substances: the role of organo-iron complexes. *Soil Science and Plant Nutrition* 50 (7): 1089–1095. https://doi.org/10.1080/00380768.2004.10408579.

Chen, X., Henriksen, T.M., Svensson, K., and Korsaeth, A. (2020). Long-term effects of agricultural production systems on structure and function of the soil microbial community. *Applied Soil Ecology* 147: 103387. https://doi.org/10.1016/j.apsoil.2019.103387.

Chiaiese, P., Corrado, G., Colla, G. et al. (2018). Renewable sources of plant biostimulation: microalgae as a sustainable means to improve crop performance. *Frontiers in Plant Science* 871: 1782. https://doi.org/10.3389/fpls.2018.01782.

Chitarra, W., Pagliarani, C., Maserti, B. et al. (2016). Insights on the impact of arbuscular mycorrhizal Symbiosis on tomato tolerance to water stress. *Plant Physiology* 171 (2): 1009–1023. https://doi.org/10.1104/pp.16.00307.

Chivenge, P.P., Murwira, H.K., Giller, K.E. et al. (2007). Long-term impact of reduced tillage and residue management on soil carbon stabilization: implications for conservation agriculture on contrasting soils. *Soil and Tillage Research* 94 (2): 328–337. https://doi.org/10.1016/j.still.2006.08.006.

Ciampitti, I.A., García, F.O., Picone, L.I., and Rubio, G. (2011). Phosphorus budget and soil extractable dynamics in field crop rotations in Mollisols. *Soil Science Society of America Journal* 75 (1): 131–142. https://doi.org/10.2136/sssaj2009.0345.

Colla, G. and Rouphael, Y. (2015). Biostimulants in horticulture. *Scientia Horticulturae* 196: 1–2. https://doi.org/10.1016/j.scienta.2015.10.044.

Colla, G., Mitchell, J.P., Joyce, B.A. et al. (2000). Soil physical properties and tomato yield and quality in alternative cropping systems. *Agronomy Journal* 92 (5): 924–932. https://doi.org/10.2134/agronj2000.925924x.

Colla, G., Rouphael, Y., Canaguier, R. et al. (2014). Biostimulant action of a plant-derived protein hydrolysate produced through enzymatic hydrolysis. *Frontiers in Plant Science* 5 (SEP): 448. https://doi.org/10.3389/fpls.2014.00448.

Colla, G., Nardi, S., Cardarelli, M. et al. (2015a). Protein hydrolysates as biostimulants in horticulture. *Scientia Horticulturae* 196: 28–38. https://doi.org/10.1016/j.scienta.2015.08.037.

Colla, G., Rouphael, Y., Di Mattia, E. et al. (2015b). Co-inoculation of Glomus intraradices and Trichoderma atroviride acts as a biostimulant to promote growth, yield and nutrient uptake of vegetable crops. *Journal of the Science of Food and Agriculture* 95 (8): 1706–1715. https://doi.org/10.1002/jsfa.6875.

Colla, G., Hoagland, L., Ruzzi, M. et al. (2017). Biostimulant action of protein hydrolysates: unraveling their effects on plant physiology and microbiome. *Frontiers in Plant Science* 8: 2202. https://doi.org/10.3389/fpls.2017.02202.

Compant, S., Clément, C., and Sessitsch, A. (2010). Plant growth-promoting bacteria in the rhizo- and endosphere of plants: their role, colonization, mechanisms involved and prospects for utilization. *Soil Biology and Biochemistry* 42 (5): 669–678. https://doi.org/10.1016/j.soilbio.2009.11.024.

Condron, L.M., Turner, B.L., and Cade-Menun, B.J. (2015). Chemistry and dynamics of soil organic phosphorus. In: *Phosphorus: Agriculture and the Environment* (ed. J.T. Sims and A.N. Sharpley), 87–121. Agronomy Monographs. https://doi.org/10.2134/agronmonogr46.c4.

Coninx, L., Martinova, V., and Rineau, F. (2017). Mycorrhiza-assisted phytoremediation. In: *Advances in Botanical Research* (ed. A. Cuypers and J. Vangronsveld), 127–188. Academic Press. https://doi.org/10.1016/bs.abr.2016.12.005.

Copetta, A., Lingua, G., and Berta, G. (2006). Effects of three AM fungi on growth, distribution of glandular hairs, and essential oil production in Ocimum basilicum L. var. Genovese. *Mycorrhiza* 16 (7): 485–494. https://doi.org/10.1007/s00572-006-0065-6.

Cordell, D., Drangert, J.-O., and White, S. (2009). The story of phosphorus: global food security and food for thought. *Global Environmental Change* 19 (2): 292–305. https://doi.org/10.1016/j.gloenvcha.2008.10.009.

Craigie, J.S. (2011). Seaweed extract stimuli in plant science and agriculture. *Journal of Applied Phycology* 23 (3): 371–393. https://doi.org/10.1007/s10811-010-9560-4.

Crittenden, S.J., Poot, N., Heinen, M. et al. (2015). Soil physical quality in contrasting tillage systems in organic and conventional farming. *Soil and Tillage Research* 154: 136–144. https://doi.org/10.1016/j.still.2015.06.018.

Crouch, I.J. and Van Staden, J. (1993). Commercial seaweed products as biostimulants in horticulture. *Journal of Home & Consumer Horticulture* 1 (1): 19–76. https://doi.org/10.1300/J280v01n01_03.

Damodar Reddy, D., Subba Rao, A., Sammi Reddy, K., and Takkar, P.N. (1999). Yield sustainability and phosphorus utilization in soybean–wheat system on Vertisols in response to integrated use of manure and fertilizer phosphorus. *Field Crops Research* 62 (2): 181–190. https://doi.org/10.1016/S0378-4290(99)00019-2.

Dardanelli, M.S., Manyani, H., González-Barroso, S. et al. (2010). Effect of the presence of the plant growth promoting rhizobacterium (PGPR) Chryseobacterium balustinum Aur9 and salt stress in the pattern of flavonoids exuded by soybean roots. *Plant and Soil* 328 (1): 483–493. https://doi.org/10.1007/s11104-009-0127-6.

Dasgan, H.Y., Aydoner, G., and Akyol, M. (2012). Use of some microorganisms as bio-fertilizers in soilless grown squash for saving chemical nutrients. In: *Acta Horticulturae*, 155–162. Leuven, Belgium: International Society for Horticultural Science (ISHS). https://doi.org/10.17660/ActaHortic.2012.927.17.

Davis, K.F., Gephart, J.A., Emery, K.A. et al. (2016). Meeting future food demand with current agricultural resources. *Global Environmental Change* 39: 125–132. https://doi.org/10.1016/j.gloenvcha.2016.05.004.

De Meyer, A., Poesen, J., Isabirye, M. et al. (2011). Soil erosion rates in tropical villages: a case study from Lake Victoria Basin, Uganda. *CATENA* 84 (3): 89–98. https://doi.org/10.1016/j.catena.2010.10.001.

De Ponti, T., Rijk, B., and Van Ittersum, M.K. (2012). The crop yield gap between organic and conventional agriculture. *Agricultural Systems* 108: 1–9. https://doi.org/10.1016/j.agsy.2011.12.004.

De Vries, F.T., Griffiths, R.I., Knight, C.G. et al. (2020). Harnessing rhizosphere microbiomes for drought-resilient crop production. *Science* 368 (6488): 270–274. https://doi.org/10.1126/science.aaz5192.

Deng, Z. and Cao, L. (2017). Fungal endophytes and their interactions with plants in phytoremediation: a review. *Chemosphere* 168: 1100–1106. https://doi.org/10.1016/j.chemosphere.2016.10.097.

Department for Environment Food and Rural Affairs (2006). *Soil and Nutrient Management on Organic Farms*. UK: Gloucestershire. https://www.organicresearchcentre.com/manage/authincludes/article_uploads/iota/technical-leaflets/soil-and-nutrient-management-on-organic-farms.pdf.

Diagne, N., Baudoin, E., Svistoonoff, S. et al. (2018). Effect of native and allochthonous arbuscular mycorrhizal fungi on Casuarina equisetifolia growth and its root bacterial community. *Arid Land Research and Management* 32 (2): 212–228. https://doi.org/10.1080/15324982.2017.1406413.

Drakopoulos, D., Scholberg, J.M.S., Lantinga, E.A., and Tittonell, P.A. (2016). Influence of reduced tillage and fertilization regime on crop performance and nitrogen utilization of organic potato. *Organic Agriculture* 6 (2): 75–87. https://doi.org/10.1007/s13165-015-0110-x.

Drinkwater, L.E., Letourneau, D.K., Workneh, F. et al. (1995). Fundamental differences between conventional and organic tomato agroecosystems in California. *Ecological Applications* 5 (4): 1098–1112. https://doi.org/10.2307/2269357.

Drinkwater, L.E., Wagoner, P., and Sarrantonio, M. (1998). Legume-based cropping systems have reduced carbon and nitrogen losses. *Nature* 396 (6708): 262–265. https://doi.org/10.1038/24376.

Durrer, A., Gumiere, T., Zagatto, M.R.G. et al. (2021). Organic farming practices change the soil bacteria community, improving soil quality and maize crop yields. *Peer J* 9: e11985. https://doi.org/10.7717/peerj.11985.

Dwivedi, M. (2020). Gluconobacter. In: *Beneficial Microbes in Agro-Ecology* (ed. N. Amaresan, M. Senthil Kumar, K. Annapurna, et al.), 521–544. Elsevier. https://doi.org/10.1016/B978-0-12-823414-3.00025-3.

EBIC (2021). European Biostimulants Industry Council. https://biostimulants.eu/issue/plant-biostimulants-contribute-to-climate-smart-agriculture (accessed 15 February 2022).

Edmeades, D.C. (2003). The long-term effects of manures and fertilisers on soil productivity and quality: a review. *Nutrient Cycling in Agroecosystems* 66 (2): 165–180. https://doi.org/10.1023/A:1023999816690.

Ertani, A., Cavani, L., Pizzeghello, D. et al. (2009). Biostimulant activity of two protein hydrolyzates in the growth and nitrogen metabolism of maize seedlings. *Journal of Plant Nutrition and Soil Science* 172 (2): 237–244. https://doi.org/10.1002/jpln.200800174.

Ertani, A., Pizzeghello, D., Francioso, O. et al. (2014). Capsicum chinensis L. growth and nutraceutical properties are enhanced by biostimulants in a long-term period: chemical and metabolomic approaches. *Frontiers in Plant Science* 5: 375. https://doi.org/10.3389/fpls.2014.00375.

Esfahani, M.R., Stretz, H.A., and Wells, M.J.M. (2015). Abiotic reversible self-assembly of fulvic and humic acid aggregates in low electrolytic conductivity solutions by dynamic light scattering and zeta potential investigation. *Science of the Total Environment* 537: 81–92. https://doi.org/10.1016/j.scitotenv.2015.08.001.

European Comission (2019). Regulation (EU) 2019/1009 of the European Parliament and of the Council of 5 June 2019 laying down rules on the making available on the market of EU fertilising products and amending regulations (EC) no 1069/2009 and (EC) no 1107/2009 and repealing Regula. *Off. J. Eur. Union* 2019: 1–114. https://eur-lex.europa.eu/legal-content/EN/TXT/PDF/?uri=CELEX:32019R1009&from=EN.

European Commission (2020). *Caring for Soil Is Caring for Life: Ensure 75% of Soils Are Healthy by 2030 for Food, People, Nature and Climate : Report of the Mission Board for Soil Health and Food*. Publications Office. https://doi.org/10.2777/821504.

European Commission (2021a). *Communication from the Commission to the European Parliament, The Council, The European Economic and Social Committee and The Committee of the Regions – EU Soil Strategy for 2030: Reaping the Benefits of Healthy Soils for People, Food, Nature and Climate*, Brussels, Belgium. https://eur-lex.europa.eu/legal-content/EN/TXT/?uri=CELEX%3A52021DC0699.

European Commission (2021b). *Communication from the Commission to the European Parliament, The Council, The European Economic and Social Committee and The Committee of the Regions – Forging a Climate-Resilient Europe: The New EU Strategy on Adaptation to Climate Change.* https://eur-lex.europa.eu/legal-content/EN/TXT/PDF/?uri=CELEX:52021DC0082&from=EN.

European Environment Agency (2000). *Down to Earth: Soil Degradation and Sustainable Development in Europe.* Copenhagen, Denmark: European Environment Agency. https://www.eea.europa.eu/publications/Environmental_issue_series_16.

European Environment Agency (2019). *The European Environment: State and Outlook 2020 – Knowledge for Transition to a Sustainable Europe.* Publications. Office https://data.europa.eu/doi/10.2800/085135.

European Environment Agency (2022). European Environment Agency. https://www.eea.europa.eu (accessed 22 February 2022).

European Union (2018). *Regulation (EU) 2018/848 of the European Parliament and of the Council of 30 May 2018 on Organic Production and Labelling of Organic Products and Repealing Council Regulation (EC) no 834/2007.* Brussels: Belgium. https://eur-lex.europa.eu/legal-content/EN/TXT/?uri=CELEX%3A32018R0848.

Eurostat (2021). Statistics Explained. https://ec.europa.eu/eurostat/statistics-explained (accessed 22 February 2022).

Eyhorn, F., Muller, A., Reganold, J.P. et al. (2019). Sustainability in global agriculture driven by organic farming. *Nature Sustainability* 2 (4): 253–255. https://doi.org/10.1038/s41893-019-0266-6.

FAO (2004). *Carbon Sequestration in Dryland Soils.* Rome: World Soils Resources Reports 102. https://www.fao.org/3/y5738e/y5738e00.

FAO (2011). *The State of the world's Land and Water Resources for Food and Agriculture (SOLAW) – Managing Systems at Risk.* (ed. Food and Agriculture Organization of the United Nations). London: Rome and Earthscan. https://www.fao.org/3/i1688e/i1688e00.htm.

FAO (2015a). FAO – Helping to Build a World Without Hunger. http://www.fao.org/ag/ca/index.html.

FAO (2015b). *Healthy Soils Are the Basis for Healthy Food Production.* Rome: Food and Agriculture Organization of the United Nations. https://www.fao.org/3/i4405e/i4405e.pdf.

FAO (2015c). *Status of the World's Soil Resources: Main Report.* Rome, Italy: FAO. https://www.fao.org/documents/card/en/c/c6814873-efc3-41db-b7d3-2081a10ede50.

FAO (2017a). *The Future of Food and Agriculture – Trends and Challenges.* Rome: Italy. https://www.fao.org/3/i6583e/i6583e.pdf.

FAO (2017b). *Voluntary Guidelines for Sustainable Soil Management Food and Agriculture Organization of the United Nations.* Rome: Italy. https://www.fao.org/3/bl813e/bl813e.pdf.

FAO (2022a). Conservation Agriculture. https://www.fao.org/conservation-agriculture/en (accessed 22 February 2022).

FAO (2022b). Food and Agriculture Data. https://www.fao.org/faostat/en (accessed 22 February 2022).

FAO, ITPS, GSBI, SCBD, and EC (2020). *State of Knowledge of Soil Biodiversity – Status, Challenges and Potentialities.* Report 202. Rome: FAO. http://www.fao.org/documents/card/en/c/cb1928en.

Faria, J.M.S., Teixeira, D.M., Pinto, A.P. et al. (2020). Arbuscular mycorrhiza inoculum type influences phosphorus subcellular distribution in shoots of wheat grown in acidic soil under sustainable agricultural practices. *Biology and Life Sciences Forum* 4 (1): 62. http://sciforum.net/conference/IECPS2020/paper/8596.

Faria, J.M.S., Pinto, A.P., Teixeira, D. et al. (2021a). Diversity of native arbuscular mycorrhiza extraradical mycelium influences antioxidant enzyme activity in wheat grown under Mn toxicity. *Bulletin of Environmental Contamination and Toxicology* 108: 451–456. https://doi.org/10.1007/s00128-021-03240-5.

Faria, J.M.S., Teixeira, D.M., Pinto, A.P. et al. (2021b). The protective biochemical properties of arbuscular mycorrhiza extraradical mycelium in acidic soils are maintained throughout the Mediterranean summer conditions. *Agronomy* 11 (4): 748. https://www.mdpi.com/2073-4395/11/4/748.

Farrell, M., Prendergast-Miller, M., Jones, D.L. et al. (2014). Soil microbial organic nitrogen uptake is regulated by carbon availability. *Soil Biology and Biochemistry* 77: 261–267. https://doi.org/10.1016/j.soilbio.2014.07.003.

Ferreira, S., Oliveira, F., Gomes da Silva, F. et al. (2020). Assessment of factors constraining organic farming expansion in Lis Valley, Portugal. *AgriEngineering* 2 (1): 111–127. https://doi.org/10.3390/agriengineering2010008.

Fess, T. and Benedito, V. (2018). Organic versus conventional cropping sustainability: a comparative system analysis. *Sustainability* 10 (1): 272. https://doi.org/10.3390/su10010272.

Fließbach, A., Oberholzer, H.R., Gunst, L., and Mäder, P. (2007). Soil organic matter and biological soil quality indicators after 21 years of organic and conventional farming. *Agriculture, Ecosystems and Environment* 118 (1–4): 273–284. https://doi.org/10.1016/j.agee.2006.05.022.

Flores-Félix, J.D., Menéndez, E., Rivas, R., and Velázquez, M.d.l.E. (2019). Future perspective in organic farming fertilization. In: *Organic Farming* (ed. S. Chandran, M.R. Unni, and S. Thomas), 269–315. Elsevier. https://doi.org/10.1016/B978-0-12-813272-2.00010-0.

Foereid, B. (2017). Phosphorus availability in residues as fertilizers in organic agriculture. *Agricultural and Food Science* 26 (1): 25–33. https://doi.org/10.23986/afsci.59307.

Fortin, J.A., Plenchette, C., and Piché, Y. (2009). *Mycorrhizas: The New Green Revolution*. Québec City, CA: Editions MultiMondes.

Fortune, S., Hollies, J., and Stockdale, E.A. (2006). Effects of different potassium fertilizers suitable for use in organic farming systems on grass/clover yields and nutrient offtakes and interactions with nitrogen supply. *Soil Use and Management* 20 (4): 403–409. https://doi.org/10.1111/j.1475-2743.2004.tb00389.x.

Friedel, J.K., Gabel, D., and Stahr, K. (2001). Nitrogen pools and turnover in arable soils under different durations of organic farming: II: source-and-sink function of the soil microbial biomass or competition with growing plants? *Journal of Plant Nutrition and Soil Science* 164 (4): 421. https://doi.org/10.1002/1522-2624(200108)164:4<421::AID-JPLN421>3.0.CO;2-P.

García, A.C., Santos, L.A., Izquierdo, F.G. et al. (2012). Vermicompost humic acids as an ecological pathway to protect rice plant against oxidative stress. *Ecological Engineering* 47: 203–208. https://doi.org/10.1016/j.ecoleng.2012.06.011.

Garcia-Gonzalez, J. and Sommerfeld, M. (2016). Biofertilizer and biostimulant properties of the microalga Acutodesmus dimorphus. *Journal of Applied Phycology* 28 (2): 1051–1061. https://doi.org/10.1007/s10811-015-0625-2.

García-Martínez, A.M. et al. (2010). Enzymatic production of an organic soil biostimulant from wheat-condensed distiller solubles: effects on soil biochemistry and biodiversity. *Process Biochemistry* 45 (7): 1127–1133. https://doi.org/10.1016/j.procbio.2010.04.005.

García-Mina, J.M. (2007). Advantages and limitations of the use of an extended polyelectrolyte model to describe the proton-binding process in macromolecular systems. Application to a poly(acrylic acid) and a humic acid. *Journal of Physical Chemistry B* 111 (17): 4488–4494. https://doi.org/10.1021/jp0689518.

Garg, N. and Chandel, S. (2012). Role of arbuscular mycorrhizal (AM) Fungi on growth, cadmium uptake, osmolyte, and Phytochelatin synthesis in Cajanus cajan (L.) Millsp. Under NaCl and cd stresses. *Journal of Plant Growth Regulation* 31 (3): 292–308. https://doi.org/10.1007/s00344-011-9239-3.

Gattinger, A., Muller, A., Haeni, M. et al. (2012). Enhanced top soil carbon stocks under organic farming. *Proceedings of the National Academy of Sciences* 109 (44): 18226–18231. https://doi.org/10.1073/pnas.1209429109.

George, T.S., Richardson, A.E., and Simpson, R.J. (2005). Behaviour of plant-derived extracellular phytase upon addition to soil. *Soil Biology and Biochemistry* 37 (5): 977–988. https://doi.org/10.1016/j.soilbio.2004.10.016.

Gichangi, E.M., Mnkeni, P.N.S., and Brookes, P.C. (2009). Effects of goat manure and inorganic phosphate addition on soil inorganic and microbial biomass phosphorus fractions under laboratory incubation conditions. *Soil Science and Plant Nutrition* 55 (6): 764–771. https://doi.org/10.1111/j.1747-0765.2009.00415.x.

Giovannini, L., Palla, M., Agnolucci, M. et al. (2020). Arbuscular mycorrhizal Fungi and associated microbiota as plant biostimulants: research strategies for the selection of the best performing Inocula. *Agronomy* 10 (1). https://doi.org/10.3390/agronomy10010106.

Glick, B.R. (2014). Bacteria with ACC deaminase can promote plant growth and help to feed the world. *Microbiological Research* 169 (1): 30–39. https://doi.org/10.1016/j.micres.2013.09.009.

Gomiero, T. (2016). Soil degradation, land scarcity and food security: reviewing a complex challenge. *Sustainability* 8 (3): 281. https://doi.org/10.3390/su8030281.

Govasmark, E., Stäb, J., Holen, B. et al. (2011). Chemical and microbiological hazards associated with recycling of anaerobic digested residue intended for agricultural use. *Waste Management* 31 (12): 2577–2583. https://doi.org/10.1016/j.wasman.2011.07.025.

Graham, R., Wortman, S., and Pittelkow, C. (2017). Comparison of organic and integrated nutrient management strategies for reducing soil N2O emissions. *Sustainability* 9 (4): 510. https://doi.org/10.3390/su9040510.

Guan, G., Tu, S., Li, H. et al. (2013). Phosphorus fertilization modes affect crop yield, nutrient uptake, and soil biological properties in the Rice–wheat cropping system. *Soil Science Society of America Journal* 77 (1): 166–172. https://doi.org/10.2136/sssaj2011.0324.

Guiñazú, L.B., Andrés, J.A., Del Papa, M.F. et al. (2010). Response of alfalfa (Medicago sativa L.) to single and mixed inoculation with phosphate-solubilizing bacteria and Sinorhizobium meliloti. *Biology and Fertility of Soils* 46 (2): 185–190. https://doi.org/10.1007/s00374-009-0408-5.

Gupta Choudhury, S., Srivastava, S., Singh, R. et al. (2014). Tillage and residue management effects on soil aggregation, organic carbon dynamics and yield attribute in rice–wheat cropping system under reclaimed sodic soil. *Soil and Tillage Research* 136: 76–83. https://doi.org/10.1016/j.still.2013.10.001.

Gupta, G., Panwar, J., and Jha, P.N. (2013a). Natural occurrence of Pseudomonas aeruginosa, a dominant cultivable diazotrophic endophytic bacterium colonizing *Pennisetum glaucum* (L.) R. Br. *Applied Soil Ecology* 64: 252–261. https://doi.org/10.1016/j.apsoil.2012.12.016.

Gupta, V., Ratha, S.K., Sood, A. et al. (2013b). New insights into the biodiversity and applications of cyanobacteria (blue-green algae) – prospects and challenges. *Algal Research* 2 (2): 79–97. https://doi.org/10.1016/j.algal.2013.01.006.

Gurav, R.G. and Jadhav, J.P. (2013). A novel source of biofertilizer from feather biomass for banana cultivation. *Environmental Science and Pollution Research* 20 (7): 4532–4539. https://doi.org/10.1007/s11356-012-1405-z.

Haider, F.U., Coulter, J.A., Cai, L. et al. (2022). An overview on biochar production, its implications, and mechanisms of biochar-induced amelioration of soil and plant characteristics. *Pedosphere* 32 (1): 107–130. https://doi.org/10.1016/S1002-0160(20)60094-7.

Halpern, M., Bar-Tal, A., Ofek, M. et al. (2015). The use of biostimulants for enhancing nutrient uptake. In: *Advances in Agronomy* (ed. D. Sparks), 141–174. Academic Press. https://doi.org/10.1016/bs.agron.2014.10.001.

Hamid, B., Zaman, M., Farooq, S. et al. (2021). Bacterial plant biostimulants: a sustainable way towards improving growth, productivity, and health of crops. *Sustainability* 13 (5). https://doi.org/10.3390/su13052856.

Harris, J.A. (2003). Measurements of the soil microbial community for estimating the success of restoration. *European Journal of Soil Science* 54 (4): 801–808. https://doi.org/10.1046/j.1351-0754.2003.0559.x.

Harvey, P.R., Warren, R.A., and Wakelin, S. (2009). Potential to improve root access to phosphorus: the role of non-symbiotic microbial inoculants in the rhizosphere. *Crop and Pasture Science* 60 (2): 144–151. https://doi.org/10.1071/CP08084.

Hatten, J. and Liles, G. (2019). A 'healthy' balance – the role of physical and chemical properties in maintaining forest soil function in a changing world. *Developments in Soil Science* 36: 373–396. https://doi.org/10.1016/B978-0-444-63998-1.00015-X.

Hauggaard-Nielsen, H., Gooding, M., Ambus, P. et al. (2009). Pea–barley intercropping for efficient symbiotic N2-fixation, soil N acquisition and use of other nutrients in European organic cropping systems. *Field Crops Research* 113 (1): 64–71. https://doi.org/10.1016/j.fcr.2009.04.009.

Hayat, R., Ali, S., Amara, U. et al. (2010). Soil beneficial bacteria and their role in plant growth promotion: a review. *Annals of Microbiology* 60 (4): 579–598. https://doi.org/10.1007/s13213-010-0117-1.

He, H., Zhihong, Y., Yang, D. et al. (2013). Characterization of endophytic Rahnella sp. JN6 from Polygonum pubescens and its potential in promoting growth and cd, pb, Zn uptake by Brassica napus. *Chemosphere* 90 (6): 1960–1965. https://doi.org/10.1016/j.chemosphere.2012.10.057.

Hellequin, E., Monard, C., Chorin, M. et al. (2020). Responses of active soil microorganisms facing to a soil biostimulant input compared to plant legacy effects. *Scientific Reports* 10 (1): 1–15. https://doi.org/10.1038/s41598-020-70695-7.

Herencia, J.F., Ruiz-Porras, J.C., Melero, S. et al. (2007). Comparison between organic and mineral fertilization for soil fertility levels, crop macronutrient concentrations, and yield. *Agronomy Journal* 99 (4): 973–983. https://doi.org/10.2134/agronj2006.0168.

Hinsinger, P. (2001). Bioavailability of soil inorganic P in the rhizosphere as affected by root-induced chemical changes: a review. *Plant and Soil* 237 (2): 173–195. https://doi.org/10.1023/A:1013351617532.

Hole, D.G., Perkins, A.J., Wilson, J.D. et al. (2005). Does organic farming benefit biodiversity? *Biological Conservation* 122 (1): 113–130. https://doi.org/10.1016/j.biocon.2004.07.018.

Holland, J.M. (2004). The environmental consequences of adopting conservation tillage in Europe: reviewing the evidence. *Agriculture, Ecosystems & Environment* 103 (1): 1–25. https://doi.org/10.1016/j.agee.2003.12.018.

Hurek, T. and Reinhold-Hurek, B. (2003). Azoarcus sp. strain BH72 as a model for nitrogen-fixing grass endophytes. *Journal of Biotechnology* 106 (2–3): 169–178. https://doi.org/10.1016/j.jbiotec.2003.07.010.

IFOAM (2014). The IFOAM Norms for Organic Production and Processing. Version 2014.

IFOAM (2016). Organic Agriculture & Food Security. https://www.ifoam.bio/organic-agriculture-food-security (accessed 22 February 2022).

IPCC (2000). *Land Use, Land Use Change and Forestry* (ed. R.T. Watson, I.R. Noble, B. Bolin, et al.). Cambridge, UK: Cambridge University Press. https://www.ipcc.ch/report/land-use-land-use-change-and-forestry.

IPCC (2019). Summary for policymakers. In: *Climate Change and Land: An IPCC Special Report on Climate Change, Desertification, Land Degradation, Sustainable Land Management, Food Security, and Greenhouse Gas Fluxes in Terrestrial Ecosystems* (ed. P.R. Shukla et al.). https://www.ipcc.ch/srccl/chapter/summary-for-policymakers.

Ito, Y., Nakanomyo, I., Motose, H. et al. (2006). Dodeca-CLE peptides as suppressors of plant stem cell differentiation. *Science* 313 (5788): 842–845. https://doi.org/10.1126/science.1128436.

Iyamuremye, F. and Dick, R.P. (1996). Organic amendments and phosphorus sorption by soils. In: *Advances in Agronomy* (ed. D. Sparks), 139–185. Academic Press. https://doi.org/10.1016/S0065-2113(08)60181-9.

Janeau, J.L., Bricquet, J.P., Planchon, O., and Valentin, C. (2003). Soil crusting and infiltration on steep slopes in northern Thailand. *European Journal of Soil Science* 54 (3): 543–554. https://doi.org/10.1046/j.1365-2389.2003.00494.x.

Jannin, L., Arkoun, M., Ourry, A. et al. (2012). Microarray analysis of humic acid effects on Brassica napus growth: involvement of N, C and S metabolisms. *Plant and Soil* 359 (1–2): 297–319. https://doi.org/10.1007/s11104-012-1191-x.

du Jardin, P. (2012). The Science of Plant Biostimulants. A Bibliographic Analysis: Final Report.

du Jardin, P. (2015). Plant biostimulants: definition, concept, main categories and regulation. *Scientia Horticulturae* 196: 3–14. https://doi.org/10.1016/j.scienta.2015.09.021.

Jindo, K., Soares, T.S., Peres, L.E.P. et al. (2016). Phosphorus speciation and high-affinity transporters are influenced by humic substances. *Journal of Plant Nutrition and Soil Science* 179 (2): 206–214. https://doi.org/10.1002/jpln.201500228.

JLI Social Media (2020). Soil Fertility in Organic Farming Systems. https://www.jliedu.com/blog/soil-fertility-organic-farming-systems.

Johnston, A.E., Poulton, P.R., Fixen, P.E., and Curtin, D. (2014). Chapter five – phosphorus: its efficient use in agriculture. In: *Advances in Agronomy* (ed. D. Sparks), 177–228. Academic Press. https://doi.org/10.1016/B978-0-12-420225-2.00005-4.

Joshi, N., Parewa, H.P., Joshi, S. et al. (2021). Chapter 5 – use of microbial biostimulants in organic farming. In: *Advances in Organic Farming* (ed. V.S. Meena, S.K. Meena, A. Rakshit, et al.), 59–73. Woodhead Publishing. https://doi.org/10.1016/B978-0-12-822358-1.00011-0.

Kapoore, R.V., Wood, E.E., and Llewellyn, C.A. (2021). Algae biostimulants: a critical look at microalgal biostimulants for sustainable agricultural practices. *Biotechnology Advances* 49: 107754. https://doi.org/10.1016/j.biotechadv.2021.107754.

Kassam, A., Friedrich, T., Shaxson, F., and Pretty, J. (2009). The spread of conservation agriculture: justification, sustainability and uptake. *International Journal of Agricultural Sustainability* 7 (4): 292–320. https://doi.org/10.3763/ijas.2009.0477.

Khan, K.S. and Joergensen, R.G. (2009). Changes in microbial biomass and P fractions in biogenic household waste compost amended with inorganic P fertilizers. *Bioresource Technology* 100 (1): 303–309. https://doi.org/10.1016/j.biortech.2008.06.002.

Khan, A.A., Jilani, G., Akhtar, M.S. et al. (2009). Phosphorus solubilizing Bacteria: occurrence, mechanisms and their role in crop production. *Journal of agriculture and biological sciences* 1 (1): 48–58.

Khosro, M. and Yousef, S. (2012). Bacterial biofertilizers for sustainable crop production: a review. *Journal of Agricultural and Bological Science* 7 (5): 307–316.

Klodd, A.E., Eissenstat, D.M., Wolf, T.K., and Centinari, M. (2016). Coping with cover crop competition in mature grapevines. *Plant and Soil* 400 (1–2): 391–402. https://doi.org/10.1007/s11104-015-2748-2.

Kolbe, H. (2019). Mineral P and K fertilization effect on crop yield response as a function of soil nutrient supply under the cultivation conditions of organic farming in Germany. *J für Kulturpflanzen* 71: 161–181.

Kondo, T., Sawa, S., Kinoshita, A. et al. (2006). A plant peptide encoded by CLV3 identified by in situ MALDI-TOF MS analysis. *Science* 313 (5788): 845–848. https://doi.org/10.1126/science.1128439.

Kong, L., Wang, Y.-B., Zhao, L.-N., and Chen, Z.-H. (2009). Enzyme and root activities in surface-flow constructed wetlands. *Chemosphere* 76 (5): 601–608. https://doi.org/10.1016/j.chemosphere.2009.04.056.

Krishnaprabu, S. (2019). Organic farming in India: concept, applications and perspectives. *Journal of Physics: Conference Series* 1362 (1): 012020. https://doi.org/10.1088/1742-6596/1362/1/012020.

Kulikova, N.A., Badun, G.A., Korobkov, V.I. et al. (2014). Accumulation of coal humic acids by wheat seedlings: direct evidence using tritium autoradiography and occurrence in lipid fraction. *Journal of Plant Nutrition and Soil Science* 177 (6): 875–883. https://doi.org/10.1002/jpln.201300648.

Kumar, S. (2016). *Characterization of Phosphate Solubilising and Mineralizing bacteria from Grain Legumes*. Ludhiana, India: Punjab Agricultural University.

Kumar, A., Suri, V.K., and Choudhary, A.K. (2014). Influence of inorganic phosphorus, VAM Fungi, and irrigation regimes on crop productivity and phosphorus transformations in okra (Abelmoschus esculentus L.)–pea (Pisum sativum L.) cropping system in an acid Alfisol.

Communications in Soil Science and Plant Analysis 45 (7): 953–967. https://doi.org/10.1080/00103624.2013.874025.

Kunicki, E., Grabowska, A., Sękara, A., and Wojciechowska, R. (2010). The effect of cultivar type, time of cultivation, and biostimulant treatment on the yield of spinach (L.). *Folia Horticulturae* 22 (2): 9–13. https://doi.org/10.2478/fhort-2013-0153.

Lal, R. (1997). Degradation and resilience of soils. (eds. Greenland, D.J., Gregory, P.J., and Nye, P.H). *Philosophical Transactions of the Royal Society of London. Series B: Biological Sciences* 352 (1356): 997–1010. https://doi.org/10.1098/rstb.1997.0078.

Lal, R. (2015). Restoring soil quality to mitigate soil degradation. *Sustainability* 7 (5): 5875–5895. https://doi.org/10.3390/su7055875.

Lally, R.D., Galbally, P., Moreira, A.S. et al. (2017). Application of endophytic pseudomonas fluorescens and a bacterial consortium to brassica napus can increase plant height and biomass under greenhouse and field conditions. *Frontiers in Plant Science* 8. https://doi.org/10.3389/fpls.2017.02193.

Latati, M., Bargaz, A., Belarbi, B. et al. (2016). The intercropping common bean with maize improves the rhizobial efficiency, resource use and grain yield under low phosphorus availability. *European Journal of Agronomy* 72: 80–90. https://doi.org/10.1016/j.eja.2015.09.015.

Leifeld, J. and Fuhrer, J. (2010). Organic farming and soil carbon sequestration: what do we really know about the benefits? *AMBIO* 39 (8): 585–599. https://doi.org/10.1007/s13280-010-0082-8.

Leung, H., Wu, F.Y., Cheung, K.C. et al. (2010a). Synergistic effects of arbuscular mycorrhizal fungi and phosphate rock on heavy metal uptake and accumulation by an arsenic hyperaccumulator. *Journal of Hazardous Materials* 181 (1): 497–507. https://doi.org/10.1016/j.jhazmat.2010.05.042.

Leung, H., Wu, F.Y., Cheung, K.C. et al. (2010b). The effect of arbuscular mycorrhizal Fungi and phosphate Amendement on arsenic uptake, accumulation and growth of Pteris Vittata in as-contaminated soil. *International Journal of Phytoremediation* 12 (4): 384–403. https://doi.org/10.1080/15226510903051740.

Leung, H.M., Wang, Z.W., Ye, Z.H. et al. (2013). Interactions between arbuscular mycorrhizae and plants in phytoremediation of metal-contaminated soils: a review. *Pedosphere* 23 (5): 549–563. https://doi.org/10.1016/S1002-0160(13)60049-1.

Li, H., Li, X., Xiang, L. et al. (2018). Phytoremediation of soil co-contaminated with cd and BDE-209 using hyperaccumulator enhanced by AM fungi and surfactant. *Science of The Total Environment* 613–614: 447–455. https://doi.org/10.1016/j.scitotenv.2017.09.066.

Liao, M., Hocking, P.J., Dong, B. et al. (2008). Variation in early phosphorus-uptake efficiency among wheat genotypes grown on two contrasting Australian soils. *Australian Journal of Agricultural Research* 59 (2): 157. https://doi.org/10.1071/AR06311.

Lienhard, P., Tivet, F., Chabanne, A. et al. (2013). No-till and cover crops shift soil microbial abundance and diversity in Laos tropical grasslands. *Agronomy for Sustainable Development* 33 (2): 375–384. https://doi.org/10.1007/s13593-012-0099-4.

Lisiecka, J., Knaflewski, M., Spizewski, T. et al. (2011). The effect of animal protein hydrolysate on quantity and quality of strawberry daughter plants cv. "Elsanta." *Acta Scientiarum Polonorum, Hortorum Cultus* 10 (1): 31–40.

Liu, X., Ko, K., Kim, S., and Lee, K. (2007). Effect of amino acid fertilization on nitrate assimilation of leafy radish and soil chemical properties in high nitrate soil. *Communications in Soil Science and Plant Analysis* 39 (1–2): 269–281. https://doi.org/10.1080/00103620701759301.

Lockeretz, W., Shearer, G., and Kohl, D.H. (1981). Organic farming in the Corn Belt. *Science* 211 (4482): 540–547. https://doi.org/10.1126/science.211.4482.540.

López-Bucio, J., Pelagio-Flores, R., and Herrera-Estrella, A. (2015). Trichoderma as biostimulant: exploiting the multilevel properties of a plant beneficial fungus. *Scientia Horticulturae* 196: 109–123. https://doi.org/10.1016/j.scienta.2015.08.043.

Lori, M., Symnaczik, S., Mäder, P. et al. (2017). Organic farming enhances soil microbial abundance and activity – a meta-analysis and meta-regression. (ed. Lehman, R.M.). *PLOS ONE* 12 (7): e0180442. https://doi.org/10.1371/journal.pone.0180442.

Lotter, D.W., Seidel, R., and Liebhardt, W. (2003). The performance of organic and conventional cropping systems in an extreme climate year. *American Journal of Alternative Agriculture* 18 (3): 146–154. https://doi.org/10.1079/AJAA200345.

Lucini, L., Rouphael, Y., Cardarelli, M. et al. (2015). The effect of a plant-derived biostimulant on metabolic profiling and crop performance of lettuce grown under saline conditions. *Scientia Horticulturae* 182: 124–133. https://doi.org/10.1016/j.scienta.2014.11.022.

Lugtenberg, B. and Kamilova, F. (2009). Plant-growth-promoting rhizobacteria. *Annual Review of Microbiology* 63 (1): 541–556. https://doi.org/10.1146/annurev.micro.62.081307.162918.

Lynch, J.P. and Brown, K.M. (2008). Root strategies for phosphorus acquisition. In: *The Ecophysiology of Plant-Phosphorus Interactions* (ed. P.J. White and J.P. Hammond), 83–116. Dordrecht: Springer Netherlands. https://doi.org/10.1007/978-1-4020-8435-5_5.

Ma, Y., Rajkumar, M., Zhang, C., and Freitas, H. (2016). Beneficial role of bacterial endophytes in heavy metal phytoremediation. *Journal of Environmental Management* 174: 14–25. https://doi.org/10.1016/j.jenvman.2016.02.047.

Maeder, P., Fliessbach, A., Dubois, D. et al. (2002). Soil fertility and biodiversity in organic farming. *Science* 296 (5573): 1694–1697. https://doi.org/10.1126/science.1071148.

Malik, M.A., Marschner, P., and Khan, K.S. (2012). Addition of organic and inorganic P sources to soil – effects on P pools and microorganisms. *Soil Biology and Biochemistry* 49: 106–113. https://doi.org/10.1016/j.soilbio.2012.02.013.

Malik, M.A., Khan, K.S., Marschner, P., and Ali, S. (2013). Organic amendments differ in their effect on microbial biomass and activity and on P pools in alkaline soils. *Biology and Fertility of Soils* 49 (4): 415–425. https://doi.org/10.1007/s00374-012-0738-6.

Mandal, A., Dhaliwal, S.S., Mani, P.K., and Toor, A.S. (2021). Conservation agricultural practices under organic farming. In: *Advances in Organic Farming* (ed. V.S. Meena, Q. Rakshit, C. Srinivasarao, et al.), 17–37. Duxford, UK: Elsevier. https://doi.org/10.1016/B978-0-12-822358-1.00014-6.

Mannino, G., Nerva, L., Gritli, T. et al. (2020). Effects of different microbial Inocula on tomato tolerance to water deficit. *Agronomy* 10 (2). https://doi.org/10.3390/agronomy10020170.

Mannino, G., Gentile, C., Ertani, A. et al. (2021). Anthocyanins: biosynthesis, distribution, ecological role, and use of biostimulants to increase their content in plant foods – a review. *Agriculture (Switzerland)* 11 (3): 1–25. https://doi.org/10.3390/agriculture11030212.

Mao, L.L., Zhang, L.Z., Zhang, S.P. et al. (2015). Resource use efficiency, ecological intensification and sustainability of intercropping systems. *Journal of Integrative Agriculture* 14 (8): 1542–1550. https://doi.org/10.1016/S2095-3119(15)61039-5.

Marinari, S., Mancinelli, R., Campiglia, E., and Grego, S. (2006). Chemical and biological indicators of soil quality in organic and conventional farming systems in Central Italy. *Ecological Indicators* 6 (4): 701–711. https://doi.org/10.1016/j.ecolind.2005.08.029.

Marschner, H. (ed.) (1996). Chapter: 8 – functions of mineral nutrients: macronutrients. In: *Marschner's Mineral Nutrition of Higher Plants*, 3e, 229–312. London: Elsevier. https://doi.org/10.1016/C2009-0-02402-7.

Marschner, P. (2012). *Marschner's Mineral Nutrition of Higher Plants*, 3e. London: Elsevier. https://doi.org/10.1016/C2009-0-63043-9.

Matsumiya, Y. and Kubo, M. (2011). Soybean peptide: novel plant growth promoting peptide from soybean. In: *Soybean and Nutrition* (ed. M.K.E.-H. El-Shemy). Rijeka: InTech, Ch. 11. https://doi.org/10.5772/19132.

Métraux, J.-P. (1987). Gibberellins and plant cell elongation. In: *Plant Hormones and their Role in Plant Growth and Development* (ed. P.J. Davies), 296–317. Dordrecht: Springer Netherlands. https://doi.org/10.1007/978-94-009-3585-3_16.

Michalak, I., Chojnacka, K., and Saeid, A. (2017). Plant growth biostimulants, dietary feed supplements and cosmetics formulated with supercritical CO_2 algal extracts. *Molecules* 22 (1): 66. https://doi.org/10.3390/molecules22010066.

Mirzabaev, A., Wu, J., Evans, J. et al. (2019). Desertification. In: *Climate Change and Land: An IPCC Special Report on Climate Change, Desertification, Land Degradation, Sustainable Land Management, Food Security, and Greenhouse Gas Fluxes in Terrestrial Ecosystems* (ed. P.R. Shukla et al.).

Miyakawa, T., Fujita, Y., Yamaguchi-Shinozaki, K., and Tanokura, M. (2013). Structure and function of abscisic acid receptors. *Trends in Plant Science* 18 (5): 259–266. https://doi.org/10.1016/j.tplants.2012.11.002.

Mohammad, A. and Mittra, B. (2013). Effects of inoculation with stress-adapted arbuscular mycorrhizal fungus Glomus deserticola on growth of *Solanum melongena* L. and Sorghum sudanese staph. Seedlings under salinity and heavy metal stress conditions. *Archives of Agronomy and Soil Science* 59 (2): 173–183. https://doi.org/10.1080/03650340.2011.610029.

Mondelaers, K., Aertsens, J., and Van Huylenbroeck, G. (2009). A meta-analysis of the differences in environmental impacts between organic and conventional farming. *British Food Journal* 111 (10): 1098–1119. https://doi.org/10.1108/00070700910992925.

Mora, V., Bacaicoa, E., Zamarreño, A.-M. et al. (2010). Action of humic acid on promotion of cucumber shoot growth involves nitrate-related changes associated with the root-to-shoot distribution of cytokinins, polyamines and mineral nutrients. *Journal of Plant Physiology* 167 (8): 633–642. https://doi.org/10.1016/j.jplph.2009.11.018.

Morton, J.B. and Redecker, D. (2001). Two new families of Glomales, Archaeosporaceae and Paraglomaceae, with two new genera Archaeospora and Paraglomus, based on concordant molecular and morphological characters. *Mycologia* 93 (1): 181–195. https://doi.org/10.1080/00275514.2001.12063147.

Mu, D., Horowitz, N., Casey, M., and Jones, K. (2017). Environmental and economic analysis of an in-vessel food waste composting system at Kean University in the U.S. *Waste Management* 59: 476–486. https://doi.org/10.1016/j.wasman.2016.10.026.

Muller, A., Schader, C., Scialabba, N.E.-H. et al. (2017). Strategies for feeding the world more sustainably with organic agriculture. *Nature Communications* 8 (1): 1290. https://doi.org/10.1038/s41467-017-01410-w.

Nadeem, S.M., Zahir, Z.A., Naveed, M. et al. (2010). Rhizobacteria capable of producing ACC-deaminase may mitigate salt stress in wheat. *Soil Science Society of America Journal* 74 (2): 533–542. https://doi.org/10.2136/sssaj2008.0240.

Nardi, S., Pizzeghello, D., Gessa, C. et al. (2000). A low molecular weight humic fraction on nitrate uptake and protein synthesis in maize seedlings. *Soil Biology and Biochemistry* 32 (3): 415–419. https://doi.org/10.1016/S0038-0717(99)00168-6.

Nardi, S., Pizzeghello, D., Muscolo, A., and Vianello, A. (2002). Physiological effects of humic substances on higher plants. *Soil Biology and Biochemistry* 34 (11): 1527–1536. https://doi.org/10.1016/S0038-0717(02)00174-8.

Nelson, N. and Mikkelsen, R. (2008). Meeting the phosphorus requirement on organic farms. *Better Crops* 92 (1): 12–14. https://ucanr.edu/sites/nm/files/76653.pdf.

Ngwira, A.R., Aune, J.B., and Mkwinda, S. (2012). On-farm evaluation of yield and economic benefit of short term maize legume intercropping systems under conservation agriculture in Malawi. *Field Crops Research* 132: 149–157. https://doi.org/10.1016/j.fcr.2011.12.014.

Novara, A., Gristina, L., Guaitoli, F. et al. (2013). Managing soil nitrate with cover crops and buffer strips in Sicilian vineyards. *Solid Earth* 4 (2): 255–262. https://doi.org/10.5194/se-4-255-2013.

Obale-ebanga, F. (2001). *Impacts of Agricultural Land Use Histories on Soil Organic Matter Dynamics and Related Properties of Savannah Soils in North Cameroon*. Universiteit van Amsterdam. https://hdl.handle.net/11245/1.185515.

Oehl, F., Sieverding, E., Ineichen, K. et al. (2005). Community structure of arbuscular mycorrhizal fungi at different soil depths in extensively and intensively managed agroecosystems. *New Phytologist* 165 (1): 273–283. https://doi.org/10.1111/j.1469-8137.2004.01235.x.

Olaetxea, M., De Hita, D., Garcia, C.A. et al. (2018). Hypothetical framework integrating the main mechanisms involved in the promoting action of rhizospheric humic substances on plant root- and shoot- growth. *Applied Soil Ecology* 123: 521–537. https://doi.org/10.1016/j.apsoil.2017.06.007.

Olsson, L., Barbosa, H., Bhadwal, S. et al. (2019). Land degradation. In: *Climate Change and Land: An IPCC Special Report on Climate Change, Desertification, Land Degradation, Sustainable Land Management, Food Security, and Greenhouse Gas Fluxes in Terrestrial Ecosystems* (ed. P.R. Shukla et al.).

Pacini, C., Wossink, A., Giesen, G. et al. (2003). Evaluation of sustainability of organic, integrated and conventional farming systems: a farm and field-scale analysis. *Agriculture, Ecosystems and Environment* 95 (1): 273–288. https://doi.org/10.1016/S0167-8809(02)00091-9.

Palm, C., Blanco-Canqui, H., DeClerck, F. et al. (2014). Conservation agriculture and ecosystem services: an overview. *Agriculture, Ecosystems & Environment* 187: 87–105. https://doi.org/10.1016/j.agee.2013.10.010.

Pampuro, N., Caffaro, F., and Cavallo, E. (2018). Reuse of animal manure: a case study on Stakeholders' perceptions about pelletized compost in northwestern Italy. *Sustainability* 10 (6): 2028. https://doi.org/10.3390/su10062028.

Panagos, P., Borrelli, P., Poesen, J. et al. (2015). The new assessment of soil loss by water erosion in Europe. *Environmental Science and Policy* 54: 438–447. https://doi.org/10.1016/j.envsci.2015.08.012.

Panke-Buisse, K., Poole, A.C., Goodrich, J.K. et al. (2015). Selection on soil microbiomes reveals reproducible impacts on plant function. *The ISME Journal* 9 (4): 980–989. https://doi.org/10.1038/ismej.2014.196.

Paradiković, N., Vinković, T., Vinković Vrček, I. et al. (2011). Effect of natural biostimulants on yield and nutritional quality: an example of sweet yellow pepper (*Capsicum annuum* L.) plants. *Journal of the Science of Food and Agriculture* 91 (12): 2146–2152. https://doi.org/10.1002/jsfa.4431.

Parrado, J., Bautista, J., Romero, E.J. et al. (2008). Production of a carob enzymatic extract: potential use as a biofertilizer. *Bioresource Technology* 99 (7): 2312–2318. https://doi.org/10.1016/j.biortech.2007.05.029.

de Pascale, S., Rouphael, Y., and Colla, G. (2018). Plant biostimulants: innovative tool for enhancing plant nutrition in organic farming. *European Journal of Horticultural Science* 82 (6): 277–285. https://doi.org/10.17660/eJHS.2017/82.6.2.

Peigné, J., Casagrande, M., Payet, V. et al. (2016). How organic farmers practice conservation agriculture in Europe. *Renewable Agriculture and Food Systems* 31 (1): 72–85. https://doi.org/10.1017/S1742170514000477.

Peltoniemi, K., Velmala, S., Fritze, H. et al. (2021). Long-term impacts of organic and conventional farming on the soil microbiome in boreal arable soil. *European Journal of Soil Biology* 104: 103314. https://doi.org/10.1016/j.ejsobi.2021.103314.

Piccolo, A. (ed.) (1996). Humus and soil conservation. In: *Humic Substances in Terrestrial Ecosystems*, 225–264. Elsevier. https://doi.org/10.1016/b978-044481516-3/50006-2.

Piccolo, A. (2002). The supramolecular structure of humic substances: a novel understanding of humus chemistry and implications in soil science. *Advances in Agronomy* 75: 57–134. https://doi.org/10.1016/S0065-2113(02)75003-7.

Piccolo, A., Conte, P., Spaccini, R., and Chiarella, M. (2003). Effects of some dicarboxylic acids on the association of dissolved humic substances. *Biology and Fertility of Soils* 37 (4): 255–259. https://doi.org/10.1007/s00374-003-0583-8.

Pimentel, D., Hepperly, P., Hanson, J. et al. (2005). Environmental, energetic, and economic comparisons of organic and conventional farming systems. *BioScience* 55 (7): 573–582. https://doi.org/10.1641/0006-3568(2005)055[0573:EEAECO]2.0.CO;2.

Pindi, P.K. (2012). Liquid microbial consortium – a potential tool for sustainable soil health. *Journal of Biofertilizers & Biopesticides* 03 (04): 1–9. https://doi.org/10.4172/2155-6202.1000124.

Pinto, A.P., de Varennes, A., Dias, C.M.B., and Lopes, M.E. (2018). Microbial-assisted phytoremediation: a convenient use of plant and microbes to clean up soils. In: *Phytoremediation* (ed. A.A. Ansari, S.S. Gill, R.R. Gill, et al.), 21–87. Cham: Springer International Publishing https://doi.org/10.1007/978-3-319-99651-6_2.

Pinto, A.P., Lopes, M.E., Dordio, A., and Castanheiro, J.E.F. (2021). Microbe and plant-assisted remediation of organic xenobiotics. In: *Handbook of Assisted and Amendment: Enhanced Sustainable Remediation Technology*, 437–475. Wiley Online Books. Wiley https://doi.org/10.1002/9781119670391.ch22.

Pinton, R., Cesco, S., De Nobili, M. et al. (1997). Water- and pyrophosphate-extractable humic substances fractions as a source of iron for Fe-deficient cucumber plants. *Biology and Fertility of Soils* 26 (1): 23–27. https://doi.org/10.1007/s003740050337.

Pinton, R., Cesco, S., Iacolettig, G. et al. (1999). Modulation of NO_3^- uptake by water-extractable humic substances: involvement of root plasma membrane H^+ ATPase. *Plant and Soil* 215 (2): 155–161. https://doi.org/10.1023/A:1004752531903.

Priyadarshini, P., Choudhury, S., Tilgam, J. et al. (2021). Nitrogen fixing cereal: a rising hero towards meeting food security. *Plant Physiology and Biochemistry* 167: 912–920. https://doi.org/10.1016/j.plaphy.2021.09.012.

Prudent, M., Dequiedt, S., Sorin, C. et al. (2020). The diversity of soil microbial communities matters when legumes face drought. *Plant Cell and Environment* 43 (4): 1023–1035. https://doi.org/10.1111/pce.13712.

Puglisi, E., Pascazio, S., Suciu, N. et al. (2013). Rhizosphere microbial diversity as influenced by humic substance amendments and chemical composition of rhizodeposits. *Journal of Geochemical Exploration* 129: 82–94. https://doi.org/10.1016/j.gexplo.2012.10.006.

Qiao, Y., Crowley, D., Wang, K. et al. (2015). Effects of biochar and arbuscular mycorrhizae on bioavailability of potentially toxic elements in an aged contaminated soil. *Environmental Pollution* 206: 636–643. https://doi.org/10.1016/j.envpol.2015.08.029.

Quaggiotti, S., Ruperti, B., Pizzeghello, D. et al. (2004). Effect of low molecular size humic substances on nitrate uptake and expression of genes involved in nitrate transport in maize (Zea mays L.). *Journal of Experimental Botany* 55 (398): 803–813. https://doi.org/10.1093/jxb/erh085.

Rabot, E., Wiesmeier, M., Schlüter, S., and Vogel, H.J. (2018). Soil structure as an indicator of soil functions: a review. *Geoderma* 314: 122–137. https://doi.org/10.1016/j.geoderma.2017.11.009.

Ramesh, A., Sharma, S.K., Joshi, O.P., and Khan, I.R. (2011). Phytase, phosphatase activity and P-nutrition of soybean as influenced by inoculation of Bacillus. *Indian Journal of Microbiology* 51 (1): 94–99. https://doi.org/10.1007/s12088-011-0104-7.

Ramesh, A., Sharma, S.K., Yadav, N., and Joshi, O.P. (2014). Phosphorus mobilization from native soil P-Pool upon inoculation with phytate-mineralizing and phosphate-solubilizing Bacillus aryabhattai isolates for improved P-acquisition and growth of soybean and wheat crops in microcosm conditions. *Agricultural Research* 3 (2): 118–127. https://doi.org/10.1007/s40003-014-0105-y.

Raphael, J.P.A., Calonego, J.C., Milori, D.M.B.P., and Rosolem, C.A. (2016). Soil organic matter in crop rotations under no-till. *Soil and Tillage Research* 155: 45–53. https://doi.org/10.1016/j.still.2015.07.020.

Reeves, P.G. and Stallard, L. (1994). Zinc deficiency reduces the activity of angiotensin-converting enzyme in testicular germ cells and sperm of adult rats. *Journal of Trace Elements in Experimental Medicine* 7 (3): 125–134.

Reganold, J.P. and Wachter, J.M. (2016). Organic agriculture in the twenty-first century. *Nature Plants* 2 (2): 15221. https://doi.org/10.1038/nplants.2015.221.

Reganold, J.P., Elliott, L.F., and Unger, Y.L. (1987). Long-term effects of organic and conventional farming on soil erosion. *Nature* 330 (6146): 370–372. https://doi.org/10.1038/330370a0.

Richardson, A.E. (1994). Soil microorganisms and phosphorus availability. In: *Soil Biota: Management in Sustainable Farming Systems* (ed. C.E. Pankhurst, B.M. Doube, V.V.S.R. Gupta, and P.R. Grace), 50–62. Melbourne, Australia: CSIRO.

Richardson, A.E., Lynch, J.P., Ryan, P.R. et al. (2011). Plant and microbial strategies to improve the phosphorus efficiency of agriculture. *Plant and Soil* 349 (1): 121–156. https://doi.org/10.1007/s11104-011-0950-4.

Ritchie, H. (2017). Is Organic Really Better for the Environment than Conventional Agriculture? https://ourworldindata.org/is-organic-agriculture-better-for-the-environment (accessed 22 February 2022).

Robačer, M., Canali, S., Kristensen, H.L. et al. (2016). Cover crops in organic field vegetable production. *Scientia Horticulturae* 208: 104–110. https://doi.org/10.1016/j.scienta.2015.12.029.

Rodríguez, H. and Fraga, R. (1999). Phosphate solubilizing bacteria and their role in plant growth promotion. *Biotechnology Advances* 17 (4–5): 319–339. https://doi.org/10.1016/S0734-9750(99)00014-2.

Rodríguez, H., Fraga, R., Gonzalez, T., and Bashan, Y. (2007). Genetics of phosphate solubilization and its potential applications for improving plant growth-promoting bacteria. In: *First International Meeting on Microbial Phosphate Solubilization* (ed. E. Velázquez and C. Rodríguez-Barrueco), 15–21. Dordrecht: Springer Netherlands.

Rose, M.T., Patti, A.F., Little, K.R. et al. (2014). A meta-analysis and review of plant-growth response to humic substances: practical implications for agriculture. In: *Advances in Agronomy*, 37–89. Elsevier. https://doi.org/10.1016/B978-0-12-800138-7.00002-4.

Rosolem, C.A. and Calonego, J.C. (2013). Phosphorus and potassium budget in the soil–plant system in crop rotations under no-till. *Soil and Tillage Research* 126: 127–133. https://doi.org/10.1016/j.still.2012.08.003.

Rouphael, Y. and Colla, G. (2020a). Editorial: biostimulants in agriculture. *Frontiers in Plant Science* 11: https://doi.org/10.3389/fpls.2020.00040.

Rouphael, Y. and Colla, G. (2020b). Toward a sustainable agriculture through plant biostimulants: from experimental data to practical applications. *Agronomy* 10 (10). https://doi.org/10.3390/agronomy10101461.

Rouphael, Y., Cardarelli, M., and Colla, G. (2015). Role of arbuscular mycorrhizal fungi in alleviating the adverse effects of acidity and aluminium toxicity in zucchini squash. *Scientia Horticulturae* 188: 97–105. https://doi.org/10.1016/j.scienta.2015.03.031.

Rouphael, Y., De Micco, V., Arena, C. et al. (2017). Effect of Ecklonia maxima seaweed extract on yield, mineral composition, gas exchange, and leaf anatomy of zucchini squash grown under saline conditions. *Journal of Applied Phycology* 29 (1): 459–470. https://doi.org/10.1007/s10811-016-0937-x.

Ruzzi, M. and Aroca, R. (2015). Plant growth-promoting rhizobacteria act as biostimulants in horticulture. *Scientia Horticulturae* 196: 124–134. https://doi.org/10.1016/j.scienta.2015.08.042.

Sacco, D., Moretti, B., Monaco, S., and Grignani, C. (2015). Six-year transition from conventional to organic farming: effects on crop production and soil quality. *European Journal of Agronomy* 69: 10–20. https://doi.org/10.1016/j.eja.2015.05.002.

Sánchez, P.A. (2010). Tripling crop yields in tropical Africa. *Nature Geoscience* 3 (5): 299–300. https://doi.org/10.1038/ngeo853.

Santander, C., Sanhueza, M., Olave, J. et al. (2019). Arbuscular mycorrhizal colonization promotes the tolerance to salt stress in lettuce plants through an efficient modification of ionic balance. *Journal of Soil Science and Plant Nutrition* 19 (2): 321–331. https://doi.org/10.1007/s42729-019-00032-z.

Santi, C., Bogusz, D., and Franche, C. (2013). Biological nitrogen fixation in non-legume plants. *Annals of Botany* 111 (5): 743–767. https://doi.org/10.1093/aob/mct048.

Santos, V.B., Araújo, A.S.F., Leite, L.F.C. et al. (2012). Soil microbial biomass and organic matter fractions during transition from conventional to organic farming systems. *Geoderma* 170: 227–231. https://doi.org/10.1016/j.geoderma.2011.11.007.

Saritha, M. and Prasad Tollamadugu, N.V.K.V. (2019). The status of research and application of biofertilizers and biopesticides: global scenario. In: *Recent Developments in Applied Microbiology and Biochemistry* (ed. V. Buddolla), 195–207. Elsevier. https://doi.org/10.1016/B978-0-12-816328-3.00015-5.

Sarkar, S., Dey, A., Kumar, V. et al. (2021). Fungal endophyte: an interactive endosymbiont with the capability of modulating host physiology in myriad ways. *Frontiers in Plant Science* 12: 701800. https://doi.org/10.3389/fpls.2021.701800.

Sarwar, N., Imran, M., Shaheen, M.R. et al. (2017). Phytoremediation strategies for soils contaminated with heavy metals: modifications and future perspectives. *Chemosphere* 171: 710–721. https://doi.org/10.1016/j.chemosphere.2016.12.116.

Savvas, D. and Ntatsi, G. (2015). Biostimulant activity of silicon in horticulture. *Scientia Horticulturae* 196: 66–81. https://doi.org/10.1016/j.scienta.2015.09.010.

Scalise, A., Tortorella, D., Pristeri, A. et al. (2015). Legume–barley intercropping stimulates soil N supply and crop yield in the succeeding durum wheat in a rotation under rainfed conditions. *Soil Biology and Biochemistry* 89: 150–161. https://doi.org/10.1016/j.soilbio.2015.07.003.

Schaafsma, G. (2009). Safety of protein hydrolysates, fractions thereof and bioactive peptides in human nutrition. *European Journal of Clinical Nutrition* 63 (10): 1161–1168. https://doi.org/10.1038/ejcn.2009.56.

Schiavon, M., Ertani, A., and Nardi, S. (2008). Effects of an alfalfa protein hydrolysate on the gene expression and activity of enzymes of the tricarboxylic acid (TCA) cycle and nitrogen metabolism in Zea mays L. *Journal of Agricultural and Food Chemistry* 56 (24): 11800–11808. https://doi.org/10.1021/jf802362g.

Seitz, S., Goebes, P., Puerta, V.L. et al. (2019). Conservation tillage and organic farming reduce soil erosion. *Agronomy for Sustainable Development* 39 (1): 4. https://doi.org/10.1007/s13593-018-0545-z.

Sessitsch, A., Brader, G., Pfaffenbichler, N. et al. (2018). The contribution of plant microbiota to economy growth. *Microbial Biotechnology* 11 (5): 801–805. https://doi.org/10.1111/1751-7915.13290.

Seufert, V. (2012). Organic Agriculture as an Opportunity for Sustainable Agricultural Development Verena – Policy Brief No. 13. https://www.mcgill.ca/isid/files/isid/seufert.pb13.pdf.

Sharma, H.S.S., Fleming, C., Selby, C. et al. (2014). Plant biostimulants: a review on the processing of macroalgae and use of extracts for crop management to reduce abiotic and biotic stresses. *Journal of Applied Phycology* 26 (1): 465–490. https://doi.org/10.1007/s10811-013-0101-9.

Sheoran, N., Nadakkakath, A.V., Munjal, V. et al. (2015). Genetic analysis of plant endophytic Pseudomonas putida BP25 and chemo-profiling of its antimicrobial volatile organic compounds. *Microbiological Research* 173: 66–78. https://doi.org/10.1016/j.micres.2015.02.001.

Shi, Y., Ziadi, N., Messiga, A.J. et al. (2013). Changes in soil phosphorus fractions for a long-term corn-soybean rotation with tillage and phosphorus fertilization. *Soil Science Society of America Journal* 77 (4): 1402–1412. https://doi.org/10.2136/sssaj2012.0427.

Siegrist, S., Schaub, D., Pfiffner, L., and Mäder, P. (1998). Does organic agriculture reduce soil erodibility? The results of a long-term field study on loess in Switzerland. *Agriculture, Ecosystems & Environment* 69 (3): 253–264. https://doi.org/10.1016/S0167-8809(98)00113-3.

Sieverding, E. and Oehl, F. (2006). Revision of Entrophospora and description of Kuklospora and Intraspora, two new genera in the arbuscular mycorrhizal Glomeromycetes. *Journal of Applied Botany and Food Quality* 80 (1): 69–81.

Singh, J.S. (2015). Microbes: the chief ecological engineers in reinstating equilibrium in degraded ecosystems. *Agriculture, Ecosystems & Environment* 203: 80–82. https://doi.org/10.1016/j.agee.2015.01.026.

Singh, M., Reddy, K.S., Singh, V.P., and Rupa, T.R. (2007). Phosphorus availability to rice (Oriza sativa L.)–wheat (Triticum estivum L.) in a vertisol after eight years of inorganic and organic fertilizer additions. *Bioresource Technology* 98 (7): 1474–1481. https://doi.org/10.1016/j.biortech.2006.02.045.

Singh, A., Vaish, B., and Singh, P. (2016). Eco-restoration of degraded lands through microbial biomass: an ecological engineer. *Acta Biomedica Scientia* 3 (1): 133–135.

Sivaranjani, S. and Rakshit, A. (2019). Organic farming in protecting water quality. In: *Organic Farming* (ed. C. Sarath Chandran, S. Thomas, and M.R. Unni), 1–9. Cham: Springer International Publishing. https://doi.org/10.1007/978-3-030-04657-6_1.

Six, J., Paustian, K., Elliott, E.T., and Combrink, C. (2000). Soil structure and organic matter I. distribution of aggregate-size classes and aggregate-associated carbon. *Soil Science Society of America Journal* 64 (2): 681–689. https://doi.org/10.2136/sssaj2000.642681x.

Smith, L.G. and Lampkin, N.H. (2019). Greener farming: managing carbon and nitrogen cycles to reduce greenhouse gas emissions from agriculture. In: *Managing Global Warming* (ed. T.M. Letcher), 553–577. United Kingdom: Academic Press. https://doi.org/10.1016/B978-0-12-814104-5.00019-3.

Smith, S. and Read, D. (2008a). Chapter 1: the symbionts forming arbuscular mycorrhizas. In: *Mycorrhizal Symbiosis*, 3e (ed. S.E. Smith and D.B.T. Read), 13–41. London: Elsevier. https://doi.org/10.1016/B978-0-12-370526-6.X5001-6.

Smith, S.E. and Read, D. (2008b). The symbionts forming arbuscular mycorrhizas. In: *Mycorrhizal Symbiosis* (ed. S.E. Smith and D.B.T. Read), 13–41. London: Elsevier. https://doi.org/10.1016/B978-012370526-6.50003-9.

Smith, O.M., Cohen, A.L., Rieser, C.J. et al. (2019). Organic farming provides reliable environmental benefits but increases variability in crop yields: a global meta-analysis. *Frontiers in Sustainable Food Systems* 3: https://doi.org/10.3389/fsufs.2019.00082.

Souza, R.C., Cantão, M.E., Vasconcelos, A.T.R. et al. (2013). Soil metagenomics reveals differences under conventional and no-tillage with crop rotation or succession. *Applied Soil Ecology* 72: 49–61. https://doi.org/10.1016/j.apsoil.2013.05.021.

Stevenson, F.J. (1994). *Humus Chemistry: Genesis, Composition, Reactions*, 2e. New York, NY: Wiley.

Stockdale, E.A., Lampkin, N.H., Hovi, M. et al. (2001). Agronomic and environmental implications of organic farming systems. In: *Advances in Agronomy*, 261–327. Elsevier. https://doi.org/10.1016/S0065-2113(01)70007-7.

Stockdale, E.A., Shepherd, M.A., Fortune, S., and Cuttle, S.P. (2006). Soil fertility in organic farming systems – fundamentally different? *Soil Use and Management* 18: 301–308. https://doi.org/10.1111/j.1475-2743.2002.tb00272.x.

Szparaga, A., Kuboń, M., Kocira, S. et al. (2019). Towards sustainable agriculture – agronomic and economic effects of biostimulant use in common bean cultivation. *Sustainability* 11 (17): 4575. https://doi.org/10.3390/su11174575.

Tang, J., Leung, A., Leung, C., and Lim, B.L. (2006). Hydrolysis of precipitated phytate by three distinct families of phytases. *Soil Biology and Biochemistry* 38 (6): 1316–1324. https://doi.org/10.1016/j.soilbio.2005.08.021.

Tarafdar, J.C. and Claassen, N. (1988). Organic phosphorus compounds as a phosphorus source for higher plants through the activity of phosphatases produced by plant roots and microorganisms. *Biology and Fertility of Soils* 5 (4): 308–312. https://doi.org/10.1007/BF00262137.

Tavares, O.C.H., Santos, L.A., Ferreira, L.M. et al. (2017). Humic acid differentially improves nitrate kinetics under low- and high-affinity systems and alters the expression of plasma membrane H^+-ATPases and nitrate transporters in rice. *Annals of Applied Biology* 170 (1): 89–103. https://doi.org/10.1111/aab.12317.

Toor, M.D., Adnan, M., Raza, A. et al. (2020). Land degradation and its management: a review. *International Journal of Environmental Sciences & Natural Resources* 25 (1): 63–66. https://doi.org/10.19080/ijesnr.2020.25.556157.

Trevisan, S., Botton, A., Vaccaro, S. et al. (2011). Humic substances affect Arabidopsis physiology by altering the expression of genes involved in primary metabolism, growth and development. *Environmental and Experimental Botany* 74 (1): 45–55. https://doi.org/10.1016/j.envexpbot.2011.04.017.

Tsvetkov, I., Atanassov, A., Vlahova, M. et al. (2018). Plant organic farming research–current status and opportunities for future development. *Biotechnology and Biotechnological Equipment* 32 (2): 241–260. https://doi.org/10.1080/13102818.2018.1427509.

Tu, C., Louws, F.J., Creamer, N.G. et al. (2006). Responses of soil microbial biomass and N availability to transition strategies from conventional to organic farming systems. *Agriculture, Ecosystems and Environment* 113 (1–4): 206–215. https://doi.org/10.1016/j.agee.2005.09.013.

Tuomisto, H.L., Hodge, I.D., Riordan, P., and Macdonald, D.W. (2012). Does organic farming reduce environmental impacts? – a meta-analysis of European research. *Journal of Environmental Management* 112: 309–320. https://doi.org/10.1016/j.jenvman.2012.08.018.

Turner, B.L., McKelvie, I.D., and Haygarth, P.M. (2002). Characterisation of water-extractable soil organic phosphorus by phosphatase hydrolysis. *Soil Biology and Biochemistry* 34 (1): 27–35. https://doi.org/10.1016/S0038-0717(01)00144-4.

Turner, B.L., Newman, S., Cheesman, A.W., and Reddy, K.R. (2007). Sample pretreatment and phosphorus speciation in wetland soils. *Soil Science Society of America Journal* 71 (5): 1538–1546. https://doi.org/10.2136/sssaj2007.0017.

Ullah, A., Mushtaq, H., Ali, H. et al. (2015). Diazotrophs-assisted phytoremediation of heavy metals: a novel approach. *Environmental Science and Pollution Research* 22 (4): 2505–2514. https://doi.org/10.1007/s11356-014-3699-5.

United Nations (2022). World Population Projected to Reach 9.8 billion in 2050, and 11.2 billion in 2100. https://www.un.org/en/desa/world-population-projected-reach-98-billion-2050-and-112-billion-2100 (accessed 10 March 2022).

United Nations. Population Division (1999). *World Population Prospects, the 1998 Revision. Volume II, the Sex and Age Distribution of the World Population*. New York, NY: United Nations Publications.

Urrutia, O., Erro, J., Guardado, I. et al. (2014). Physico-chemical characterization of humic-metal-phosphate complexes and their potential application to the manufacture of new types of phosphate-based fertilizers. *Journal of Plant Nutrition and Soil Science* 177 (2): 128–136. https://doi.org/10.1002/jpln.201200651.

Vaccaro, S., Musculo, A., Pizzeghello, D. et al. (2009). Effect of a compost and its water-soluble fractions on key enzymes of nitrogen metabolism in maize seedlings. *Journal of Agricultural and Food Chemistry* 57 (23): 11267–11276. https://doi.org/10.1021/jf901808s.

Vaccaro, S., Ertani, A., Nebbioso, A. et al. (2015). Humic substances stimulate maize nitrogen assimilation and amino acid metabolism at physiological and molecular level. *Chemical and Biological Technologies in Agriculture* 2 (1): 1–12. https://doi.org/10.1186/s40538-015-0033-5.

Varanini, Z. and Nannipieri, P. (2000). Direct versus indirect effects of soil humic substances on plant growth and nutrition. In: *The Rhizosphere* (ed. S. Willig, Z. Varanini, and P. Nannipieri), 157–174. Boca Raton: CRC Press https://doi.org/10.1201/9780849384974-11.

Varanini, Z., Pinton, R., De Biasi, M.G. et al. (1993). Low molecular weight humic substances stimulate H+-ATPase activity of plasma membrane vesicles isolated from oat (Avena sativa L.) roots. *Plant and Soil* 153 (1): 61–69. https://doi.org/10.1007/BF00010544.

Vasseur-Coronado, M., du Boulois, H.D., Pertot, I., and Puopolo, G. (2021). Selection of plant growth promoting rhizobacteria sharing suitable features to be commercially developed as biostimulant products. *Microbiological Research* 245: 126672. https://doi.org/10.1016/j.micres.2020.126672.

Verbruggen, E., Röling, W.F.M., Gamper, H.A. et al. (2010). Positive effects of organic farming on below-ground mutualists: large-scale comparison of mycorrhizal fungal communities in agricultural soils. *New Phytologist* 186 (4): 968–979. https://doi.org/10.1111/j.1469-8137.2010.03230.x.

Viaene, J., Van Lancker, J., Vandecasteele, B. et al. (2016). Opportunities and barriers to on-farm composting and compost application: a case study from northwestern Europe. *Waste Management* 48: 181–192. https://doi.org/10.1016/j.wasman.2015.09.021.

Vujinović, T., Zanin, L., Venuti, S. et al. (2020). Biostimulant action of dissolved humic substances from a conventionally and an organically managed soil on nitrate acquisition in maize plants. *Frontiers in Plant Science* 10. https://doi.org/10.3389/fpls.2019.01652.

Walker, C., Vestberg, M., Demircik, F. et al. (2007). Molecular phylogeny and new taxa in the Archaeosporales (Glomeromycota): Ambispora fennica gen. Sp. nov., Ambisporaceae fam. Nov., and emendation of Archaeospora and Archaeosporaceae. *Mycological Research* 111 (2): 137–153. https://doi.org/10.1016/j.mycres.2006.11.008.

Wang, L., Ji, B., Hu, Y. et al. (2017a). A review on in situ phytoremediation of mine tailings. *Chemosphere* 184: 594–600. https://doi.org/10.1016/j.chemosphere.2017.06.025.

Wang, Y., Li, C., Tu, C. et al. (2017b). Long-term no-tillage and organic input management enhanced the diversity and stability of soil microbial community. *Science of the Total Environment* 609: 341–347. https://doi.org/10.1016/j.scitotenv.2017.07.053.

Watson, C.A., Bengtsson, H., Ebbesvik, M. et al. (2002). A review of farm-scale nutrient budgets for organic farms as a tool for management of soil fertility. *Soil Use and Management* 18 (3): 264–273. https://doi.org/10.1079/SUM2002127.

Watson, C.A., Atkinson, D., Gosling, P. et al. (2006). Managing soil fertility in organic farming systems. *Soil Use and Management* 18: 239–247. https://doi.org/10.1111/j.1475-2743.2002.tb00265.x.

Weil, R. and Magdoff, F. (2004). Significance of soil organic matter to soil quality and health. In: *Soil Organic Matter in Sustainable Agriculture* (ed. F. Magdoff and R. Weil), 1–43. Boca Raton: CRC Press Inc. https://doi.org/10.1201/9780203496374.ch1.

Weyens, N., van der Lelie, D., Taghavi, S. et al. (2009a). Exploiting plant–microbe partnerships to improve biomass production and remediation. *Trends in Biotechnology* 27 (10): 591–598. https://doi.org/10.1016/j.tibtech.2009.07.006.

Weyens, N., van der Lelie, D., Taghavi, S., and Vangronsveld, J. (2009b). Phytoremediation: plant–endophyte partnerships take the challenge. *Current Opinion in Biotechnology* 20 (2): 248–254. https://doi.org/10.1016/j.copbio.2009.02.012.

Willer, H., Moeskops, B., Busacca, E., and De La Vega, N. (2019). Organic in Europe: recent developments. In: *The World of Organic Agriculture: Statistics and Emerging Trends 2019* (ed. H. Willer and J. Lernoud), 208–216. Research Institut of Organic Agriculture FiBL, IFOAM – Organics International.

Yang, G., Roy, J., Veresoglou, S.D., and Rillig, M.C. (2021). Soil biodiversity enhances the persistence of legumes under climate change. *New Phytologist* 229 (5): 2945–2956. https://doi.org/10.1111/nph.17065.

Zafar, M., Abbasi, M.K., Khan, M.A. et al. (2012). Effect of plant growth-promoting rhizobacteria on growth, nodulation and nutrient accumulation of lentil under controlled conditions. *Pedosphere* 22 (6): 848–859. https://doi.org/10.1016/S1002-0160(12)60071-X.

Zamboni, A., Zanin, L., Tomasi, N. et al. (2016). Early transcriptomic response to Fe supply in Fe-deficient tomato plants is strongly influenced by the nature of the chelating agent. *BMC Genomics* 17 (1): 35. https://doi.org/10.1186/s12864-015-2331-5.

Zandonadi, D.B., Santos, M.P., Caixeta, L.S. et al. (2016). Plant proton pumps as markers of biostimulant action. *Scientia Agricola* 73 (1): 24–28. https://doi.org/10.1590/0103-9016-2015-0076.

Zanin, L., Tomasi, N., Rizzardo, C. et al. (2015). Iron allocation in leaves of Fe-deficient cucumber plants fed with natural Fe complexes. *Physiologia Plantarum* 154 (1): 82–94. https://doi.org/10.1111/ppl.12296.

Zanin, L., Tomasi, N., Zamboni, A. et al. (2018). Water-extractable humic substances speed up transcriptional response of maize roots to nitrate. *Environmental and Experimental Botany* 147: 167–178. https://doi.org/10.1016/j.envexpbot.2017.12.014.

Zanin, L., Tomasi, N., Cesco, S. et al. (2019). Humic substances contribute to plant iron nutrition acting as chelators and biostimulants. *Frontiers in Plant Science* 10. https://doi.org/10.3389/fpls.2019.00675.

Zhang, T., Hu, Y., Zhang, K. et al. (2018). Arbuscular mycorrhizal fungi improve plant growth of Ricinus communis by altering photosynthetic properties and increasing pigments under drought and salt stress. *Industrial Crops and Products* 117 (5268): 13–19. https://doi.org/10.1016/j.indcrop.2018.02.087.

Zhen, Z., Liu, H., Wang, N. et al. (2014). Effects of manure compost application on soil microbial community diversity and soil microenvironments in a temperate cropland in China. *PLoS ONE* 9 (10): e108555. https://doi.org/10.1371/journal.pone.0108555.

6

Ecological Restoration of Degraded Soils Through Protective Afforestation

Marcin Pietrzykowski, Bartłomiej Woś, and Marek Pająk

Department of Ecological Engineering and Forest Hydrology, Faculty of Forestry, University of Agriculture in Krakow, al. 29 Listopada 46, 31–425, Krakow, Poland

6.1 Introduction

Moving away from fossil fuels and the implementation of the concept of a wide-scale European program, the so-called Green Deal, which will be done consistently for climatic and technological reasons as well as due to the inevitable depletion of hydrocarbon deposits. However, this does not change the fact that the problem of reclamation of post-mining sites will continue to exist and will remain a challenge in terms of mitigating the effects of obtaining minerals, because it also applies to raw materials that will always be necessary for the development of the industry and the economy (metal ores, raw materials for construction, chemical industry, etc.). On the one hand, post-mining sites which must be located around raw material mines, are a negative environmental and landscape consequence, and on the other, with appropriate reclamation, they may develop into the so-called novel ecosystems, fulfilling a number of ecosystem services (Macdonald et al. 2015). This is due to the fact that it is reclamation to forest and afforestation that allow us to provide these services fully. However, in order for reclamation and afforestation not to be accidental and chaotic, and in order for the ecosystem to develop dynamically in the long term, these processes need to be studied and suitable reclamation patterns must be developed (Favas et al. 2018; Pietrzykowski 2019).

Reclamation to forest and afforestation of post-mining sites are considered the most optimal way of managing and restoring environmental stability (Zipper et al. 2011; Pietrzykowski 2014). In ecological terms, reclamation to forest is defined as the process of restoring bioecological conditions where the exchange between the soil and plant cover ensures a positive course of plant succession and intensive development of soil formation (Bradshaw 1983; Pietrzykowski and Krzaklewski 2018). Mine soils which as a result of mining operations consist of freshly uncovered substrates from various geological epochs typically display unfavorable physico-chemical properties, insufficiency of soil organic matter, and soil microorganisms. As mentioned before, the performed reclamation

Agroecological Approaches for Sustainable Soil Management, First Edition.
Edited by Majeti Narasimha Vara Prasad and Chitranjan Kumar.

measures significantly affect the soil formation process and spontaneous succession. Intensive agrotechnical cultivation and fertilization change habitat conditions for invading plants, while afforestation is not used in the first stages of succession and is intended to accelerate the habitat-forming process and transform the "raw" substrates of post-mining sites into soil (Pietrzykowski 2014). The introduced tree species forming compact stands are a factor which modify forest habitats in interaction with the soil and climate. The mechanism of the habitat-forming effect of stands is expressed through crown cover and the regulation of access of light and microclimatic relations, the biochemical impact of organic canopy litter, the activity of the roots and biochemically active secretions. These factors modify habitat conditions in various ways, which in turn leads to the development of soils with different properties (Hagen-Thorn et al. 2004).

This chapter presents an overview of the issues related to afforestation as a way to protect post-mining sites, with particular emphasis on soil reconstruction and the selection of species composition for afforestation in the reclaimed post-mining sites based on the assessment of tree adaptation taking into account the criteria of: (i) growth features, (ii) nutrient supply, (iii) impact on the properties of soil formation.

6.2 The Importance of Reclamation for the Protection of Post-Mining Sites

Reclamation of degraded sites to terrestrial ecosystems, according to most international legislation includes measures aimed at restoring or giving land and sites use value by carrying out the necessary treatments such as: improving the physical properties of the soil; regulation of water relations; soil restoration by technical or biological methods; enforcement of slopes and road construction. In a broader sense, reclamation should include not only restoring the use value of degraded land through a series of treatments, but also of entire ecosystems and landscapes (Hüttl and Weber 2001; Pietrzykowski and Krzaklewski 2018). The ecological definition of reclamation (Bradshaw 1983; Pietrzykowski and Krzaklewski 2018) describes it as the process of restoring the ecosystem and its functions. To conclude it may be said that reclamation is considered completed when a self-sufficient system is formed, i.e., a bioecological system with circulation of matter and energy, i.e., the ecosystem.

Afforestation is potentially the best sustainable strategy for reclaiming mine lands to their former state (Pietrzykowski and Krzaklewski 2007). In temperate climate zones of the northern hemisphere, the forest is a climax biome, the most organized, and dominant among terrestrial ecosystems. For this reason, a large part of post-mining landscapes are reclaimed to forest for instance in Europe and in North America (Hüttl and Weber 2001). When expanding on the above-mentioned ecological definition of reclamation in relation to forest ecosystems, it may be assumed that the reclamation procedure and the assessment of its effects depend on the objectives (Figure 6.1). The greater the expenditure of energy, labor, technical means and agrotechnical treatments, the greater the effects in the form of growth of stands and biomass (Pietrzykowski and Socha 2011). This may be achieved by developing the habitat from homogeneous, potentially most productive substrates when introducing monocultures. Its disadvantages include low biodiversity and as well as possible susceptibility to fungal and insect outbreaks.

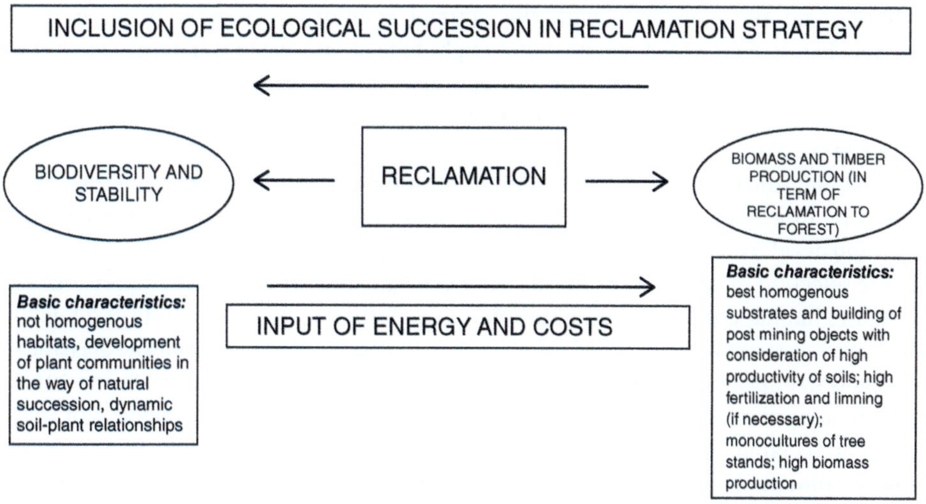

Figure 6.1 Development of an ecological concept of the reclamation strategy depends on the aim and approaches. *Source:* Originally by M. Pietrzykowski.

On the other hand, by incorporating the process of natural succession in the developing ecosystems and using varied geological formations of the overburden, which are frequently dumped with no selection, we get greater biodiversity due to a mosaic of various geological formations, but also smaller tree growth and biomass (Photo 6.1). Thus, the course of action in specific sites should be the result of careful analysis of natural and economic

Photo 6.1 Succession communities as the important supplement to afforestation on hard coal mine spoil heap. *Source:* Upper Silesia, Poland; photo by M. Pietrzykowski.

conditions and the objectives, with the optimal solution somewhere between these extremes (Figure 6.1), which means that it ought to take into account natural succession, avoid introducing monocultures, but also optimize biomass production (an appropriate form of mixing species, a sufficient amount of neutralizers or mineral fertilizers; Pietrzykowski 2015).

Correct identification of habitat conditions at mining sites and planning species composition in afforestation both have a fundamental impact on the stability of reclaimed stands and developed forest ecosystems (Mukhopadhyay et al. 2014; Pietrzykowski 2014). Suitable selection of tree species means their ecological requirements are appropriately adjusted to habitat conditions (climate conditions, fertility, and soil humidity). Thus, in recent years, interest has grown regarding soil reconstruction methods and site classification versus tree species selection (Kuznetsova et al. 2010; Pietrzykowski and Socha 2011). Understanding the rate of forest stand development and assessing the response of trees growing in reclaimed areas to environmental stress may also be useful in the context of the development of forest ecosystems in the event of extreme weather conditions and natural large-scale disturbances which intensify along with climate change (Malmsheimer et al. 2008).

6.3 Soil Reconstruction in Varied Post-Mine Site Conditions

The chief aim of reclamation is to restore soil as the basic element of the terrestrial ecosystem (Hüttl and Bradshaw 2000). The methods of soil restoration may be divided into technical ones, where the full profile of the reconstructed soil is restored with the use of organic and organic-mineral horizons previously removed from mining sites (Photo 6.2) and into

Photo 6.2 Spreading forest humus on a former sand mine cast to improve its properties. *Source:* Photo by M. Pietrzykowski.

biological ones with the use of cariogenic plants. The technology of restoring a complete soil profile is limited by the availability of overburden with suitable soil-forming properties. An example is reclamation in the Rhine River Basin in Germany, where, for the purposes of forest reclamation, loess is mixed with sand and gravel in a 1 : 3 ratio, creating an excellent substrate for afforestation, the so-called forest gravel (*Forstkies*). In order to increase the diversity of the developed biotopes, small fragments of even the worst, acidic Miocene substrates are left on the spoil heaps (Topp et al. 2001).

When there is no good quality soil in the overburden, as is the case with extremely poor sandy substrates on sand mine casts, the techniques of soil restoration in the biological reclamation phase make use of green manuring, i.e., the cultivation of atmospheric nitrogen-fixing cariogenic plants. In the reclamation of former sand mine casts in Upper Silesia in southern Poland, before afforestation, yellow lupine (*Lupinus luteus L.*) is grown, the biomass of which is plowed in in a two-year cycle of cultivation combined with N-P-K mineral fertilization (Pietrzykowski et al. 2017). Additionally, after the afforestation of the former sand mine casts, garden lupin (*Lupinus polyphyllus L.*) is planted between the rows of trees, which enriches the soil with nitrogen (Pietrzykowski and Krzaklewski 2009).

Special treatment with neutralizers is required for deposits in areas made of toxic Miocene acidic sandy deposits which occur with lignite deposits (Katzur and Haubold-Rosar 1996). Forms which naturally occur with the mined mineral such as Neogene Lake lime may be used as neutralizers in the case of acidic formations (Pietrzykowski 2014). In the Lower Lusatian Basin, alkaline ash from lignite combustion from nearby power plants was used as a neutralizer (Katzur and Haubold-Rosar 1996). Regardless of the above-mentioned conditions, mine soils typically lack soil organic matter, display poor air-water relations, compaction, extreme acidification or alkalization, and frequently also excessive salinity (Brevik and Lazari 2014; Pietrzykowski 2014). These soils typically show a deficiency of biogenic elements or disturbances in their quantitative relations. In some regions, mine soils are reported to have an excessive concentration of heavy metals and sometimes also of sulfur. A characteristic feature is significant spatial and profile variability of the aforementioned soil properties (Pietrzykowski et al. 2014). The listed features of reclaimed mine soils make the emerging habitat conditions for the introduced forest trees difficult and significantly different from natural forest habitats (Photo 6.3). Poor soil properties result in low supply of tree nutrients (Heinsdorf 1996; Pietrzykowski et al. 2013). This, in turn, impacts the growth, health and stability of the introduced stands (Katzur and Haubold-Rosar 1996; Pietrzykowski et al. 2013).

An unfavorable factor in the case of RMS (reclaimed mine soils) are potential rapid changes in their properties, especially acidification (pH), which takes place, for example, in areas made up of carbonized Carboniferous deposits stored on coal spoil heaps (Aldag and Strzyszcz 1980) and in Neogene formations of lignite mine spoil heaps. In these deposits, there is strong acidification as a result of weathering of pyrites and marcasites (Katzur and Haubold-Rosar 1996). In such conditions, afforestation planning should take into account risk reduction through diversified species composition, including the introduction of pioneering species which may later be replaced with climax species (Pietrzykowski 2014). However, the afforestation strategy and the question of the role and timing of introducing

Photo 6.3 Open pit lignite mining and different sediments (mix of Quaternary and Neogene rocks) as the substrate for reclaimed mine soil and site conditions for introduced forest tree species. *Source:* Lignite Mining Bełchatów, Central Poland; photo by M. Pietrzykowski.

early-succession and climax species in reclaimed post-industrial sites remain the subject of debate (Zipper et al. 2011; Frouz et al. 2015).

6.4 Criteria for Assessing the Adaptation of Tree Species to the Conditions of Reclaimed Areas

The basic element of the restored forest ecosystems, apart from to the soil, are tree stands, which transform the biotope most dynamically, creating a specific microclimate over time and changing the properties of the developing soil (Photo 6.4). The assessment of the growth and productivity of stands is one of the measures of the response and adaptation of forest tree species to the habitat conditions of reclaimed sites. The assessment of the results of forest reclamation based on the biomass of communities and the growth of stands is important both in the ecological dimension, for the assessment of the potential and current productivity of ecosystems, as well as economically, i.e., for example, production timber and biomass used as biofuel (Rodrigue et al. 2002). It is also important to estimate the potential of carbon sequestration in the biomass of new ecosystems in order to minimize the greenhouse effect and the impact of mining for minerals (Shrestha and Lal 2006; Pietrzykowski and Daniels 2014). Knowing the biomass condition also makes it possible to assess the energy balance of reclamation as one of the ecological and economic criteria for measuring its effects (Pietrzykowski and Krzaklewski 2007). It is possible to quite accurately determine

Photo 6.4 Afforestation of hard coal spoil heap with different tree species – a first phase of novel forest ecosystem development on extremely hard site conditions on carboniferous spoiled rocks with topsoiling. *Source:* Upper Silesia, Poland; photo by M. Pietrzykowski.

the state of tree stand biomass on the basis of dendrometric measurements and appropriate empirical formulas. For example, in the study of Pietrzykowski and Socha (2011), an analysis of biomass productivity of "new pine forests" in various post-mining sites was presented. The authors emphasized significant variability in the productivity of mine soil ecosystems, which depends mainly on the abundance of nutrients in soil substrates. The lowest productivity of stands was recorded on sandy formations of sand mine casts (the average annual increase in biomass was 2.78 $Mg\,ha^{-1}yr^{-1}$), while the highest which was similar to the control plots in natural habitats (4.34 $Mg\,ha^{-1}yr^{-1}$) was reported on Quaternary sands mixed with Neogene clays on a spoil heap of a former opencast sulfur mine.

The tree stands introduced on post-mining sites may have similar growth parameters compared to natural, undisturbed forest ecosystems (e.g., Vares et al. 2004; Kuznetsova et al. 2010; Table 6.1). On the other hand, studies conducted in the Czech Republic indicate that pine stands in areas reclaimed after lignite and sand mining show stand volume which is lower by 22% and biomass production by 19% (Vacek et al. 2021a). In Poland, it was indicated that the Scots pine shows good growth parameters on various soil substrates (Table 6.1; Pietrzykowski and Socha 2011). The achieved growth parameters confirm a wide ecological amplitude and high adaptation potential to the new habitats in the post-mining areas of this tree species. The growth of the European larch strongly depends on the grain size of the soil. This species grows well on soils with grainy clay sands and sandy loams (Krzaklewski et al. 2011). Similarly, the common birch grows well in soils with grainy loamy sands and sandy loams, while it grows rather poorly in soils with grainy loose

Table 6.1 Growth parameters of tree species in post-mining sites of Central Europe.

Tree species	Tree stand age (years)	Mine soil substrate and sediments	Type of mining extraction	Stand density (pcs. ha^{-1})	Mean DBH [cm]	Mean height (m)	Volume (m^3 ha^{-1})	References
Scots pine (*Pinus sylvestris*)	12	Acid Miocene sands	Lignite, open pit mining	6375	2.9	3.3	–	Pietrzykowski et al. 2015
	17	Quaternary loamy sands, loam, boulder clay	Lignite, open pit mining	7500	5.6	5.7	–	Pietrzykowski et al. 2015
	22	Clays	Lignite, open pit mining	4193	8.6	5.2	–	Frouz et al. 2009
	22–23	Oil shales	Oil shales, open pit mining	2000	10.5	8.7	73.0	Kuznetsova et al. 2010
	28	Quaternary sands	Sand, open pit mining	2800	11.3	11.8	160.3	Woś 2015
	27	Quaternary loamy sands and sandy loams	Sand, open pit mining	2067	13.6	14.3	193.1	Woś and Pietrzykowski 2019
	30	Carboniferous deposits, mudstones, sandstones and carbonaceous shales	Hard coal, underground mining	1800	13.8	12.3	–	Pietrzykowski et al. 2015
	40	Quaternary sands	Sulfur, open pit mining	1000	18.1	16.2	173.7	Woś 2015
	40	Mixture of Quaternary sands and Neogene clays	Sulfur, open pit mining	733	22.3	20.9	233.7	Woś 2015
	40	Sand	Sand, open pit mining	2822	14.6	16.9	335.0	Vacek et al. 2021a
	48	Clays	Lignite, open pit mining	2034	18.5	17.5	364.0	Vacek et al. 2021b
European larch (*Larix decidua*)	25	Quaternary loamy sands and sandy loams	Sand, open pit mining	1450	17.7	17.2	297.3	Woś and Pietrzykowski 2019
	30	Clays	Lignite, open pit mining	2162	14.5	15.4	–	Frouz et al. 2009
	40	Mixture of Quaternary sands and Neogene clays	Sulfur, open pit mining	525	27.4	20.1	292.7	Woś 2015
	48	Clays	Lignite, open pit mining	1969	14.8	16.1	200.0	Vacek et al. 2021b

Norway Spruce (*Picea abies*)	27	Clays	Lignite, open pit mining	1050	14.5	8.9	–	Frouz et al. 2009
	48	Clays	Lignite, open pit mining	2034	14.3	13.7	191.0	Vacek et al. 2021b
Common birch (*Betula pendula*)	28	Quaternary sands	Sand, open pit mining	1070	8.6	10.7	33.2	Woś 2015
	25	Quaternary loamy sands and sandy loams	Sand, open pit mining	1480	13.3	15.2	146.6	Woś and Pietrzykowski 2019
	40	Mixture of Quaternary sands and Neogene clays	Sulfur, open pit mining	600	20.5	20.0	178.0	Woś 2015
Black alder (*Alnus glutinosa*)	17	Quaternary loamy sands and sandy loams	Sand, open pit mining	3850	6.0	7.0	–	Chodak et al. 2019
	21	Oil shales	Oil shales, open pit mining	2300	12.4	13.4	189.0	Vares et al. 2004
	28	Clays	Lignite, open pit mining	2000	9.5	9.7	–	Frouz et al. 2009
	33	Quaternary loamy sands and sandy loams	Sand, open pit mining	1457	13.5	10.9	108.0	Woś and Pietrzykowski 2019
	40	Neogene clays	Sulfur, open pit mining	695	18.6	17.2	140.2	Woś 2015
European Oak (*Quercus robur*)	28	Clays	Lignite, open pit mining	1881	7.9	6.9	–	Frouz et al. 2009
	40	Mixture of Quaternary sands and Neogene clays	Sulfur, open pit mining	791	17.1	16.8	157.4	Woś 2015
	40	Neogene clays	Sulfur, open pit mining	688	16.9	15.9	104.3	Woś 2015

sands (Woś 2015; Table 6.1). The European oak is a climax species with high habitat requirements which can be introduced only in the most fertile parts of post-mining sites. Introduced on a spoil heap of an open-cast sulfur mine and a spoil heap of an open-cast lignite mine into spots with dominant clay, it displayed good growth parameters (Frouz et al. 2009; Pietrzykowski et al. 2015; Table 6.1).

Despite the fact that in Estonia it was shown that the black alder stands may show similar parameters as in natural habitats (Vares et al. 2004), most studies indicate poor growth parameters of this tree species (Woś 2015). Observations from reclaimed post-mining areas show spontaneous retreat of the alder (after about 30–40 years) from the introduced stands. This is due both to the habitat conditions that are far from its ecological optimum and the failure to compete with the fast-growing target tree species. Poorer growth of the alder and its retreat is not the most important indicator for assessing the suitability of this species for afforestation of reclaimed sites. The most important element is the forecrop and phytomelioration functions of the alder, i.e., the production of a large amount of easily decomposable organic matter with a high content of nutrients, the ability to assimilate atmospheric nitrogen by symbiotic bacteria of the genus *Frankia*, and thus a positive effect on the growth of the target species (Pietrzykowski 2019).

Compared to other terrestrial ecosystems, in forest ecosystems with considerable aboveground biomass, there is also significant underground biomass which plays an important role in the circulation of matter and energy (Wardle et al. 2004). Research on tree root systems, including biomass broken down into fractions of coarse and fine roots, assessment of their viability, range and density, as well as distribution in the soil profile are important for assessing the condition of trees and their adaptability to various habitat conditions (Nielsen and Hansen 2006). Recognizing the share and role of roots in the biomass of terrestrial ecosystems is also important in the context of assessing the ability to sequester CO_2 by new forest ecosystems (Rodrigue et al. 2002; Pietrzykowski and Daniels 2014). While the aboveground biomass of phytocoenoses and the productivity of ecosystems are well understood, determining the underground biomass of trees and stands is much more difficult from the methodological point of view and very labour-intensive (Pietrzykowski et al. 2010).

For these reasons, there are still very few studies devoted to the development of roots in post-mining habitats. Based on research conducted on a reclaimed sand mine cast in Upper Silesia, Poland, Pietrzykowski (2008) showed an increase in the vertical range of roots in chronological sequence from 5 to 25 years in reclaimed sites and in sites with natural succession. Research by Pietrzykowski et al. (2010) showed deformations and alterations of the Scots pine root systems in post-mining areas, including in a reclaimed spoil heap of a lignite opencast mine called KWB Bełchatów and in a backfilling sand mine cast called Szczakowa. The cited study indicated that the deformations of the pine tap root in the spoil heaps were due to zones of extreme soil compaction. On the other hand, the root systems of pines growing in a sand mine cast in zones with shallow groundwater showed alterations consisting of the disappearance of the main tap root and an increased share of fine roots in the form of bunches in the area of the groundwater table. Moreover, under these conditions, lateral roots which provided main support for tree stability were better developed (Pietrzykowski et al. 2010).

The reaction and adaptive strategy of tree species to reclaimed post-mine sites must take into account nutrient bio-geochemical cycles and tree-stand nutrient supply and stability

(Baumann et al. 2006; Pietrzykowski et al. 2013). The basic criterion for assessing the nutritional status of trees is the content of individual elements in the foliage (Pietrzykowski et al. 2013). Despite the importance of this subject, there are still no studies on the adaptive characteristics of woody species, including the assessment of nutrition strategies in post-mine habitats. Data on the Scots pine, one of the main species used in the afforestation of post-industrial sites in Central Europe, was published by Heinsdorf (1996), Baumann et al. (2006), Pietrzykowski et al. (2013), among others. The cited studies clearly show that the given thresholds for the content of macroelements in the needles of the Scots pine growing in natural habitats should be complemented with data from post-mine sites. The reported macronutrient contents in these areas was often lower than deficit values, however the pine did not show any signs of growth inhibition and needle chlorosis. Thus, this confirms significant adaptability of the species to extreme habitat conditions in post-mining sites. It was shown that nitrogen and phosphorus were the most scarce elements in new forests as compared to the thresholds provided in the literature on the subject (Pietrzykowski et al. 2013). It would seem that the deficiencies in the supply of nitrogen would be alleviated by mineral fertilization. However, this treatment in reclaimed post-mining sites should be considered only in cases of the greatest deficits, and individual single amounts must not be too big due to the risk of leaching and nitrogen losses and the possibility of disturbing quantitative relations of macroelements in the initial soils. In this context, the most important objective should be for the emerging ecosystem to achieve efficient circulation of nutrients and stability of tree stands which are the main modifying elements of forest habitats (Knoche et al. 2002). Currently, it is even believed that attempts to achieve maximum increase in the biomass of stands through "luxurious" tree nutrition in reclaimed sites is unfavorable (Knoche et al. 2002; Pietrzykowski et al. 2013). Following a period of fast tree growth and biomass increase in the juvenile stages, a decrease in the increment in further development stages may occur. Therefore, there is a need to monitor the emerging ecosystems.

6.5 The Impact of Tree Species on Soil Properties

The soil properties of developing forest ecosystems on degraded land depend on the climate, topography, time, vegetation (mainly tree species in forest ecosystems), and parent material, with feedback occurring between these factors (Frouz et al. 2008; Kooch et al. 2018). The main factor which determines the impact of tree species on soil properties is varying accumulation of organic litter and the rate of its decomposition (Hagen-Thorn et al. 2004). The substances leached from canopy litter and the organic horizon are a factor influencing the course of the soil-development process by transporting humus compounds, nutrients, and elements within the soil profile (Kalbitz et al. 2000). Another factor determining the impact of litter on the soil formation process is the release of organic acids during its decomposition, which lower soil pH, stimulate the processes of weathering of primary minerals, and increase the content of aluminum and iron in soil solutions (Binkley and Giardina 1998).

In addition to canopy litter, roots, and particularly fine roots ($d < 2$ mm) play an important role in the soil formation process (Rosenvald et al. 2011; Świątek and Pietrzykowski 2021).

Though fine roots constitute only about 10% of the underground biomass of trees (Miller et al. 2006), due to natural dying and decomposition processes, they can sometimes supply the soil with several-fold more nutrients and organic carbon than the litter from the biomass of the aboveground parts. Partly as a result of the secretion of their own substances, and partly due to decomposition of dead cells, fine roots form a microhabitat (rhizosphere) around them. In this zone, the number and activity of microorganisms is greater than in the soil further away from the roots (Rosenvald et al. 2011). In addition, the activity of the roots may significantly affect the physicochemical properties of the developing soil, for instance the activity of the alder roots and symbiotic bacteria of the genus *Frankia* sp. not only leads to an increase in nitrogen content, but also in soil acidity (Woś et al. 2022). An increase in nutrient bioavailability may occur as a result of intensification of the biological process of parent rock weathering due to various types of root secretions (Binkley and Giardina 1998).

Another mechanism of tree stand impact on the chemical properties of soils is the effect of crown cover and features of tree crowns. The nature of forest tree crowns, which depends, among others, on their spatial distribution, the size, shape, and thickness of leaves, modifies the hydrological factors of the habitat through the ability to intercept atmospheric precipitation. These changes reduce nutrient leaching to the deeper horizons of parent rock in the soil (Prescott 2002). The amount of rainfall and chemicals dissolved in it going through tree canopy and then reaching the soil varies depending on the species composition of the stand (Schrijver et al. 2007). Based on a literature review, Schrijver et al. (2007) found that soils under coniferous stands (consisting of: *Picea abies*, *Pinus sp.*, *Abies balsamea*, and *Pseudotsuga menzesii*) get a bigger amount of N and S from atmospheric precipitation and more leaching of NO_3^-, SO_4^{2-}, K^+, Ca^{2+}, Mg^{2+}, and Al^{3+} into the profile compared to soils under deciduous stands (consisting of: *Fagus sylvatica*, *Betula pendula*, *Quercus sp.*, *Acer sp.*, and others). This phenomenon may cause greater susceptibility of soils under coniferous stands to acidification and eutrophication due to atmospheric precipitation compared to soils under deciduous stands.

On degraded sites, the introduced tree species have features which may alter ecosystem characteristics (e.g., resource turnover, nutrient distribution, and C sequestration), thereby changing the properties of the developing soil (Frouz et al. 2009). According to a meta-analysis, Laganiere et al. (2010) suggested that deciduous tree species have a greater potential for SOM accumulation than coniferous species. The average increase in soil C, two to three decades after the afforestation of agricultural land with deciduous tree species, was 25% compared to 2% with conifers (Laganiere et al. 2010). However, some studies have shown that soils under coniferous species display higher total (in organic + mineral horizons) C_{stock} than soils under deciduous tree species (Oostra et al. 2006; Gurmesa et al. 2013). Studies have also indicated that N-fixing species change soil properties more rapidly than non-N-fixing species, with consistent increases in the C_{stock} and N_{stock} (Mayer et al. 2020). On other hand, a number of studies have found that tree species had no impact on the total C_{stock} in soil but differed in the distribution of C between the organic and mineral horizons (Vesterdal et al. 2013; Gruba and Socha 2019). Tree species producing higher C_{stock} in the organic horizons of their soils have been found to have lower C_{stock} in their mineral horizons (Vesterdal et al. 2013). In general, coniferous species tend to accumulate more SOM in the forest floor, but less in the mineral horizon, compared with deciduous trees (Jandl et al. 2007). The impact of tree species on soil properties determined, for instance in natural, undisturbed soils cannot always be directly translated into conditions of disturbed soils. For example, a study by Hüblová and Frouz (2021) showed that deciduous tree species sequester more C in alkaline, higher-clay-content post-mining soils than conifers do. Conversely, conifers sequester more C in mature, undisturbed forest soils. Greater difference in C stock is reported to occur in young soils such as mine soils at an early stage of development (Hüblová and Frouz 2021; Woś et al. 2022). In their study, Frouz et al. (2009) analyzed the impact of the stands of the Norway spruce, Scots pine, European larch, European oak, black alder (with an admixture of gray alder), and small-leaved lime on the content and accumulation of carbon in soil organic matter on a spoil heap of an opencast lignite mine in the Sokolov Basin (The Czech Republic). The presented research showed that the content of carbon in soil organic matter ranged from 4.5 to 38.0 $Mg\,ha^{-1}$, while its accumulation was from 0.15 to 1.28 $Mg\,ha^{-1}$ $year^{-1}$ for soils under the stands (in rising order): spruce < pine, oak < larch < alder < linden.

The Scots pine (*Pinus sylvestris*) and the common birch (*B. pendula*) are commonly used for the afforestation of degraded lands because of their low habitat requirements (Photo 6.5). Compared to the pine, the birch is recognized as a soil improver, as it increases soil pH and base saturation, positively affects the content and availability of nutrients (Jonczak et al. 2020) and improves the microbial properties of soils (Chodak and Niklińska 2010a). N-fixing tree species, such as the black alder (*Alnus glutinosa*), are planted in disturbed sites (Pietrzykowski 2019). These species not only boost the amount of

Photo 6.5 Extremely poor sites on sand mine excavation and afforestation of mixed pioneer tree species (Scots pine, common birch and black alder) as the key to dynamic and biodiverse new ecosystems. *Source:* Szczakowa sand pit, Upper Silesia, Poland; photo by M. Pietrzykowski.

soil N but also stimulate C accumulation in afforested soils (Chodak et al. 2019; Woś and Pietrzykowski 2019).

In natural forest ecosystems, the biomass and the activity of soil microorganisms and macrofauna depend on the properties of the soil and the impact of the dominant tree species in the stand (Priha et al. 2001). Similar correlations occur in the reclaimed mine soils (Šourková et al. 2005; Chodak and Niklińska 2010a; Chodak and Niklińska 2010b). The question remains to what extent the type of parent rock compared to the introduced tree species impact the biological properties of mine soils, and what role in altering these properties is played by individual tree species used in afforestation from pioneering, climax, and phytomelioration species. Šourková et al. (2005) showed that the activity of microbial biomass in mine soils is more influenced by the tree species and litter quality than by the type of soil substrate. They reported higher microbial biomass in soils under deciduous species such as the black and gray alder and the red oak compared to soils under the Scots pine. On the other hand, Chodak and Niklińska (2010a) showed greater impact of the type of soil substrate on the microbiological properties measured with the biomass of microorganisms, soil respiration index and soil enzyme activity than the tree species introduced as part of reclamation in the spoil heap of KWB Bełchatów. However, this study only included pioneering species (the pine, birch, and mixed pine-birch stands). This study also indicated that the soils under the pine trees displayed lower biomass and microorganism activity compared to the soils under the birch and mixed pine-birch stands. In other studies, carried out in Szczakowa backfill sand mine cast, the authors (Chodak and Niklińska 2010b) reported a positive effect of deciduous species (the birch, alder) on the development and activity of soil microorganisms compared to coniferous species (the larch, pine). According to the authors, the black alder was the species with the biggest impact on the chemical properties of soils. The soils under the alder had the highest carbon and nitrogen content and the lowest pH compared to the other species (Chodak and Niklińska 2010b).

6.6 Conclusion

Afforestation is the best method of reclamation and restoration of ecosystems in degraded post-mining sites as it alters the microclimate, soil protection functions, and water retention features; it is also important for its landscaping properties and a number of other ecosystem services. The main element in the success of afforestation, however, is correct selection of the species composition for the newly emerging habitat conditions, incomparable to natural ones. Tree stands are the main layer of recreated phytocoenoses; under their canopy and the impact of canopy litter and a number of complicated processes taking place belowground, natural succession of other species takes place as well as the formation of novel forest ecosystems. Based on previous research on the usefulness and growth of tree species in afforested post-mining sites on the basis of a case study of the temperate zone of Central Europe, a satisfactory increase in above-ground biomass of the introduced afforestation was found, in some cases even exceeding the values obtained in the compared natural forest habitats in the vicinity of post-industrial facilities. However, there is a noticeable problem of deformation of the root systems in relation to the typical morphology of the roots of a given species in natural conditions, which may in the future result in disturbances in the stability of stands. An important issue is also the quantitative correlation of nutrients in soils and their bioavailability as well as efficient circulation in the emerging ecosystems. This applies in particular to the deficiency of nitrogen and phosphorus and the mutual quantitative correlations of these elements. As a consequence, problems may occur with suitable nutrition of stands and disturbance of their stability in reclaimed post-industrial sites. A complex assessment of the suitability for reclamation of tree species should take into account not only the assessment of growth characteristics and biomass growth, but also their varying impact on the properties of the developing soil.

Acknowledgments

The study was financed by The National Science Centre, Poland, Grant No. 2018/31/D/ST10/02137.

References

Aldag, R.W. and Strzyszcz, Z. (1980). Inorganic and organic nitrogen compounds in carboniferous phyllosilicates on spoils with regard to forest reclamation. *Reclam. Rev.* 3: 69–73.

Baumann, K., Rumpelt, A., Schneider, B.U. et al. (2006). Seedling biomass and element content of *Pinus sylvestris* and *Pinus nigra* grown in sandy substrates with lignite. *Geoderma* 136: 573–578. https://doi.org/10.1016/j.geoderma.2006.04.013.

Binkley, D. and Gardina, C. (1998). Why do tree species affect soils? The warp and woof of tree–soil interactions. *Biogeochemistry* 42: 89–106. https://doi.org/10.1023/A:1005948126251.

Bradshaw, A.D. (1983). The reconstruction of ecosystems. *J. Appl. Ecol.* 20: 1–17.

Brevik, E.C. and Lazari, A.G. (2014). Rates of pedogenesis in reclaimed lands as compared to rates of natural pedogenesis. *Soil Horizons* 55: 1–6. https://doi.org/10.2136/sh13-06-0017.

Chodak, M. and Niklińska, M. (2010a). Effect of texture and tree species on microbial properties of mine soils. *Appl. Soil Ecol.* 46 (2): 268–275. https://doi.org/10.1016/j.apsoil.2010.08.002.

Chodak, M. and Niklińska, M. (2010b). The effect of different tree species on chemical and microbial properties of reclaimed mine soils. *Biol. Fertil. Soils* 46 (6): 555–566. https://doi.org/10.1007/s00374-010-0462-z.

Chodak, M., Sroka, K., Woś, B., and Pietrzykowski, M. (2019). Effect of green alder (*Alnus viridis*) and black alder (*Alnus glutinosa*) on chemical and microbial properties of sandy mine soils. *Geoderma* 356: 113924. https://doi.org/10.1016/j.geoderma.2019.113924.

Favas, P.J.C., Martino, L.E., and Prasad, M.N.V. (2018). Abandoned mine land reclamation – challenges and opportunities (holistic approach). In: *Bio-Geotechnologies for Mine Site Rehabilitation* (ed. M.N.V. Prasad, P. Favas, and S.K. Maiti), 3–31. Amsterdam-Oxford-Cambridge: Elsevier.

Frouz, J., Prach, K., Pižl, V. et al. (2008). Interactions between soil development, vegetation and soil fauna during spontaneous succession in post mining sites. *Eur. J. Soil Biol.* 44 (1): 109–121. https://doi.org/10.1016/j.ejsobi.2007.09.002.

Frouz, J., Pižl, V., Cienciala, E., and Kalčík, J. (2009). Carbon storage in post-mining forest soil, the role of tree biomass and soil bioturbation. *Biogeochemistry* 94: 111–121. https://doi.org/10.1007/s10533-009-9313-0.

Frouz, J., Voborilova, V., Janusova, I., and Kodochova, S. (2015). Spontaneous establishment of late successional tree species English oak (*Quercus robur*) and European beech (*Fagus sylvatica*) at reclaimed alder plantation and unreclaimed post mining sites. *Ecol. Eng.* 77: 1–8. https://doi.org/10.1016/j.ecoleng.2015.01.001.

Gruba, P. and Socha, J. (2019). Exploring the effects of dominant forest tree species, soil texture, altitude, and pH_{H2O} on soil carbon stocks using generalized additive models. *For. Ecol. Manag.* 447: 105–114. https://doi.org/10.1016/j.foreco.2019.05.061.

Gurmesa, G.A., Schmidt, I.K., Gundersen, P., and Vesterdal, L. (2013). Soil carbon accumulation and nitrogen retention traits of four tree species grown in common gardens. *For. Ecol. Manag.* 309: 47–57. https://doi.org/10.1016/j.foreco.2013.02.015.

Hagen-Thorn, A., Callesen, I., Armolaitis, K., and Nihlgård, B. (2004). The impact of six European tree species on the chemistry of mineral topsoil in forest plantations on former agricultural land. *For. Ecol. Manag.* 195 (3): 373–384. https://doi.org/10.1016/j.foreco.2004.02.036.

Heinsdorf, D. (1996). Development of forest stands in the Lusatian lignite Mining District after mineral fertilization adopted to site and tree species. *Water Air Soil Pollut.* 91: 33–42. https://doi.org/10.1007/BF00280921.

Hüblová, L. and Frouz, J. (2021). Contrasting effect of coniferous and broadleaf trees on soil carbon storage during reforestation of forest soils and afforestation of agricultural and post-mining soils. *J. Environ. Manag.* 290: 112567. https://doi.org/10.1016/j.jenvman.2021.112567.

Hüttl, R.F. and Bradshaw, A. (2000). Aspect of reclamation ecology. *Landsc. Urban Plan.* 51 (2): 73–74. https://doi.org/10.1016/S0169-2046(00)00120-1.

Hüttl, R. and Weber, E. (2001). Forest ecosystem development in post-mining landscapes: a case study of the Lusatian lignite district. *Naturwissenschaften* 88: 322–329.

Jandl, R., Lindner, M., Vesterdal, L. et al. (2007). How strongly can forest management influence soil carbon sequestration? *Geoderma* 137 (3–4): 253–268. https://doi.org/10.1016/j.geoderma.2006.09.003.

Jonczak, J., Jankiewicz, U., Kondras, M. et al. (2020). The influence of birch trees (*Betula* spp.) on soil environment – a review. *For. Ecol. Manag.* 477: 118486. https://doi.org/10.1016/j.foreco.2020.118486.

Kalbitz, K., Solinger, S., Park, J.-H. et al. (2000). Controls on the dynamics dissolved organic matter in soils: a review. *Soil Sci.* 165 (4): 277–304.

Katzur, J. and Haubold-Rosar, M. (1996). Amelioration and reforestation of sulfurous mine soils in Lusatia (eastern Germany). *Water Air Soil Pollut.* 91: 17–32. https://doi.org/10.1007/BF00280920.

Knoche, D., Embacher, A., and Katzur, J. (2002). Water and element fluxes of red oak ecosystems during stand development on post-mining sites (Lusatian Lignite District). *Water Air Soil Pollut.* 141: 219–231. https://doi.org/10.1023/A:1021350321058.

Kooch, Y., Tavakoli, M., and Akbarinia, M. (2018). Tree species could have substantial consequences on topsoil fauna: a feedback of land degradation/restoration. *Eur. J. For. Res.* 137: 793–805. https://doi.org/10.1007/s10342-018-1140-1.

Krzaklewski, W., Woś, B., Stachnik, B., and Pietrzykowski, M. (2011). Influence of some properties of initial soils on the growth of European larch (*Larix decidua*). *Sylwan* 155: 653–659.

Kuznetsova, T., Mandre, M., Klõšeiko, J., and Pärn, H. (2010). A comparison of the growth of Scots pine (*Pinus sylvestris* L.) in a reclaimed oil shale post-mining area and in a *Calluna* site in Estonia. *Environ. Monit. Assess.* 166: 257–265. https://doi.org/10.1007/s10661-009-0999-1.

Macdonald, S.E., Landhäusser, S.M., Skousen, J. et al. (2015). Forest restoration following surface mining disturbance: challenges and solutions. *New For.* 46: 703–732. https://doi.org/10.1007/s11056-015-9506-4.

Malmsheimer, R.W., Heffernan, P., Brink, S. et al. (2008). Forest management solutions for mitigating climate change in the United States. *J. For.* 106: 115–173.

Mayer, M., Prescott, C.E., Abaker, W.E.A. et al. (2020). Tamm review: influence of forest management activities on soil organic carbon stocks: a knowledge synthesis. *For. Ecol. Manag.* 466: 118127. https://doi.org/10.1016/j.foreco.2020.118127.

Miller, A.T., Allen, H.L., and Maier, C.A. (2006). Quantifying the coarse-root biomass of intensively managed loblolly pine plantations. *Can. J. For. Res.* 36: 12–22.

Mukhopadhyay, S., Maiti, S.K., and Masto, R.E. (2014). Development of mine soil quality index (MSQI) for evaluation of reclamation success: a chronosequence study. *Ecol. Eng.* 71: 10–20. https://doi.org/10.1016/j.ecoleng.2014.07.001.

Nielsen, C.C.N. and Hansen, J.K. (2006). Root CSA-root biomass prediction models in six tree species and improvement of models by inclusion of root architectural parameters. *Plant Soil* 280: 339–356. https://doi.org/10.1007/s11104-005-3503-x.

Oostra, S., Majdi, H., and Olsson, M. (2006). Impact of tree species on soil carbon stocks and soil acidity in southern Sweden. *Scand. J. For. Res.* 21: 364–371. https://doi.org/10.1080/02827580600950172.

Pietrzykowski, M. (2008). Soil and plant communities development and ecological effectiveness of reclamation on a sand mine cast. *J. For. Sci.* 54: 554–565.

Pietrzykowski, M. (2014). Soil quality index as a tool for Scots pine (*Pinus sylvestris*) monoculture conversion planning on afforested, reclaimed mine land. *J. For. Res.* 25: 63–74. https://doi.org/10.1007/s11676-013-0418-x.

Pietrzykowski, M. (2015). Reclamation and reconstruction of terrestrial ecosystems on mine sites - ecological effectiveness assessment. In: *Energy Science and Technology Series* (ed. J.N. Govil et al.), 121–151. Houston, New Delhi: Studium Press LLC.

Pietrzykowski, M. (2019). Tree species selection and reaction to mine soil reconstructed at reforested post-mine sites: central and eastern European experiences. *Ecol. Eng.* 142: 100012. https://doi.org/10.1016/j.ecoena.2019.100012.

Pietrzykowski, M. and Daniels, W.L. (2014). Estimation of carbon sequestration by pine (*Pinus sylvestris* L.) ecosystems developed on reforested post-mining sites in Poland on differing mine soil substrates. *Ecol. Eng.* 73: 209–218. https://doi.org/10.1016/j.ecoleng.2014.09.058.

Pietrzykowski, M. and Krzaklewski, W. (2007). An assessment of energy efficiency in reclamation to forest. *Ecol. Eng.* 30: 341–348. https://doi.org/10.1016/j.ecoleng.2007.04.003.

Pietrzykowski, M. and Krzaklewski, W. (2009). *Reclamation to Forest in Sand-Filing Mining Exemplified on Szczakowa Sand Mine Cast. Monography*. Krakow: Department of Forest Ecology, University of Agriculture in Krakow and Monos.

Pietrzykowski, M. and Krzaklewski, W. (2018). Reclamation of mine lands in Poland. Chapter 27. In: *Bio-Geotechnologies for Mine Site Rehabilitation* (ed. M.N.V. Prasad, P. Favas, and S.K. Maiti), 493–513. Amsterdam-Oxford-Cambridge: Elsevier.

Pietrzykowski, M. and Socha, J. (2011). An estimation of Scots pine (*Pinus sylvestris* L.) ecosystem productivity on reclaimed post-mining sites in Poland (Central Europe) using of allometric equations. *Ecol. Eng.* 37: 381–386. https://doi.org/10.1016/j.ecoleng.2010.10.006.

Pietrzykowski, M., Socha, J., and Woś, B. (2010). Biomass and deformation of the Scots pine (*Pinus sylvestris* L.) root systems in reclaimed open–cast mining pit and dumping ground conditions. *Sylwan* 154: 107–116.

Pietrzykowski, M., Woś, B., and Haus, N. (2013). Scots pine needles macronutrient (N, P, K, CA, MG, and S) supply at different reclaimed mine soil substrates – as an indicator of the stability of developed forest ecosystems. *Environ. Monit. Assess.* 185: 7445–7457. https://doi.org/10.1007/s10661-013-3111-9.

Pietrzykowski, M., Socha, J., and van Doorn, N.S. (2014). Linking heavy metal bioavailability (Cd, Cu, Zn and Pb) in Scots pine needles to soil properties in reclaimed mine areas. *Sci. Total Environ.* 470-471: 501–510. https://doi.org/10.1016/j.scitotenv.2013.10.008.

Pietrzykowski, M., Socha, J., and van Doorn, N.S. (2015). Scots pine (*Pinus sylvestris* L.) site index in relation to physico-chemical and biological properties in reclaimed mine soils. *New For.* 46: 247–266. https://doi.org/10.1007/s11056-014-9459-z.

Pietrzykowski, M., Gruba, P., and Sproull, G. (2017). The effectiveness of yellow lupine (*Lupinus luteus* L.) green manure cropping in sand mine cast reclamation. *Ecol. Eng.* 102: 72–79. https://doi.org/10.1016/j.ecoleng.2017.01.026.

Prescott, C.E. (2002). The influence of the forest canopy on nutrient cycling. *Tree Physiol.* 22: 1193–1200. https://doi.org/10.1093/treephys/22.15-16.1193.

Priha, O., Grayston, S.J., Hiukka, R. et al. (2001). Microbial community structure and characteristics of the organic matter in soils under *Pinus sylvestris, Picea abies and Betula pendula* at two forest sites. *Biol. Fertil. Soils* 33 (1): 17–24. https://doi.org/10.1007/s003740000281.

Rodrigue, J.A., Burger, J.A., and Oderwald, R.G. (2002). Forest productivity and commercial value of pre-law reclaimed mined land in the eastern United States. *North. J. Appl. For.* 19: 106–114. https://doi.org/10.1093/njaf/19.3.106.

Rosenvald, K., Kuznetsova, T., Ostonen, I. et al. (2011). Rhizosphere effect and fine-root morphological adaptations in a chronosequence of silver birch stands on reclaimed oil shale post-mining areas. *Ecol. Eng.* 37 (7): 1027–1034. https://doi.org/10.1016/j.ecoleng.2010.05.011.

Schrijver, A.D., Geudens, G., Augusto, L. et al. (2007). The effect of forest type on throughfall deposition and seepage flux: a review. *Oecologia* 153 (3): 663–674. https://doi.org/10.1007/s00442-007-0776-1.

Shrestha, R.K. and Lal, R. (2006). Ecosystem C budgeting and soil C sequestration in reclaimed mine soil. *Environ. Int.* 32: 781–796. https://doi.org/10.1016/j.envint.2006.05.001.

Šourková, M., Frouz, J., and Šantrůčková, H. (2005). Accumulation of carbon, nitrogen and phosphorus during soil formation on alder spoil heaps after brown-coal mining, near

Sokolov (Czech Republic). *Geoderma* 124: 203–214. https://doi.org/10.1016/j.geoderma.2004.05.001.

Świątek, B. and Pietrzykowski, M. (2021). Soil factors determining the fine-root biomass in soil regeneration after a post-fire and soil reconstruction in reclaimed post-mining sites under different tree species. *Catena* 204: 105449. https://doi.org/10.1016/j.catena.2021.105449.

Topp, W., Simon, M., Kautz, G. et al. (2001). Soil fauna of a reclaimed lignite open-cast mine of the Rhineland: improvement of soil quality by surface pattern. *Ecol. Eng.* 17: 307–322. https://doi.org/10.1016/S0925-8574(00)00147-6.

Vacek, Z., Linda, R., Cukor, J. et al. (2021a). Scots pine (*Pinus sylvestris* L.), the suitable pioneer species for afforestation of reclamation sites? *For. Ecol. Manag.* 485: 118951. https://doi.org/10.1016/j.foreco.2021.118951.

Vacek, Z., Cukor, J., Vacek, S. et al. (2021b). Production potential, biodiversity and soil properties of forest reclamations: opportunities or risk of introduced coniferous tree species under climate change? *Eur. J. For. Res.* 140: 1243–1266. https://doi.org/10.1007/s10342-021-01392-x.

Vares, A., Lohmus, K., Truu, M. et al. (2004). Productivity of black alder (*Alnus glutinosa* (L.) Gaertn.) plantations on reclaimed oil-shale mining detritus and mineral soils in relation to rhizosphere conditions. *Oil Shale* 21 (1): 43–58.

Vesterdal, L., Clarke, N., Sigurdsson, B.D., and Gundersen, P. (2013). Do tree species influence soil carbon stocks in temperate and boreal forests? *For. Ecol. Manag.* 309: 4–18. https://doi.org/10.1016/j.foreco.2013.01.017.

Wardle, D.A., Bardgett, R.D., Klironomos, J.N. et al. (2004). Ecological linkages between aboveground and underground biota. *Science* 304: 1629–1633. https://doi.org/10.1126/science.1094875.

Woś, B. (2015). Impact of selected forest tree species on soil-forming process in conditions of lithologically diversified post-mining sites. PhD thesis. University of Agriculture in Krakow, Faculty of Forestry, Krakow, Manuscript.

Woś, B. and Pietrzykowski, M. (2019). Impact of tree species on macroelements content and properties of the initial soils in condition of reclaimed sand pit. *Sylwan* 163 (5): 407–414. https://doi.org/10.26202/sylwan.2018151.

Woś, B., Chodak, M., Józefowska, A., and Pietrzykowski, M. (2022). Influence of tree species on carbon, nitrogen, and phosphorus stocks and stoichiometry under different soil regeneration scenarios on reclaimed and afforested mine and post-fire forest sites. *Geoderma* 415: 115782. https://doi.org/10.1016/j.geoderma.2022.115782.

Zipper, C.E., Burger, J.A., McGrath, J.M. et al. (2011). Forest restoration potentials of coal-mined lands in the eastern United States. *J. Environ. Qual.* 40: 1567–1577. https://doi.org/10.2134/jeq2011.0040.

7

Biochar Applications for Sustainable Agriculture and Environmental Management

Majeti Narasimha Vara Prasad

School of Life Sciences, University of Hyderabad, Hyderabad, India

7.1 Introduction

Recent years have observed growing problems related to the need to sustainable agriculture. Management of biodegradable waste (this includes a portion of municipal solid waste), sewage sludge is posing a problem. However, several success stories are reported for energy recovery improving soil properties, reducing the use of mineral fertilizers and plant protection products, reducing greenhouse gas emissions, and carbon sequestration in soil. In order to achieve this successfully at global scale new solutions are sought which will enable finding a common ground to solve these problems in accordance with the principles of sustainable development.

Applications of biochar in environmental and agricultural management, uses of biochar made from trash and biomass, the use of biochar in agriculture, treatment of water and wastewater to further sustainable development goals have gained considerable importance in recent times (Mohan et al. 2022; Ok et al. 2018; Singh and Singh 2020; Tsang et al. 2022).

Terra preta soils along Amazon River Basin of South America more than 2500 years ago, charcoal is added to soil as a corrective measure to boost agricultural productivity. Archeological proof suggests that natives collected biomass viz., charcoal, manure, and animal bones in earthen pits, and converted to biochar in a Kiln technology with controlled air supply. This process resulted in poor tropical soils enriched by the higher carbon (C) content (9% OC), higher nitrogen (N), and phosphorus (P) (Brewer et al. 2009; Cao et al. 2009).

Since its inception, several scientists have contributed to biochar technology, and it is proven technology and accepted practice that biochar can address multiple challenges related to sustainable development (Figure 7.1).

Agroecological Approaches for Sustainable Soil Management, First Edition.
Edited by Majeti Narasimha Vara Prasad and Chitranjan Kumar.

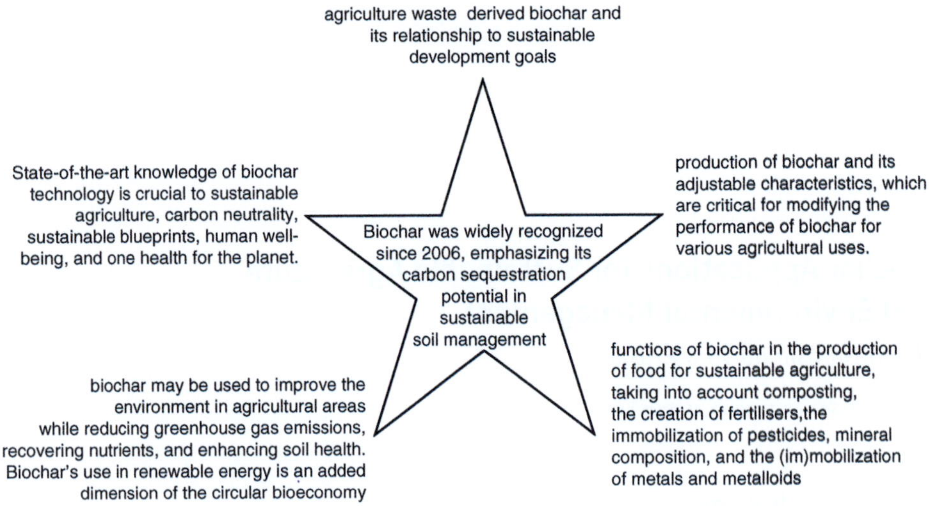

Figure 7.1 Biochar application in agriculture is a proven technology and accepted practice to address multiple challenges related to sustainable development.

7.2 Resume of Biochar for Sustainable Soil Management

Biochar application to contaminated agricultural soil reduces the availability and plant uptake of potentially toxic elements viz., lead, cadmium, arsenic, nickel, etc. (Puga et al. 2015, Qian et al. 2022, Radziemska et al. 2022, Rehman et al. 2016, Rizwan et al. 2018a,b; Shen et al. 2016). Hence, biochar research gained momentum significantly (Singh and Singh 2020, Tsang and Ok 2022). Biochar applications for climate change mitigation, waste management, and wastewater treatment is subject to contemporary interest (Qambrani et al. 2017, Sasongko 2021). Biochar is made from different sources such as wood waste, bamboo, poultry manure, sewage sludge, coconut shell, white clover waste (Abbasi and Anwar 2015, Agegnehu et al. 2017, Ahmad et al. 2022, Liu et al. 2018a,b). Sewage sludge produced biochar improved soil nutrient status, microbial diversity, and enhanced plant bioproductivity (Ahmad et al. 2022). Biochar in agroecosystems enhance soil nutrient status and plant growth (Abbasi and Anwar 2015). Agegnehu et al. (2017) noticed soil quality and crop performance in biochar amended agricultural land. Hardwood produced biochar exercised influence on soil bacterial community as confirmed from high-throughput PCR sequencing (Cole et al. 2019).

Different biochars derived from various sources have exhibited different water dynamics (Conte and Nestle 2015). El Rasafi et al. (2020a) in diverse experiments concluded that biochar play an important role in soil hydraulic properties. Biochar and animal manure amended mine soil enhanced growth, survival, and bioproductivity of barley (El Rasafi et al. 2020b, 2021b). Biochar increases plant water use efficiency and biomass production while reducing Cu concentration in *Brassica juncea* L. in Cu-contaminated soil (Silva Gonzaga et al. 2019). Drought and salt stress to plants was ameliorated by the addition of biochar to soil (Ali et al. 2017b). Hemp, paper mulch, and biochar applications improve weed management in strawberry production (Gramig et al. 2022). Biochar application to soil irrigated with saline waste improves cotton-wheat system performance and soil functions in a study conducted in northwest India (Singh et al. 2021).

It is demonstrated that biochar ameliorated the toxicity of Pb, Cd, and Cr and the development of maize in artificially contaminated soil (Alaboudi et al. 2019, Assa et al. 2021), bamboo char combined with compost reduced phytotoxicity of heavy metals in China's mine-contaminated soils (Ali et al. 2017a). Pot and field experiments with Douglas fir biochar-supported phosphate lead immobilization in polluted soil (Beatrice et al. 2022); adsorption and stabilization of Pb in soil amended with biochar and phosphorus rich biochar for immobilization of Pb have been reported (Chen et al. 2021, 2022a,b).

Microplastic contaminated agricultural soil microbial community activity was studied in a combined biochar and soil + moisture and soil chemical functions (Dissanayake et al. 2022). Such an approach would help to solve microplastics application to soil and management. Microplastic and biochar co-existence on paddy soil improved bacterial community structure, stimulated the methane emission but suppressed nitrous oxide emission from a rice paddy soil (Han et al. 2022a,b). Fahad et al. 2015 demonstrated that biochar application protects rice pollen from high temperature stress; a combined application of biochar and phosphorus alleviates heat stress and improves agronomical attributes of rice (Fahad et al. 2016). Biochar and manure amendments had an impact on soil nutrients and microbial enzymatic activity in irrigated maize crop (Foster et al. 2016).

Cd^{2+} biosorption by viable *Bacillus cereus* RC-1 immobilized on different biochars was reported by Huang et al. (2020). Kamran et al. (2019), reported biochar alleviation of Cd phytotoxicity. Keerthanan et al. (2022), reported that cinnamon wood biochar reducd the bioavailability of sulfamethoxazole. It was observed that the availability of soil organic pollutants, pesticides were detoxified/immobilized by biochar (Lin et al. 2022b, Liu et al. 2018a,b, Tan et al. 2022, Tsang and Ok 2022, Uchimiya et al. 2010, Włodarczyk and Kwarciak-Kozłowska 2021, Xu et al. 2019, Yang et al. 2022c). Slag immobilized As and Cd contaminated paddy soil (Liu et al. 2020). Rice husk biochar impacts soil phosphorous availability, phosphatase activities and bacterial community characteristics in three different soil types (Liu et al. 2017).

Khan et al. (2018) observed mitigation of greenhouse gas emissions from a red acidic soil by using magnesium-modified wheat straw biochar. Kiran and Prasad (2019) reported biochar and rice husk ash assisted phytoremediation. Kumar et al. (2022) observed biochar as "environmental armor" and its diverse role toward protecting soil, water, and air.

Li et al. (2022a) research finding revealed the passivation of lead and cerium in soil was facilitated by biochar-supported phosphate-doped ferrihydrite including microbial community evolution in the root system of soybean. Li et al. (2022b) found the impact of biochar on soil environment, e.g. co-pyrolysis of sewage sludge and metal-free/metal-loaded polyvinyl chloride (PVC) microplastics improved biochar properties and reduced environmental risk of heavy metals (Li et al. 2022c). Li et al. (2022d) also observed microplastics and biochar coexistence decreases global warming resulting from soil greenhouse gas emissions (Li et al. 2022a-d). In a greenhouse experiment using maize crop, biochar altered the physicochemical properties of sandy soil (Lima et al. 2018). Yrjälä et al. (2022) in a recent review suggested that utilizing biochar will have a favorable impact on the following:

1) soil nutrient fluxes, which could reduce eutrophication;
2) soil status and fertility, including porosity, pH, ion exchange, and root growth;
3) microbial processes. Food security and health are positively impacted by plant or crop productivity, quality, and physiology;
4) soil contamination can be remedied to affect land use;

5) markets for biomass and land can favorably affect land-use changes, for instance, by driving up biomass demand;
6) the reduction of industrial agriculture inputs such as agrochemicals, water, and equipment;
7) supply bioenergy and biochemical products (such as heat, power, car fuel, and lubricants) from pyrolysis gases and tars;
8) offer a service for treating biomass waste (farm waste, garden garbage); and
9) equipment manufacturing market (e.g., pyrolysis).

Some of the recent publications highlighting the production of biochar, biochar properties, biochar for soil sustainability and crop production, biochar for improving soil properties and for removal of impurities and cleanup of contaminated soils by the addition of biochar are shown in Table 7.1 and Figure 7.2.

Table 7.1 Biochar assistance for sustainable agriculture and environmental management.

Metalloid and toxic metals cleanup with the help of biochar	Dhiman et al. (2021)
In problematic oils, biochar supplementation increases crop productivity.	Dhir (2021)
Production of biochar, its attributes, and its use in environmental protection against harmful metals.	Gałwa-Widera (2021)
Using biochar to treat resistant soils will improve agronomic performance.	Grobelak and Jaskulak (2021)
Wastewater treatment with biochar.	Kwarciak-Kozłowska and Włodarczyk (2021)
Prospects and difficulties of using biochar as a soil supplement to clean up heavy metal contamination.	Maiti and Ghosh (2021), Lahori et al. (2017)
Water treatment systems based on biochar for the provision of clean water.	Sasongko et al. (2021)
Biochar for bioremediation of toxic metals.	Włodarczyk and Kwarciak-Kozłowska (2021)

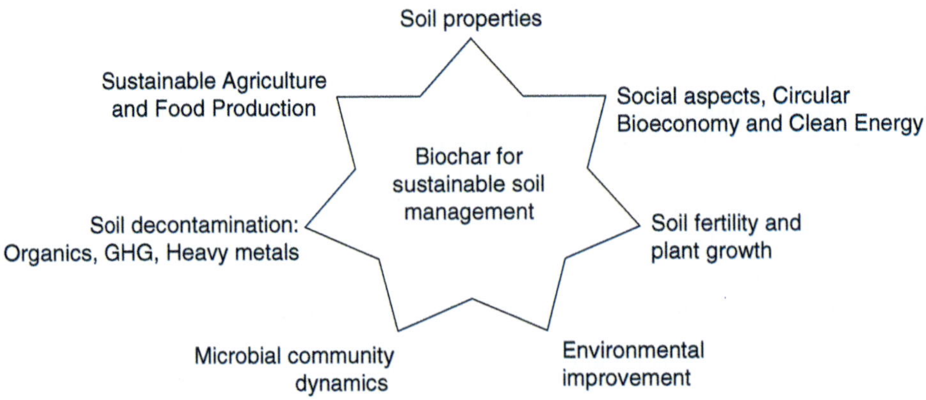

Figure 7.2 Biochar for sustainable soil management. *Source:* Kamali et al. (2022) / with permission of Elsevier.

7.3 Biochar Advantages for Sustainable Soil Management

A great diversity of biodegradable waste (Table 7.2) can be used as a starting point for biofertilizer production for application on degraded soil, introducing valuable phytochemicals such as polysaccharides, proteins, polyphenols, lipids, macro- and micronutrients, among many others, during the process of decomposition. Key causes of soil degradation and biochar benefits to soil and atmosphere are depicted in Figure 7.3.

Table 7.2 The most frequently used biodegradable waste and products for degraded soils improvement. Advantages and challenges for bio-based waste substrates for degraded soil improvement "emerging contaminants" such microplastics, pharmaceuticals waste (Kacprzak et al. (2022) / MDPI / CC BY 4.0).

Product	Biodegradable waste
Compost/Vermicompost	Food, agricultural, forest waste Vermicompost
Digestate	Pig, cow, poultry manure
Organic mineral fertilizers	Sewage sludge
Soil substitutes	Organic fraction of municipal solid waste
Plant growth promoting substrates bakeries	Waste from the food processing units such slaughterhouses, dairies, breweries
Biochar/Ash	Pulp and paper mill by products
Struvite	Domestic wastewater

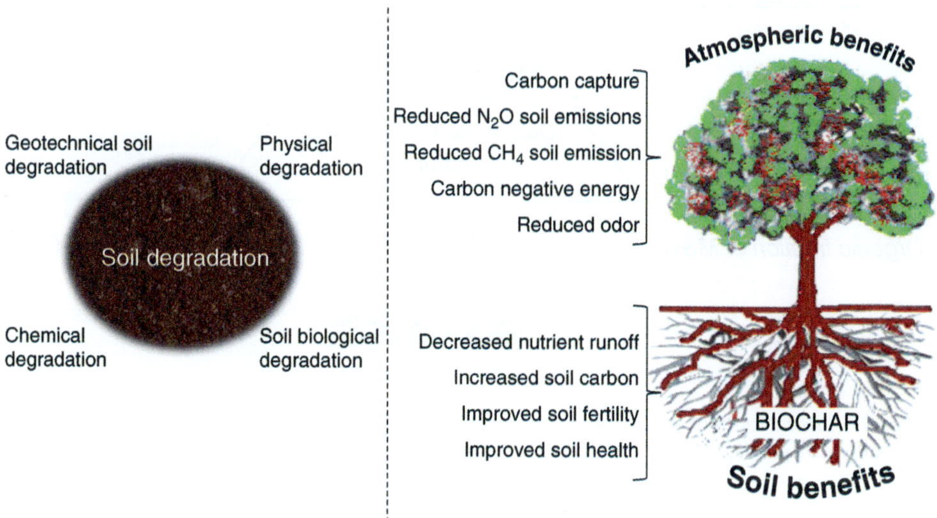

Figure 7.3 Soil degradation factors and benefits of biochar to soil and atmosphere.

7.4 Feedstock for Production of Biochar

Increased carbon storage, improved soil fertility, and a reduction in greenhouse gas emissions are all possible benefits of biochar. Biochar amended soils helps in carbon sequestration, alter physico-chemical properties of the soil favorably, play a significant role in nutrient dynamics and biological properties and productivity of soil.

Lehmann (2007) commented that "Instead of storing carbon in plants and trees that eventually degrade, it makes more sense to lock a little amount of carbon in soil. Biochar also helps in reduced emissions and carbon neutrality" (Figure 7.4).

C/N, or the mass ratio of organic carbon to total nitrogen, is the fundamental factor for a biofertilizer such as compost or a biochar. Therefore, proper progression of the biomass or biowaste conversion process decided the quality of the final product. Microorganisms need carbon as a source of energy, whereas nitrogen affects how quickly their populations develop (Figure 7.5). Compost has a more immediate and long-lasting influence on the pH and cation exchange capacity of the soil than biochar has on the soil's carbon pool. Using biochar and compost together for sustainable soil management has been suggested for degraded soils.

This chapter offers a resume agricultural application of various biochars. As a complex organic compound, biochar's properties depend on the feedstock utilized and the technology applied for production. The use of certain forms of biochar with the correct field

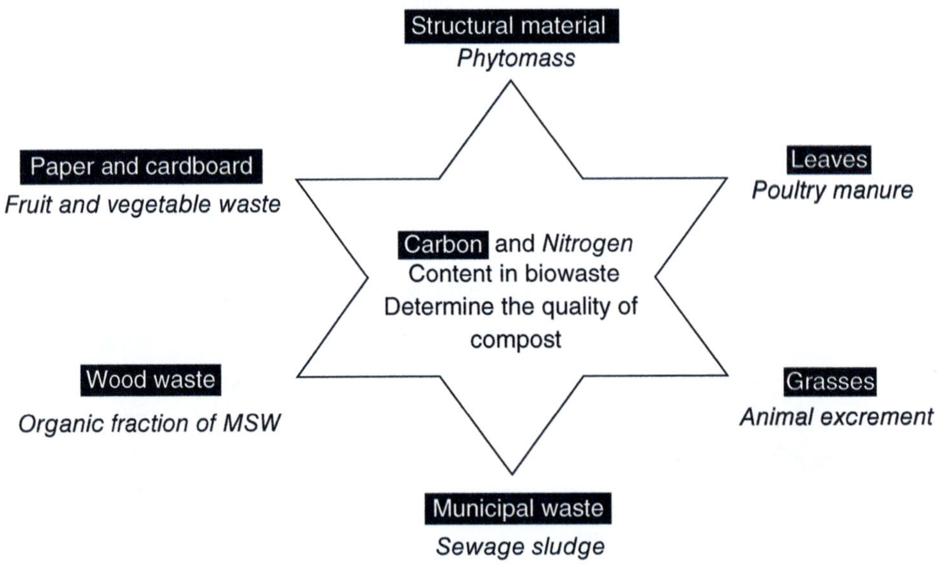

Figure 7.4 Advantages of a handful of carbon for productivity, reduced emissions and carbon neutrality. *Source:* Adapted from Lehmann (2007) and Kacprzak et al. (2022).

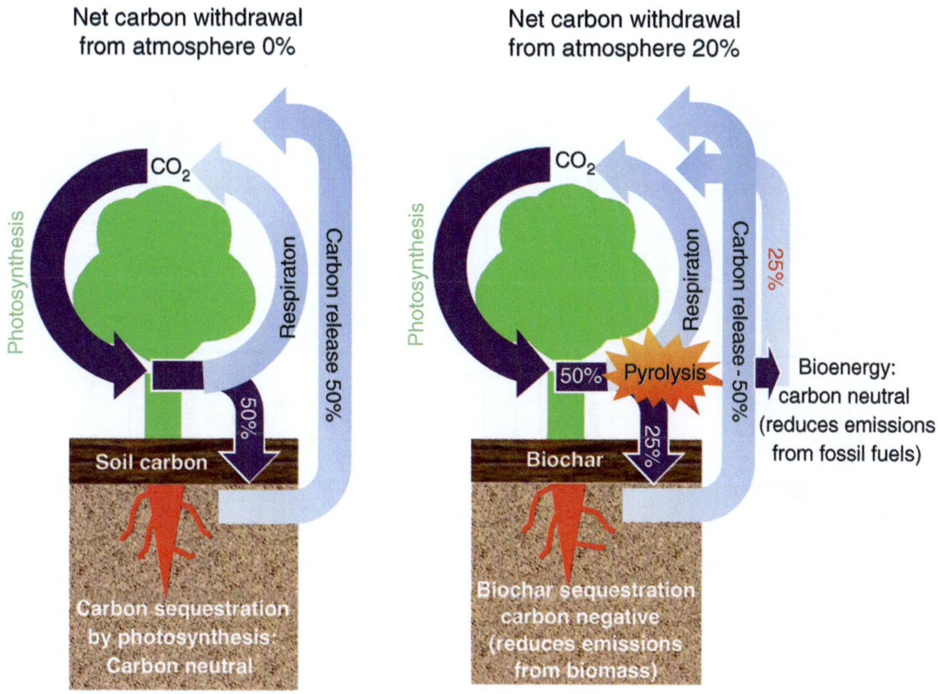

Figure 7.5 Potential sources of carbon (highlighted in black with white letters) and nitrogen rich compost mixes (in italics) utilized in the composting process.

application rate under the relevant soil types and circumstances is the only way to employ biochar for its agronomic advantages (Figure 7.6).

7.5 Soil Carbon Storage/Sequestration

For sustainable agriculture and environmentally friendly farming, the function of soil microorganisms in mineralization and boosting nutrient availability to plants is of the utmost importance (Figure 7.7).

The core of soil carbon sequestration (SCS) is the so-called soil microbial community (SMC). SMC controls the deterioration and change of soil organic matter (SOM). The rate of SCS is determined by soil aggregates and SOM. Reintroducing decaying SMC under various agricultural systems would guarantee long-term C storage sustainability as shown in Figure 7.8 (Bhattacharyya et al. 2022).

An annotated synopsis of research findings are tabulated for Biochar influence on detoxification of potentially toxic elements in soil (Table 7.3), Biochar mitigates salinity in different crop fields (Table 7.4), Biochar amendment alleviate drought in croplands (Table 7.5) and Miscellaneous functions of biochar for soil sustainability (Table 7.6).

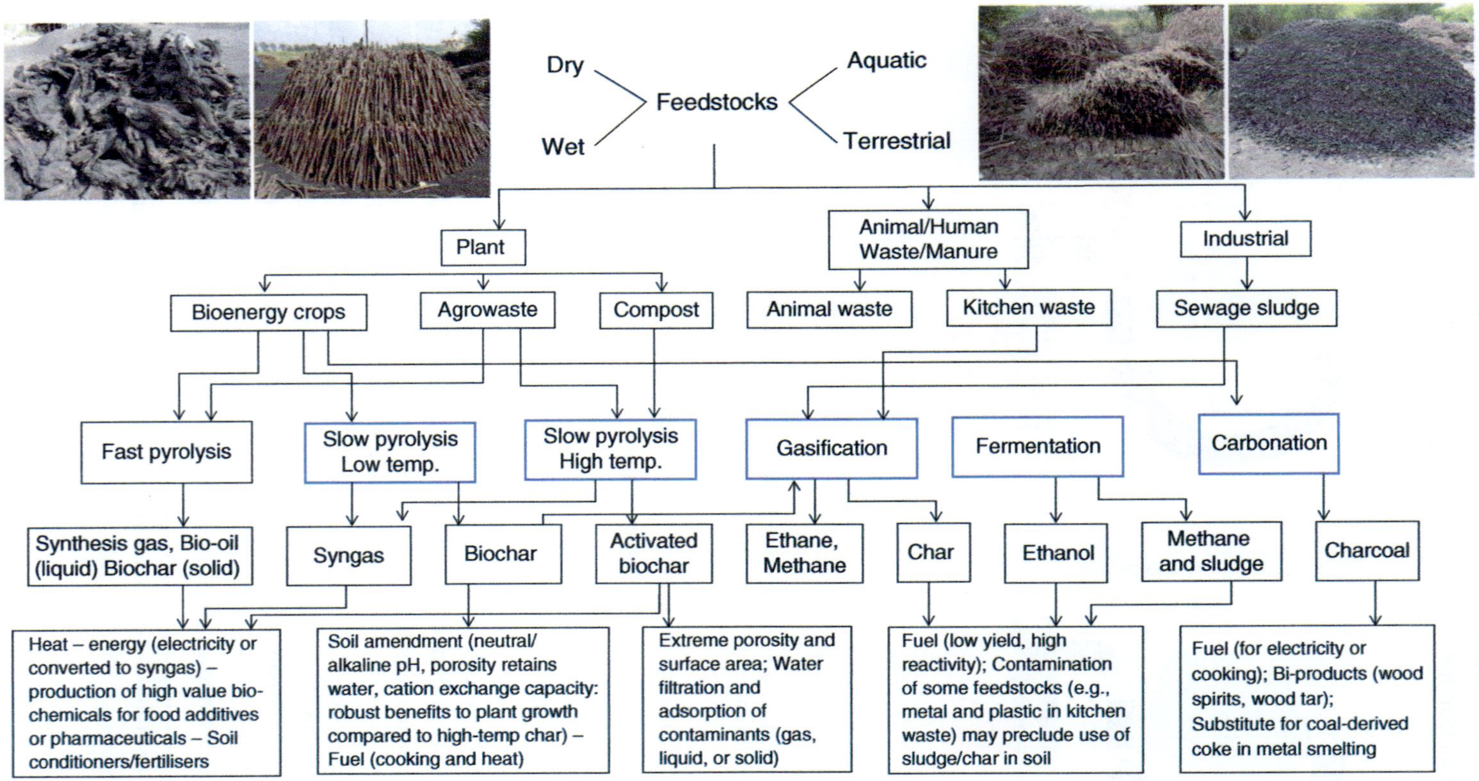

Figure 7.6 Feedstocks, biochar, and other byproducts of the thermal conversion of biomass in accordance with diverse sources of feed stock and methods.

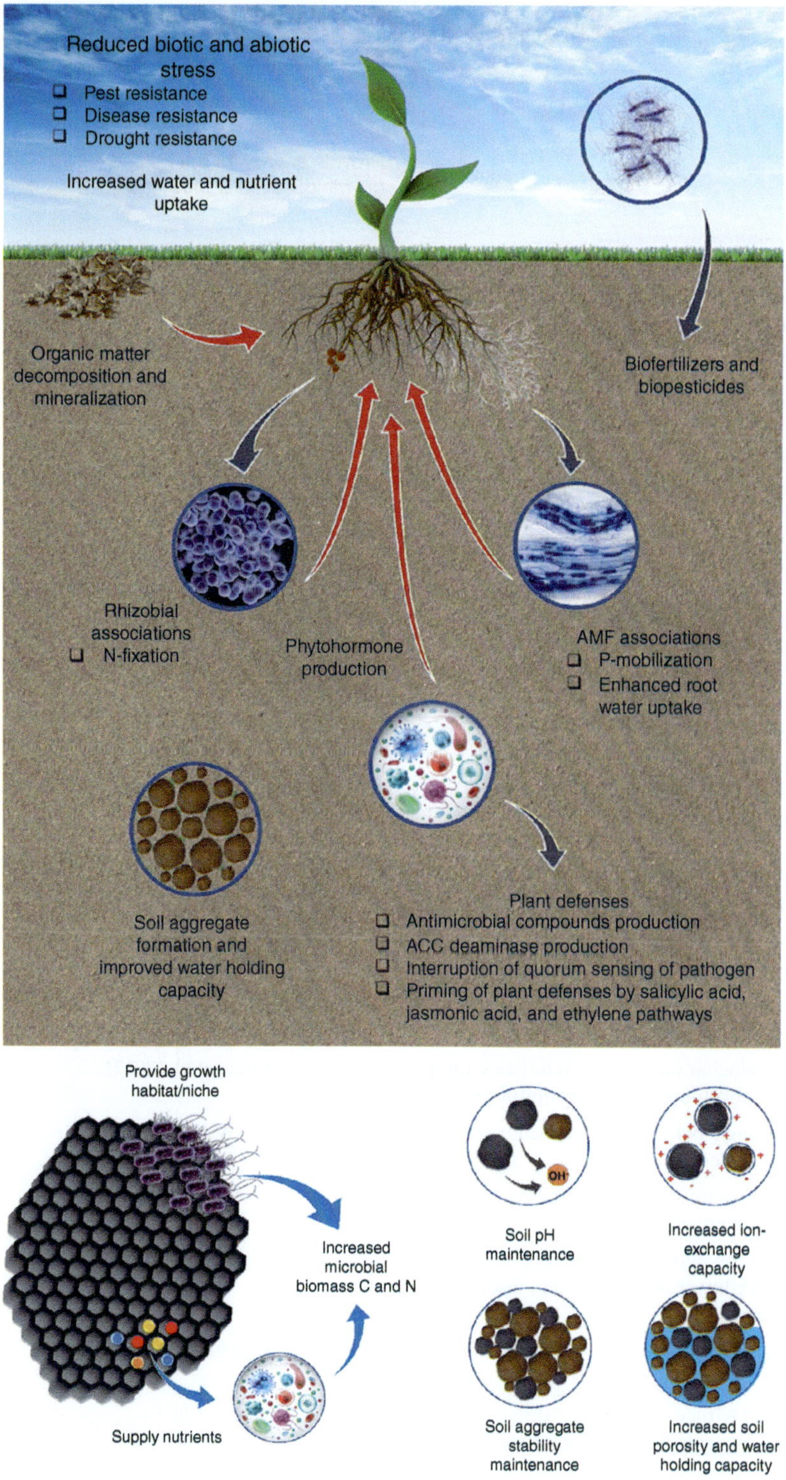

Figure 7.7 Soil microbiology plays a significant role in biogeochemical cycling of nutrients for sustainable agriculture and eco-friendly farming. Biochar augmentation to soil in agroecosystems influence soil biota and physicochemical properties.

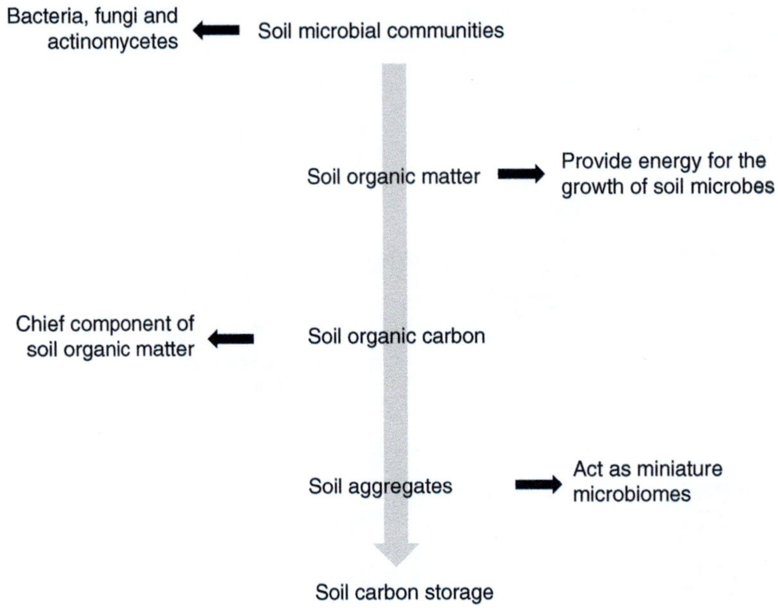

Figure 7.8 Soil carbon sequestration–An interplay between soil microbial community and soil organic matter dynamics. *Source:* Adapted from Bhattacharyya et al. (2022).

7.6 Biochar Influence on Detoxification of Potentially Toxic Elements in Soil

Table 7.3 Biochar influence on detoxification of potentially toxic elements in soil.

A review of heavy metal sorbents that could be inexpensive	Bailey et al. (1999)
Biochar is used to immobilize and retain soluble arsenic, cadmium, and zinc.	Beesley and Marmiroli (2011)
Effect of applying biochar on soil characteristics and lettuce (*Lactuca sativa*) nitrogen uptake in chromium-polluted soils	Nigussie et al. (2012)
In sandy soils, adding biochar to the soil improves tomato seedling drought resilience.	Mulcahy et al. (2013)
Using biochar to clean up soils contaminated with organic contaminants and heavy metals.	Xiaokai et al. (2013)
Effects of a biochar amendment on the root characteristics and contaminant availability of maize plants in a soil with high levels of copper and arsenic.	Brennan et al. (2014)
The effects of hydrophobicity and porosity on water absorption in biochars.	Gray et al. (2014)
Effect of biochar on lettuce (*L. sativa* L.) uptake and immobilization of heavy metals in agricultural soil.	Kim et al. (2015)
In an effort to phytostabilize an abandoned copper mine, native trees are being evaluated for their metal absorbing potential.	Nirola et al. (2015)

Table 7.3 (Continued)

Applying biochar to contaminated soil decreases the availability and uptake of zinc, lead, and cadmium by plants.	Puga et al. (2015)
In cadmium-contaminated soil, biochar impacts the development and metabolic processes of fenugreek (*Trigonella corniculata*).	Younis et al. (2015)
Mechanisms and future directions for the physical–chemical characteristics and microbial responses in soils modified with biochar.	Gul et al. (2015)
Effect of biochar amendment on sunflower plants' morphology, production, and water relations in the absence of irrigation.	Paneque et al. (2016)
contrasting the effects of biochar, compost, and farm manure on plant growth, photosynthesis, and metal uptake in relation to the reduction of nickel toxicity in maize (*Zea mays* L.).	Rehman et al. (2016)
The impact of biochar and crop straws on the bioavailability of heavy metals and plant accumulation in a soil contaminated with Cd and Pb.	Shen et al. (2016)
The effect of biochar and crop straws on heavy metal bioavailability and plant accumulation in a Cd and Pb polluted soil.	Xu et al. (2016)
Effect of biochar on soil enzyme activity and the capacity of heavy metals (Cd, Cu, Pb, and Zn) to be extracted	Yang et al. (2016)
Through changes to Cd absorption and other physiological and biochemical characteristics, biochar improves the cadmium tolerance in spinach (*Spinacia oleracea*) and physiological and biochemical attributes.	Younis et al. (2016)
Influence of biochar on cadmium absorption and bioavailability in wheat (*Triticum aestivum* L.) cultivated in soil with historical contamination	Abbas et al. (2017)
Mechanisms of metal sorption by biochars: Features and changes of biochar.	Li et al. (2017)
Effects of old and fresh biochar in situ on soil hydraulic properties and microbial C consumption in drought conditions	Paetsch et al. (2018)
By reducing nutrient stress in a moderately acidic low-input Nepalese soil, biochar enhances maize growth.	Pandit et al. (2018)
Contrasting effects of biochar that has aged before and after application on the immobilization of Cd in polluted soils.	Xu et al. (2018)
When drought-stressed wheat cultivated in an old, contaminated soil was treated with biochar, growth and yield increased while cadmium levels decreased.	Abbas et al. (2018a)
Biochar's residual effects on rice (*Oryza sativa* L.) grown in soil contaminated with cadmium.	Rizwan et al. (2018a)
Wheat (*T. aestivum* L.) grown on cadmium-contaminated salty soil is less hazardous because to biochar.	Abbas et al. (2018b)
The effects of biochar on rice (*O. sativa* L.) growth, photosynthesis, and cadmium uptake under Cd stress with various water conditions.	Rizwan et al. (2018b)
Metal behavior, plant toxicity, and microbial activity all contributed to the phytoremediation of cadmium-contaminated sediments made easier by the use of biochar.	Gong et al. (2019)

(Continued)

Table 7.3 (Continued)

Rice grown in a salty, Cd-contaminated soil has a residual effect of biochar on its growth, antioxidant defence, and cadmium (Cd) accumulation.	Hafeez et al. (2019)
Mechanisms, potential hazards, and applications of biochar in the remediation of heavy metal-contaminated soils in China.	He et al. (2019)
Impact of beeswax waste biochar on saffron's growth, physiology, and uptake of cadmium	Moradi et al. (2019)
Mobility of Pb, Zn, Ba, As, and Cd from a mine soil modified with biochar toward soil pore water and plants (willow and ryegrass).	Norini et al. (2019)
By lowering the bioavailability of Cd and Zn, biochar compost mixes promote the establishment of switchgrass in mine soils.	Novak et al. (2019)
Effect of biochar on heavy metal buildup in potatoes after irrigation with wastewater.	Nzediegwu et al. (2019)
A critical analysis of how microbial populations react to soils altered with biochar.	Palansooriya et al. (2019)
Soil's nutrient mineralization and microbial activity are improved by recently aged biochar.	Yadav et al. (2019)
Effects of biochar on moso bamboo (*Phyllostachy pubescens*) growth, heavy metal deposition, soil physical characteristics, and soil solubility	Ying et al. (2019)
Under various water regimes, the use of co-composted agricultural manure and biochar boosted wheat growth while reducing cadmium buildup in plants.	Bashir et al. (2020)
Metal-immobilizing biochar and Liquefacient Serratia Wheat grain metal buildup was decreased by CL-1 in a soil with metal contamination.	Cheng et al. (2020)
Review of the function of biochar in the improvement of soil and the composting of organic waste.	Guo et al. (2020)
Plant-bacteria-biochar partnerships: eco-friendly ways to combat heavy metal contamination	Harindintwali et al. (2020)
Insight into organic carbon release, surface physicochemical characteristics, and cadmium sorption from the simulation of photocatalytic aging of biochar in soil ecosystem.	Quan et al. (2020)
Managing growth, physiological, and oxidative abnormalities in *Brassica napus* caused by cadmium by using biochar and Enterobacter sp. MN17 together.	Sabir et al. (2020)
Endogenous copper and cadmium in biochar are activated by freeze–thaw and dry-wet aging processes.	Cui et al. (2021)
In croplands, the impacts of biochar and its manure-based feedstock differ in terms of soil organic carbon and greenhouse gas emissions.	Gross et al. (2022)
An overview of biochar production, its effects, and the mechanisms through which soil and plant properties are improved by biochar.	Haider et al. (2022)
An evaluation of the effects of biochar-based soil replacement and integrated amendments on the phytotoxicity of metals (loids) in radish, soya bean, and amaranth rotations in acidic mine soil.	Jiang et al. (2022)

Table 7.3 (Continued)

Effects and processes of mixed bacteria-loaded biochar for the immobilization of arsenic, lead, and cadmium in a polluted soil system.	Liang and Zhang (2022)
A strategy for cleaning up chromium-tainted water and soil that uses designed biochar made from peanut shells.	Murad et al. (2022)
Microplastic-contaminated soil's chemical and microbiological characteristics are changed by biochar.	Palansooriya et al. (2022)
Toward estimating the effects of biochar on soil nitrogen availability for plants	Phillips et al. (2022)
A nanoscale zero-valent iron compound supported by porous biochar that is particularly effective for the simultaneous cleanup of cadmium and lead-contaminated soil.	Qian et al. (2022)
Under various temperature circumstances, the soil additives nanosized halloysite and biochar helped to phytostabilize metal (-oid) s-contaminated soil.	Radziemska et al. (2022)
In *Camellia oleifera* plantings, biochar made from waste mushroom substrate decreased N_2O emissions with lower water content but increased CH_4 emissions when the soil was inundated.	Xu et al. (2022)
The short-term immobilization of lead and cadmium in soil and aqueous solutions by nano-hydroxyapatite/biochar.	Zhou et al. (2022)
Under varying regulated redox conditions, biochar made from pig carcasses had inconsistent effects on the mobilization of arsenic in a polluted paddy soil.	Yang et al. (2022a)
Using geochemical and spectroscopic methods, elucidating the redox-driven dynamic interactions between arsenic and iron-impregnated biochar in a paddy soil.	Yang et al. (2022b)
Effects of biochar type and aging method on the aging characteristics of metals (loid) in soil treated with biochar.	Yang et al. (2022b)

7.7 Biochar Mitigates Salinity in Different Crop Fields

Table 7.4 Biochar mitigates salinity in different crop fields.

Potato salinity stress is reduced by biochar.	Akhtar et al. (2015)
Conocarpus under saline irrigation, biochar changes, soil nutrient availability, and tomato growth.	Usman et al. (2016)
Rice grown in a salty, Cd-contaminated soil has a residual effect of biochar on its growth, antioxidant defense, and cadmium (Cd) accumulation.	Kibria et al. (2017)
Effect of biochar on the ion content and development of the bean plant in a salty environment.	Farhangi-Abriz and Torabian (2018)
Utilizing biochar to reduce the harmful effects of salt stress on wheat (*Triticum aestivum* L.).	Kanwal et al. (2018)
Benefits of adding biochar to the soil to improve tomato growth when using irrigation with salt water.	She et al. (2018)

(*Continued*)

Table 7.4 (Continued)

Rice grown in a salty, Cd-contaminated soil has a residual effect of biochar on its growth, antioxidant defense, and cadmium (Cd) accumulation.	Hafeez et al. (2019)
Effect of straw biochar on soil characteristics and wheat yield when irrigation with salt water is used.	Huang et al. (2019a)
Effect of biochar on sweet corn and soil salinity in coastal saline soil under simultaneous watering with brackish water.	Huang et al. (2019b)
The biochemical features of rice seedlings with varying salt tolerance were enhanced and the salt stress of produced saline paddy soil was reduced.	Zhang et al. (2019)
Under saline circumstances, biochar affects the development of sweet basil (*Ocimum basilicum* L.) and the kinetics of NH_3-volatilization.	Ding et al. (2020)
The use of biochar reduces the detrimental effects of salinity and drought stress on soybean yield and water use effectiveness.	Yaojun et al. (2020)
In the Songnen Plain, use biochar to remediate soda saline-alkaline land, enhance soil productivity, and boost maize nutrient availability.	Zhao et al. (2020)
Optimizing biochar application to enhance saline-alkaline soils' physical and hydraulic qualities	Liang et al. (2021)
In northwest India, applying rice straw biochar to soil that receives salty irrigation in a cotton-wheat system enhances crop performance and soil functionality.	Singh et al. (2021)
Effects of irrigation and biochar on pearl millet and ammonia volatilization kinetics from saline sandy soils	Ding et al. (2022)

Table 7.5 Biochar amendment alleviates drought in croplands.

In sandy soils, adding biochar to the soil improves tomato seedling drought resilience.	Mulcahy et al. (2013)
Application of biochar for *Silybum marianum* physiological responses to drought stress	Keshavarz Afshar et al. (2016)
Review of the effectiveness of biochar as a soil addition for reducing plant stress caused by salt and drought	Ali et al. (2017a,b)
The effects of in situ aged and fresh biochar on the hydraulics and microbiological utilization of carbon during a drought	Paetsch et al. (2018)
Biochar–potential tool against drought and climate change	Zbigniew et al. (2018)
Does biochar mitigate the adverse effects of drought on the agronomic traits and yield components of soybean?	Gavili et al. (2019)
In drought-stricken soybean plants, applying biochar at a higher rate did not increase plant water availability.	Mannan et al. (2021)

7.8 Miscellaneous Benefits of Biochar for Soil Sustainability

Table 7.6 Miscellaneous functions of biochar for soil sustainability.

Sequestration of biochar in terrestrial environments	Lehamnn et al. (2006)
Preliminary assessment of biochar in a Canadian commercial farm.	Husk (2008)
Pyrolysis platform for biofuels, bioproducts, and biorefining, as well as co-producing charcoal and biooil	Laird et al. (2009)
Science and technology of biochar for environmental management	Lehman et al. (2011)
Biofuels and biochar for soil management are produced and characterized through the pyrolysis of agricultural leftovers from oilseed rape and sunflower seeds.	Sànchez et al. (2009)
Potential methods for applying biochar to temperate soils in order to promote agriculture	Atkinson et al. (2010)
The significance of biochar in crop nutrition	Krull (2010)
Roles of natural organic matter in the immobilization of contaminants and the release of nutrients by biochar soil amendment.	Uchimiya et al. (2010)
Possible use of biochar in restoring polluted soils through revegetation and remediation	Beesely et al. (2011)
Arsenic, cadmium, and zinc that are soluble are immobilized and retained by biochar.	Beesley and Marmiroli (2011)
Temperature during pyrolysis has an impact on the creation and nutritional characteristics of biochar made from wastewater sludge.	Hossain et al. (2011)
Review of the impacts of biochar on soil biota	Lehmann et al. (2011)
From a global and Canadian viewpoint, biochar is a viable solution for sequestering carbon from a global and Canadian viewpoint, biochar is a viable solution for sequestering carbon while also increasing the production of durum wheat.	Matovic (2011)
Biochar as a strategy to sequester carbon and increase yield in durum wheat	Vaccari et al. (2011)
Effect of applying biochar on soil characteristics and lettuce (*Lactuca sativa*) nitrogen uptake in chromium-polluted soils	Nigussie et al. (2012)
Biochar has an impact on seed germination and the first stages of seedling growth.	Solaiman et al. (2012)
In sandy soils, adding biochar to the soil improves tomato seedling drought resilience.	Mulcahy et al. (2013)
Using biochar to clean up soils contaminated with organic contaminants and heavy metals	Xiaokai et al. (2013)
Impact of biochar's structural characteristics on the catalytic hydrolysis of carbaryl and atrazine on pig manure-derived biochar	Zhang et al. (2013)
Tomato output and quality are improved by biochar when irrigation is reduced.	Akhtar et al. (2014)
Effects of a biochar amendment on the root characteristics and contaminant availability of maize plants in a soil with high levels of copper and arsenic.	Brennan et al. (2014)

(*Continued*)

Table 7.6 (Continued)

The effects of hydrophobicity and porosity on water absorption in biochars	Gray et al. (2014)
Effects of biochar on *Malus hupehensis*'s photosynthetic and antioxidant system Rehd. seedlings in replanting circumstances	Wang et al. (2014)
Rice pollen is shielded from heat stress by biochar.	Fahad et al. (2015)
Mechanisms and future directions for the physical–chemical characteristics and microbial responses in soils modified with biochar	Gul et al. (2015)
Application of biochar does not enhance a sandy soil's hydrological function.	Jeffery et al. (2015)
In a soil contaminated with Cd, combining phytoextraction and biochar addition enhances soil biochemical characteristics.	Lu et al. (2015)
The effects of biochar on enhancing wheat under salt stress in terms of growth, physiology, and yield	Saleem et al. (2015)
Following the addition of both fresh and used biochar to an unproductive soil, Catena Carbon mineralization.	Zhao et al. (2015)
Biochar is a potential long-term soil additive for phytostabilization of TE-contaminated soils.	Bopp et al. (2016)
When biochar and phosphorus are applied together, heat-induced negative effects on the physiological, agronomical, and qualitative characteristics of rice are lessened.	Fahad et al. (2016)
Application of biochar has an impact on the sandy soil's physical attributes and hydrological characteristics.	Głąb et al. (2016)
Effect of biochar amendment on sunflower plants' morphology, production, and water relations in the absence of irrigation	Paneque et al. (2016)
Contrasting the effects of biochar, compost, and farm manure on plant growth, photosynthesis, and metal uptake in relation to the reduction of nickel toxicity in maize (*Zea mays* L.)	Rehman et al. (2016)
The phytoavailability of Cd and Pb in crop straw biochar-amended soil is correlated with the soil and biochar heavy metal contents.	Shen et al. (2016)
Polycylic aromatic hydrocarbons (PAHs) and their affinity for binding to dissolved organic matter produced from various types of biochars are characterized using spectroscopy	Tang et al. (2016)
Under salty irrigation, *Conocarpus* biochar causes modifications in the availability of soil nutrients and tomato growth.	Usman et al. (2016)
The impact of biochar and crop straws on the bioavailability of heavy metals and plant accumulation in a soil contaminated with Cd and Pb.	Xu et al. (2016)
Effects of new and old biochars on the amount of water accessible to plants and their ability to use it.	Aller et al. (2017)
Effects of argan shell biochar on the germination of barley and lettuce.	Bouqbis et al. (2017)
Changes in quantity and quality of biochar aged for 5 yr in soil under field circumstances	Dong et al. (2017)
Under salt stress, biochar has an impact on alterations in antioxidant enzymes and osmotic adjustment in bean seedlings.	Farhangi-Abriz and Torabian (2017)
In calcareous sandy and clayey soils, reactions of microbial performance and community to corn biochar.	Khadem and Raiesi (2017)

Table 7.6 (Continued)

In three different soil types, rice husk biochar affects the phosphorus availability in the soil, phosphatase activity, and bacterial community features.	Liu et al. (2017)
Applications for waste management, wastewater treatment, and climate change mitigation using biochar	Qambrani et al. (2017)
Returning biochar to fields	Tan et al. (2017)
A review of the impact that biochar has on the composting of organic wastes and the cleanup of contaminated soils.	Wu et al. (2017)
The organic compounds in biochar leacheates increase the cold tolerance of rice seedlings.	Yuan et al. (2017)
Comparison of the effects of biochar produced from chicken dung on the growth of barley in soils contaminated with petroleum	Barati et al. (2018)
Effect of biochar on the ion content and development of the bean plant in a salty environment	Farhangi-Abriz and Torabian (2018)
Utilizing biochar to reduce the harmful effects of salt stress on wheat (*Triticum aestivum* L.)	Kanwal et al. (2018)
Use of biochar has an effect on the sorption, desorption, and degradation of pesticides	Liu et al. (2018b)
Effects of old and fresh biochar in situ on soil hydraulic properties and microbial C consumption in drought conditions	Paetsch et al. (2018)
By reducing nutrient stress in a moderately acidic low-input Nepalese soil, biochar enhances maize growth.	Pandit et al. (2018)
Biochar's residual effects on rice (*Oryza sativa* L.) grown in soil contaminated with cadmium	Rizwan et al. (2018a)
The effects of biochar on rice (*O. sativa* L.) growth, photosynthesis, and cadmium uptake under Cd stress with various water conditions.	Rizwan et al. (2018b)
A succinct analysis of the use of biochar in agricultural soils to enhance soil quality and combat pollution.	Shaaban et al. (2018)
Benefits of adding biochar to the soil to improve tomato growth when using irrigation with salt water	She et al. (2018)
Contrasting effects of biochar that has aged before and after application on the immobilization of Cd in polluted soils	Xu et al. (2018)
Biochar potential against climate change	Zbigniew et al. (2018)
Nitrate leaching and N_2O emissions are reduced through interactions between biochar, soil, and land use	Borchard et al. (2019)
Environmental risks, carbon mineralization, and soil nutrient release as a function of biochar composition	El-Naggar et al. (2019)
Applications of biochar and inorganic fertilizers have an impact on the growth, yield, and water use effectiveness of maize when irrigation is insufficient.	Faloye et al. (2019)
Enhancing the effectiveness of agricultural water use with biochar: a summary of the impacts of biochar on water fluxes and storage at various scales	Fischer et al. (2019)
Soil hydraulic characteristics before and after freezing–thawing in response to the addition of biochar	Fu et al. (2019)

(Continued)

Table 7.6 (Continued)

Brassica napus phytoextraction and biochar addition combined for Riotinto's mining soil remediation (Spain).	Gascó et al. (2019)
Does biochar lessen the negative impact of drought on soybean yield components and agronomic traits?	Gavili et al. (2019)
Safflower (*Carthamus tinctorius* L.) seedlings are less prone to oxidative stress and fluoride poisoning when treated with biochar.	Ghassemi-Golezani and Farhangi-Abriz (2019)
Amaranthus mangostanus (L.physiological)'s response and mineral element absorption as influenced by biochar.	Jia et al. (2019)
Can biochar increase the physiological traits and productivity of pumpkins grown with less water?	Langeroodi et al. (2019)
Using biochar as an amendment, assisted phytoremediation of a co-contaminated soil: removal of pollutants and characteristics of the bacterial community	Manyun et al. (2019)
Effect of biochar made from beeswax waste on saffron's growth, physiology, and uptake of cadmium	Moradi et al. (2019)
A critical analysis of how microbial populations react to soil altered with biochar.	Palansooriya et al. (2019)
A sustainable method for enhancing soil quality and plant growth is biochar.	Rawat et al. (2019)
Influence of rice straw biochar on ramie (*Boehmeria nivea* L.) planted as forage in old copper-contaminated soil in terms of growth, antioxidant capacity, and copper uptake.	Rehman et al. (2019)
Brassica juncea L. in a Cu-contaminated soil produces more biomass and uses less water when exposed to biochar, which also lowers its content of copper.	Silva Gonzaga et al. (2019)
Sandy loam soil's mineralization and microbial activity are improved by aged biochar.	Yadav et al. (2019)
Review of the preparation, use, and prospects of biochar for the management of polluted soil.	Yuan et al. (2019)
Under various water regimes, the use of co-composted agricultural manure and biochar boosted wheat growth while reducing cadmium buildup in plants.	Bashir et al. (2020)
Barley growth, survival, and biomass production in biochar and animal manure-amended polluted mine soil	El Rasafi and Haddioui (2020)
Can biochar reuse the waste from coal mines?	Ghosh and Maiti (2020)
Ecotoxicological characteristics of biochar and biochar-amended soils: the shadow side of black gold	Godlewska et al. (2020)
Review of the function of biochar in the improvement of soil and the composting of organic waste	Guo et al. (2020)
Soil-biochar-plant interactions: discrepancies between agricultural and artificial soils	Hussain et al. (2020)
Effects of lodestone, charcoal, and carbide slag on As and Cd immobilization, plant uptake, and translocation in polluted paddy soil	Liu et al. (2020)
Improvement of summer savory growth and mineral components under salt stress and biochar use in alkaline soil	Mehdizadeh et al. (2020)

Table 7.6 (Continued)

Effects of various biochar application techniques on the variability of the soybean growth indicator in a soil area with seasonal soil freezing	Qinglin et al. (2020)
Biochar's simulated photocatalytic aging in a soil ecosystem: insight into the release of organic carbon, the physicochemical makeup of surfaces, and cadmium sorption	Quan et al. (2020)
Managing growth, physiological, and oxidative abnormalities in *B. napus* caused by cadmium by using biochar and Enterobacter sp. MN17 together	Sabir et al. (2020)
Mechanisms, physical and chemical changes, evaluation, and field application consequences of biochar aging	Wang et al. (2020)
Response of maize to application of sheep dung and biochar on contaminated mine soil in coniferous tree woodlands	EL Rasafi et al. (2021a)
Optimizing biochar application to enhance saline-alkaline soils' physical and hydraulic qualities	Liang et al. (2021)
When soybean plants are under drought stress, applying biochar at a higher rate does not increase their access to water.	Mannan et al. (2021)
In north-west India, applying rice straw biochar to soil that receives salty irrigation in a cotton-wheat system enhances crop performance and soil functionality.	Singh et al. (2021)
The effects of adding biochar on the physicochemical characteristics of soil	Zhang et al. (2021)
Evidence from a global meta-analysis shows that biochar improves soil carbon reservoirs.	Chagas et al. (2022)
Effects of irrigation and biochar on pearl millet and the rate at which ammonia is released from salty sandy soils	Ding et al. (2022)
The improvement of soil hydraulic characteristics by biochar particle size and hydrophobicity	Edeh and Mašek (2022)
Improved nitrobenzene removal in soil using sulfated nano zerovalent iron accompanied by biochar: Solubilization impact and mechanism	Gao et al. (2022)
A field experiment looked at the effects of biochar on the chemistry of the soil, greenhouse gas emissions, and forage productivity.	Ginebra et al. (2022)
Effect of adding biomass before pyrolyzing sewage sludge on the persistence and bioavailability of polycyclic aromatic hydrocarbons in biochar-amended soil.	Godlewska and Oleszczuk (2022)
In croplands, the impacts of biochar and its manure-based feedstock differ in terms of soil organic carbon and greenhouse gas emissions.	Gross et al. (2022)
An overview of biochar production, its effects, and the mechanisms through which soil and plant properties are improved by biochar	Haider et al. (2022)
Application of biochar in compacted and uncompacted soil to reduce greenhouse gas emissions	Horák et al. (2022)
A review of the effects of several feedstocks-based biochars on soil remediation	Ji et al. (2022)
Applications of biochar in soil: problems, opportunities, and sustainability	Kamali et al. (2022)
Use of magnesium-modified wheat straw biochar to reduce greenhouse gas emissions from a red acidic soil	Khan et al. (2022)

(Continued)

Table 7.6 (Continued)

Biochar serves as environmental armor and plays a variety of roles in defending soil, water, and the atmosphere.	Kumar et al. (2022)
By enhancing the health of the soil and the functional diversity of the rhizosphere microorganisms, biochar inhibits the bacterial wilt disease of flue-cured tobacco.	Li et al. (2022)
The effects of biochar on the soil environment of soybean root systems.	Li et al. (2022)
A potential conditioner for the secure repair of Pb-contaminated alkaline soil is vinegar residue biochar.	Li et al. (2022)
Effects and processes of mixed bacteria-loaded biochar for the immobilization of arsenic, lead, and cadmium in a polluted soil system.	Liang and Zhang (2022)
The temperature of the pyrolysis process may have an impact on how biochar affects the activity of the enzymes that help soils absorb nutrients: iterative analysis	Liao et al. (2022)
A strategy for cleaning up chromium-tainted water and soil that uses designed biochar made from peanut shells	Murad et al. (2022)
Microplastic-contaminated soil's chemical and microbiological characteristics are changed by biochar.	Palansooriya et al. (2022)
Toward estimating the effects of biochar on soil nitrogen availability for plants	Phillips et al. (2022)
Using biochar substrates to enhance the physical qualities of soil and the connections between soil parameters in agricultural soils with varying textures	Šimanský et al. (2022)
Influence of biochar remediation on *Eisenia fetida* in Pb-contaminated soils	Sun (2022a,b)
A quality assessment of poultry litter biochar's nutrient preservation and pollutant immobilization in multi-contaminated soil	Tsai and Chang (2022)
Response of soil enzyme activity to the addition of various kinds of biochar	Wojewódzki et al. (2022)
By changing the soil's C/N component when N fertilizer is placed deeply, the interaction between biochar and N fertilizers improves the efficiency of N consumption.	Xia et al. (2022)
In *Camellia oleifera* plantings, biochar made from waste mushroom substrate decreased N_2O emissions with lower water content but increased CH_4 emissions when the soil was inundated.	Xu et al. (2022)
Evidence from a global meta-analysis shows that biochar improves soil carbon reservoirs.	Chagas et al. (2022)
Effects of biochar-based materials on the biological effects and bioavailability of soil organic contaminants.	Lin et al. (2022a)
Under varying regulated redox conditions, biochar made from pig carcasses had inconsistent effects on the mobilization of arsenic in a polluted paddy soil.	Yang et al. (2022a)
In a maize-soil system, a biochar amendment promotes the retention of 15N fertilizer and rhizosphere N enrichment.	Zhang et al. (2022a)
Effects of biochar and slag additives on soil organic carbon fractions and microorganisms during floods in a rice field in southeast China after 2 yr.	Lin et al. (2022b)

Table 7.6 (Continued)

Using geochemical and spectroscopic methods, elucidating the redox-driven dynamic interactions between arsenic and iron-impregnated biochar in a paddy soil.	Yang et al. (2022b)
A 4-yr study on biochar citrus growth responding favorably to biochar in an acidic soil environment when potassium levels are low.	Zhang et al. (2022b)
Metolachlor recovery from soil and adsorption using a new magnetic illite-biochar	Liu (2022c)
Understanding the microbially-driven nitrogen dynamics that reduce nitrogen losses in paddy soil with the application of reed-biochar	Zhang et al. (2022c)
The effects of biochar soil amendment on rice growth may vary greatly with rice genotypes.	Liu (2022d)
The impact of a biochar soil amendment on rice development can range significantly between various rice genotypes.	Liu (2022b)
By changing the root development strategy in Ferralsol and Anthrosol, biochar and alternating wetting-drying cycles improve the availability of nutrients in the rhizosphere soil and tobacco growth.	Liu (2022d)

References

Abbas, T., Rizwan, M., Ali, S. et al. (2017). Effect of biochar on cadmium bioavailability and uptake in wheat (*Triticum aestivum* L.) grown in a soil with aged contamination. *Ecotoxicol. Environ. Saf.* 140: 37–47. https://doi.org/10.1016/j.ecoenv.2017.02.028.

Abbas, T., Rizwan, M., Ali, S. et al. (2018a). Biochar application increased the growth and yield and reduced cadmium in drought stressed wheat grown in an aged contaminated soil. *Ecotoxicol. Environ. Saf.* 148: 825–833. https://doi.org/10.1016/j.ecoenv.2017.11.063.

Abbas, T., Rizwan, M., Ali, S. et al. (2018b). Effect of biochar on alleviation of cadmium toxicity in wheat (*Triticum aestivum* L.) grown on Cd-contaminated saline soil. *Environ. Sci. Pollut. Res.* 25: 25668–25680. https://doi.org/10.1007/s11356-017-8987-4.

Abbasi, M.K. and Anwar, A.A. (2015). Ameliorating effects of biochar derived from poultry manure and white clover residues on soil nutrient status and plant growth promotion – greenhouse experiments. *PLoS One* 10: 1–18. https://doi.org/10.1371/journal.pone.0131592.

Agegnehu, G., Srivastava, A.K., and Bird, M.I. (2017). The role of biochar and biochar-compost in improving soil quality and crop performance: a review. *Appl. Soil Ecol.* 119: 156–170. https://doi.org/10.1016/j.apsoil.2017.06.008.

Ahmad, A., Chowdhary, P., Khan, N. et al. (2022). Effect of sewage sludge biochar on the soil nutrient, microbial abundance, and plant biomass: a sustainable approach towards mitigation of solid waste. *Chemosphere* 287: 132112.

Akhtar, S.S., Li, G., Andersen, M.N., and Liu, F. (2014). Biochar enhances yield and quality of tomato under reduced irrigation. *Agric. Water Manag.* 138: 37–44. https://doi.org/10.1016/j.agwat.2014.02.016.

Akhtar, S.S., Andersen, M.N., and Liu, F. (2015). Biochar mitigates salinity stress in potato. *J. Agron. Crop Sci.* 201: 368–378. https://doi.org/10.1111/jac.12132.

Alaboudi, K.A., Ahmed, B., and Brodie, G. (2019). Effect of biochar on Pb, Cd and Cr availability and maize growth in artificial contaminated soil. *Ann. Agric. Sci.* 64: 95–102. https://doi.org/10.1016/j.aoas.2019.04.002.

Ali, A., Guo, D., Zhang, Y. et al. (2017a). Using bamboo biochar with compost for the stabilization and phytotoxicity reduction of heavy metals in mine-contaminated soils of China. *Sci. Rep.* 7: 1–12. https://doi.org/10.1038/s41598-017-03045-9.

Ali, S., Rizwan, M., Qayyum, M.F. et al. (2017b). Biochar soil amendment on alleviation of drought and salt stress in plants: a critical review. *Environ. Sci. Pollut. Res.* 24: 12700–12712. https://doi.org/10.1007/s11356-017-8904-x.

Aller, D., Rathke, S., Laird, D. et al. (2017). Impacts of fresh and aged biochars on plant available water and water use efficiency. *Geoderma* 307: 114–121. https://doi.org/10.1016/j.geoderma.2017.08.007.

Assa, H., Li, X., Jeyakumar, P. et al. (2021). Influence of biochar and soil properties on soil and plant tissue concentrations of Cd and Pb: a meta-analysis. *Sci. Total Environ.* 755: 142582. https://doi.org/10.1016/j.scitotenv.2020.142582.

Atkinson, C.J., Fitzgerald, J.D., and Hipp,s, N.A. (2010). Potential mechanisms for achieving agricultural benefits from biochar application to temperate soils: review. *Plant Soil* 337: 1–18.

Bailey, S.E., Olin, T.J., Bricka, R.M., and Adrian, D.D. (1999). A review of potentially low-cost sorbents for heavy metals. *Water Res.* 33 (11): 2469–2479.

Barati, M., Bakhtiari, F., Mowla, D., and Safarzadeh, S. (2018). Comparison of the effects of poultry manure and its biochar on barley growth in petroleum-contaminated soils. *Int. J. Phytorem.* 20: 98–103. https://doi.org/10.1080/15226514.2017.1337069.

Bashir, A., Rizwan, M., Zia ur Rehman, M. et al. (2020). Application of co-composted farm manure and biochar increased the wheat growth and decreased cadmium accumulation in plants under different water regimes. *Chemosphere* 246: 125809. https://doi.org/10.1016/j.chemosphere.2019.125809.

Bhattacharyya, S.S., Ros, G.H., Furtak, K., et al. (2022). Soil carbon sequestration–an interplay between soil microbial community and soil organic matter dynamics. *Science of The Total Environment*, 152928.

Beatrice, A., Varco, J.J., Dygert, A. et al. (2022). Lead immobilization in simulated polluted soil by Douglas fir biochar-supported phosphate. *Chemosphere* 292: 133355.

Beesely, L., Moreno-Jiménez, E., Gomez-Eyles, J.L. et al. (2011). A review of biochars' potential role in the remediation, revegetation and restoration of contaminated soils. *Environ. Pollut.* 159: 3269–3282.

Beesley, L. and Marmiroli, M. (2011). The immobilisation and retention of soluble arsenic, cadmium and zinc by biochar. *Environ. Pollut.* 159: 474–480. https://doi.org/10.1016/j.envpol.2010.10.016.

Bopp, C., Christl, I., Schulin, R., and Evangelou, M.W.H. (2016). Biochar as possible long-term soil amendment for phytostabilisation of TE-contaminated soils. *Environ. Sci. Pollut. Res.* 23: 17449–17458. https://doi.org/10.1007/s11356-016-6935-3.

Borchard, N., Schirrmann, M., Cayuela, M.L. et al. (2019). Biochar, soil and land-use interactions that reduce nitrate leaching and N_2O emissions: a meta-analysis. *Sci. Total Environ.* 651: 2354–2364. https://doi.org/10.1016/j.scitotenv.2018.10.060.

Bouqbis, L., Daoud, S., Koyro, H.W. et al. (2017). Phytotoxic effects of argan shell biochar on salad and barley germination. *Agric. Nat. Resour.* 51: 247–252. https://doi.org/10.1016/j.anres.2017.04.001.

Brennan, A., Jiménez, E.M., Puschenreiter, M. et al. (2014). Effects of biochar amendment on root traits and contaminant availability of maize plants in a copper and arsenic impacted soil. *Plant Soil* 379: 351–360. https://doi.org/10.1007/s11104-014-2074-0.

Brewer, C.E., Schmidt-Rohr, K., Satrio, J.A., and Brown, R.C. (2009). Characterization of biochar from fast pyrolysis and gasification systems. *Environ. Prog. Sustain. Energy* 28 (3): 386–396.

Cao, X., Ma, L., Gao, B., and Harris, W. (2009). Dairy-manure derived biochar effectively sorbs lead and atrazine. *Environ. Sci. Technol.* 43 (9): 3285–3291.

Chagas, J.K.M., de Figueiredo, C.C., and Ramos, M.L.G. (2022). Biochar increases soil carbon pools: evidence from a global meta-analysis. *J. Environ. Manage.* 305: 114403.

Chen, D., Liu, W., Wang, Y., and Lu, P. (2021). Effect of biochar aging on the adsorption and stabilization of Pb in soil. *J. Soil. Sediment.* 22. https://doi.org/10.1007/s11368-021-03059-x.

Chen, D., Liu, W., Wang, Y., and Lu, P. (2022a). Effect of biochar aging on the adsorption and stabilization of Pb in soil. *J. Soil. Sediment.* 22 (1): 56–66.

Chen, H., Feng, Y., Yang, X. et al. (2022b). Assessing simultaneous immobilization of lead and improvement of phosphorus availability through application of phosphorus-rich biochar in a contaminated soil: a pot experiment. *Chemosphere* 296: 133891.

Cheng, C., Han, H., Wang, Y. et al. (2020). Biochar and metal-immobilizing *Serratia liquefaciens* CL-1 synergistically reduced metal accumulation in wheat grains in a metal-contaminated soil. *Sci. Total Environ.* 740: 139972. https://doi.org/10.1016/j.scitotenv.2020.139972.

Cole, E.J., Zandvakili, O.R., Blanchard, J. et al. (2019). Investigating responses of soil bacterial community composition to hardwood biochar amendment using high-throughput PCR sequencing. *Appl. Soil Ecol.* 136: 80–85. https://doi.org/10.1016/j.apsoil.2018.12.010.

Conte, P. and Nestle, N. (2015). Water dynamics in different biochar fractions. *Magn. Reson. Chem.* 53: 726–734. https://doi.org/10.1002/mrc.4204.

Cui, H., Li, D., Liu, X. et al. (2021). Dry-wet and freeze-thaw aging activate endogenous copper and cadmium in biochar. *J. Clean. Prod.* 288: 125605. https://doi.org/10.1016/j.jclepro.2020.125605.

Dhiman, S., Ibrahim, M., Devi, K. et al. (2021). Biochar assisted remediation of toxic metals and metalloids. In: *Handbook of Assisted and Amendment: Enhanced Sustainable Remediation Technology* (ed. M.N.V. Prasad), 131–162.

Dhir, B. (2021). Biochar amendment improves crop production in problematic soils. In: *Handbook of Assisted and Amendment: Enhanced Sustainable Remediation Technology* (ed. M.N.V. Prasad), 189–204.

Ding, Z., Zhou, Z., Lin, X. et al. (2020). Biochar impacts on NH_3 -volatilization kinetics and growth of sweet basil (*Ocimum basilicum* L.) under saline conditions. *Ind. Crops Prod. J.* 15. https://doi.org/10.1016/j.indcrop.2020.112903.

Ding, Z., Majrashi, M.A., Ghoneim, A.M. et al. (2022). Irrigation and biochar effects on pearl millet and kinetics of ammonia volatilization from saline sandy soils. *J. Soil Sci. Plant Nutr.* 22: 1–13.

Dissanayake, P.D., Palansooriya, K.N., Sang, M.K. et al. (2022). Combined effect of biochar and soil moisture on soil chemical properties and microbial community composition in microplastic-contaminated agricultural soil. *Soil Use Manage.* 38 (3): 1446–1458.

Dong, X., Li, G., Lin, Q., and Zhao, X. (2017). Quantity and quality changes of biochar aged for 5 years in soil under field conditions. *Catena* 159: 136–143. https://doi.org/10.1016/j.catena.2017.08.008.

Edeh, I.G. and Mašek, O. (2022). The role of biochar particle size and hydrophobicity in improving soil hydraulic properties. *Eur. J. Soil Sci.* 73 (1): e13138.

El Rasafi, T. and Haddioui, A. (2020). Growth, survival and biomass production of barley in a polluted mine soil amended with biochar and animal manure. *Bull. Environ. Contam. Toxicol.* 105: 155–165. https://doi.org/10.1007/s00128-020-02914-w.

EL Rasafi, T., Bouda, S., Nouri, M., and Haddioui, A. (2020a). Assessment of metals (Cu, Ni) and metalloids (As) induced stress responses in Barley (*Hordeum vulgare*) and wheat (*Triticum aestivum*). *J. Mater. Environ. Sci.* 11: 795–807.

El Rasafi, T., Oukarroum, A., Haddioui, A. et al. (2020b). Cadmium stress in plants: a critical review of the effects, mechanisms, and tolerance strategies. *Crit. Rev. Environ. Sci. Technol.* 52: 1–52. https://doi.org/10.1080/10643389.2020.1835435.

EL Rasafi, T., Oukarroum, A., and Haddioui, A. (2021a). Response of maize to coniferous tree woods biochar and sheep manure application to contaminated mine soil. *Biomass Convers. Biorefin.* 1–11. https://doi.org/10.1007/s13399-021-01390-4.

EL Rasafi, T., Pereira, R., Pinto, G. et al. (2021b). Potential of *Eucalyptus globulus* for the phytoremediation of metals in a Moroccan iron mine soil – a case study. *Environ. Sci. Pollut. Res.* 28: 15782–15793. https://doi.org/10.1007/s11356-020-11494-5.

El-Naggar, A., El-Naggar, A.H., Shaheen, S.M. et al. (2019). Biochar composition-dependent impacts on soil nutrient release, carbon mineralization, and potential environmental risk: a review. *J. Environ. Manage.* 241: 458–467. https://doi.org/10.1016/j.jenvman.2019.02.044.

Fahad, S., Hussain, S., Saud, S. et al. (2015). A biochar application protects rice pollen from high-temperature stress. *Plant Physiol. Biochem.* 96: 281–287. https://doi.org/10.1016/j.plaphy.2015.08.009.

Fahad, S., Hussain, S., Saud, S. et al. (2016). A combined application of biochar and phosphorus alleviates heat-induced adversities on physiological, agronomical and quality attributes of rice. *Plant Physiol. Biochem.* 103: 191–198. https://doi.org/10.1016/j.plaphy.2016.03.001.

Faloye, O.T., Alatise, M.O., Ajayi, A.E., and Ewulo, B.S. (2019). Effects of biochar and inorganic fertiliser applications on growth, yield and water use efficiency of maize under deficit irrigation. *Agric Water Manag* 217: 165–178. https://doi.org/10.1016/j.agwat.2019.02.044.

Farhangi-Abriz, S. and Torabian, S. (2017). Antioxidant enzyme and osmotic adjustment changes in bean seedlings as affected by biochar under salt stress. *Ecotoxicol. Environ. Saf.* 137: 64–70. https://doi.org/10.1016/j.ecoenv.2016.11.029.

Farhangi-Abriz, S. and Torabian, S. (2018). Effect of biochar on growth and ion contents of bean plant under saline condition. *Environ. Sci. Pollut. Res.* 25: 11556–11564. https://doi.org/10.1007/s11356-018-1446-z.

Fischer, B.M.C., Manzoni, S., Morillas, L. et al. (2019). Improving agricultural water use efficiency with biochar – a synthesis of biochar effects on water storage and fluxes across scales. *Sci. Total Environ.* 657: 853–862. https://doi.org/10.1016/j.scitotenv.2018.11.312.

Foster, E.J., Hansen, N., Wallenstein, M., and Cotrufo, M.F. (2016). Biochar and manure amendments impact soil nutrients and microbial enzymatic activities in a semi-arid irrigated maize cropping system. *Agric. Ecosyst. Environ.* 233: 404–414. https://doi.org/10.1016/j.agee.2016.09.029.

Fu, Q., Zhao, H., Li, T.X. et al. (2019). Effects of biochar addition on soil hydraulic properties before and after freezing-thawing. *Catena* 176: 112–124. https://doi.org/10.1016/j.catena.2019.01.008.

Gałwa-Widera, M. (2021). Biochar–production, properties, and service to environmental protection against toxic metals. In: *Handbook of Assisted and Amendment: Enhanced Sustainable Remediation Technology* (ed. M.N.V. Prasad), 53–75.

Gao, F., Ahmad, S., Tang, J. et al. (2022). Enhanced nitrobenzene removal in soil by biochar supported sulfidated nano zerovalent iron: solubilization effect and mechanism. *Sci. Total Environ.* 826: 153960.

Gascó, G., Álvarez, M.L., Paz-Ferreiro, J., and Méndez, A. (2019). Combining phytoextraction by Brassica napus and biochar amendment for the remediation of a mining soil in Riotinto (Spain). *Chemosphere* 231: 562–570. https://doi.org/10.1016/j.chemosphere.2019.05.168.

Gavili, E., Moosavi, A.A., and Kamgar Haghighi, A.A. (2019). Does biochar mitigate the adverse effects of drought on the agronomic traits and yield components of soybean? *Ind. Crop Prod.* 128: 445–454. https://doi.org/10.1016/j.indcrop.2018.11.047.

Ghassemi-Golezani, K. and Farhangi-Abriz, S. (2019). Biochar alleviates fluoride toxicity and oxidative stress in safflower (*Carthamus tinctorius* L.) seedlings. *Chemosphere* 223: 406–415. https://doi.org/10.1016/j.chemosphere.2019.02.087.

Ghosh, D. and Maiti, S.K. (2020). Can biochar reclaim coal mine spoil? *J. Environ. Manage.* 272: 111097. https://doi.org/10.1016/j.jenvman.2020.111097.

Ginebra, M., Muñoz, C., Calvelo-Pereira, R. et al. (2022). Biochar impacts on soil chemical properties, greenhouse gas emissions and forage productivity: a field experiment. *Sci. Total Environ.* 806: 150465.

Głąb, T., Palmowska, J., Zaleski, T., and Gondek, K. (2016). Effect of biochar application on soil hydrological properties and physical quality of sandy soil. *Geoderma* 281: 11–20. https://doi.org/10.1016/j.geoderma.2016.06.028.

Godlewska, P. and Oleszczuk, P. (2022). Effect of biomass addition before sewage sludge pyrolysis on the persistence and bioavailability of polycyclic aromatic hydrocarbons in biochar-amended soil. *Chem. Eng. J.* 429: 132143.

Godlewska, P., Ok, Y.S., and Oleszczuk, P. (2020). The dark side of black gold: ecotoxicological aspects of biochar and biochar-amended soils. *J. Hazard. Mater.* 403: 123833. https://doi.org/10.1016/j.jhazmat.2020.123833.

Gong, X., Huang, D., Liu, Y. et al. (2019). Biochar facilitated the phytoremediation of cadmium contaminated sediments: metal behavior, plant toxicity, and microbial activity. *Sci. Total Environ.* 666: 1126–1133. https://doi.org/10.1016/j.scitotenv.2019.02.215.

Gramig, G.G., Hogstad, S.K., and Carr, P.M. (2022). Can hemp hurd or paper mulch and biochar application improve weed management in matted-row strawberry production systems? *Renewable Agric. Food Syst.* 37 (3): 257–267.

Gray, M., Johnson, M.G., Dragila, M.I., and Kleber, M. (2014). Water uptake in biochars: the roles of porosity and hydrophobicity. *Biomass Bioenergy* 61: 196–205. https://doi.org/10.1016/j.biombioe.2013.12.010.

Grobelak, A. and Jaskulak, M. (2021). Biochars for remediation of recalcitrant soils to enhance agronomic performance. In: *Handbook of Assisted and Amendment: Enhanced Sustainable Remediation Technology* (ed. M.N.V. Prasad), 179–188.

Gross, C.D., Bork, E.W., Carlyle, C.N., and Chang, S.X. (2022). Biochar and its manure-based feedstock have divergent effects on soil organic carbon and greenhouse gas emissions in croplands. *Sci. Total Environ.* 806: 151337.

Gul, S., Whalen, J.K., Thomas, B.W. et al. (2015). Physico-chemical properties and microbial responses in biochar-amended soils: mechanisms and future directions. *Agric. Ecosyst. Environ.* 206: 46–59. https://doi.org/10.1016/j.agee.2015.03.015.

Guo, X., Liu, H., and Zhang, J. (2020). The role of biochar in organic waste composting and soil improvement: a review. *Waste Manag.* 102: 884–899. https://doi.org/10.1016/j.wasman.2019.12.003.

Hafeez, F., Rizwan, M., Saqib, M. et al. (2019). Residual effect of biochar on growth, antioxidant defence and cadmium (Cd) accumulation in rice in a Cd contaminated saline soil. *Pakistan J. Agric. Sci.* 56: 197–204. https://doi.org/10.21162/PAKJAS/19.7546.

Haider, F.U., Coulter, J.A., Liqun, C.A.I. et al. (2022). An overview on biochar production, its implications, and mechanisms of biochar-induced amelioration of soil and plant characteristics. *Pedosphere* 32 (1): 107–130.

Han, L., Chen, L., Li, D. et al. (2022a). Influence of polyethylene terephthalate microplastic and biochar co-existence on paddy soil bacterial community structure and greenhouse gas emission. *Environ. Pollut.* 292: 118386.

Han, L., Zhang, B., Li, D. et al. (2022b). Co-occurrence of microplastics and hydrochar stimulated the methane emission but suppressed nitrous oxide emission from a rice paddy soil. *J. Clean. Prod.* 337: 130504.

Harindintwali, J.D., Zhou, J., Yang, W. et al. (2020). Biochar-bacteria-plant partnerships: eco-solutions for tackling heavy metal pollution. *Ecotoxicol. Environ. Saf.* 204: 111020. https://doi.org/10.1016/j.ecoenv.2020.111020.

He, L., Zhong, H., Liu, G. et al. (2019). Remediation of heavy metal contaminated soils by biochar: mechanisms, potential risks and applications in China. *Environ. Pollut.* 252: 846–855. https://doi.org/10.1016/j.envpol.2019.05.151.

Horák, J., Šimanský, V., Kotuš, T. et al. (2022). Mitigation of greenhouse gas emissions with biochar application in compacted and uncompacted soil. *Agronomy* 12 (3): 546.

Hossain, M.K., Strezov, V., Chan, K.Y. et al. (2011). Influence of pyrolysis temperature on production and nutrient properties of wastewater sludge biochar. *J. Environ. Manage.* 92: 223–228.

Huang, M., Zhang, Z., Zhai, Y. et al. (2019a). Effect of straw biochar on soil properties and wheat production under salinewater irrigation. *Agronomy* 9: 1–15. https://doi.org/10.3390/agronomy9080457.

Huang, M., Zhang, Z., Zhu, C. et al. (2019b). Effect of biochar on sweet corn and soil salinity under conjunctive irrigation with brackish water in coastal saline soil. *Sci. Hortic. (Amsterdam)* 250: 405–413. https://doi.org/10.1016/j.scienta.2019.02.077.

Huang, F., Li, K., Wu, R.R. et al. (2020). Insight into the Cd^{2+} biosorption by viable Bacillus cereus RC-1 immobilized on different biochars: roles of bacterial cell and biochar matrix. *J. Clean. Prod.* 272: 122743. https://doi.org/10.1016/j.jclepro.2020.122743.

Husk B., (2008). Preliminary evaluation of biochar in a commercial farming operation in Canada. http://www.blue-leaf.ca/main-en/files/BlueLeaf_Biochar_Field_Trial_2008_fv.pdf.

Hussain, R., Garg, A., and Ravi, K. (2020). Soil-biochar-plant interaction: differences from the perspective of engineered and agricultural soils. *Bull. Eng. Geol. Environ.* 79. https://doi.org/10.1007/s10064-020-01846-3.

Jeffery, S., Meinders, M.B.J., Stoof, C.R. et al. (2015). Biochar application does not improve the soil hydrological function of a sandy soil. *Geoderma* 251–252: 47–54. https://doi.org/10.1016/j.geoderma.2015.03.022.

Ji, M., Wang, X., Usman, M. et al. (2022). Effects of different feedstocks-based biochar on soil remediation: a review. *Environ. Pollut.* 294: 118655.

Jia, W., Wang, C., Ma, C. et al. (2019). Mineral elements uptake and physiological response of *Amaranthus mangostanus* (L.) as affected by biochar. *Ecotoxicol. Environ. Saf.* 175: 58–65. https://doi.org/10.1016/j.ecoenv.2019.03.039.

Jiang, S., Dai, G., Zhou, J. et al. (2022). An assessment of integrated amendments of biochar and soil replacement on the phytotoxicity of metal (loid) s in rotated radish-soya bean-amaranth in a mining acidy soil. *Chemosphere* 287: 132082.

Kacprzak, M., Kupich, I., Jasinska, A., and Fijalkowski, K. (2022). Bio-based waste' substrates for degraded soil improvement – advantages and challenges in European context. *Energies* 15 (1): 385.

Kamali, M., Sweygers, N., Al-Salem, S. et al. (2022). Biochar for soil applications-sustainability aspects, challenges and future prospects. *Chem. Eng. J.* 428: 131189.

Kamran, M., Malik, Z., Parveen, A. et al. (2019). Biochar alleviates Cd phytotoxicity by minimizing bioavailability and oxidative stress in pak choi (*Brassica chinensis* L.) cultivated in Cd-polluted soil. *J. Environ. Manage.* 250: 109500. https://doi.org/10.1016/j.jenvman.2019.109500.

Kanwal, S., Ilyas, N., Shabir, S. et al. (2018). Application of biochar in mitigation of negative effects of salinity stress in wheat (*Triticum aestivum* L.). *J. Plant Nutr.* 41: 526–538. https://doi.org/10.1080/01904167.2017.1392568.

Keerthanan, S., Jayasinghe, C., Bolan, N. et al. (2022). Retention of sulfamethoxazole by cinnamon wood biochar and its efficacy of reducing bioavailability and plant uptake in soil. *Chemosphere* 297: 134073.

Keshavarz Afshar, R., Hashemi, M., DaCosta, M. et al. (2016). Biochar application and drought stress effects on physiological characteristics of *Silybum marianum*. *Commun. Soil Sci. Plant Anal.* 47: 743–752. https://doi.org/10.1080/00103624.2016.1146752.

Khadem, A. and Raiesi, F. (2017). Responses of microbial performance and community to corn biochar in calcareous sandy and clayey soils. *Appl. Soil Ecol.* 114: 16–27. https://doi.org/10.1016/j.apsoil.2017.02.018.

Khan, M.A., Ding, X., Khan, S. et al. (2018). The influence of various organic amendments on the bioavailability and plant uptake of cadmium present in mine-degraded soil. *Sci. Total Environ.* 636: 810–817. https://doi.org/10.1016/j.scitotenv.2018.04.299.

Khan, M.N., Huang, J., Li, D. et al. (2022). Mitigation of greenhouse gas emissions from a red acidic soil by using magnesium-modified wheat straw biochar. *Environ. Res.* 203: 111879.

Kibria, M.G., Hossain, M., Murata, Y., and Hoque, M.A. (2017). Antioxidant defense mechanisms of salinity tolerance in rice genotypes. *Ric. Sci.* 24: 155–162. https://doi.org/10.1016/j.rsci.2017.05.001.

Kim, H.S., Kim, K.R., Kim, H.J. et al. (2015). Effect of biochar on heavy metal immobilization and uptake by lettuce (*Lactuca sativa* L.) in agricultural soil. *Environ. Earth Sci.* 74: 1249–1259. https://doi.org/10.1007/s12665-015-4116-1.

Kiran, B.R. and Prasad, M.N.V. (2019). Biochar and rice husk ash assisted phytoremediation potentials of *Ricinus communis* L. for lead-spiked soils. *Ecotoxicol. Environ. Saf.* 183: 109574. https://doi.org/10.1016/j.ecoenv.2019.109574.

Krull, E. (2010). Biochar – Its Role in Crop Nutrition. www.grdc.com.au/ResearchandDevelopment/Research-Updates/2010/02/Biochar-its-role-in-crop-nutrition.

Kumar, A., Singh, E., Mishra, R., and Kumar, S. (2022). Biochar as environmental armour and its diverse role towards protecting soil, water and air. *Sci. Total Environ.* 806: 150444.

Kwarciak-Kozłowska, A. and Włodarczyk, R. (2021). Biochar for wastewater treatment. In: *Handbook of Assisted and Amendment: Enhanced Sustainable Remediation Technology* (ed. M.N.V. Prasad), 103–117.

Lahori, A.H., Zhanyu, G.U.O., Zhang, Z. et al. (2017). Use of biochar as an amendment for remediation of heavy metal-contaminated soils: prospects and challenges. *Pedosphere* 27 (6): 991–1014.

Laird, D.A., Brown, R.C., Amonette, J.E., and Lehmann, J. (2009). Review of the pyrolysis platform for coproducing bio-oil and biochar. *Biofuels Bioprod. Biorefin.* 3: 547–562.

Langeroodi, A.R.S., Campiglia, E., Mancinelli, R., and Radicetti, E. (2019). Can biochar improve pumpkin productivity and its physiological characteristics under reduced irrigation regimes? *Sci. Hortic. (Amsterdam)* 247: 195–204. https://doi.org/10.1016/j.scienta.2018.11.059.

Lehamnn, J., Gaunt, J., and Rondon, M. (2006). Bio-char sequestration in terrestrial ecosystems. *Mitig. Adapt. Strat. Glob. Chang.* 11: 395–419.

Lehmann, J. (2007). A handful of carbon. *Nature* 447: 143–144.

Lehmann, J., Rilling, M.C., Thies, J. et al. (2011). Biochar effects on soil biota – a review. *Soil Biotechnol. Biochem.* 43: 1812–1836.

Li, H., Dong, X., da Silva, E.B. et al. (2017). Mechanisms of metal sorption by biochars: biochar characteristics and modifications. *Chemosphere* 178: 466–478. https://doi.org/10.1016/j.chemosphere.2017.03.072.

Li, C., Ahmed, W., Li, D. et al. (2022a). Biochar suppresses bacterial wilt disease of flue-cured tobacco by improving soil health and functional diversity of rhizosphere microorganisms. *Appl. Soil Ecol.* 171: 104314.

Li, Q., Fu, Q., Li, T. et al. (2022b). Biochar impacts on the soil environment of soybean root systems. *Sci. Total Environ.* 153421.

Li, H., Jiang, Q., Li, R. et al. (2022c). Passivation of lead and cerium in soil facilitated by biochar-supported phosphate-doped ferrihydrite: mechanisms and microbial community evolution. *J. Hazard. Mater.* 436: 129090.

Li, W., Meng, J., Zhang, Y. et al. (2022d). Co-pyrolysis of sewage sludge and metal-free/metal-loaded polyvinyl chloride (PVC) microplastics improved biochar properties and reduced environmental risk of heavy metals. *Environ. Pollut.* 302: 119092.

Liang, W. and Zhang, W. (2022). Mixed bacteria-loaded biochar for the immobilization of arsenic, lead, and cadmium in a polluted soil system: effects and mechanisms. *Sci. Total Environ.* 811: 152112.

Liang, J., Li, Y., Si, B. et al. (2021). Optimizing biochar application to improve soil physical and hydraulic properties in saline-alkali soils. *Sci. Total Environ.* 771. https://doi.org/10.1016/j.scitotenv.2020.144802.

Liao, X., Kang, H., Haidar, G. et al. (2022). The impact of biochar on the activities of soil nutrients acquisition enzymes is potentially controlled by the pyrolysis temperature: a meta-analysis. *Geoderma* 411: 115692.

Lima, J.R.d.S., de Moraes Silva, W., de Medeiros, E.V. et al. (2018). Effect of biochar on physicochemical properties of a sandy soil and maize growth in a greenhouse experiment. *Geoderma* 319: 14–23. https://doi.org/10.1016/j.geoderma.2017.12.033.

Lin, Q., Tan, X., Almatrafi, E. et al. (2022a). Effects of biochar-based materials on the bioavailability of soil organic pollutants and their biological impacts. *Sci. Total Environ.* 826: 153956.

Lin, S., Wang, W., Sardans, J. et al. (2022b). Effects of slag and biochar amendments on microorganisms and fractions of soil organic carbon during flooding in a paddy field after two years in southeastern China. *Sci. Total Environ.* 824: 153783.

Liu, S., Meng, J., Jiang, L. et al. (2017). Rice husk biochar impacts soil phosphorous availability, phosphatase activities and bacterial community characteristics in three different soil types. *Appl. Soil Ecol.* 116: 12–22. https://doi.org/10.1016/j.apsoil.2017.03.020.

Liu, H., Xu, F., Xie, Y. et al. (2018a). Effect of modified coconut shell biochar on availability of heavy metals and biochemical characteristics of soil in multiple heavy metals contaminated soil. *Sci. Total Environ.* 645: 702–709. https://doi.org/10.1016/j.scitotenv.2018.07.115.

Liu, Y., Lonappan, L., Brar, S.K., and Yang, S. (2018b). Impact of biochar amendment in agricultural soils on the sorption, desorption, and degradation of pesticides: a review. *Sci. Total Environ.* 645: 60–70. https://doi.org/10.1016/j.scitotenv.2018.07.099.

Liu, G., Meng, J., Huang, Y. et al. (2020). Effects of carbide slag, lodestone and biochar on the immobilization, plant uptake and translocation of As and Cd in a contaminated paddy soil. *Environ. Pollut.* 266: 115194. https://doi.org/10.1016/j.envpol.2020.115194.

Lu, H., Li, Z., Fu, S. et al. (2015). Combining phytoextraction and biochar addition improves soil biochemical properties in a soil contaminated with Cd. *Chemosphere* 119: 209–216. https://doi.org/10.1016/j.chemosphere.2014.06.024.

Maiti, S.K. and Ghosh, D. (2021). Use of biochar as an amendment for remediation of heavy metal-contaminated soils. In: *Handbook of Assisted and Amendment: Enhanced Sustainable Remediation Technology* (ed. M.N.V. Prasad), 163–177.

Mannan, M.A., Mia, S., Halder, E., and Dijkstra, F.A. (2021). Biochar application rate does not improve plant water availability in soybean under drought stress. *Agric. Water Manag.* 253: 106940. https://doi.org/10.1016/j.agwat.2021.106940.

Manyun, Z., Wang, J., Bai, S.H. et al. (2019). Assisted phytoremediation of a co-contaminated soil with biochar amendment: contaminant removals and bacterial community properties. *Geoderma* 348: 115–123. https://doi.org/10.1016/j.geoderma.2019.04.031.

Matovic, D. (2011). Biochar as a viable carbon sequestration option: global and Canadian perspective. *Energy* 36: 2011–2016.

Mehdizadeh, L., Moghaddam, M., and Lakzian, A. (2020). Amelioration of soil properties, growth and leaf mineral elements of summer savory under salt stress and biochar application in alkaline soil. *Sci. Hortic. (Amsterdam)* 267: 109319. https://doi.org/10.1016/j.scienta.2020.109319.

Mohan, D., Pittman, C. Jr., and Mlsna, T. (2022). *Sustainable Biochar for Water and Wastewater Treatment*. Elsevier.

Moradi, R., Pourghasemian, N., and Naghizadeh, M. (2019). Effect of beeswax waste biochar on growth, physiology and cadmium uptake in saffron. *J. Clean. Prod.* 229: 1251–1261. https://doi.org/10.1016/j.jclepro.2019.05.047.

Mulcahy, D.N., Mulcahy, D.L., and Dietz, D. (2013). Biochar soil amendment increases tomato seedling resistance to drought in sandy soils. *J. Arid Environ.* 88: 222–225. https://doi.org/10.1016/j.jaridenv.2012.07.012.

Murad, H.A., Ahmad, M., Bundschuh, J. et al. (2022). A remediation approach to chromium-contaminated water and soil using engineered biochar derived from peanut shell. *Environ. Res.* 204: 112125.

Nigussie, A., Kissi, E., Misganaw, M., and Ambaw, G. (2012). Effect of biochar application on soil properties and nutrient uptake of lettuces (Lactuca sativa) grown in chromium polluted soils. *Am. Eurasian J. Agric. Environ. Sci.* 12 (3): 369–376.

Nirola, R., Megharaj, M., Palanisami, T. et al. (2015). Evaluation of metal uptake factors of native trees colonizing an abandoned copper mine – a quest for phytostabilization. *J. Sustain. Min.* 14: 115–123. https://doi.org/10.1016/j.jsm.2015.11.001.

Norini, M.P., Thouin, H., Miard, F. et al. (2019). Mobility of Pb, Zn, Ba, As and Cd toward soil pore water and plants (willow and ryegrass) from a mine soil amended with biochar. *J. Environ. Manage.* 232: 117–130. https://doi.org/10.1016/j.jenvman.2018.11.021.

Novak, J.M., Ippolito, J.A., Watts, D.W. et al. (2019). Biochar compost blends facilitate switchgrass growth in mine soils by reducing Cd and Zn bioavailability. *Biochar* 1: 97–114. https://doi.org/10.1007/s42773-019-00004-7.

Nzediegwu, C., Prasher, S., Elsayed, E. et al. (2019). Effect of biochar on heavy metal accumulation in potatoes from wastewater irrigation. *J. Environ. Manage.* 232: 153–164. https://doi.org/10.1016/j.jenvman.2018.11.013.

Ok, Y.S., Tsang, D.C.W., Bolan, N., and Novak, J.M. (ed.) (2018). *Biochar from Biomass and Waste: Fundamentals and Applications*. Elsevier.

Paetsch, L., Mueller, C.W., Kögel-Knabner, I. et al. (2018). Effect of in-situ aged and fresh biochar on soil hydraulic conditions and microbial C use under drought conditions. *Sci. Rep.* 8: 1–11. https://doi.org/10.1038/s41598-018-25039-x.

Palansooriya, K.N., Wong, J.T.F., Hashimoto, Y. et al. (2019). Response of microbial communities to biochar-amended soils: a critical review. *Biochar* 1: 3–22. https://doi.org/10.1007/s42773-019-00009-2.

Palansooriya, K.N., Sang, M.K., Igalavithana, A.D. et al. (2022). Biochar alters chemical and microbial properties of microplastic-contaminated soil. *Environ. Res.* 209: 112807.

Pandit, N.R., Mulder, J., Hale, S.E. et al. (2018). Biochar improves maize growth by alleviation of nutrient stress in a moderately acidic low-input Nepalese soil. *Sci. Total Environ.* 625: 1380–1389. https://doi.org/10.1016/j.scitotenv.2018.01.022.

Paneque, M., De la Rosa, J.M., Franco-Navarro, J.D. et al. (2016). Effect of biochar amendment on morphology, productivity and water relations of sunflower plants under non-irrigation conditions. *Catena* 147: 280–287. https://doi.org/10.1016/j.catena.2016.07.037.

Phillips, C.L., Meyer, K.M., Garcia-Jaramillo, M. et al. (2022). Towards predicting biochar impacts on plant-available soil nitrogen content. *Biochar* 4 (1): 1–15.

Puga, A.P., Abreu, C.A., Melo, L.C.A., and Beesley, L. (2015). Biochar application to a contaminated soil reduces the availability and plant uptake of zinc, lead and cadmium. *J. Environ. Manage.* 159: 86–93. https://doi.org/10.1016/j.jenvman.2015.05.036.

Qambrani, N.A., Rahman, M.M., Won, S. et al. (2017). Biochar properties and eco-friendly applications for climate change mitigation, waste management, and wastewater treatment: a review. *Renew. Sustain. Energy Rev.* 79: 255–273. https://doi.org/10.1016/j.rser.2017.05.057.

Qian, W., Liang, J.Y., Zhang, W.X. et al. (2022). A porous biochar supported nanoscale zero-valent iron material highly efficient for the simultaneous remediation of cadmium and lead contaminated soil. *J. Environ. Sci.* 113: 231–241.

Qinglin, L., Li, H., Fu, Q. et al. (2020). Effects of different biochar application methods on soybean growth indicator variability in a seasonally frozen soil area. *Catena* 185. https://doi.org/10.1016/j.catena.2019.104307.

Quan, G., Fan, Q., Cui, L. et al. (2020). Simulated photocatalytic aging of biochar in soil ecosystem: insight into organic carbon release, surface physicochemical properties and cadmium sorption. *Environ. Res.* 183: 109241. https://doi.org/10.1016/j.envres.2020.109241.

Radziemska, M., Gusiatin, Z.M., Kumar, V., and Brtnicky, M. (2022). Co-application of nanosized halloysite and biochar as soil amendments in aided phytostabilization of metal (-oid) s-contaminated soil under different temperature conditions. *Chemosphere* 288: 132452.

Rawat, J., Saxena, J., and Sanwal, P. (2019). Biochar: a sustainable approach for improving plant growth and soil properties. In: *Biochar – An Imperative Amendment for Soil and the Environment* (ed. V. Abrol and P. Sharma), 1–17. IntechOpen.

Rehman, M.Z.U., Rizwan, M., Ali, S. et al. (2016). Contrasting effects of biochar, compost and farm manure on alleviation of nickel toxicity in maize (*Zea mays* L.) in relation to plant growth, photosynthesis and metal uptake. *Ecotoxicol. Environ. Saf.* 133: 218–225. https://doi.org/10.1016/j.ecoenv.2016.07.023.

Rehman, M., Liu, L., Bashir, S. et al. (2019). Influence of rice straw biochar on growth, antioxidant capacity and copper uptake in ramie (*Boehmeria nivea* L.) grown as forage in aged copper-contaminated soil. *Plant Physiol. Biochem.* 138: 121–129. https://doi.org/10.1016/j.plaphy.2019.02.021.

Rizwan, M., Ali, S., Abbas, T. et al. (2018a). Residual impact of biochar on cadmium uptake by rice (*Oryza sativa* L.) grown in Cd-contaminated soil. *Arab. J. Geosci.* 11: 1–8. https://doi.org/10.1007/s12517-018-3974-8.

Rizwan, M., Ali, S., Abbas, T. et al. (2018b). Residual effects of biochar on growth, photosynthesis and cadmium uptake in rice (*Oryza sativa* L.) under Cd stress with different water conditions. *J. Environ. Manage.* 206: 676–683. https://doi.org/10.1016/j.jenvman.2017.10.035.

Sabir, A., Naveed, M., Bashir, M.A. et al. (2020). Cadmium mediated phytotoxic impacts in *Brassica napus*: managing growth, physiological and oxidative disturbances through combined use of biochar and Enterobacter sp. MN17. *J. Environ. Manage.* 265. https://doi.org/10.1016/j.jenvman.2020.110522.

Saleem, S., Neumann, M., and Liu, F. (2015). Residual effects of biochar on improving growth, physiology and yield of wheat under salt stress. *Agric. Water Manag.* 158: 61–68. https://doi.org/10.1016/j.agwat.2015.04.010.

Sànchez, M.E., Lindao, E., Margaleff, D. et al. (2009). Pyrolysis of agricultural residues from rape and sunflower: production and characterization of bio-fuels and biochar soil management. *J. Anal. Appl. Pyrolysis* 85: 142–144.

Sasongko, D., Gunawan, D., and Indarto, A. (2021). Biochar-based water treatment systems for clean water provision. In: *Handbook of Assisted and Amendment: Enhanced Sustainable Remediation Technology* (ed. M.N.V. Prasad), 77–101.

Shaaban, M., Van Zwieten, L., Bashir, S. et al. (2018). A concise review of biochar application to agricultural soils to improve soil conditions and fight pollution. *J. Environ. Manage.* 228: 429–440. https://doi.org/10.1016/j.jenvman.2018.09.006.

She, D., Sun, X., Gamareldawla, A.H.D. et al. (2018). Benefits of soil biochar amendments to tomato growth under saline water irrigation. *Sci. Rep.* 8: 1–10. https://doi.org/10.1038/s41598-018-33040-7.

Shen, X., Huang, D.Y., Ren, X.F. et al. (2016). Phytoavailability of Cd and Pb in crop straw biochar-amended soil is related to the heavy metal content of both biochar and soil. *J. Environ. Manage.* 168: 245–251. https://doi.org/10.1016/j.jenvman.2015.12.019.

Silva Gonzaga, M.I., Oliveira da Silva, P.S., Carlos de Jesus Santos, J., and Ganassali de Oliveira Junior, L.F. (2019). Biochar increases plant water use efficiency and biomass production while reducing Cu concentration in Brassica juncea L. in a Cu-contaminated soil. *Ecotoxicol. Environ. Saf.* 183: 109557. https://doi.org/10.1016/j.ecoenv.2019.109557.

Šimanský, V., Horák, J., and Bordoloi, S. (2022). Improving the soil physical properties and relationships between soil properties in arable soils of contrasting texture enhancement using biochar substrates: case study in Slovakia. *Geoderma Reg.* 28: e00443.

Singh, J.S. and Singh, C. (ed.) (2020). *Biochar Applications in Agriculture and Environment Management*. Springer International Publishing.

Singh, G., Mavi, M.S., Choudhary, O.P.N.G., and Singh, Y. (2021). Rice straw biochar application to soil irrigated with saline water in a cotton-wheat system improves crop performance and soil functionality in north-west India. *J. Environ. Manage.* 295: 113277. https://doi.org/10.1016/j.jenvman.2021.113277.

Solaiman, Z.M., Murphy, D.V., and Abbott, L.K. (2012). Biochars influence seed germination and early growth of seedlings. *Plant and Soil* 353: 273–287. https://doi.org/10.1007/s11104-011-1031-4.

Sun, F., Chen, J., Chen, F. et al. (2022a). Influence of biochar remediation on Eisenia fetida in Pb-contaminated soils. *Chemosphere* 295: 133954.

Sun, Y., Shaheen, S.M., Ali, E.F. et al. (2022b). Enhancing microplastics biodegradation during composting using livestock manure biochar. *Environ. Pollut.* 306: 119339.

Tan, Z., Lin, C.S.K., Ji, X., and Rainey, T.J. (2017). Returning biochar to fields: a review. *Appl. Soil Ecol.* 116: 1–11. https://doi.org/10.1016/j.apsoil.2017.03.017.

Tan, M., Zhang, H., and Chi, J. (2022). Responses of bioavailability and degradation of phenanthrene in soils with or without earthworms to the addition of mixed particles of biochar and polyethylene. *J. Soil. Sediment.* 22 (1): 185–195.

Tang, J., Li, X., Luo, Y. et al. (2016). Spectroscopic characterization of dissolved organic matter derived from different biochars and their polycylic aromatic hydrocarbons (PAHs) binding affinity. *Chemosphere* 152: 399–406. https://doi.org/10.1016/j.chemosphere.2016.03.016.

Tsang, C.W. and Ok, Y.S. (ed.) (2022). *Biochar in Agriculture for Achieving Sustainable Development Goals*, 422. Academic Press.

Tsang, D.C.W., Ok, Y.S., and Tsang, D.C.W. (ed.) (2022). *Biochar in Agriculture for Achieving Sustainable Development Goals.* Elsevier.

Tsai, C.C. and, Chang, Y.F. (2022). Poultry litter biochar as a gentle soil amendment in multi-contaminated soil: quality evaluation on nutrient preservation and contaminant immobilization. *Agronomy*, 12(2), 405.

Uchimiya, M., Lima, I.M., Klasson, K.T., and Wartelle, L.H. (2010). Contaminant immobilization and nutrient release by biochar soil amendment: roles of natural organic matter. *Chemosphere* 80: 935–940. https://doi.org/10.1016/j.chemosphere.2010.05.020.

Usman, A.R.A., Al-Wabel, M.I., Ok, Y.S. et al. (2016). Conocarpus biochar induces changes in soil nutrient availability and tomato growth under saline irrigation. *Pedosphere* 26: 27–38. https://doi.org/10.1016/S1002-0160(15)60019-4.

Vaccari, F.P., Baronti, S., Lugato, E. et al. (2011). Biochar as a strategy to sequester carbon and increase yield in durum wheat. *Eur. J. Agron.* 34: 231–238. https://doi.org/10.1016/j.eja.2011.01.006.

Wang, Y., Pan, F., Wang, G. et al. (2014). Effects of biochar on photosynthesis and antioxidative system of Malus hupehensis Rehd. seedlings under replant conditions. *Sci. Hortic. (Amsterdam)* 175: 9–15. https://doi.org/10.1016/j.scienta.2014.05.029.

Wang, L., O'Connor, D., Rinklebe, J. et al. (2020). Biochar aging: mechanisms, physicochemical changes, assessment, and implications for field applications. *Environ. Sci. Technol.* 54: 14797–14814. https://doi.org/10.1021/acs.est.0c04033.

Włodarczyk, R. and Kwarciak-Kozłowska, A. (2021). Biochar for bioremediation of toxic metals. In: *Handbook of Assisted and Amendment: Enhanced Sustainable Remediation Technology* (ed. M.N.V. Prasad), 119–130.

Wojewódzki, P., Lemanowicz, J., Debska, B., and Haddad, S.A. (2022). Soil enzyme activity response under the amendment of different types of biochar. *Agronomy* 12 (3): 569.

Wu, S., He, H., Inthapanya, X. et al. (2017). Role of biochar on composting of organic wastes and remediation of contaminated soils – a review. *Environ. Sci. Pollut. Res.* 24: 16560–16577. https://doi.org/10.1007/s11356-017-9168-1.

Xia, H., Riaz, M., Zhang, M. et al. (2022). Biochar-N fertilizer interaction increases N utilization efficiency by modifying soil C/N component under N fertilizer deep placement modes. *Chemosphere* 286: 131594.

Xiaokai, Z., Wang, H., He, L. et al. (2013). Using biochar for remediation of soils contaminated with heavy metals and organic pollutants. *Environ. Sci. Pollut. Res.* 20: 8472–8483. https://doi.org/10.1007/s11356-013-1659-0.

Xu, P., Sun, C.X., Ye, X.Z. et al. (2016). The effect of biochar and crop straws on heavy metal bioavailability and plant accumulation in a Cd and Pb polluted soil. *Ecotoxicol. Environ. Saf.* 132: 94–100. https://doi.org/10.1016/j.ecoenv.2016.05.031.

Xu, Z., Xu, X., Tsang, D.C.W., and Cao, X. (2018). Contrasting impacts of pre- and post-application aging of biochar on the immobilization of Cd in contaminated soils. *Environ. Pollut.* 242: 1362–1370. https://doi.org/10.1016/j.envpol.2018.08.012.

Xu, X., Huang, H., Zhang, Y. et al. (2019). Biochar as both electron donor and electron shuttle for the reduction transformation of Cr(VI) during its sorption. *Environ. Pollut.* 244: 430–423. https://doi.org/10.1016/j.envpol.2018.10.068.

Xu, X., Yuan, X., Zhang, Q. et al. (2022). Biochar derived from spent mushroom substrate reduced N_2O emissions with lower water content but increased CH_4 emissions under flooded condition from fertilized soils in Camellia oleifera plantations. *Chemosphere* 287: 132110.

Yadav, V., Jain, S., Mishra, P. et al. (2019). Amelioration in nutrient mineralization and microbial activities of sandy loam soil by short term field aged biochar. *Appl. Soil Ecol.* 138: 144–155. https://doi.org/10.1016/j.apsoil.2019.01.012.

Yang, X., Liu, J., McGrouther, K. et al. (2016). Effect of biochar on the extractability of heavy metals (Cd, Cu, Pb, and Zn) and enzyme activity in soil. *Environ. Sci. Pollut. Res.* 23: 974–984. https://doi.org/10.1007/s11356-015-4233-0.

Yang, X., Hinzmann, M., Pan, H. et al. (2022a). Pig carcass-derived biochar caused contradictory effects on arsenic mobilization in a contaminated paddy soil under fluctuating controlled redox conditions. *J. Hazard. Mater.* 421: 126647.

Yang, X., Shaheen, S.M., Wang, J. et al. (2022b). Elucidating the redox-driven dynamic interactions between arsenic and iron-impregnated biochar in a paddy soil using geochemical and spectroscopic techniques. *J. Hazard. Mater.* 422: 126808.

Yang, X., Wang, L., Guo, J. et al. (2022c). Aging features of metal(loid)s in biochar-amended soil: effects of biochar type and aging method. *Sci. Total Environ.* 815: 152922.

Yaojun, Z., Ding, J., Wang, H. et al. (2020). Biochar addition alleviate the negative effects of drought and salinity stress on soybean productivity and water use efficiency. *BMC Plant Biol.* 20: 1–11.

Ying, W., Zhong, B., Shafi, M. et al. (2019). Effects of biochar on growth, and heavy metals accumulation of moso bamboo (*Phyllostachy pubescens*), soil physical properties, and heavy metals solubility in soil. *Chemosphere* 219: 510–516. https://doi.org/10.1016/j.chemosphere.2018.11.159.

Younis, U., Malik, S.A., Qayyum, M.F. et al. (2015). Biochar affects growth and biochemical activities of fenugreek (*Trigonella corniculata*) in cadmium polluted soil. *J. Appl. Bot. Food Qual.* 88: 29–33. https://doi.org/10.5073/JABFQ.2015.088.006.

Younis, U., Malik, S.A., Rizwan, M. et al. (2016). Biochar enhances the cadmium tolerance in spinach (*Spinacia oleracea*) through modification of Cd uptake and physiological and biochemical attributes. *Environ. Sci. Pollut. Res.* 23: 21385–21394. https://doi.org/10.1007/s11356-016-7344-3.

Yrjälä, K., Ramakrishnan, M. and Salo, E. (2022). Agricultural waste streams as resource in circular economy for biochar production towards carbon neutrality. *Curr. Opin. Eenviron. Sci. Health*, 26: 100339.

Yuan, J., Meng, J., Liang, X. et al. (2017). Organic molecules from biochar leacheates have a positive effect on rice seedling cold tolerance. *Front. Plant Sci.* 8: 1–13. https://doi.org/10.3389/fpls.2017.01624.

Yuan, P., Wang, J., Pan, Y. et al. (2019). Review of biochar for the management of contaminated soil: preparation, application and prospect. *Sci. Total Environ.* 659: 473–490. https://doi.org/10.1016/j.scitotenv.2018.12.400.

Zbigniew, B., Kobyłecki, R., Scisłowska, M., and Zarzycki, R. (2018). Biochar – potential tool to combat climate change and drought. *Ecohydrol. Hydrobiol. J.* 18: https://doi.org/10.1016/j.ecohyd.2018.11.005.

Zhang, P., Sun, H., Yu, L., and Sun, T. (2013). Adsorption and catalytic hydrolysis of carbaryl and antrazine on pig manure-derived biochars: impact of structural properties of biochars. *J. Hazard. Mater.* 244–245: 217–224.

Zhang, J., Bai, Z., Huang, J. et al. (2019). Biochar alleviated the salt stress of induced saline paddy soil and improved the biochemical characteristics of rice seedlings differing in salt tolerance. *Soil Tillage Res.* 195: 104372. https://doi.org/10.1016/j.still.2019.104372.

Zhang, Y., Wang, J., and Feng, Y. (2021). The effects of biochar addition on soil physicochemical properties: a review. *Catena* 202. https://doi.org/10.1016/j.catena.2021.105284.

Zhang, J., Zhang, L., and Qiu, S. (2022a). Biochar amendment benefits 15N fertilizer retention and rhizosphere N enrichment in a maize-soil system. *Geoderma* 412: 115713.

Zhang, M., Riaz, M., Xia, H. et al. (2022b). Four-year biochar study: positive response of acidic soil microenvironment and citrus growth to biochar under potassium deficiency conditions. *Sci. Total Environ.* 813: 152515.

Zhang, Y., Zhao, H., Hu, W. et al. (2022c). Understanding how reed-biochar application mitigates nitrogen losses in paddy soil: insight into microbially-driven nitrogen dynamics. *Chemosphere* 295: 133904.

Zhao, R., Coles, N., and Wu, J. (2015). Catena carbon mineralization following additions of fresh and aged biochar to an infertile soil. *Catena* 125: 183–189. https://doi.org/10.1016/j.catena.2014.10.026.

Zhao, W., Zhou, Q., Tian, Z. et al. (2020). Apply biochar to ameliorate soda saline-alkali land, improve soil function and increase corn nutrient availability in the Songnen Plain. *Sci. Total Environ.* 722: 137428. https://doi.org/10.1016/j.scitotenv.2020.13742.

Zhou, C., Song, X., Wang, Y. et al. (2022). The sorption and short-term immobilization of lead and cadmium by nano-hydroxyapatite/biochar in aqueous solution and soil. *Chemosphere* 286: 131810.

8

Restoring Ecosystems: Guidance from Agroecology for Sustainability in Thailand

Woranan Nakbanpote[1], Pranee Srihaban[2], Wutthisat Chokkuea[1], Winya Dungkaew[1], Uraiwan Taya[2], Piyanutt Khanema[1], Ruttanakorn Munjit[1], Ponlakit Jitto[3], Piyapatr Busababodhin[1], Surasak Khankhum[1], Khanitta Somtrakoon[1], and Majeti Narasimha Vara Prasad[4]

[1] *Isan Saline Soil Research Unit (ISSRU), Faculty of Science, Mahasarakham University, Khamriang, Kantarawichi, Mahasarakham, 44150, Thailand*
[2] *Land Development Regional Office 5, Department of Land, Development, Ministry of Agriculture and Cooperatives, Khon Kaen, 40000, Thailand*
[3] *Faculty of Environment and Resource Studies, Mahasarakham University, Khamriang, Kantarawichi District, Maha Sarakham, 44150, Thailand*
[4] *School of Life Sciences, University of Hyderabad, Hyderabad, India*

8.1 Introduction

Ecological restoration is the process of restoring ecosystems that have been degraded, damaged, or destroyed and unable to repair themselves. The main purpose of restoring ecosystems is to restore resilient ecological processes and functions (UNEP 2021). Agriculture causes widespread degradation of ecosystems. This is because the agricultural processes in the past have focused on the practice of maximizing productivity by using machines and overloading the production inputs into the area. In addition, the product price is generally established without valuing the lost natural cost (Crossman et al. 2016). Ecosystem restoration is also an important part of sustainable agricultural production and brings many benefits to society Collaborative land restoration and rehabilitation is one of the key strategies for achieving SDG 15 (Life on Land) of the Sustainable Development Goals (SDGs) and in particular for achieving the neutrality goal of land degradation (15.3) (IRP 2019).

Restoration means solving problems that occurred in terms of stability and soil conditions, surface and groundwater quality, biodiversity, and stability of micro-and-global climate in accordance to people's livelihoods and cultures (Crossman et al. 2016). There may be no actual or historical analogy to guide ecological restoration. Therefore, renewal needs to be forward-looking, focusing on the future trajectory of climate, land use, demographic and socioeconomic changes, and species dynamics. A key goal of revitalization is economic, social, and environmental sustainability, so adopting new environmentally friendly

Agroecological Approaches for Sustainable Soil Management, First Edition.
Edited by Majeti Narasimha Vara Prasad and Chitranjan Kumar.

technologies to increase productivity may be necessary, especially in dry and degraded areas. The term "climate-smart agriculture" focuses on encouraging farming techniques that increase yields, reduce vulnerability to climate change, and reduce emissions of greenhouse gas (FAO 2011). Although there are not a specific practice for agroecology, agroecology is a comprehensive approach to management that considers the ecosystem, health, society, and economy of agriculture and food, as well as agricultural practices at the field, farm and landscape level (Wezel et al. 2020; Kerr et al. 2021).

The current development of Thai agriculture is based on the 20-year Agriculture and Cooperatives Strategy (2017–2036) in line with SDGs (MOAC 2017). This chapter presents an overview of sustainable agricultural development in both large-scale farming, small-scale farming, and cooperative farming in Thailand. Farm-scale agroecological approaches in Thailand are focusing on the integrated farming system, organic farming, natural farming, agroforest, and the new theory agriculture (Jitsanguan 2001). Since 2021, Thailand has launched the Bio-Circular-Green (BCG) Economy model as a strategy to build sustainability and inclusiveness for Thailand's economy, society, and environment (NSTDA 2021). Case studies on the development of salinity areas for agriculture, reforestation of drylands for tourism sites and agroforest, integrated farming learning center, and cooperative farming through development of farmers are presented.

8.2 Importance of Agricultural Strategy and Ecological Restoration in Thailand

The importance of agriculture in Thailand can be considered at both national and household levels. Agriculture affects the country in four areas of economics, society, policy, and the environment (Figure 8.1a). From an economic perspective, the cost of agricultural production influences the price of imported-exported agricultural goods. Agricultural produce is a raw material for many industries, especially the food industry. Therefore, agricultural commodity prices drive the cost of various commodities in the supply and demand chain, including the cost of living and the level of inflation in the country (Sukprasert 2019). In society, farming brings people together and creates a community and a society that reflects the activities and lifestyle of people in the area. Agriculture is also a tool for defining the position or development zone of global communities (Zugravu et al. 2017). In politics, agriculture encourages and supports communities, society, and the nation to be strong and stable through communication, cooperation, and co-existence, such as support for the harvesting of rice and agriculture in the community, agricultural cooperatives, the establishment of farmers' councils and agricultural industrial estates, etc. Agriculture also ensures the welfare and peace of the country, e.g., the presence of agricultural agreements, regulations, and legislation that address the quality standards of agricultural products, inputs, processing systems, and impacts of agricultural operations (Zugravu et al. 2017; Sukprasert 2019). From an environmental point of view, agriculture can support or damage the environment. Agricultural activities support human subsistence, supporting biodiversity and maintaining ecosystem services. The negative impacts of agriculture are the conversion of forests, grasslands, and other habitats for agricultural use, degradation of soil quality soil and surface water, aquifers and coastal wetlands by excessive or inappropriate

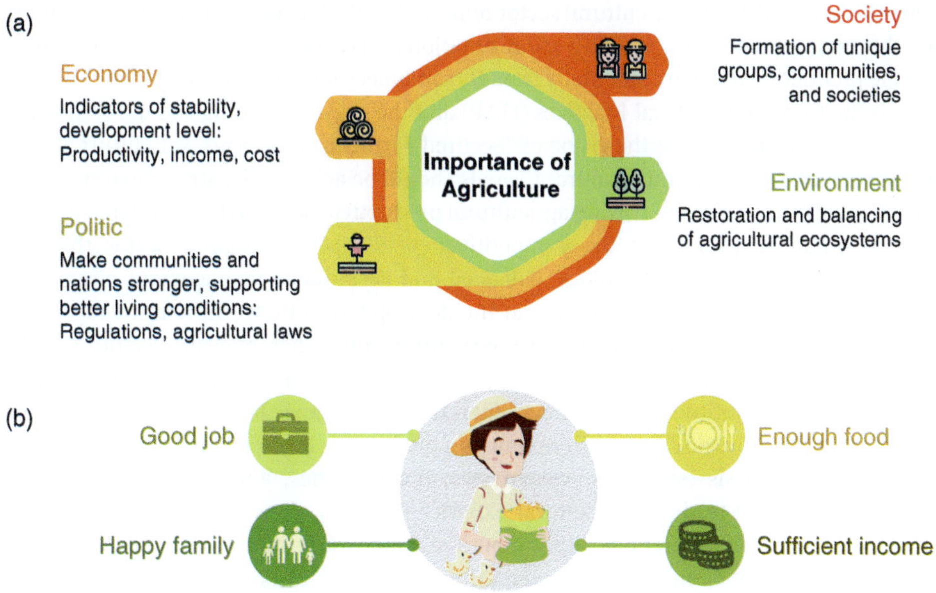

Figure 8.1 The importance of agriculture in Thailand on politics, environment, society, and economy (a), and (b) Basic life needs of farmers.

use of pesticides and fertilizers, significant loss of crop and livestock genetic diversity through the spread of industrial monocultures, reducing resilience to climatic and other changes (Rohila et al. 2017). As a result, agriculture is now focusing on the environment that helps to restore, treat, and/or create ecological and environmental balance, including maintaining the ecosystem. At the household level, the needs of the farmer are the basic needs of a stable job, sufficient and healthy food, an adequate income, and a happy family (Figure 8.1b). Most farmers in Thailand still have small farms not exceeding 15 Rai. In addition, the lack of scale on individual farms has left farmers with uneconomic investments in mechanization, fertilizer application, and the purchase of seeds and other agricultural inputs. Farmers without land and tenants do not have sufficient incentives to increase their production efficiency, many of whom turn to farm workers (MOAC 2017). These problems can be alleviated by applying the sufficiency economy philosophy. If farmers, who are the majority in Thailand, are strong and self-sufficient, it will lead to a strong and sustainable country (The Chaipattana Foundation 2017).

Thailand's agricultural sector has always played a vital role as a major source of agricultural production for domestic consumption and exports to other countries worldwide. However, the Thai agricultural sector still faces a number of problems, namely the instability of agricultural commodity prices, the inappropriate use of agricultural inputs, the depletion of natural resources, and high debt. The MOAC has prepared the 20-year Agriculture and Cooperatives Strategy (2017–2036) to use it as an agenda for the continued and effective operation of development in the agricultural sector. The strategy is consistent with the National Strategy and the National Economic and Social Development Plans, which are the twelfth plan (2017–2021) and the thirteen plan (2022–2026), as well as the

SDGs (MOAC 2017). The agricultural sector remains highly dependent on natural resources and the environment. Consequently, the restoration and conservation of natural resources and the production of environmentally friendly products should be encouraged at the same time as the Good Agricultural Practices (GAP) and the organic farming standard. In addition, the 20-year strategy has the vision of "Secured Farmers, Prosperous Agriculture Sector and Sustainable Agricultural Resources," goals should be achieved by strengthening farmers and their institutions, increasing agricultural productivity and quality standards in agricultural products, promoting the competitiveness of the agricultural sector through improved and innovative technology, balanced and sustainable management of agricultural resources and the environment, and the development of government management (MOAC 2017). Organizations related to agriculture and other relevant sectors should emphasize the promotion of the next generation of farmers by educating, raising awareness, and changing attitudes for the new generation to recognize both the importance and benefits of agriculture and the environment as a profession (Rayasawath 2018).

The SWOT analysis (strengths, weaknesses, opportunities, and threats) shows that Thailand is a world-leading agricultural exporter in terms of rice, rubber, and cassava. Thailand's farmers in different localities inherited their local wisdoms from their ancestors in agricultural production. They also adapt the Sufficiency Economy Philosophy and the New Theory Agriculture initiated by the King Rama IX as their self-immunity. Geographically, Thailand has gained its strength to be located within the land of the appropriate environment for diversified agricultural production. The real weaknesses are the majority of Thai farmers are smallholder farmers, many of whom are landless, poor, and lack technology in managing their production. Among the fluctuations in the world economic situation, rougher and more frequent natural disasters, and problems caused by the COVID-19 epidemic, Thailand still has an opportunity in a strong agricultural section. In addition, the increase in the world's population has led to an explosion in global demand for food, especially food security, as Thailand has a high production potential (MOAC 2017; Sapbamrer et al. 2022).

Focusing only the 20-year Agriculture and Cooperatives Strategy no. 4: Balanced and Sustainable Management of Agricultural Resources and the Environment. The targets are 10 million Rai of sustainable agricultural areas, 2 million rains per year of conserved, improved, and rehabilitated agricultural land, and 49.52 million Rai of irrigated agricultural land. Development guidelines are agricultural resources managed in a sustainable manner, balanced, and revitalized in a sustainable manner, and agricultural resources preserved (MOAC 2017).

8.3 Management of Thailand's Restoration of Agricultural Areas

Based on a 2012 survey by the International Center for Tropical Agriculture (CIAT), the total economic impact of climate change on agriculture in Thailand is forecasted to be between \$300 million and \$420 million (9.8–13.9 billion Thai Baht), with field crop yields such as rice, maize, and sugarcane declining by 3.6% – and as much as 15% – by 2050 (FAO 2022). In particular, average temperatures above 35 °C affect sterile stamens and pollen from rice flowers, resulting in a decrease in rice yield (Zhang et al. 2018; Chen et al. 2020;

Hu et al. 2021). Besides the problem of climate change, the deterioration of natural resources is due to natural depletion and human activity. The ecological balance and biodiversity are lost due to repeated cultivation on the same plot of land, deforestation, improper use of farm chemicals, and misuse of resources without proper conservation. Agricultural adaptation and management are needed to reduce unavoidable impacts, support environmental sustainability, and emphasize social and economic dimensions (Rohila et al. 2017; Soulé et al. 2021). Although large-scale agriculture can have a negative impact on the environment, there is a need to continue exporting and supplying food to the population. As a result, large-scale agriculture would have to be adapted to society, the environment and modern technology. Agricultural system in Thailand under the 20-year agriculture and cooperative strategy (2017–2036) is divided into large-scale farming and small-scale farming for sustainable agriculture systems (MOAC 2017), as shown in Figure 8.2. The bioeconomy is the production of renewable biological resources and converting them into value-added products. The circular economy is designed to reuse and recycle materials to maximize the value of limited resources. The green economy is committed to balancing the economy, society, and the environment to achieve sustainable development. Therefore, the BCG model supports food and agriculture management throughout the chain for high productivity, value-added products, and effective and appropriate investments (NSTDA 2021; Tontisirin 2021).

8.3.1 Large-Scale Agriculture and Modern Agricultural Technology

The Large-Scale Agricultural Extension System project is an important government policy and aligns with the current needs of farmers, in addition to large plots of agribusiness. Cooperative agriculture can solve the problem of high production costs and

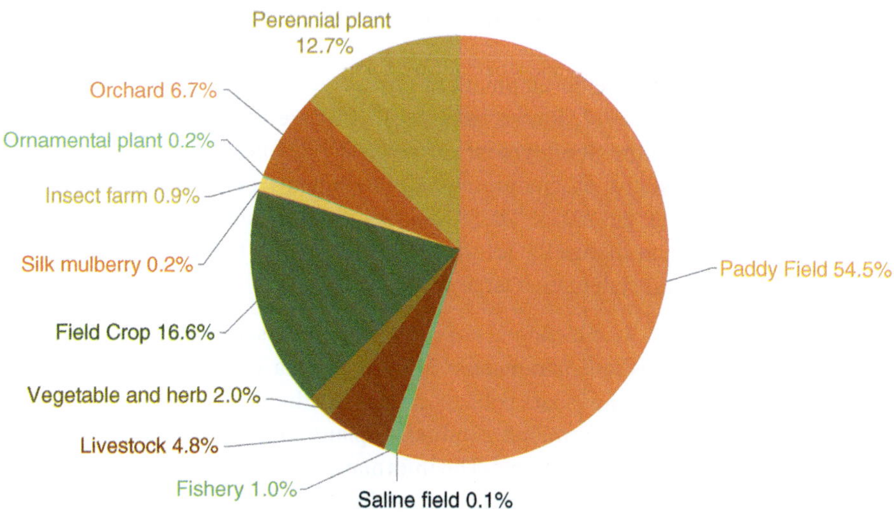

Figure 8.2 Agricultural system in Thailand according to the 20-year agriculture and cooperative strategy (2017–2036) which is divided into large scale farming and small- scale farming for sustainable agriculture systems, supporting the Bio-Circular-Green economic (BCG) model.

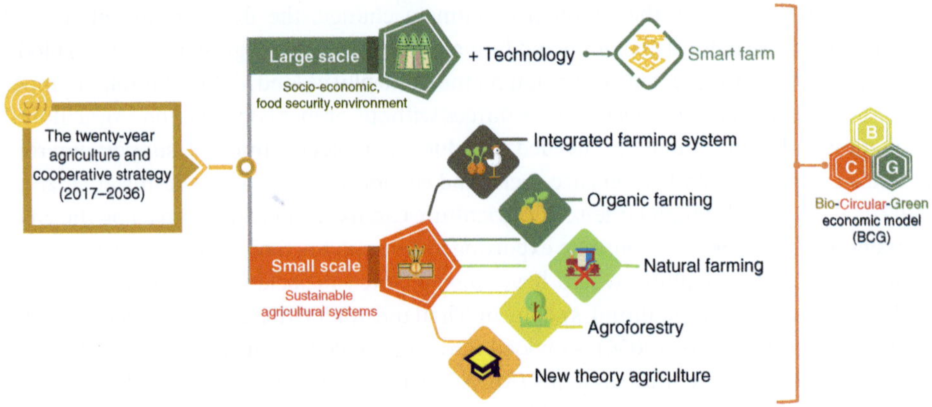

Figure 8.3 The percentage of large-scale farming classified by product group (100% = 6 785 994.01 Rai), in 2021 data from Large Scale Agricultural Extension System, Department of Agricultural Extension (DOAE) (Adapted from https://co-farm.doae.go.th/graph/Dashboard1dsb.php, 19 September 2022).

the absence of negotiating power in the markets for individual farmers. A key principle of this project is an emphasis on farmer groups and participatory management through collaborative area-based production and distribution. Large-scale agriculture categorized by agricultural products in 2021 (Figure 8.3) indicates that the majority of Thai agriculture is rice and field crops (rubber, cassava, sugarcane, maize, and oil palm) (DOAE 2022). The large-scale system has moved Thai agriculture from traditional practices in the past to become more marketed and mechanized. It helps to address agricultural labor shortages, higher wages, and a growing trend of aging in the agricultural sector (Sekkhunthod 2020). Furthermore, large-scale farming should be encouraged to reduce the water footprint of rice production and mitigate the effects of climate change (Arunrat et al. 2020).

The MOAC has prioritized the development of farmers to be smart farmers, especially the new generation (MOAC 2017; Rayasawath 2018). Climate Smart Agriculture (CSA) contributes significantly to the development of cost-effective and refined solutions that improve crops and the status of environmental assessment, field management, and decision-making (Gupta et al. 2022). There are six trends in agricultural technology (AgTech), which are going to change the future of Thai agriculture to become a smart farm and support the agricultural system on a large scale (NIA 2020):

i) *Digital agriculture (DA)*: A collection of near-field data from sensors such as measurements of soil conditions and parameters; Mid-range data collection from cameras attached to drones; and remote data collection from satellite imagery that can be used to determine the details of farmland conditions, plant species, growing stages, and problems from diseases and insects, etc. The information obtained can be coupled with artificial intelligence (AI) to help farmers control their crop fields. In addition, big data can reflect the details of topography, climate, farming at the plot level, and farmers both now and in the past, which will help understand the problems and different needs of farmers.

ii) *Agricultural machinery, robots, drones, and automation*: Agricultural machinery, robots, drones, and automation can help farmers to save time and labor, produce more product, reduce costs, and improve farmers' quality of life, such as using drones to spray fertilizers and other bio-based chemicals instead of labor.
iii) *Agricultural biotechnology*: The agricultural biotechnology is the application of diverse knowledges of biology, chemistry, and agriculture. Biotechnology is becoming widely used in agriculture, for example by exploiting soil microorganisms to encourage growth and drought tolerance, resist weed, and reduce the use of fertilizers and pesticides. Biotechnology also includes gene modification, improve plants or animals to become products, businesses, and solutions to agricultural problems.
iv) *Modern farm management*: A new trend in farm management enhances the efficiency of the production process, the regulation of plants and animals' growth factors, and helping farmers not to worry about the damage of their produce. New type of farms such as "plant factory" that can control the environment and provide water, light, and appropriate temperature and humidity. Then, plants grow well and can also grow crops out of season. In addition, "vertical farming" is a suitable trend of urban farming because it solves the problem of lacking arable land. Farm management with a control system can be applied to "Insect farming," a new source of protein that is starting to attract consumers.
v) *Post-harvest handling and transportation*: Thailand is a tropical country, various agricultural products with thin skins are prone to perishables. Innovative developments related to post-harvesting processes have been interested such as packaging with extending shelf life and environmentally friendly, chemical-free life extension processes, non-destructive product quality monitoring systems, product coating, and innovations in fast transportation, etc. In addition, (v) *Agribusiness services*: Agro-industrial structure in Thailand mostly consists of medium-to-small-sized farming; therefore, farmers unable to invest in the innovations. Agribusiness services and supporting application tend to increase so that farmers can borrow agricultural innovations such as drones and agricultural machinery, etc.

8.3.2 Small-Scale Agriculture and Sustainable Agricultural Systems

Alternative agriculture is an agriculture system that emphasizes the management of existing resources and actual production inputs. It is an eco– friendly system with minimum utilization of resources. Agriculturalists tend to be in line with the alternative agriculture system are small-holder farmers who own not over 15 Rai and landless farmers. However, there are entrepreneur farmers who have better understanding and knowledge on the importance of ecosystems, environment, quality of living and social demands, willingly entering into the world of alternative agriculture. Thus, the known appearance of alternative agriculture is "eco-friendly management" agriculture (Sukprasert 2019). Thailand has diverse agricultural resources, along with rural culture and traditional wisdom. In which sustainable farming should be adapted to different contexts. Sustainable agriculture systems, according to the policy of the MOAC are integrated farming, organic farming, natural farming, Agroforestry, and new theory agriculture (Jitsanguan 2001). Table 8.1 presents the characterization of the five sustainable agricultural systems and their effects on the environment.

Table 8.1 Sustainable agriculture systems, according to the policy of the Ministry of Agriculture and Cooperatives, characterization, and advantage.

Agricultural systems	Characterization	Advantage
Integrated farming	- Two or more agriculture activities are undertaken to maximize benefits rather than profits. - There is ongoing cooperation in terms of support between plants and plants, animals and fish, plants and animals, animals, and animals.	- Reduce the risk of weather variability - Reduce the spread of plant diseases and pests - Reduce the risk of product price volatility - Reduce household expenses - Reduce dependence on external inputs
Organic farming	- An agricultural system that focuses on environmental, social, and economic sustainability, with a focus on improving soils. - Reduce the use of external inputs - In principle: not use chemical fertilizers and pesticides, not use genetically modified crops, use birds and insects to kill pests in the food chain, use organic fertilizers, bio-fertilizers, and microorganisms to improve soil.	- Conservation and improving the environment - Restore the soil's fertility. - Secure production processes for farmers and consumers. - Reduce the cost of agricultural production - Increase the value of products for farmers.
Nature farming	- Farming in accordance with the ecology - Cultivation of healthy plants in a balanced environment. - In principles: no tillage, no use of chemicals and fertilizers, no weeding, no pesticides, or herbicides.	- Create a natural balance of the ecosystem - Do not pollute people or livestock. - Restore the degraded natural resources. - Produce quality products free from toxic substances. - Reduce production costs and use less labor
Agroforestry	- Forestry agriculture, such as the cultivation of plants in the forest, raising animals in the forest and collecting forest products for sustainable use. - Create an agriculture system that imitates the natural forest ecosystem. - Farmers produce without compromising the forest of origin, which contributes to preserving the environment, increasing the abundance of resources, and conserving the diversity of flora and fauna. - Focus on the relative height of each plant to ensure compatibility with nutrients and exposure to sunlight.	- Reduce the problem of forest encroachment - Agriculture in harmony with nature for sustainability
New theory farming	- There are a variety of mixed agricultural activities in four areas: digging ponds, growing rice, growing perennial fruits and plants, buildings and stables (30:30 : 30:10). - Three main objectives: food security creates self-sufficiency, water resources management to maximize benefits and income security by focusing on household consumption and selling the rest as income.	- Self-sufficiency based on the principles of sufficiency economy - Reduce the risk of drought - Generate income from the remaining products for consumption - Allocation of existing land to create meaningful benefits - Promote sustainable agricultural development

8.3.2.1 Integrated Farming

Integrated agriculture comprises many agricultural activities utilizing complementary renewable resources (symbiosis) in the same space leading to the empowerment of benefits, e.g., planting multiple crops and raising animals in the same space, relying on each other to share renewable resources coherently and appropriately. This type of farming is a prototype that contains a dynamic balance indicating the stability of the production system and the abundance of a diversified environment. The Integrated crop-livestock systems contribute to food security, social and economic benefits, and resilience, and provide adaptation strategies in low-, medium-, and high-income countries (Sekaran et al. 2021).

8.3.2.2 Organic Farming

Organic farming can be applied in other farming systems. The system does not allow chemicals to be used at every stage of production. Inputs, processes, and production techniques should be free from all chemicals. It is possible to use fungicides, fertilizers, and insecticides that are biological or biological agents and that are not harmful to the ecosystem or to users and consumers. The objective is to improve food security and restore soil fertility and water quality that have been damaged through the use of chemicals (Jitsanguan 2001). It can be expected that improvements in the environmental impact associated with organic agriculture will include improvements in surface and groundwater quality, conservation of natural enemies, and biodiversity (Tscharntke et al. 2021). Therefore, organic farming tends to become the main agricultural activity in the future due to increasing consumer confidence and social trends in achieving the health benefits.

The MOAC launched a national Programme (2017–2021) for the Promotion of Organic Rice Production (hereafter PPORP), which work together with Rice Research Centers and the rice mills. The stated objectives of the PPROP are to enhance exports of organic rice, to improve the health of farmers and consumers, to stabilize agricultural prices for organic rice and farmers' incomes, and to protect the environment. The PPORP provides training courses in organic farming, distributes organic rice seeds to farmers free of charge, pays subsidies to compensate for lower yields during the conversion period, and grants Organic Thailand certification (Hérique and Faysse 2020).

8.3.2.3 Natural Farming

Nature farming is close to conservation agriculture. It is likely the ultimate model of sustainable agriculture in terms of resource and environmental conservation. As far as possible, natural agriculture tries to avoid the use of inputs from outside sources and reduces any physical methods that affect the ecological balance of farmland. The concept of natural agriculture focuses on preserving the natural ecology, planning, and controlling the balance between the various life cycles of crops, animals, and weeds (Chauhan et al. 2012; Nichols et al. 2015). Natural farming is based on four principles: no tillage, no chemical and fertilizer use, no weeding, no pesticides, or herbicides. The practice of pure natural agriculture remains relatively uncommon in Thailand (Jitsanguan 2001).

8.3.2.4 Agroforestry

The agroforestry system combines agricultural production and forestry to create sustainable agriculture. Commercial crops and livestock are grown in forest areas. This preserves, forests, as well as biodiversity and natural ecology. The objective of agroforestry is not just to produce food and generate income for farmers, but also to increase forest resources. Agroforestry sustains the economic benefits of forests, such as soil fertility and nutrients from trees, topsoil protection, windbreaks, watersheds, and recreational value. The compromise between agricultural and environmental needs is a solution to the widespread problem of deforestation to provide land for agricultural purposes. In addition, the incorporating trees on at least 10% of agricultural land in agroforestry provides an alternative cultural practice that is a way to adapt to climate change (Burgess et al. 2022).

In the north of Thailand, the Thai government wants to replace maize with more sustainable options. For instance, agricultural systems such as integrated agriculture, perennial fruit trees, and agroforestry have been promoted as better options. Once the upland environment and social context were taken into account, agroforestry and perennial fruit trees such as lychee and avocado, including coffee, were better choices (Talerngsri-Teerasuwannajak and Pongkijvorasin 2021). Furthermore, the Forest Landscape Restoration (FLR349) Fund has been developed based on the King's Philosophy of "Three Forests, Four Benefits" and the development of a value chain, which could become a model for farmers living in watershed areas (WWF 2022). The "three forests, four benefits" that are (i) usable forest for self-sufficiency in terms of housing, (ii) economic forest for self-sufficiency in terms of income, (iii) edible forest for self-sufficiency in terms of foods and herbs, and (iv) well-being and sustainability. The FLR349 Fund helps farmers turn their agricultural operation from forest-encroaching mono-agriculture with intensive use of chemicals into the "Three Forests, Four Benefits" agricultural system, which helps to restore the environment by stopping the destruction of topsoil. This project reduces the living cost of farmers and promotes community agriculture. Their products will be purchased by a social enterprise based on a strong marketing plan and an advanced production plan. The efforts of the FLR349 funding program have already helped farmers succeed on project sites of Mae Chaem district, Chiang Mai Province, and Nan Province, in Northern Thailand to replace maize plantations into sustainable agriculture. The agroforest can be further developed into to attract eco-tourism. According to Thailand's northeast tradition, the ancestral spiritual forest (Don Pu Ta) has become a way to effectively manage community resources. The area of the Pu Ta Shrine is covered with lush vegetation, full of trees and wildlife. The villagers have to harmonize with the forest conditionally, for example, collecting products or making economical use of the forest with respect, and planting trees to replace cutting for use (Kettate 2000).

8.3.2.5 New Theory Agriculture

The New Theory is the most distinct and concrete example of the application of the Philosophy of Sufficient Economics in the agricultural sector. His Majesty King Bhumibol initiated this theory to help Thai farmers who suffer from the impacts of economic crisis, natural disasters, and other unproductive natural conditions. In order to adopt the new agricultural theory, farmers are invited to follow the three phases (The Chaipattana Foundation 2016; Porpeang 2021).

Phase I: Self-reliance. The land is split in four parts with a ratio of 30:30:30:10 as shown in Figure 8.4a. The division of land to obtain the highest benefit is calculated and considered from the land ownership rate of 15 Rai per household. However, farmers with fewer or more land may also apply this ratio to their land. The first 30% is designated for

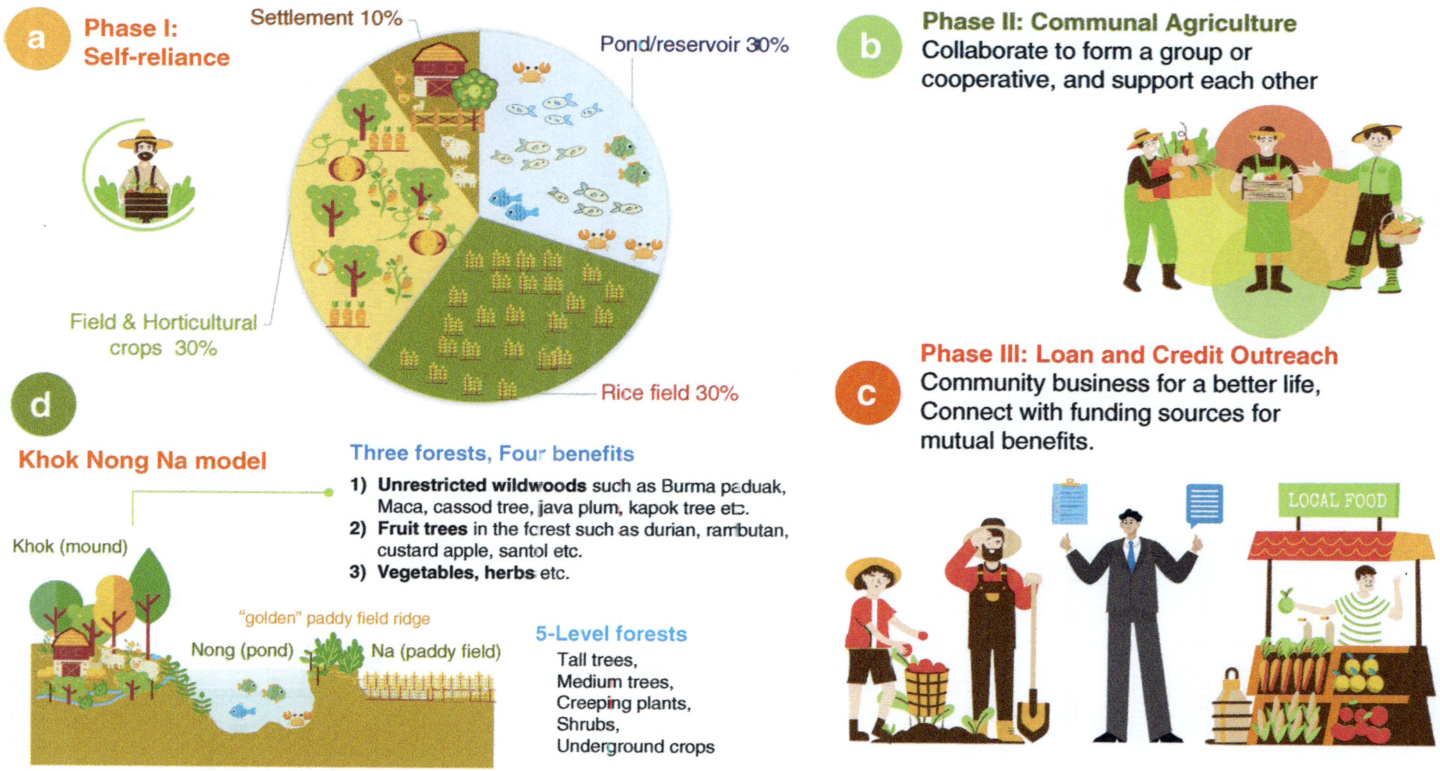

Figure 8.4 King's Philosophy for Sustainable livelihoods, safe food, reforestation and collaboration under "New theory agriculture," which consists of three steps: (a) Phase I, Self-reliance; (b) Phase II, Communal agriculture; (c) Phase III, Loan and credit outreach; and (d) Khok Nong Na model with three forests, four benefits.

a pond to hold rainwater during the rainy season. While during the dry season, it is used to supply water for the cultivation and breeding of aquatic animals and plants. The second 30% is spent in the cultivation of rice during the rainy season for the daily consumption of the family throughout the year. The third 30% is used to cultivate fruit and perennial trees, vegetables, field crops, and herbs for daily consumption. If there is any surplus, it will be sold. The remaining 10% is reserved for housing, livestock, roads, and other facilities. The above proportion of land use can be changed or improved according to topography, rainfall, and environment. Rice and edible products enable farmers to become self-sufficient. The products left over from the consumption can be sold to generate income

Phase II: Communal Agriculture (Figure 8.4b). After self-sufficiency, farmers are encouraged to form groups or cooperatives to conduct farming activities in the area. Farmers must cooperate in crop production, starting with soil preparation, crop species, fertilizers, water supply, etc. for cultivation marketing (rice drying space, silos, rice mills, the sale of crops). In particular, when rice crops are harvested, the preparations of drying space, silos, rice mills, must be made to derive the greatest profits, as well as joint co-operation in the sale of products to obtain good prices and reduce expenses.

Phase III: Loan and credit outreach (Figure 8.4c). Farmers should move into the third phase by making contact with banks or private companies to obtain funds to help them invest or develop their quality of life. In that way, farmers and banks or private companies benefit each other. Banks or private businesses can buy rice directly from farmers at low prices without going through middlemen. Farmers unite to sell rice, so they don't have to be forced to sell it at a cheap price. The establishment of cooperatives made it possible to purchase consumer goods at lower prices at wholesale prices.

"Khok Nong Na model" (Figure 8.4d) combines with the "Three forests, four benefits" can be applied to the land division. Khok (mound) should contain five levels of plants that are tall trees (iron woods, rubber trees, teak, etc.), medium trees (coconut trees, banana trees, etc.), creeping plants (watermelons, pumpkins, etc.), shrubs (eggplants, holy basil, chilies, etc.), and underground crops (potatoes, taros, onions, etc.). The water system of the Khok-Nong-Na model consists of Nong (pond), Lum Khanom Krok, Khong Sai Kai and Check dam. "Nong" is a water storage area to be used during the dry season and a watershed area during floods, as well as habitat for fish and other aquatic animals. "Khanom Krok Basin" is a water storage area both on the ground and underground. It helps to drain water into the canal, disperses moisture throughout the area, and is a sediment resting point. "Khong Sai Kai" is a small canal that stretches from upstream or water source in a meandering manner to deliver water to remote spots, enhancing the diversity and abundance of trees and maintaining moisture in the area. "Check dams" are built along the "Khong Sai Kai" or water flow to slow water velocity and prevent sediment from flowing into the pond. Na (rice fields) are designed and cultivated according to the social landscape and rice ecosystem such as a lowland rice field, upland rice field, floating rice field. Not just a place to plant rice, but also a habitat for aquatic animals like fish and shrimp. In addition, "golden paddy field ridge", which should be at least 1 m in width, height 2 m, and base width 2 m, is an area for planting food plants. The raising ridge increases the water storage area in the paddy fields, and vetiver grass should be planted together to prevent soil erosion (CDD 2022).

8.4 Special Cases of Restoration and Sustainable Agriculture in Thailand

Rehabilitation and development of agricultural areas in Thailand to be sustainable are presented in four special cases of (i) rice cultivation in inland saline soil of Northeast Thailand, (ii) restoring arid areas to become a floating market in forest with the King's philosophy, (iii) integrated agricultural learning center for sustainability, and (iv) large community organic rice fields and promotion of sustainability by up-skills and re-skills to members.

8.4.1 Rice Cultivation in Inland Saline Soil of Northeast Thailand

Northeast region of Thailand is the most affected area of saline soil problem about 17.8 million Rai. The map of saline soil in the Northeast region of Thailand shown in Figure 8.5a. Most areas about 12.6 million Rai have slightly to moderately saline soil, which their electrical conductivity in saturated paste extracts (ECe) is 2–8 dS/m. In which, the slight to moderate salt-affected soil are typically used for rice cultivation as shown in Figure 8.5b. Therefore, an appropriate agronomic management is applied to the inland salt for rice cultivation. The management of several soils affected by salt requires rehabilitation, leaching and drainage, mulching and modification of crop models, in particular the participation of local people. Suitable agronomic practices include the selection of salt-tolerant species, the use of organic amendments and green fertilizers. In addition, mechanical method is one of the main approaches used for saline soil management in Northeast Thailand. Figure 8.6a,b show the restoration of a salinity area in Non-Sung Sub-district, Kalasin province, Thailand. Soil leaching by surface water is another method leaches a contaminated salt from the soil and then drained out from the area via a tubular sub-drainage system (Figure 8.6c,d).

Green manure maintains soil moisture during the rice planting period and reduces the salinity effect on rice growth (Phaithong and Pagdee 2013; Naher et al. 2019; Pongwichian 2016). *Sesbania* spp. can be used as a green manure for rice planting systems in rainy lowlands for both transplanted and directly seeded rice systems in moderate saline soil (Anuluxtipun et al. 1992; Nawaz et al. 2022). *Sesbania rostrata* L., a nodular legume, is recommended on account of its salt tolerance in the range of 4–12 dS/m, flood tolerance and high nitrogen fixation (Arunin et al. 1987; Arunin et al. 1988). In addition, *S. rostrata* inoculated with rhizobium ORS571 produced higher-yielding biomass and a higher accumulation of nitrogen (Arunin and Pongwichian 2015). Restoring saline soils by growing halophyte crops is an effective strategy. The mechanisms of salinity tolerance are osmotic modulation in plants, root salt leaching, tissue water retention, salt avoidance, ionization, and glandular salt retention or excretion (Hasanuzzaman et al. 2014). Some halophytes can also be used for fodder plants. *Sporobolus virginicus*, *Spartina patens*, and *Distichlis spicata* can survive in severely salt-affected soils of 42 dS/m (Yuvaniyama and Arunin 1992; Pongwichian 2016). For salt-tolerant trees, *Acacia ampliceps*, *Casuarina glauca*, and *Melaleuca acaciodes* have been successfully cultivated in soils seriously affected by salt. Native species such as *Azadirachta indica*, *Cassia siamea*, *Tamarindus indica*, and *Pithecellobium dulce* could also tolerate at a high level of salinity (Arunin and Pongwichian 2015). Interaction with salt-tolerant grasses further reduced salt buildup in the tree root area.

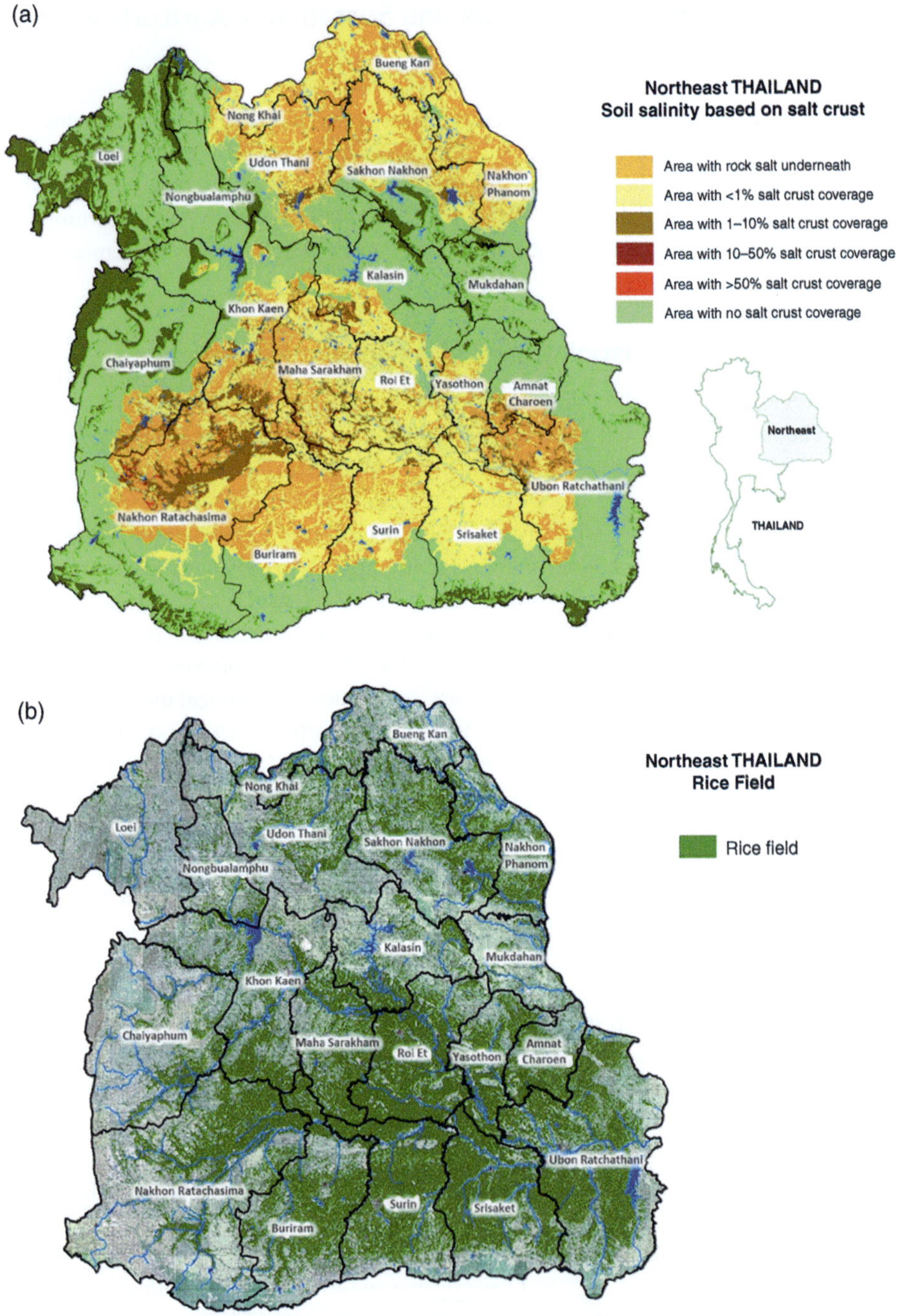

Figure 8.5 The saline soil map of northeastern Thailand, estimating from the percentage of salt covering the area during the dry season (a) and (b) rice planting area in the northeastern region.

Figure 8.6 Large polder and drainage system for soil salinity restoration in Kalasin province, Northeast of Thailand (16°27′29.952″ N, 7°13′41.858″E), (a) the agricultural area on 28 May 2021 indicating to polder and drainage system, (b) the restored agricultural area on 28 August 2022, (c) and (d) a tubular sub-drainage system.

As a consequence, integrated management practices to increase rice yields follow; (i) selected salt tolerant of rice cultivar of KDML 105, (iii) transplant seedlings that are 30–35 days old, spacing closer to 15×20 cm, with more seedlings (6–8 seedlings/hill), (ii) apply organic soil improvers such as green manure, farm manure or compost to increase soil fertility and improve soil physical properties, and (iv) leveling the land using salt-tolerant trees, e.g., *A. ampliceps* on the crest has been successfully sanitized soils, moderately affected by salt. In addition, plant growth promoting bacteria (PGPB) have roles in the alleviation of salinity effect on rice, promoting the rice growth under salt stress, and increasing aroma compound, 2-acetyl-1-pyrroline (2AP) as shown in Table 8.2. Biochar is

Table 8.2 Role of plant growth promoting bacteria (PGPB) and biochar amendment in the alleviation of salinity effect on rice and induce the rice growth under the salt stress.

Bio-amendment	Plant	Treatment	Properties and results	References
B. megaterium A12ag	*O. sativa* L cv. KDML105	0, 5, 10, 15, and 20 dS/m NaCl (Laboratory), 3.3 dS/m NaCl (Greenhouse)	- Contain the plant growth properties of phosphate solubilization - Promote germination, growth and yield of rice - Increase shoot wet-dry weight and N-P content in shoots	Sapsirisopa et al. (2009)
Pseudomonas sp. PDMZnCd2003	*O. sativa* L.cv. RD6	0, 4, 8, and 16 dS/m NaCl	- Contain the plant growth properties of IAA production, nitrogen fixation, and phosphate solubilization, under salt stress condition - This isolate might have a negative effect on rice growth in cases where the soil lacks a carbon source. - Coating rice seeds with carboxymethyl cellulose (CMC) could decrease the salt stress	Nakbanpote et al. (2014)
Bacillus amyloliquefaciens Y35-1, P1 (unknown), and P2 (unknown)	*O. sativa* L.cv. KDML 105	171 mM NaCl	- Promote rice seedling growth under salt stress, both fresh weight (29–35%) and height (18–22%).	Putaporntip et al. (2014)
Nocardiopsis yanglingensis A18, *Streptomyces jiujiangensis* JXJ 0074, *Streptomyces psammoticus* NBRC13971, and *Pseudonocardia oroxyli* D10	Rice seedling	200 mM NaCl	- Contain the plant growth properties of 1-aminocyclopropane-1-carboxylate (ACC) deaminase, IAA production, phosphate solubilization, and siderophore production - Increase shoot and root weight, and root length of rice seedling	Suksaard et al. (2017)
Streptomyces sp. GMKU 336	*O. sativa* L.cv. KDML 105	150 mM NaCl	- Contain the plant growth properties of ACC-deaminase, IAA production, phosphate solubilization, and siderophore production - Increase plant growth, chlorophyll, proline, K^+, Ca^{2+}, and water contents of rice seedling - Decrease ethylene, reactive oxygen species (ROS), Na^+, and Na^+/K^+ ratio in rice	Jaemsaeng et al. (2018)

Streptomyces venezuelae ATCC 10712, and *S. venezuelae*/pIJ85-acdS	*O. sativa* L.cv. KDML 105	150 mM NaCl	- Contain the plant growth properties of ACC deaminase overexpressing, and IAA production - Increase proline, total chlorophyll, relative water content (RWC), malondialdehyde (MDA), and K^+ in rice - Decrease ethylene, ROS and Na^+ contents in rice - Promote plant growth and salt tolerance of rice	Yoolong et al. (2019)
Streptomyces sp. SEF11	Glutinous rice cv. RD6	2.37 dS/m, and 7.25 dS/m NaCl	- Contain the plant growth properties of IAA production - Promote germination of rice seed - Increase shoot and root length of rice seedling	Somthakoon et al. (2020)
Sinomonas sp. ORF15–23	*O. sativa* L.cv. KDML 105	50 mM NaCl	- Contain the plant growth properties of phosphate solubilization, potassium solubilization, and siderophore production - Increase 2-acetyl-1-pyrroline (2AP) content in leaves of rice seedling	Chinachanta et al. (2021)
Pseudomonas putida ATCC 17484	*O. sativa* L. cv. KDML105	0, 75, and 150 mM NaCl	- Contain the plant growth properties of ACC deaminase, IAA production, and siderophore production - Increase root length, shoot length, fresh and dry weight of shoot of rice seedling	Sitisuanjik et al. (2021)

another soil amendment material reported in the area for saline soil management. Rice husk biochar (RHB) mixed with a cow manure improve saline soil quality to enable jasmine rice cultivation, especially the RHB significantly reduce soil salinity, sodium content, and sodium adsorption ratio (Prasertsuk and Wijitkosum 2021). Wijitkosum (2020) also reported that adding RHB with organic fertilizer into saline sodic soil increases the amount of exchangeable magnesium and calcium, which could replace the sodium ion in the salty soil making the soil less salty. Application of RHB (2.0 kg/m^2) yielded of rice with the highest number of ears of paddy per area.

Thung Kula Rong-Hai, the span area about 2.1 million Rai and five provinces of the northeastern of Thailand, this dry area is mostly contained within a Roi Et province and followed by Surin, Sisaket, Maha Sarakham, and Yasothon province, respectively. Due to the arid condition, especially in the dry season, the land is parched and cracked, and 84.86% of the area has low fertility. Also, the soil is rather sandy with more sodium and silicon (Saetung and Trelo-ges 2017). The land lacks freshwater resources and underground is a salt deposition layer, but most areas are paddy fields. In the rainy season, during August, the limiting abiotic factors alleviate, and the soil property tend to be suitable for rice cultivation. Two cultivars of Thai jasmine rice (*Oryza sativa* L.) that are favorites in this area are namely Khao Dawk Mali 105 (KDML 105) and RD 15. Although the yields are mostly lower than other areas due to the limiting factors, they are considered to be of high quality, especially for the fragrance. The rice has been registered as the geographical indication (GI) with the Department of Intellectual Property of Thailand on April 28, 2006, with the name "Thung Kula Rong-Hai Thai Hom Mali Rice (TKR)." The fragrance of TKR is specifically similar to a pandan-like odor, but actually is activated from an appropriate level of salinity stress. In general, jasmine rice is a salt susceptible variety, but for TKR the salt-tolerant property generating from the genetically local climate adaptation is beneficial in the specific order. 2AP is the main chemical compound in TKR fragrance which is synthesized under slight salinity (Poonlaphdecha et al. 2012; Funsueb et al. 2016; Kongpun et al. 2020; Chinachanta et al. 2021). The strong fragrance is an important character to determine the economic value of rice in the export market and consumer acceptance. The TKR has been registered for the protected geographical indication (PGI) in the European Union, protected by Council Regulation (EC) No. 510/2006 on February 12, 2013 (Napasintuwong 2019).

8.4.2 Restoring Arid Areas to Become a Floating Market in the Forest with the King's Philosophy

Guangzhou Waterfall is a floating market in the middle of a forest that has been restored from arid areas in Nong Ya Plong District, Phetchaburi Province. The area has been restored to an integrated forest using the King's philosophy since 2004 (MGR Online 2017). The restoration has begun with adjusting the area into the stairs, then planting vetiver grass to improve soil structure and prevent soil erosion. Bamboo, which is a fast-growing plant, is planted for ecological and functional plants (Figure 8.7a). The water retention (Figure 8.7c) was accomplished by digging a large Nong (pond) and pumping water from nearby water resources to store it. Then the water has been distributed to be stored in various small pounds, e.g. the monkey cheek pond, Khanom Krok Pond, and Puang Pond, as well as

Figure 8.7 Restoring degraded and arid areas in Nong Ya Plong District, Phetchabun Province, which used to be pineapple plantations, into tourist attractions "Floating Market in the Forest of Guangzhou Waterfall" using the theory of His Majesty King Rama IX. (a) and (c) Started managing the water system by digging ponds and planting bamboo forests in 2004, (b) and (d) the areas in 2022, (e) distributing water into dehydrated areas using check dams and (f) Khong Sai Kai, (g) areas around the bamboo forests and the floating markets are being restored and rehabilitated using bananas as a leading crop, and (h) the area connecting to the restored area which is awaiting restoration.

distributing water throughout the area with check dams (Figure 8.7g) and Khong Sai Kai (Figure 8.7h). Nowadays, the area beside the bamboo tunnel is arranged for selling various products from the community (Figure 8.7b), and the larger pond becomes a floating market in the middle of a forest (Figure 8.7d) to attract tourists and generate income for people in the community. Moreover, the area surrounding the floating market and bamboo forest is still restored and rehabilitated with "three forests, four benefits" by growing bananas as a pioneer crop (Figure 8.7e). As comparison between the recovery area (Figure 8.7e and the awaiting regeneration area (Figure 8.7f), the restoration of this degraded area to be agroforest is ongoing.

8.4.3 Integrated Agricultural Learning Center for Sustainability

Khao Krapuk Agricultural Learning Center, Tha Yang District, Phetchaburi Province, has been created according to a royal initiative by Her Royal Highness Princess Maha Chakri Sirindhorn. This learning center has an area of 34.29 Rai for presenting an integrated farming system. This learning center has an area of 34.29 Rai for presenting an integrated farming system. The plot area is a demonstration plot of planting a variety of crops: vegetables such as pumpkins, lemons, corn, zucchini, asparagus, kale, eggplant, yard long beans, cabbage, corn, etc., fruits such as bananas, guava, bilimbi, pomegranate, etc., Napier grass for animals' feed, and edible mushrooms. Figure 8.8a–d show part of the entire plot. Catfish, pit pigs, goats, ducks, and chickens are raised in the learning center (Figure 8.8e-g). Weeds and plant residues from pruning are composted with manure as fertilizer for reuse in the demonstration plots (Figure 8.8j). In addition, there was also a demonstration of culturing earthworms to produce earthworm manure (Figure 8.8k), while the earthworms would be used to feed the catfish. Bio-fertilizer is composted from agricultural waste and microorganisms to promote plant growth, such as photosynthetic bacteria (Figure 8.8i). Banana plants and vegetable scraps left over from trimming are used to feed pigs, ducks, and chickens. The pig manure enters the biogas fermentation system (Figure 8.8h), then the methane gas will be used for warming chicks and ducklings. The learning center digs wells to store water and waters the plants using a sprinkler system and pump, which uses the electrical power from solar cells. In addition, this training center can generate income from selling vegetables, fruits, processed agricultural products, quality seeds, compost, vermicompost, duck eggs, chicken eggs, etc. The agricultural products grown in the learning center are also certified in Good Agricultural Practice for food crops, Thai Agricultural Standard TAS 9001-2013 (MOAC 2013). All activities are practical, so this agricultural learning center is a place for farmers, students, academic staff, and people interested in the integrated farming for sustainable development.

8.4.4 Large Community Organic Rice Fields

Nong Saeng Subdistrict, Wapi Pathum District Maha Sarakham Province has a total rice cultivation area of 29 854 Rai, or 97% of the total agricultural area. Farmers can grow rice once a year, so farmers earn income from selling rice only once a year. The grown rice cultivars are market rice, namely RD 6 and Hom Mali 105, with only some farmers collecting the seeds for the next growing season. Problems with rice cultivation are: (i) most

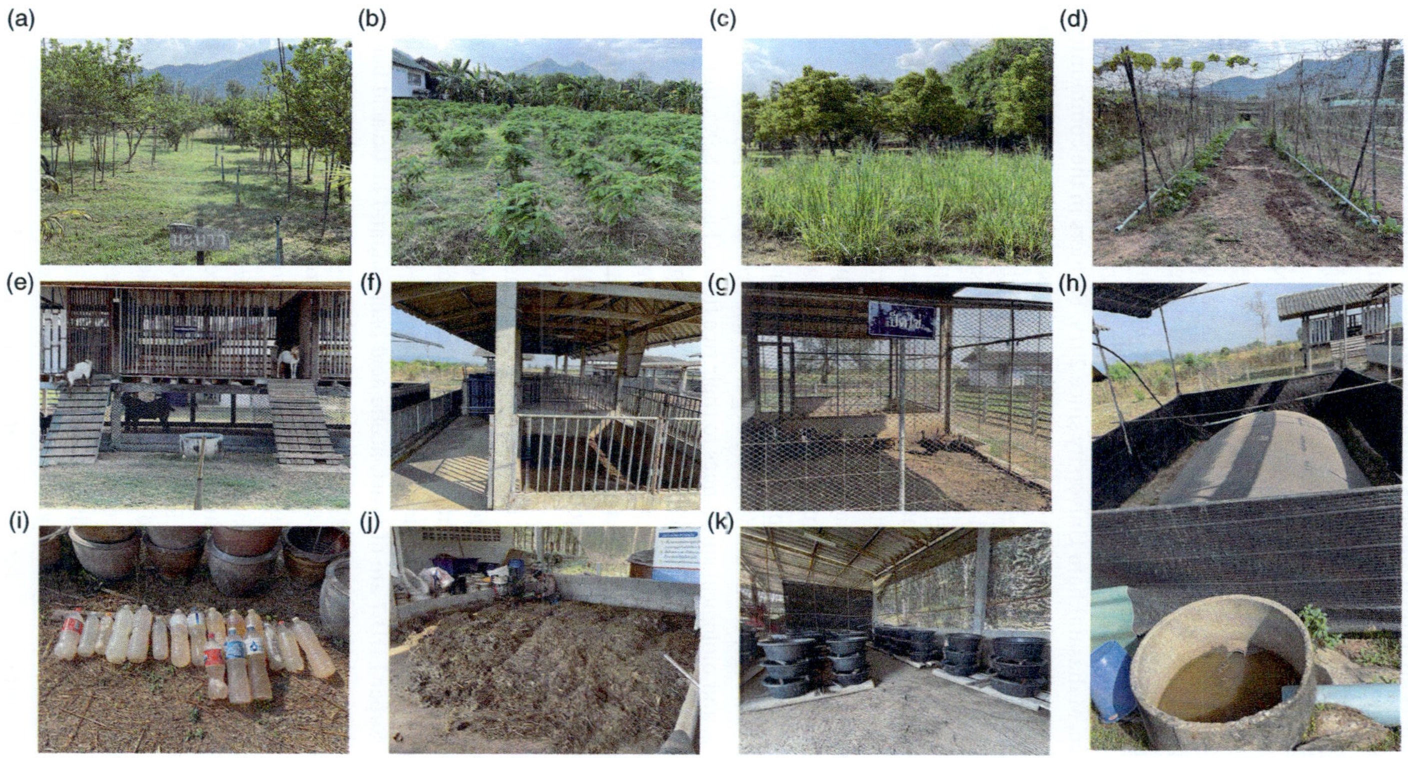

Figure 8.8 Khao Krapuk Agricultural Learning Center, Thayang District, Phetchaburi Province, created according to a royal initiative by Her Royal Highness Princess Maha Chakri Sirindhorn. demonstration of integrated farming in areas 34 rai. (a)–(d) Examples of plants are lemon, Cha-om, Napier grass and Sponge gourd, (e)–(g) examples of animals are goats, pigs, ducks and chickens, (h) methane fermentation plant from manure near the pig house, (i) propagation of photosynthetic bacteria; (j) composting from weed residues and manure; (k) cultivating earthworms from vegetable waste for fertilizer and feed.

agricultural areas are sandy loam with moderate to very saline conditions; (ii) lack of soil maintenance and improvement resulting in rice yields not meeting the demand farmers; and (iii) most farmers are retired and lack substitute workers. Farmers face higher investment challenges than farmers in other areas of farming in a non-self-sufficient way. Emphasis on growing rice in a quantitative manner without emphasizing seed selection and reliance on external inputs makes farming costs high.

The farmers' problems led to the grouping of farmers in Nong Saeng Subdistrict, six villages (Chaithong Village, Sok Yang Village, Praboo Village, Dan Kham Village, and Bom Thung Village) groups together to change their farming methods to become self-reliant farming and refined organic farming that focused on selling quality rice seeds and organic premium rice, which can generate higher income than selling paddy to a mill. The change has been supported by the Sub-district Administrative Organization and the Faculty of Science, Mahasarakham University since 2019. Farmers in the community are continually encouraged and trained in both upskilling and reskilling to enable farmers to be self-reliant and have the skills of modern agricultural technology by integrating folk science with scientific knowledge and innovation. The analysis of soil quality, soil nutrients, and salinity levels leads to the planning of fertilizer application appropriate to the area (Figure 8.9a,b). Farmer's information: soil quality data, plant species, and arable land are linked to geographic and climatic data, including rainfall through the "Water and Soil Salinity in Agricultural Area (WSSA) ver. 2.0" application, developed by the Faculty of Science. The application can also be used to make appointments to hire or ask people in the community to transplant rice seedlings and/or harvest paddy. To address the current labor shortage, communities have begun to use drones to apply biofertilizers, organic fertilizers, and biological substances to their rice fields (Figure 8.9c). In the case of the peanut planting experiment after farming, part of the plant can be used as a green manure to add nutrients, especially nitrogen to the soil, and generate income from bean sales (Figure 8.9d). Bio-fermentation liquid composted from Super LDD 2, a microbial activator derived from the Department of Land Development (LDD), was applied during the organic peanut cultivation instead of chemical fertilizer. Pest control was carried out by using local fresh herbs composted from Super LDD 7. The microorganisms in Super LDD 2 consist of yeast, lactic acid bacteria, proteolytic bacteria, lipolytic bacteria, and phosphate solubilizing bacteria, that have highly efficient to digest organic waste such as residues from fruit, vegetable, fish, and snail to become bio-extract liquid (LDD 2014). The microbes in Super LDD 7 are of yeast, acetic acid bacteria, and lactic acid bacteria. The alcohol and organic acids produced by the microbes can extract the active compounds of Azadirachtin from neem (*A. indica*), Stemonine from Non taai yak (*Stemona collinsiae*), saponin from white yam seed (*Pachyrhizus erosus*), limonene, pinene, and naphthoquinone from the Sap Sea (*Osmia odorata*) etc. to control aphid and worm (LDD 2014). Figure 8.9e shows the herb fermentation process. From the gathering of farmers for organic farming, the rice grown in Nong Saeng can be sold at a high price, especially the cultivation of registered seed for sale as certified seed is promoted by the Nong Saeng subdistrict Administrative Organization (Figure 8.9f,g). In order to sustain and promote continuous learning, the Faculty of Science has trained and educated farmers to analyze soil quality composition and make biofertilizer, and also cultivate *Trichoderma asperellam*, a biocontrol and plant growth promoting fungus

Figure 8.9 The promotion of large organic farming areas, which resulting from the collaboration of villagers in Nong Saeng Subdistrict, Wapi Pathum District, Maha Sarakham Province, together with Maha Sarakham University. (a) Soil sample collection location for nutrient and salinity level analysis of plot management by (b) Wssa ver 1.0 applications, (c) the use of drones to compost bio-fertilizers in the Member's rice field, (d) Peanut planting demonstration plot to increase soil nutrients during rice planting break, (e) Fermentation of pesticides from folk herbs and microbial inoculum, (f-g) Cultivation of registered seed for sale as certified seed, (h-j) Training of community farmers to analyze soil quality composting and making biofertilizer, (k) *Trichoderma asperellum*, a biocontrol and plant growth promoting fungus.

(Figure 8.9h–k). Nong Saeng subdistrict is therefore one of the successful model areas in the cooperation of farmers, organic farming, and safe agriculture, as well as collaborative learning between the communities, governments, and university.

8.5 Conclusions

The importance of Thai agriculture at both national and household levels has led to the 20-year Agriculture and Cooperatives Strategy (2017–2036) and the BCG model, which is consistent with the SDGs. The Strategy has the vision of "Secured Farmers, Prosperous Agriculture Sector and Sustainable Agricultural Resources." The management of Thai agriculture is divided into large agricultural fields of industry and export and small-scale agriculture for households and farmers' self-sufficiency. Agricultural technology (AgTech) will transform the future of Thai agriculture into smart farms and support large-scale farming systems, helping to reduce labor shortages and entering an aging society, as well as helping to build a stable farming career for the new generation. According to diverse agricultural resources, rural culture and traditional wisdom, the sustainable agriculture systems are integrated farming, organic farming, natural farming, agroforestry, and new theory agriculture. From the consumer confidence and social trends in achieving the health benefits, the organic farming tends to become the main agricultural activity in the future. Integrated agriculture contributes to food safety, socio-economic benefits, and resilience. Agricultural forests are notable for the reduction of encroachment on forests and the promotion of watershed conservation. Natural farming emphasizes trust in the power of the natural balance and self-renewal of the ecosystem. Finally, the New Theory Agriculture, together with "Kok Nong Na Model" and "Three Benefit Four Forests," is the clearest and most practical implementation of the application of the Sufficiency Economy Philosophy in the agricultural sector for the sustainable development of the household and the country, as well as the restoration of the ecosystem.

Acknowledgements

The authors would like to thank Mahasarakham University (MSU) for funding the projects entitled "Potential development of saline soil and brackish water area at Pla-Boo village, Nong-Saeng sub-district, Wapipathum district, Maha Sarakham province," and to acknowledge the Community Incubator (SCI) Platform, Office of the Ministry of Higher Education, Science, Research and Innovation for supporting the project "Nong-Saeng super organic fertilizer," and to recognize the Equitable Education Fund (EEF) for funding the project "Upgrading the careers of farmers who are underfunded and underprivileged through development occupational skills to be a self-reliant farmer in a new normal way Nong Saeng Subdistrict, Wapi Pathum District Maha Sarakham Province." In addition, this publication is a part of a Memorandum of Understanding (MoU) between the Department of Land Development and the Faculty of Science, MSU. Nakbanpote, W. would like to thank Mr. Suthep Trairat for the interview and the past pictures of Guangzhou Waterfall.

References

Anuluxtipun, Y., Pongwichian, P., Thamphiam, P., and Arunin, S. (1992). Nitrogen mineralization from green manure in saline soils. In: *Agriculture Research Strategies and Development in Northeast Thailand*, 285–300. Thailand: Agriculture Development Research Center in Northeast Khonkaen.

Arunin, S. and Pongwichian, P. (2015). Salt-affected soils and management in Thailand. *Bull. Soc. Sea Water Sci.* 69: 319–325.

Arunin, S., Yuvaniyama, A., Rungsangchan, P. et al. (1987). Effect of upland and lowland saline soils on 4 species of *Sesbania*. *Agric. Sci. J.* 20 (4): 237–250. (In Thai).

Arunin, S., Dissatapom, C., Anuluxtipan, Y., and Nana, D. (1988). Potential of *Sesbania rostrata* as green manure in saline rice soil in Thailand. In: *Green Manure in Rice Farming* (ed. IRRI), 83–95. Manila, Philippines: Rice Res. Inst.

Arunrat, N., Pumijumnong, N., Sereenonchai, S. et al. (2020). Assessment of climate change impact on rice yield and water footprint of large-scale and individual farming in Thailand. *Sci. Total Environ.* 726: 137864. https://doi.org/10.1016/j.scitotenv.2020.137864.

Burgess, A.J., Cano, M.E.C., and Parkes, B. (2022). The deployment of intercropping and agroforestry as adaptation to climate change. *Crop Environ.* 1: 145–160. https://doi.org/10.1016/j.crope.2022.05.001.

Chauhan, B.S., Singh, R.G., and Mahajan, G. (2012). Ecology and management of weeds under conservation agriculture: a review. *Crop Prot.* 38: 57–65. https://doi.org/10.1016/j.cropro.2012.03.010.

Chen, J., Xu, Y., Fei, K. et al. (2020). Physiological mechanism underlying the effect of high temperature during anthesis on spikelet-opening of photo-thermo-sensitive genic male sterile rice lines. *Sci. Rep.* 10: 2210. https://doi.org/10.1038/s41598-020-59183-0.

Chinachanta, K., Shutsrirung, A., Herrmann, L. et al. (2021). Enhancement of the aroma compound 2-acetyl-1-pyrroline in Thai jasmine rice (*Oryza sativa*) by rhizobacteria under salt stress. *Biology* 10: 1065. https://doi.org/10.3390/biology10101065.

Community Development Department (CDD). (2022). Khok-Nong-Na, Documents Published in the Category of "Sustainable Local Economy & Self-reliant Communities," World Soil Day 2022, Change for Good. http://cddata.cdd.go.th/apps/knn/info/km.php (accessed 19 September 2022).

Crossman, N.D., Bernard, F., Egoh, B. et al. (2016). The role of ecological restoration and rehabilitation in production landscapes: an enhanced approach to sustainable development. Working Paper for the UNCCD Global Land Outlook, Australia.

Department of Agricultural Extension (DOAE) (2022). Large-scale agricultural extension system. Digital Innovation Development Group, Information and Communication Technology Center, Ministry of Agriculture and Cooperatives (in Thai). https://co-farm.doae.go.th/graph/Dashboard1dsb.php (accessed 19 September 2022).

Food and Agriculture Organization of the United Nations (FAO) (2011). *Climate-Smart Agriculture: Managing Ecosystems for Sustainable Livelihoods*. Rome: FAO. https://www.fao.org/3/an177e/an177e.pdf (accessed 19 September 2022).

Food and Agriculture Organization of the United Nations (FAO) (2022). *Adapting to Climate Impacts in Thailand's Agricultural Sector – What to Do First?* Rome: FAO. https://www.fao.org/in-action/naps/news-events/detail/ru/c/1118438 (accessed 19 September 2022).

Funsueb, S., Krongchai, C., Mahatheeranont, S., and Kittiwachana, S. (2016). Prediction of 2-acetyl-1-pyrroline content in grains of Thai Jasmine rice based on planting condition, plant growth and yield component data using chemometrics. *Chemom. Intel. Lab. Syst.* 156: 203–210.

Gupta, D., Gujre, N., Singha, S., and Mitra, S. (2022). Role of existing and emerging technologies in advancing climate-smart agriculture through modeling: a review. *Eco. Inform.* 71: 101805. https://doi.org/10.1016/j.ecoinf.2022.101805.

Hasanuzzaman, M., Nahar, K., Alam, M.M. et al. (2014). Potential use of halophytes to remediate saline affect. *Biomed. Res. Int.* 589341. http://dx.doi.org/10.1155/2014/589341.

Hérique, O. and Faysse, N. (2020). A large-scale public programme to promote organic rice farming in Thailand: building solid foundations to enable farmers to engage. *Organ. Agric.* 11: 27–40. https://doi.org/10.1007/s13165-020-00320-4.

Hu, Q., Wang, W., Lu, Q. et al. (2021). Abnormal anther development leads to lower spikelet fertility in rice (*Oryza sativa* L.) under high temperature during the panicle initiation stage. *BMC Plant Biol.* 21: 428. https://doi.org/10.1186/s12870-021-03209-w.

Jaemsaeng, R., Jantasuriyarat, C., and Thamchaipenet, A. (2018). Molecular interaction of 1-aminocyclopropane-1-carboxylate deaminase (ACCD)-producing endophytic Streptomyces sp. GMKU 336 towards salt-stress resistance of *Oryza sativa* L. cv. KDML105. *Sci. Rep.* 8: 1950. https://doi.org/10.1038/s41598-018-19799-9.

Jitsanguan, T. (2001). *Sustainable Agricultural System for Small-Scale Farmers in Thailand: Implications for the Environment.* Taipei (Taiwan): Food and Fertilizer Technology Center for the Asian and Pacific Region. https://agris.fao.org/agris-search/search.do?recordID=XF2015013864 (accessed 19 September 2022).

Kerr, R.B., Madsen, S., Stuber, M. et al. (2021). Can agroecology improve food security and nutrition? A review. *Glob. Food Sec.* 29: 100540. https://doi.org/10.1016/j.gfs.2021.100540.

Kettate, B. (2000). The ancestral spirit forest (Don Pu Ta) and the role behavior of elders (Thao Cham) in Northeastern Thailand. *J. Siam Soc.* 88: 96–110. https://thesiamsociety.org/wp-content/uploads/2000/03/JSS_088_0l_BoonyongKettate_AncestralSpiritForest.pdf (accessed 19 September 2022).

Kongpun, A., Jaisiri, P., Rerkasem, B., Prom-u-thai, C. (2020). Impact of soil salinity on grain yield and aromatic compound in Thai Hom Mali rice cv. Khao Dawk Mali 105. *Agric. Nat. Resour.* 54: 74–78. https://doi.org/10.34044/j.anres.2020.54.1.10.

Land Development Department (LDD) (2014). Soil Biotechnology: Microbial Activator LDD 1, LDD 2, LDD 6, LDD 12. http://www1.ldd.go.th/ldd_en (accessed 19 September 2022).

MGR Online (2017). The First in Thailand! “Floating Market on Guangzhou Waterfall” Growing with the King’s Philosophy (in Thai). https://mgronline.com/onlinesection/detail/9600000125531 (accessed 18 September 2022).

Ministry of Agriculture and Cooperatives (MOAC) (2013). Good Agricultural Practice for food crops, Thai Agricultural Standard TAS 9001–2013, Published in the Royal Gazette Vol.130 Section 76D Special, Dated 25 June B.E.2556.

Ministry of Agriculture and Cooperatives (MOAC) (2017). *The Twenty-Year Agriculture and Cooperative Strategy (2017–2036) and the Five-Year Agriculture Development Plant under the Twenty National Economic and Social Development Plant (2017–2021).* Thailand, Bangkok: Office of Agricultural Economics.

Naher, U.A., Ahmed, M.N., Sarkar, M.I.U. et al. (2019). Fertilizer management strategies for sustainable rice production. In: *Organic Farming, Woodhead Publishing Series in Food Science, Technology and Nutrition*. India: Elsevier.

Nakbanpote, W., Panitlurtumpai, N., Sangdee, A. et al. (2014). Salt-tolerant and plant growth-promoting bacteria isolated from Zn/Cd contaminated soil: identification and effect on rice under saline conditions. *J. Plant Interact.* 9 (1): 379–387.

Napasintuwong, O. (2019). PGI TKR Hom Mali Rice in Thailand. In: *Sustainability of European Food Quality Schemes*. Switzerland: Springer Nature.

National Innovation Agency (Public Organization) (NIA). (2020). 6 "AgTech" Trends Changing the Future of "Agriculture" in Thailand (in Thai). https://www.nia.or.th/AgTechTrends (accessed 19 September 2022).

National Science and Technology Development Agency (NSTDA). (2021). BCG Model: Fostering Sustainable Development in Thai Economy. https://www.nstda.or.th/home/knowledge_post/bcg-booklet-english-version (accessed 19 September 2022).

Nawaz, A., Rehman, A.U., Rehman, A. et al. (2022). Increasing sustainability for rice production systems. *J. Cereal Sci.* 103: 103400. https://doi.org/10.1016/j.jcs.2021.103400.

Nichols, V., Verhulst, N., Cox, R., and Govaerts, B. (2015). Weed dynamics and conservation agriculture principles: a review. *Field Crop Res.* 183: 56–68. http://dx.doi.org/10.1016/j.fcr.2015.07.012.

Phaithong, C. and Pagdee, A. (2013). Factors affecting rice production in northeastern Thailand: the relationship between soil salinity and vegetative cover. *Int. J. Environ. Rural Dev.* 4 (1): 25–30.

Pongwichian, P. (2016). Agronomic management of saline soil in agricultural lands of Thailand. Dissertations. Nihon University. https://doi.org/10.15006/32665B7165.

Poonlaphdecha, J., Maraval, I., Roques, S. et al. (2012). Effect of timing and duration of salt treatment during growth of a fragrant rice variety on yield and 2-acetyl-1-pyrroline, proline, and GABA levels. *J. Agric. Food Chem.* 60: 3824–3830.

Porpeang (2021). New theory agriculture (in Thai). https://www.porpeang.org/content/5475/new-theory-agriculture (accessed 18 September 2022).

Prasertsuk, S. and Wijitkosum, S. (2021). Innovative use of rice husk biochar for rice cultivation in salt-affected soils with alternated wetting and drying irrigation. *Eng. J.* 25 (9). https://doi.org/10.4186/ej.2021.25.9.19.

Putaporntip, N., Mosaleeyanon, K., Pichakum, A., and Songnuan, W. (2014). Screening for plant growth-promoting bacteria to alleviate salt effect on growth of rice (*Oryza sativa* L.) seedling. *Thai J. Bot.* 6: 191–198.

Rayasawath, C. (2018). Factors affecting the household succession in agricultural occupation in Nakhon Ratchasima Province, Thailand. *Agriculture* 8: 109. https://doi.org/10.3390/agriculture8070109.

Rohila, A.K., Maan, A.D., Kumar, A., and Kumar, K. (2017). Impact of agricultural practices on environment. *Asian J. Microbiol. Biotechnol. Environ. Sci.* 19 (2): 145–148.

Saetung, W. and Trelo-ges, V. (2017). Monitoring in soil fertility change in Tung Kula Rong Hai using geographic information systems. *Int. Res. J. Adv. Eng. Sci.* 2 (4): 189–193.

Sapbamrer, R., Chitrakul, J., Sirikul, W. et al. (2022). Impact of COVID-19 pandemic on daily lives, agricultural working lives, and mental health of farmers in Northern Thailand. *Sustainability* 14: 1189. https://doi.org/10.3390/su14031189.

Sapsirisopa, S., Chookietwattana, K., Maneewan, K., and Khaengkhan, P. (2009). Effect of salt-tolerant *Bacillus* inoculum on rice KDML 105 cultivated in saline soil. *Asian-J. Food Agro-Ind.* 2: S69–S74.

Sekaran, U., Lai, L., Ussiri, D.A.N. et al. (2021). Role of integrated crop-livestock systems in improving agriculture production and addressing food security – a review. *J. Agric. Food Res.* 5: 100190. https://doi.org/10.1016/j.jafr.2021.100190.

Sekkhunthod, K. (2020). Operational development of large-scale agricultural extension system to implementation. *Baddhana J.* 7 (2): 109–122. (In Thai).

Sitisuanjik, J., Konokchan, T., Chamsarrt, S., and Indananda, C. (2021). Promoting rice seedling Khao Dawk Mali 105 growth by Pseudomonas putida grown under saline condition. *J. Sci. Technol. Mahasarakham Univ.* 40 (1): 21–29.

Somthakoon, K., Thiangsanthia, S., Sangdee, A., and Chouychai, W. (2020). Effect of Streptomyces sp. strain SEF11 on germination and seedling growth of glutinous rice cv. RD6 and peanut cv. Kasetsart 50 under salt solution. *J. Agric. Res. Extension* 37 (2): 36–46. (In Thai).

Soulé, E., Michonneau, P., Michel, N., and Bockstaller, C. (2021). Environmental sustainability assessment in agricultural system: a conceptual and methodological review. *J. Clean. Prod.* 325: 129291. https://doi.org/10.1016/j.jclepro.2021.129291.

Sukprasert, P. (2019). *The Passages from the New Theory Agriculture to the Sufficiency Economy for Sustainable Development (English Version).* Thailand: Department of Agricultural Extension and Communication, Faculty of Agriculture, Kasetsart University.

Suksaard, P., Pathom-aree, W., and Duangmal, K. (2017). Diversity and plant growth promoting activities of Actinomycetes from mangroves. *Chiang Mai J. Sci.* 44 (4): 1210–1223.

Talerngsri-Teerasuwannajak, K. and Pongkijvorasin, S. (2021). Agricultural business model and upland sustainability: evidence from northern Thailand. *Curr. Res. Environ. Sustain.* 3: 100085. https://doi.org/10.1016/j.crsust.2021.100085.

The Chaipattana Foundation (2016). The Concept of Development of Self-reliant Farmers due to the Royal Initiative (Self Reliance) (in Thai). https://www.chaipat.or.th/concept-and-theory-development/to-develop-self-reliance-of-farmers.html (accessed 19 September 2022).

The Chaipattana Foundation (2017). Philosophy of Sufficiency Economy. https://www.chaipat.or.th/eng/concepts-theories/sufficiency-economy-new-theory.html (accessed 19 September 2022).

The International Resource Panel (IRP) (2019). *Land Restoration for Achieving the Sustainable Development Goals: An International Resource Panel Think Piece.* Nairobi, Kenya: United Nations Environment Programme.

Tontisirin, K. (2021). Application of Bio-Circular-Green (BCG) economy in food and agriculture. A presentation at the Presidents' Forum 2021 "Bio-Circular-Green Economy: Tuning Challenges into Opportunities in the Post-COVID-19 World" 30th Anniversary of the NSTDA, Thailand (26 March 2021). https://www.nstda.or.th/nac/2021/wp-content/uploads/2021/03/SS17-PresidentsForum_food-and-ag_Prof_Kraisid_Final.pdf (accessed 19 September 2022).

Tscharntke, T., Grass, I., Wanger, T.C. et al. (2021). Beyond organic farming – harnessing biodiversity-friendly landscapes. *Trends Ecol. Evol.* 36 (10): 919–930. https://doi.org/10.1016/j.tree.2021.06.0109.

United Nations Environment Programme (UNEP) (2021). *Becoming #GenerationRestoration: Ecosystem Restoration for People, Nature and Climate.* Nairobi.

Wezel, A., Herren, B.G., Kerr, R.B. et al. (2020). Agroecological principles and elements and their implications for transitioning to sustainable food systems. A review. *Agron. Sustain. Dev.* 40: 40. https://doi.org/10.1007/s13593-020-00646-z.

Wijitkosum, S. (2020). Applying rice husk biochar to revitalise saline sodic soil in Khorat Plateau area – a case study for food security purposes. In: *Biochar Applications in Agriculture and Environment Management* (ed. J.S. Singh and C. Shingh), 1–31. Switzerland: Springer Nature.

World Wide Fund for Nature (WWF) (2022). Forest Landscape Restoration Fund (FLR349). https://www.wwf.or.th/en/scp/reforestation_activity/flr_349 (accessed 19 September 2022).

Yoolong, S., Kruasuwan, W., Pham, U.T.T. et al. (2019). Modulation of salt tolerance in Thai jasmine rice (*Oryza sativa* L. cv. KDML105) by Streptomyces venezuelae ATCC 10712 expressing ACC deaminase. *Sci. Rep.* 9: 1275. https://doi.org/10.1038/s41598-018-37987-5.

Yuvaniyama, A. and Arunin, S. (1992). Physiological changes of *Sporoborus virginicus* (L) Kunth under various salinity levels. In: *Strategies for Utilizing Salt-Affected Lands*, 426–432. Bangkok, Thailand.

Zhang, C., Li, G., Chen, T. et al. (2018). Heat stress induces spikelet sterility in rice at anthesis through inhibition of pollen tube elongation interfering with auxin homeostasis in pollinated pistils. *Rice* 11: 14. https://doi.org/10.1186/s12284-018-0206-5.

Zugravu, G.A., Camelia-Costela (Lungeanu), F.G., Rahoveanu, M.T. et al. (2017). Social agricultural services – review. *J. East. Eur. Res. Bus. Econ.* 477474. https://doi.org/10.5171/2017.477474.

9

Emergy Approach to the Sustainable Use of Ecosystems toward Better Land Management

Joana Marinheiro, Ana Fonseca, João Serra, and Cláudia Marques-dos-Santos

School of Agriculture, University of Lisbon, CEF – Forest Research Centre, Tapada da Ajuda, 1349-017, Lisbon, Portugal

9.1 Introduction

The necessity for increasing food production, under all means, must be stopped being seen as the final and only goal. World population is, in fact, increasing. Specially in developing countries, where two billion suffer from micronutrients deficiencies (Godfray and Garnett 2014) due to inaccessibility to food or price volatilities. Meanwhile, developed countries faces the issue of exacerbated per capita consumption. Sustainable intensification of agriculture is showing to be an important step to ensure the needs of growing population whilst maintaining a prudent use of natural resources. In fact, from the 17 sustainable development goals (SDG) settled in the United Nation's agenda, there are two directly related with these issues. SDG2 focusses on "end hunger, achieve food security and improved nutrition and promote sustainable agriculture"; and SDG15 aims to "Protect, restore, and promote sustainable use of terrestrial ecosystems, sustainably manage forests, combat desertification, and halt and reverse land degradation and halt biodiversity loss" (UN2022). Both SDG act as incentive to promote a better use of the existing resources to prevent further degradation of ecosystems in the process through a thorough land management.

According to the Economics of Land Degradation (ELD 2015), 52% of agricultural land use is already affected by soil degradation. Heavy mechanical equipment, synthetic fertilizers and pesticides can have a direct impact on soil health (Tsiafouli et al. 2015), mainly in its erosion. Topsoil loss reduces the soil's capacity for retaining nutrients and water, compromising the soils' biodiversity, hence its productivity. Soil erosion rate varies highly both in time and space. In croplands, it is estimated to be c.a. 5 t/ha/yr (Panagos et al. 2015) globally, while pastures subjected to overgrazing can go up to 6 t/ha/yr (Pimentel and Burgess 2013). Nevertheless, it varies with land's topography and use. For example, tillage practices, in particular sloping sites, has a tendency to promote erosion (FAO 2016). Production types such as integrated crop-livestock have different, more

Agroecological Approaches for Sustainable Soil Management, First Edition.
Edited by Majeti Narasimha Vara Prasad and Chitranjan Kumar.

sustainable, approaches which have the potential to limit soil erosion (i.e., cover crops, rotational systems).

Sustainable land management is defined by productivity, soil and water quality, and minimization of risks in production. This sustainability should also meet economic viability. In other words, even if the production shows to be sustainable from an environmental point of view, if it is economically unsustainable the production will cease to exist. Ecosystem services involved in agricultural production systems are of major importance, even though not fully acknowledged monetarily. Multifunctional agroecosystems for instance, are at the foundation of several ecosystem services, covering a wide range of benefits: from food production, genetic resources conservation, raw materials production, biological control, control soil erosion, nutrient cycling, pollination, providing refugee for animals, soil formation, waste detoxification, water regulation, and supply (Kremsa 2021). Its compensation is not usually present in the cost of the consumed resources. Prices are generally a result of the market supply and demand, of the purchases from producers or the willingness to pay of consumers. Thus, the importance of developing a tool for both natural and economic analysis in agroecosystems can be of key importance for policymakers on resource production and consumption.

Emergy Assessment (EmA), can contribute with a biocentric point of view of both ecological and economic progress. This tool performs a quality assessment of the energy flows of the required resources to produce the outcome. Free resources from biogeophysical systems are assessed, on the same unit of measurement as purchased resources. Emergy evaluation can be very helpful to assess agricultural production on its challenge to increase production in parallel with sustainability increase. Therefore, the following review focuses on mixed agricultural systems (i.e., integrated crop-livestock production) and how EmA reflects its land use toward resilience in the face of climate change and economic events (Martin et al. 2016).

9.2 Emergy Methodology

Emergy methodology, developed by H. T. Odum in 1983, accounts for all types of energy required to produce a service or a product. It is a tool grounded on thermodynamics and general ecological system's theory used to assess the performance and sustainability of a system, in this case agroecosystems. This method uses solar emjoules(sej) as the common energy base. All flows of money, mass, and energy (€, kg, J) are transformed into solar emergy by multiplying by a conversion factor called the unit emergy value (UEV). UEVs represent the amount of solar emergy required to produce one unit of the output under study (sej/unit).

In an emergy evaluation, the main components driving the geobiosphere are the solar, tidal, and internal Earth energy (Brown and Ulgiati 2004). The associated three primary exergies form the sum of all energy input to earth and is referred as the solar emergy baseline. The total emergy of a process is the sum of all required resources: local renewables, local non-renewables and external to the system's boundaries (purchases or feedback from the economy).

By accounting all the resources embodied on products, it is possible to assess the real value rather than attribute its money value. Moreover, for a better assessment of the system's performance and sustainability, a set of emergy indices and ratios can be used (Table 9.1).

Table 9.1 Emergy indices description for evaluating agricultural production systems.

Emergy indicator	Formula	Description
Tr – Transformity	Tr = Y/E[a]	Ratio of the emergy of the output divided by the energy of the products. Useful for comparing systems with the same output, the higher the transformity, lower the production efficiency
EYR – Emergy yield ratio	EYR = Y/(P + S)[a]	Measures the ability of the system to use local natural resources (renewable and non-renewable) by investing in external resources
Environmental loading ratio (ELR)	ELR = (NR + P)/R[a]	Indicates the pressure the system generates on the local environment based on external resources. The lower the value, the lower its impact.
Emergy sustainable index	ESI = EYR/ELR	Indicates if the production system is able to obtain a high yield with a low environmental loading. A high ESI indicates higher sustainability.
Emergy investment ratio	EIR = (P + S)/(R + NR)[a]	Indicates the dependency of the system on external purchased resources

[a] Yield (Y), Energy of product (E), Purchases (P), Services (S), Local Non-renewables (NR), Local Renewables (R).
Source: Brown and Ulgiati 1997.

9.3 Review Methodology

Emergy publications on mixed production were compiled and assessed to gain a better insight on the sustainability of these systems. The databases used for this study were Web of Science (WoS) and Scopus, with a search a limit from the year 2000 until 2022 to ensure the most recent methodologies were used in the studies (Table 9.2).

Table 9.2 Collection of papers with emergetic assessments on integrated crop-livestock productions from 2006 to 2022.

Study	Type system	Production	Goals	References
1.1	Fully combined	Fully integrated mixed sheep-permanent crops	Evaluate Mediterranean sheep and sheep-crop farming systems with diverse degrees of specialization, integration and intensification of production.	(Rodríguez-Ortega et al. 2017)
1.2	Fully combined	Partially integrated mixed sheep-arable crops		
2	Fully combined	Cattle oak production	Evaluate cattle rearing Montado farm in Portugal.	(Fonseca et al. 2016)

(*Continued*)

Table 9.2 (Continued)

Study	Type system	Production	Goals	References
3	Spatially separated	Integrated production of grains, pig, and fish	Evaluate the environmental aspects of integrated production system of grains, pig and fish in small farms in the South region of Brazil.	(Cavalett et al. 2006)
4	Rotational system	Livestock component within an 11-yr rotation of mixed production (6 yrs cropping–5 yrs livestock)	Evaluate the grazing cattle system in Argentina's Pampas.	(Rotolo et al. 2007)
5	Fully combined	Organic rice-duck mutualism	Evaluate rice-duck mutualism organic farming system versus conventional wheat rice rotation	(Xi and Qin 2009)
6	Spatially separated	Eco-agricultural integrated production system of cattle, biogas, and greenhouse vegetables	Evaluate integration production system of cattle, greenhouse vegetables and biogas in China.	(Wu et al. 2013)
7	Rotational system	Integrated crop-livestock system	Evaluate the three most common commercial crop and livestock systems observed in the Cerrado and Amazon (integrated crop-livestock, continuous crop and continuous livestock); and provide results for two innovative emergy indices: emergy footprint and carbon-emergy output intensity.	(dos Reis et al. 2021)
8	Rotational system	Integrated swine–crop–pasture–eucalyptus system (year IV of study)	Evaluate an integrated crop-livestock system in the Brazilian Cerrado at a river's headwaters	(Buller et al. 2016)
9	Fully combined	Goose raising within an organic grape production	Evaluate a goose raising system with an organic grape production	(Patrizi et al. 2018)
10	Spatially separated	Circular agriculture system (Grain, vegetables, fruits, and cattle)	Evaluate a circular agricultural system with cattle, fruit, grain, vegetables, and biogas production	(Zadgaonkar et al. 2022)

These were then selected based on its system boundary, indices and ratios assessment, and production specificities. Data collection concerns the resources used by each system and its main common indices and novelties. Once gathered, main goals and outcomes of each study were identified and aligned with the data in order to determine the improvements and technicalities in this type of production's management to restore the sustainability of agricultural systems.

9.4 Mixed Farming

Separation of crop and livestock systems rose along with the industrialization, narrowing and standardizing agricultural management practices. Environmental services were altered by synthetic external inputs (i.e., herbicides, fertilizers) and machinery to ensure higher yields. Consequentially these practices create a huge disturbance on soil's health if not followed by a proper management (de Graaff et al. 2019). Integrated production is seen as a standard to achieve good ecological performance. It has the capacity to use more local resources (e.g., organic fertilizer), improve soil quality with diversified crop rotation, enhance biodiversity with natural weed or pest controllers, reduce soil erosion, and consequentially nitrate leaching (Hilimire 2011; Nemecek et al. 2006). In short, it has the capacity to enhance the land use efficiency by producing several products while restoring environmental services. This enlargement of the farm's products scope provides more economic opportunities for the farmer and local community. Considering there is a lower reliance on heavy machinery, the workload on rural areas has the potential to increase. In short, mixed systems promote biodiversity and ecosystems restoration, with associated environmental and economic benefits.

Even though they present a wide of interesting advantages, there are some challenges that need to be accounted in these systems. Mixed farming is based on antique practices, to adapt to nowadays there must be symbiosis of knowledge and resource management to avoid their depletion. For instance, the equilibrium between supply and demand of nutrients is not easy to establish, especially in a farm with animals rearing in crop fields (Martin et al. 2016). However, the mixture of animal and crops production opens a whole range of benefits such as better nutrient and water management.

System boundaries considered in this analysis are defined at farm level scale considering this is the level where management decisions can be evaluated to understand the performance of the system. The types of integration were defined by time and space: spatially separated, fully combined, and rotational. Spatially separated farms have livestock either grazing in pastures or housed, separated from crop or supplemental feed production fields. Nevertheless, there can still be good's exchange when it comes to manure or crop residues between systems. Fully combined systems involve animals in crop pastures. Crop residues left on the field after harvest can represent a large source of biomass for ruminant feed and at the same time animals fertilize soils with manure. Rotational systems apply to both crop and animal production on the same field but at different times.

9.5 Emergy Applied to Mixed Farming

Agricultural production systems are a much-used subject for emergy assessment around the world. For example: There is a study of an agroecosystem in China (Liu et al. 2018) comprising several different subsystems of mulberry dyke, fish pond, rape dyke, and silkworm; Italy (Cristiano 2021) with a small-scale community-supported agriculture; (Castro David et al. 2018) assess a system of tilapia cage farming in Brazil; (Fonseca et al. 2019) explores an integrated production of cattle and cork oak in Portugal; whereas in India and Pakistan (Ali et al. 2019) an investigation on drivers and consequences of changes in crop production sustainability was carried. Agricultural EmA can also be adopted country-wise (Chen et al. 2006) and regionally (Principi et al. 2003). Although is mostly used at farm scale (Hercher-Pasteur et al. 2020), considering the large amount of data required for this method.

Conventional agriculture has more sustainable oriented guidelines in the last few decades through new government regulations. Common Agriculture Policy, or CAP, uses subsidies as incentives for farmers to go green if they undergo certain management practices (e.g., crop diversification, maintaining permanent grasslands, and establishing ecological focus areas). These practices are a starting point for a transition to a sustainable type of agriculture, similarly like integrated agriculture. Where a low-input farming is encouraged, quite the opposite of common conventional agriculture from an emergetic perspective, which requires more purchased inputs: fertilizers, pesticides, machinery, electricity, and fuels (La Rosa et al. 2008). Fossil fuel energy consumption on conventional farming can reach the double of what is consumed by an organic farming type (Gomiero et al. 2008).

Total emergy inputs of a system can vary notably between different types of production, e.g., conventional chicken production (4.20 E + 17 sej/yr) vs organic (3.87 E + 16 sej/yr) (Castellini et al. 2006), and a conventional blood orange (2.3 E + 16 sej/yr) vs organic (1.2 E + 16 sej/yr) (La Rosa et al. 2008). A frequent approach to give a better perspective on the total emergy required for multifunctional agricultural production systems–apart from using the dimensionless indices– is to assess, for instance, the full mixed systems and compare it with its subsystems. Assuming these subsystems were specialized production (e.g., animal or crop production only). Other common EmA of mixed agroecosystem involve anaerobic digestion for biogas production (Chen et al. 2019; Ciotola et al. 2011), hydroponics (Han et al. 2016) and aquaculture (Williamson et al. 2015). These are examples of different management techniques, site and product specific, aiming to achieve a sustainable ecosystem able to provide services on its own.

9.6 Emergy Indices and Scope Widening

From the selected literature (Table 9.2), a collection of indices was made. In a first step, the total emergy required for the system to generate its output is compared between studies together with the amount of local renewable resources and non-renewables. In most studies the non-renewable local resource consists of topsoil loss and groundwater, considering their recharge is longer than a year (time span used in these assessments). For this study, topsoil loss was collected to represent the local non-renewables. Another indicator present for comparison is the feedback. The feedback here reflects the amount of resources that are generated by the system and used afterwards by the same. This component is still not widely used as these flows are, in turn, accounted under other flows. For example, the manure from livestock applied in the crop field might not be accounted as a feedback flow but as less artificial fertilizer input. Secondly, most frequent used emergetic indices were also compared between studies to reflect the economic and environmental impact on the systems.

Figure 9.1 shows renewable resources have a high variability between the selected studies, going from 0.05% to 64.54% of the total emergy required for production. The integrated production of cattle with a greenhouse of vegetables and biogas production

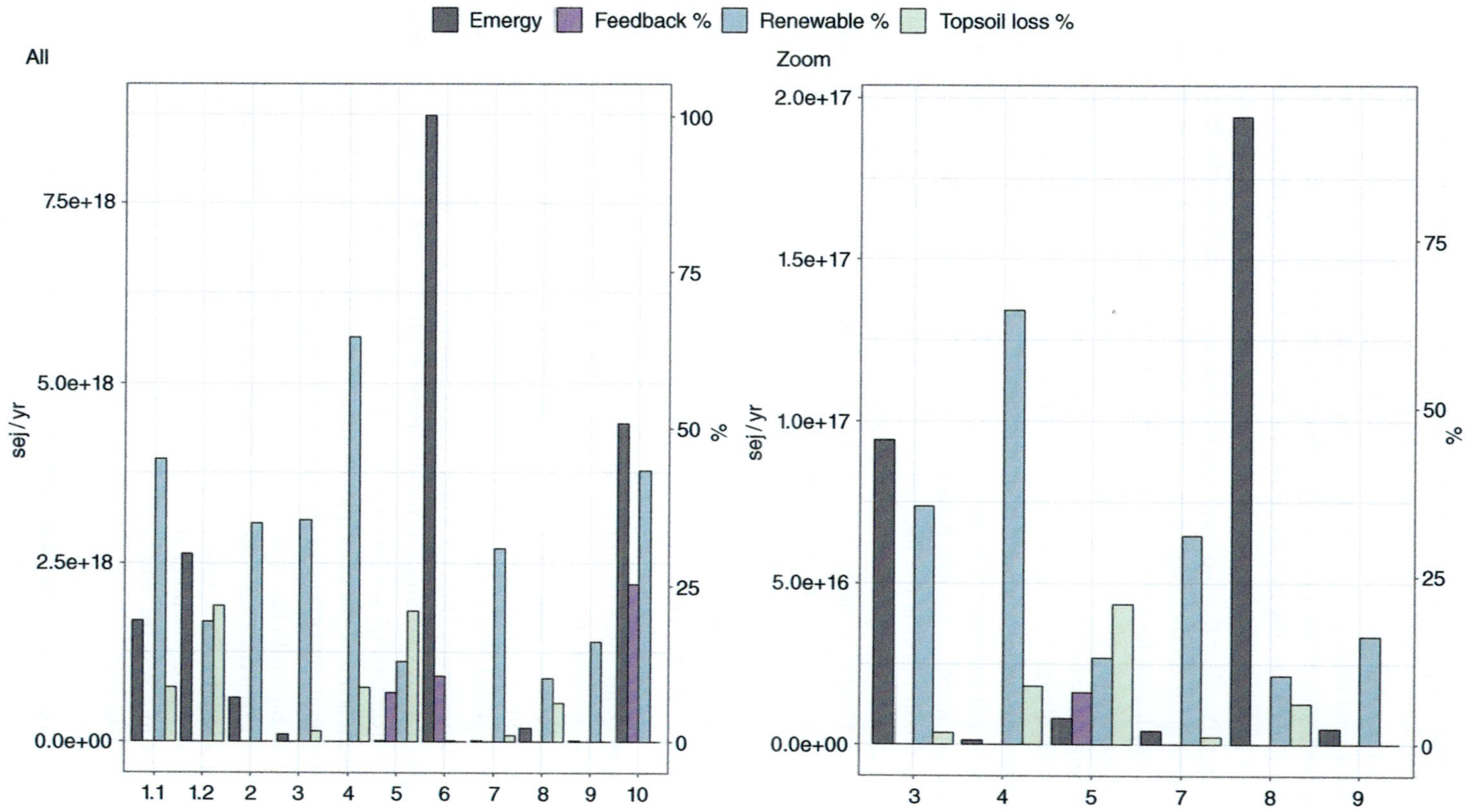

Figure 9.1 Plot of emergetic flow categories: total emergy (sej/yr); feedback from within the system (%) available only for studies 5, 6, and 10 (see Table 9.2); renewable resources (%) and topsoil loss (%) as the local non-renewable flow.

(Table 9.2, study 6), has the smallest percentage of renewables (0.05%) and the highest total emergy. Its main emergy flow is allocated in cattle feed (68.5%) which is not produced locally, thus considered as purchased inputs. Its feedback is 10.5% of the total emergy of the system. The same applies for study 10, where there is also biogas production, and it has the second highest total emergy. Even though its renewable resources contribution is higher (43.26%), almost 80% of this input is solely discarded non-milking cows, which are left to rear the farm. This forms the major contributor to the total emergy of the system, around 34%. Followed by purchased fodder and feedback, where the former represents 25% of the system's emergy. It is clear that biogas production is a major contributor for feedback inputs and for increasing the total emergy of the system. Regarding study 8, anaerobic digestor is used to generate biogas for electricity and fertilizers. This has a big impact on reducing purchased inputs, although animal's feed continues to be one of the major contributors.

Contrastingly, there are systems that benefit from locally produced feed to complement the grazing. This diminishes substantially the amount of purchased emergy. Such as study 2, where cattle rears within cork and holm oaks and the farm produces fodder for the pasture shortage in hot summers. The production for bales, involving labor, fertilizer, fuels, mechanical equipment, and plastic represent a total of 10% of the total emergy.

Another common result is the increase of labor when comparing specialized systems with integrated ones. Study 5 showed an increase from 1.03 E + 15 sej/yr in the conventional to 1.67E + 15 sej/yr in the integrated, and in study 6 a total labor emergy of 2.16E + 17 for the integrated production, while livestock and crop individual subsystems were 1.25E + 17 sej/yr and 8.65E + 16 sej/yr, respectively.

Study 1 assess three sheep pasture systems, whereas two with mixed production were considered for this study. Fully integrated mixed system's renewable resources (study 1.1) represent almost half of total emergy (45.2%), followed by purchased inputs (34.5%). The partially integrated mixed system (study 1.2) has fewer renewable resources involved, 19.3%. The latest's area was occupied mostly by cereals rather than fodder, requiring a higher purchase amount of feed.

A direct consequence from relying less on renewables is the increase of topsoil loss. Topsoil loss in study 1.1, fully integrated system, represented 8.6% of total emergy, 21.7% in the partially integrated. The same applies to other studies. Figure 9.1 reflects this increase in topsoil loss when renewables are lower (e.g., studies 1.2, 5, 8). Studies 6 and 10 topsoil loss is not, quantitively speaking, far from the other studies results. Although, it represents a very small percentage of system's total emergy. Using less land to produce the same amount of output is one of the principles of integrated crop production (Hilimire 2011). In study 9, Patrizi et al. (2018) stated the integrated system produced the same amount of goose meat and grape with less emergy (4.85E + 15 sej/yr) than the specialized, individual, system (6.43E + 15 sej/yr). There was a 33% decrease on feed emergy and complete decrease in fertilizer application. These synergies, beside economic profitability, promotes soil fertility, and erosion control (Martin et al. 2016).

Main indices to assess the emergetic performance are the Emergy Yield Ratio (EYR), Environment Load Ratio (ELR), Emergy Sustainability Index (ESI), Emergy Investment Ratio (EIR), and Transformity (Tr) Table 9.1.

From Table 9.3 we can see the main indices used in the selected studies. Study 8 has the highest ELR compared with the remain. Even though it is an integrated production, it has

Table 9.3 Emergetic indices for the selected literature.

Emergy indicators	1.1	1.2	2	3	4	5	6	7	8	9	10
Tr – sej/J	8.54E+06	4.79E+05	3.58E+05	–	1.73E+06	–	1.46E+06	1.29E+05	–	–	–
EIR – dimensionless	0.86	1.44	1.85	1.87	0.37	1.71	–	2.11	8.3	–	–
ELR – dimensionless	0.95	3.22	1.86	3.19	0.55	0.87	3.31	2.21	8.8	–	1.31
EYR – dimensionless	2.16	1.69	1.54	1.54	3.73	1.82	1	1.47	–	–	1.76
ESI – dimensionless	2.27	0.53	0.83	0.48	6.8	2.09	0.302	0.67	–	–	1.34

an intensive animal production. Resulting in a higher environmental pressure, still not reaching ELR from conventional production (Amiri et al. 2019, 2021; Asgharipour et al. 2019). Study 6, 1.2, and 3 have ELR between 3 and 4, and according to (Brown and Ulgiati 2004), ELRs between 3 and 10 indicate moderate environmental impacts. Study 6 also presents an EYR of 1, which is the lowest possible value. This indicates that this system in particular needs further improvement as it relies mostly on external resources and does not contribute with net emergy to the economy. Study 4 on the other hand shows the highest EYR and lowest ELR reflecting the system's high dependence on local renewable resources–also indicated by the ESI.

The lower the EIR, the lower the system's dependence on external purchased material. Considering EIR index is very similar to EYR, there are some studies omitting this indicator for this reason. Its presence in a table helps, nonetheless, to understand more easily the system under study. Based on what was aforementioned, study 4 would be the most likely to have the lowest EIR as it is the case, with study 1.1 showing similar results. The former has the highest transformity of all studies, or from those available. This supports the other emergy indices results, showing this integrated production is more efficient in energy conversion.

Apart from these traditional indices, there have been appearing other indicators to assess externalities which were not being accounted such as greenhouse gas emissions, infiltration of contaminants, or nutrient excess. Topsoil loss energy is calculated with the following equation:

$$\text{Topsoil loss energy} = 2\times(\text{area})\times(\text{soil loss rate})\times(\text{organic matter content})\times (5400\,\text{kcal/kg})\times(4186\,\text{J/kcal})$$

This result reflects the rate at what the soil is being eroded within the system. Two new indicators were developed to quantify, at an emergetic level, the losses of stocks of soil natural capital (Cohen et al. 2006). Fraction of use soil erosion, or FUSE, is an indicator which gives the amount of soil loss per total emergy value required for production:

$$\text{FUSE} = \text{Soil loss emergy/Total emergy input}$$

And Soil Intensity of Agriculture (SIA), represents the cost benefit of the soil use:

$$\text{SIA} = (\text{EM\$ of agricultural sales/Soil loss emergy})$$

Study 8 assesses an integrated production system and its soil improvement in 4 years with these indicators. Within 4 years of receiving the previously discussed benefits of this type of system, its FUSE went from c.a. 10 to 1. Meaning that the erosion decreased about 9% on the total resource basis. Regarding the SIA indicator, it increased from 10 to 280 which reflects the economic profitability an integrated system can have over time.

Regarding GHG, there are also some developments being made to account this externality. Carbon-emergy output Intensity (CemI) measures net ton CO_2-eq emissions per unit yields measured in emergy (Dong et al. 2018). The premiss would be to reduce this value to obtain a sustainable system. And for this case, study 7, had a CemI of −2.71E-11 tonCO_{2eq}. This negative value indicates the capacity of the integrated system to reduce CO_2 emissions.

Similarly, Greenhouse gases Emission Mitigation Intensity (GEMI) measures the net emission mitigation (GEM) between alternative production systems in terms of greenhouse gas emissions per unit of emergy input (U) (Chen and Chen 2012):

$$GEMI = GEM / U$$

This indicator was used in study 6, where it took into consideration the difference between greenhouse gases emissions in different procedures (i.e. biogas project construction, biogas use, coal substitution, chemical fertilizer substitution, and manure utilization). In total, the total mitigation was of 2.12 kg CO_{2eq}/10E+14 sej.

9.7 Main Findings and Gaps in Literature

One of the main procedures when doing an EmA to this type of production is to compare it with an EmA of the subsystems involved. There was always the consensus that mixed production involves less environment stress and it can be more self-sustaining in the long-term considering is less dependent on external purchased inputs. Although there are several studies indicating an increase in labor when passing to integrated production. This signifies a higher cost for the farmer, but it is also beneficial for the local community due to the requirement of more knowledge and/or work. High productivity is directly linked to large number of fossil-based resources, making productive systems dependable of external purchased resources.

Using a common unit basis for measuring nature and human contributions to the system is the main advantage all studies find for using this method. The studies (dos Reis et al. 2021; Rodríguez-Ortega et al. 2017; Xi and Qin 2009) concluded conventional production showed higher productivity and economic profitability. Nonetheless, the traditional economic valuation does not take environmental costs into consideration and in the long-term these systems tend to deplete the founding flow, being the soil.

Moreover, the intensification leads to a higher feed consumption. It was stated in these studies feed purchase involves a large amount of emergy versus local feeding. Feed seems to be negatively affecting the sustainability of these productions when it is external. Reducing in the intensification of animals per area is also a key result in the studies to achieve environmentally friendly productions and to protect against soil erosion. Mixed farming can aid in reducing the risk of getting economically compromised with the intensification reduction In other words, it is possible to ensure a wider selection of products within the same land along with residues valorization. Manure application in the soils is a huge contribution for instance. Especially if we are considering systems with an anaerobic digestor with the capacity to generate not only energy through biogas but also organic fertilizers.

Purchased resources and services have an associated renewability factor, which aids in identifying the renewable and non-renewable fractions. Most of these fractions are obtained from previous works. Nevertheless, there are still few studies not implementing these factors which help to reflect the complexity of these systems. Moreover, it may significantly affect the indices results considering they are functions of the renewables and non-renewables resources. Leading to an unfair comparison of different system's indices (Lu et al. 2010).

Transformity values, or UEVs, also present a similar issue. Calculating a transformity of a product or service needs to take into consideration all the flows involved for its elaboration. It is very time consuming, and it is a study on its own, to calculate a product's emergy and respective transformity. This is the reason why most of the studies use previously calculated transformities. As shown in these studies, emergy of a product may differ at region, or even, process level. A database for transformities (https://www.emergysociety.com/emergy-society-database) is being developed by the emergy society, so there are several transformities for the same products which could apply best for each particular case.

9.8 Future Advises

Further research is required, in particular to develop more knowledge in recognizing the ecosystem services provided by agroecosystems and how to valorize them monetarily. An aspect where emergy could become a useful tool together with other environmental footprints methodologies. Taking also into consideration climate change into these assessments is also vital to understand what the best approach for land use management of integrated crop-livestock production systems under extreme weather changes would be. New indicators addressing externalities, such as mentioned in this review, are interesting solutions for this issue.

Incorporation of renewability factors has the potential to reflect the complexity of these systems and to enhance the importance of renewable inputs, being purchased or from within the system. This would enable further support, through policymaking, to incentive farmers on using local renewable resources and pursue sustainable production systems such as mixed farming.

References

Ali, M., Marvuglia, A., Geng, Y. et al. (2019). Accounting emergy-based sustainability of crops production in India and Pakistan over first decade of the 21st century. *J. Clean. Prod.* 207: 111–122. https://doi.org/10.1016/j.jclepro.2018.09.236.

Amiri, Z., Asgharipour, M.R., Campbell, D.E., and Armin, M. (2019). A sustainability analysis of two rapeseed farming ecosystems in Khorramabad, Iran, based on emergy and economic analyses. *J. Clean. Prod.* 226: 1051–1066. https://doi.org/10.1016/j.jclepro.2019.04.091.

Amiri, Z., Asgharipour, M.R., Campbell, D.E. et al. (2021). Conservation agriculture, a selective model based on emergy analysis for sustainable production of shallot as a medicinal-industrial plant. *J. Clean. Prod.* 292. https://doi.org/10.1016/j.jclepro.2021.126000.

Asgharipour, M.R., Shahgholi, H., Campbell, D.E. et al. (2019). Comparison of the sustainability of bean production systems based on emergy and economic analyses. *Environ. Monit. Assess.* 191. https://doi.org/10.1007/s10661-018-7123-3.

Brown, M.T. and Ulgiati, S. (1997). Emergy-based indices and ratios to evaluate sustainability: monitoring economies and technology toward environmentally sound innovation. *Ecol. Eng.* 9(1–2): 51–69.

Brown, M.T. and Ulgiati, S. (2004). Emergy Analysis and Environmental Accounting.

Buller, L.S., Bayma-Silva, G., Ortega, E., and Bergier, I. (2016). Soil loss as a negative externality in the Emergy accounting: case study of an agricultural commodities municipality in the Brazilian Savannah. *J. Environ. Account. Manag.* 4: 129–147. https://doi.org/10.5890/JEAM.2016.06.004.

Castellini, C., Bastianoni, S., Granai, C. et al. (2006). Sustainability of poultry production using the emergy approach: comparison of conventional and organic rearing systems. *Agric. Ecosyst. Environ.* 114: 343–350. https://doi.org/10.1016/j.agee.2005.11.014.

Castro David, L.H., Pinho, S.M., and Garcia, F. (2018). Improving the sustainability of tilapia cage farming in Brazil: an emergy approach. *J. Clean. Prod.* 201: 1012–1018. https://doi.org/10.1016/j.jclepro.2018.08.124.

Cavalett, O., Queiroz, J.F.D., and Ortega, E. (2006). Emergy assessment of integrated production systems of grains, pig and fish in small farms in the South Brazil. *Ecol. Model.* 193: 205–224. https://doi.org/10.1016/j.ecolmodel.2005.07.023.

Chen, S. and Chen, B. (2012). Sustainability and future alternatives of biogas-linked agrosystem (BLAS) in China: an emergy synthesis. *Renew. Sustain. Energy Rev.* 16: 3948–3959. https://doi.org/10.1016/j.rser.2012.03.040.

Chen, G.Q., Jiang, M.M., Chen, B. et al. (2006). Emergy analysis of Chinese agriculture. *Agric. Ecosyst. Environ.* 115: 161–173. https://doi.org/10.1016/j.agee.2006.01.005.

Chen, L., Cui, Z., Dong, R. et al. (2019). Ecological evaluation of large-scale biogas engineering with biogas slurry recycle mode based on emergy analysis method [基于能值分析的规模化沼气工程沼液回流工艺生态效益评价]. *Nongye Gongcheng Xuebao/Transactions of the Chinese Society of Agricultural Engineering* 35: 228–234. https://doi.org/10.11975/j.issn.1002-6819.2019.14.029.

Ciotola, R.J., Lansing, S., and Martin, J.F. (2011). Emergy analysis of biogas production and electricity generation from small-scale agricultural digesters. *Ecol. Eng.* 37: 1681–1691. https://doi.org/10.1016/j.ecoleng.2011.06.031.

Cohen, M.J., Brown, M.T., and Shepherd, K.D. (2006). Estimating the environmental costs of soil erosion at multiple scales in Kenya using emergy synthesis. *Agric. Ecosyst. Environ.* 114: 249–269. https://doi.org/10.1016/j.agee.2005.10.021.

Cristiano, S. (2021). Organic vegetables from community-supported agriculture in Italy: Emergy assessment and potential for sustainable, just, and resilient urban-rural local food production. *J. Clean. Prod.* 292: 126015. https://doi.org/10.1016/j.jclepro.2021.126015.

Dong, G., Wang, Z., and Mao, X. (2018). Production efficiency and GHG emissions reduction potential evaluation in the crop production system based on emergy synthesis and nonseparable undesirable output DEA: a case study in Zhejiang Province. *China. PLoS One* 13: e0206680. https://doi.org/10.1371/journal.pone.0206680.

ELD (2015). ELD Initiative. https://www.eld-initiative.org/en/.

FAO (2016). Chapter 6: Global status, processes, and trends in soil erosion (Main Report). The Status of the World's Soil Resources.

Fonseca, A.M.P., Marques, C.A.F., Pinto-Correia, T., and Campbell, D.E. (2016). Emergy analysis of a silvo-pastoral system, a case study in southern Portugal. *Agrofor. Syst.* 90: 137–157. https://doi.org/10.1007/s10457-015-9888-5.

Fonseca, A.M.P., Marques, C.A.F., Pinto-Correia, T. et al. (2019). Emergy evaluation for decision-making in complex multifunctional farming systems. *Agric. Syst.* 171: 1–12. https://doi.org/10.1016/j.agsy.2018.12.009.

Godfray, H. and Garnett, T. (2014). Food security and sustainable intensification. *Philos. Trans. R. Soc. B* 369. http://dx.doi.org/10.1098/rstb.2012.0273.

Gomiero, T., Paoletti, M., and Pimentel, D. (2008). Energy and environmental issues in organic and conventional agriculture. *Crit. Rev. Plant Sci.* 27: 239–254. https://doi.org/10.1080/07352680802225456.

de Graaff, M.-A., Hornslein, N., Throop, H.L. et al. (2019). Effects of agricultural intensification on soil biodiversity and implications for ecosystem functioning: a meta-analysis. In: *Advances in Agronomy*, 1–44. Elsevier. https://doi.org/10.1016/bs.agron.2019.01.001.

Han, B., Ouyang, Z., Liu, H. et al. (2016). Courtyard integrated ecological system: an ecological engineering practice in China and its economic-environmental benefit. *J. Clean. Prod.* 133: 1363–1370. https://doi.org/10.1016/j.jclepro.2016.06.061.

Hercher-Pasteur, J., Loiseau, E., Sinfort, C., and Helias, A. (2020). Energetic assessment of the agricultural production system. A review. *Agron. Sustain. Dev.* 40: 29. https://doi.org/10.1007/s13593-020-00627-2.

Hilimire, K. (2011). Integrated crop/livestock agriculture in the United States: a review. *J. Sustain. Agric.* 35: 376–393. https://doi.org/10.1080/10440046.2011.562042.

Kremsa, V.Š. (2021). 5 – sustainable management of agricultural resources (agricultural crops and animals). In: *Sustainable Resource Management* (ed. C.M. Hussain and J.F. Velasco-Muñoz), 99–145. Elsevier. https://doi.org/10.1016/B978-0-12-824342-8.00010-9.

La Rosa, A.D., Siracusa, G., and Cavallaro, R. (2008). Emergy evaluation of Sicilian red orange production. A comparison between organic and conventional farming. *J. Clean. Prod.* 16: 1907–1914. https://doi.org/10.1016/j.jclepro.2008.01.003.

Liu, S., Min, Q., Jiao, W. et al. (2018). Integrated Emergy and economic evaluation of Huzhou mulberry-dyke and fish-pond systems. *Sustainability* 10: 3860. https://doi.org/10.3390/su10113860.

Lu, H., Bai, Y., Ren, H., and Campbell, D.E. (2010). Integrated emergy, energy and economic evaluation of rice and vegetable production systems in alluvial paddy fields: implications for agricultural policy in China. *J. Environ. Manag.* 91: 2727–2735. https://doi.org/10.1016/j.jenvman.2010.07.025.

Martin, G., Moraine, M., Ryschawy, J. et al. (2016). Crop–livestock integration beyond the farm level: a review. *Agron. Sustain. Dev.* 36: 1–21. https://doi.org/10.1007/s13593-016-0390-x.

Nemecek, T., Dubois, D., Huguenin-Elie, O., and Gaillard, G. (2006). Life cycle assessment of Swiss organic farming systems. *Aspects Appl. Biol.* 79: 15–18.

Odum, H.T. (1983). *Systems Ecology; An Introduction*. New York, USA: Wiley.

Panagos, P., Borrelli, P., Poesen, J. et al. (2015). The new assessment of soil loss by water erosion in Europe. *Environ. Sci. Policy* 54: 438–447. https://doi.org/10.1016/j.envsci.2015.08.012.

Patrizi, N., Niccolucci, V., Castellini, C. et al. (2018). Sustainability of agro-livestock integration: implications and results of emergy evaluation. *Sci. Total Environ.* 622: 1543–1552. https://doi.org/10.1016/j.scitotenv.2017.10.029.

Pimentel, D. and Burgess, M. (2013). Soil erosion threatens food production. *Agriculture* 3: 443–463. https://doi.org/10.3390/agriculture3030443.

Principi, I., Fugaro, L., and Borsa, S. (2003). Assessing sustainability of the chianti area: the role of agriculture. In: *Ecosystems and Sustainable Development Iv*, vol. 1 and 2 (ed. E. Tiezzi, C. Brebbia, and J. Uso), 723–729. Advances in Ecological Sciences.

dos Reis, J.C., Rodrigues, G.S., de Barros, I. et al. (2021). Integrated crop-livestock systems: a sustainable land-use alternative for food production in the Brazilian Cerrado and Amazon. *J. Clean. Prod.* 283. https://doi.org/10.1016/j.jclepro.2020.124580.

Rodríguez-Ortega, T., Bernués, A., Olaizola, A.M., and Brown, M.T. (2017). Does intensification result in higher efficiency and sustainability? An emergy analysis of Mediterranean sheep-crop farming systems. *J. Clean. Prod.* 144: 171–179. https://doi.org/10.1016/j.jclepro.2016.12.089.

Rotolo, G.C., Rydberg, T., Lieblein, G., and Francis, C. (2007). Emergy evaluation of grazing cattle in Argentina's pampas. *Agric. Ecosyst. Environ.* 119: 383–395. https://doi.org/10.1016/j.agee.2006.08.011.

Tsiafouli, M.A., Thébault, E., Sgardelis, S.P. et al. (2015). Intensive agriculture reduces soil biodiversity across Europe. *Glob. Chang. Biol.* 21: 973–985. https://doi.org/10.1111/gcb.12752.

Williamson, T.R., Tilley, D.R., and Campbell, E. (2015). Emergy analysis to evaluate the sustainability of two oyster aquaculture systems in the Chesapeake Bay. *Ecol. Eng.* 85: 103–120. https://doi.org/10.1016/j.ecoleng.2015.09.052.

Wu, X., Wu, F., Tong, X., and Jiang, B. (2013). Emergy-based sustainability assessment of an integrated production system of cattle, biogas, and greenhouse vegetables: insight into the comprehensive utilization of wastes on a large-scale farm in Northwest China. *Ecol. Eng.* 61: 335–344. https://doi.org/10.1016/j.ecoleng.2013.09.060.

Xi, Y.-G. and Qin, P. (2009). Emergy evaluation of organic rice-duck mutualism system. *Ecol. Eng.* 35: 1677–1683. https://doi.org/10.1016/j.ecoleng.2007.11.006.

Zadgaonkar, L.A., Darwai, V., and Mandavgane, S.A. (2022). The circular agricultural system is more sustainable: emergy analysis. *Clean Technol. Environ. Policy* https://doi.org/10.1007/s10098-021-02245-2.

10

Agroecological Transformation for Sustainable Food Systems

Ayoub Haouas[1], Anas Tallou[1], Soumia Amir[1], Abdelmajid Haddioui[2], Abdallah Oukarroum[3], and Taoufik El Rasafi[3,4]

[1] *Centre d'Agrobiotechnologie et Bioingénierie, Unité de Recherche Labellisée CNRST (Centre AgroBiotech-URL-CNRST-05), Cadi Ayyad University, Marrakech, Morocco*
[2] *Laboratory of Agro-Industrial and Medical Biotechnologies, Faculty of Science and Techniques, University of Sultan Moulay Slimane, Beni Mellal, Morocco*
[3] *High Throughput Multidisciplinary Research Laboratory, University Mohammed VI Polytechnic (UM6P), Ben Guerir, 43150, Morocco*
[4] *Laboratory of health and environment, Hassan II University, Faculty of Sciences Ain Chock, Casablanca, 5366, Morocco*

10.1 Introduction

The current agri-food system has failed to mitigate adverse hazards, provide healthy food, share equitable gains, or ensure food security for many countries throughout the world for future generations (Béné et al. 2019). For example, biodiversity (the diversity of species, across species, and within ecosystems) is declining faster than at any other point in human history (IPBES 2019; Corson et al. 2022). At this time, there are a number of scientific and political initiatives that are declaring the urgent need to move toward more sustainable, profitable, and resilient agri-food systems that are able to re-establish an organic interaction between society and nature as well as provide a variety of ecosystem services (Gerber and Gerber 2017; Padró et al. 2020). This activity is essential for making progress toward the Sustainable Development Goals set by the United Nations for the year 2030 (United Nations 2016).

Organic agriculture, agroforestry, and permaculture have all emerged as the most effective methods for ensuring the agriculture sustainability. In agriculture, organic farming refers to the use of methods of pest management and fertilizers that are less harmful to the environment and that are obtained in large part from animal and plant wastes, as well as cover crops that fix nitrogen (Willer et al. 2010). The total amount of land used for organic agriculture around the world went from 11 million ha in 1999 to 75 million ha in 2020 (i.e., to 1.6% of agricultural land). By the year 2020, there were 14.9 million ha of organic land in the European Union (EU), which was equivalent to 9.2% of its total agricultural farmlands (Corson et al. 2022).

Agroecological Approaches for Sustainable Soil Management, First Edition.
Edited by Majeti Narasimha Vara Prasad and Chitranjan Kumar.

However, simply expanding the number of organic farms will not be sufficient. A regime change at the systemic level is required for the primary agroecology live funds, such as fertile soils, land covers, livestock, farm-associated biodiversity, farmers, and consumers, to become tightly interlinked through the closure of their fundamental biophysical cycles, as well as through socially equitable and culturally meaningful food chains connected with cities (Moragues-Faus and Morgan 2015; Padró et al. 2020).

The goal of agroecological practices is to minimize socioecological costs associated with agriculture, such as soil degradation, water contamination, greenhouse gas emissions, exhaustion of non-renewable resources, and unequal social structures, while simultaneously maximizing ecological processes, the health of the environment and the well-being of all people (Wezel 2017). Agroecological farming methods differ according to the agroecosystem and the socio-cultural environment in which they are implemented. These systems are guided by principles rather than standardized procedures (Bezner Kerr et al. 2021).

Agroecological livestock systems can be considered biodiversity-based systems, which depend far more on ecosystem services provided by biological processes than they might on human interventions. This is because agroecological livestock systems are built on regenerative agriculture (Therond et al. 2017). Due to the fact that these processes are dependent on the biodiversity found in agroecosystems, making the shift to agroecology necessitates increasing the biodiversity found on farms, particularly that which is beneficial to agricultural productivity, and managing it using a landscape perspective (Duru et al. 2015). Even in natural environments, traditional methods of conserving and restoring biodiversity frequently include ongoing human involvement, such as the development of new habitats or the alteration of existing ones, the extermination of invasive species, or the implementation of captive breeding programs (Corson et al. 2022). Agri-environment systems are based on the observation that some types of intervention can generate better biodiversity and supply of specific ecosystem services than regions that do not have these actions. This observation is the foundation of the agri-environment movement. Agri-environment systems are popular in agricultural regions, notably in Europe (Batáry et al. 2015; Corson et al. 2022). This chapter discusses the fundamental elements that support the adoption of sustainable transformation of agri-food systems and the most adopted agroecological approaches on a real scale. The real challenges undermine the scaling up of these approaches.

10.2 Agroecology

The most basic definition of agroecology describes it as the application of ecological concepts and principles to agricultural practices. This is a complicated concept, however, due to the fact that agriculture and ecology are both dynamic fields that are undergoing rapid change in response to mounting evidence that the global food system is unsustainable due to the fact that agriculture, as it is currently practiced, is a major driver of climate change (Anderson et al. 2021b). Both of these fields are undergoing these changes in order to respond to mounting evidence that the global food system is unsustainable. The term "ecology" may be used to refer to both a scientific field and political groups that are concerned with the preservation of the natural environment (Barrios et al. 2020). Agriculture, which encompasses the collection of methods by which humans create food and to which other scientific disciplines are applied, is a topic emphasis rather than a scientific field. Agriculture is defined as the production of food by people. While agroecology takes into account all of these facets of complexity (Sinclair et al. 2019).

10.2.1 Agroecology and Food Systems

The Food and Agriculture Organization of the United Nations (FAO) has given its stamp of approval to the 10 elements of agroecology as an analytical framework to support the design of different approaches for agriculture and food systems progression (FAO 2019). This will allow for improved decision-making by policymakers, practitioners, and other stakeholders (Barrios et al. 2020). In addition, FAO's Common Vision for Sustainable Food and Agriculture is based on five main ideas: making better use of resources; (ii) preserving, protecting, and improving natural ecosystems; (iii) protecting and improving rural livelihoods, equity, and social well-being; (iv) making people, communities, and ecosystems more resilient; and (v) promoting good governance of both natural and human systems (FAO 2014a).

10.2.2 Principles of Agroecology

Agroecology extends beyond a one-dimensional study of agroecosystems, genetics, agronomy, edaphology, and so on, to include an understanding of ecological and social levels of co-evolution, structure, and function. Rather of focusing on a single component of the agroecosystem, agroecology highlights the interdependence of all agroecosystem components as well as the complex dynamics of ecological processes (Stanojevic and Stanojevic 2021). It has evolved from an ecology-based discipline to develop food production system and it is defined by five principles (efficiency, diversity, synergies, natural regulation, and recycling) to a broader, multidimensional concept requiring the definition of additional principles in the social, political, and economic disciplines and dimensions (Figure 10.1) (Anderson et al. 2021a; Wezel 2017). Agroecosystems are communities of plants and animals that interact with their physical and chemical surroundings in order to generate food, fiber, fuel, and other goods for human use and process.

A field, for example, is viewed as a complex system in which ecological phenomena observed in natural circumstances, such as nutrient cycling, predator/prey interactions, competition, symbiosis, and successional changes, occur. The premise that through understanding these ecological interactions and processes, agroecosystems can be managed to increase output and produce more sustainably, with fewer negative environmental or social consequences and less external inputs, is implicit in agroecological research (Chabert and Sarthou 2020). The following ecological concepts are used in the design of such systems:

- Improve biomass recycling while also maximizing nutrient availability and balancing nutrient flow.
- Creating suitable soil conditions for plant development, notably through the management of organic matter and the enhancement of soil biotic activity.

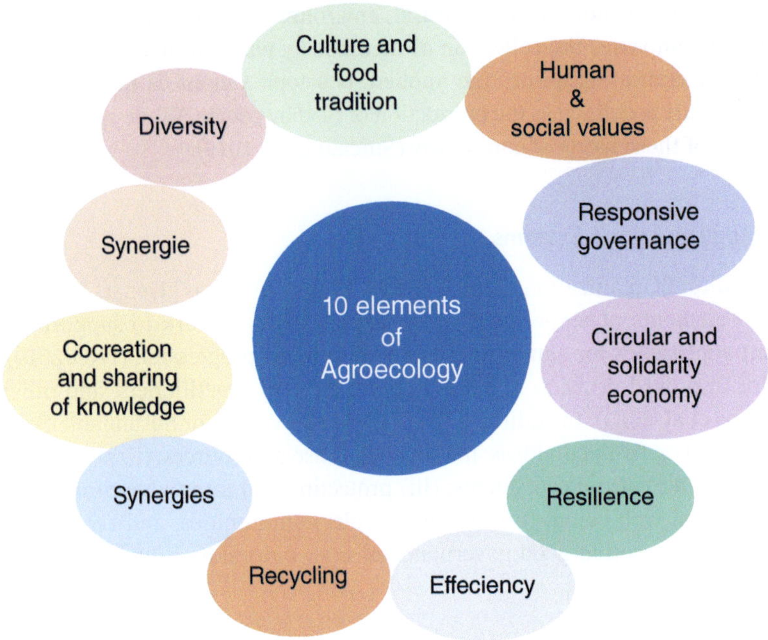

Figure 10.1 The 10 principal elements of agroecology.

- Minimizing solar radiation, air, and water losses through microclimate control, water harvesting, and soil management through enhanced soil cover.
- The evolution of species and genetic diversity in the agroecosystem across time and place.
- Improve favorable biological interactions and synergisms among agrobiodiversity components, resulting in the development of important ecological processes and services.

These concepts may be put into practice through a variety of tactics and procedures. Each of them will have a varied impact on farm production, stability, and resiliency, depending on local opportunities, resource restrictions, and, in most cases, market conditions (Siebrecht 2020). The ultimate purpose of agroecological design is to integrate components in order to increase overall biological efficiency, conserve biodiversity, and maintain agroecosystem production and self-sustaining capability. The idea is to create a patchwork of agroecosystems within a landscape unit, each of which mimics the structure and function of natural ecosystems (Béné et al. 2019).

10.2.3 In Farm Practices

Crop production research focuses on techniques to restrict off-farm fertilizers, such as mixed crops, intercropping systems to better use crop diversity, and N fixation from legumes, or to improve pest management through the use of natural processes, such as allopathy or natural enemies for plant protection (Duchene et al. 2017). Additionally, in animal agriculture, for example, research looks at natural alternatives to antibiotics such as plant extracts or the adaption of animal densities and pasture rotation to increase feed quality and availability (Anderson et al. 2021a).

10.2.3.1 Intercropping

Intercropping is a multiple cropping method in which two or more crop species are planted in the same field at the same time throughout the growing season (Duchene et al. 2017). Obviously, this does not imply that plants can be planted simultaneously in intercropping, but rather that two or more crops are together in one location, during their growth season. Therefore, it is possible that the planting times of the plants vary, and that one plant is planted after the first (Dariush et al. 2006).

10.2.3.2 Biological Control of Pests

The employment of non-chemical and environmentally friendly means of managing insect pests and illnesses through the activity of natural control agents is referred to as biological control (Ali et al. 2022). The increased use of biological management in recent decades is owing to its safety, species specificity, and long-term effect on the target pests. Unlike chemical methods, which harm non-target animals, have negative health impacts on humans, and damage the environment (Stenberg et al. 2021). Conservation biological control, importation, also known as classical biological control, is the introduction of a non-native biocontrol agent into a new location or nation for long-term pest control. Furthermore, augmentation biological control is the additional release of natural enemies by flooding or inoculation. The former involves the discharge of a high number of natural enemies to reduce pest numbers immediately, while the later involves the release of a modest amount at intervals over the pest season (Landis and Orr 2023). However, predators, parasitoids, and microbial antagonists are the most commonly utilized biological control agents in pest control, with an emphasis on entomopathogenic nematodes, fungi, *Bacillus thuringiensis*, and *Baculovirus* (Gurr and You 2016). They are modified and used in agricultural integrated pest control strategies. Unfortunately, the key variables influencing biological control adoption are effectiveness, predictability, and high cost (Stenberg et al. 2021).

10.2.3.3 Recycling into Biofertilizers

Sustainability in agriculture might be achieved without jeopardizing the environment's resources or future generations' capacity to fulfill their own needs (Calabi-Floody et al. 2018; Sartori et al. 2021). Excessive usage of chemical fertilizers depletes suitable living circumstances since residues that function as secondary pollutants might enter food chains/food webs and eventually infect humans. Secondary pollutant may stay in the surrounding environment for a longer amount of time due to health hazards (Haouas et al. 2021b). The use of biofertilizers instead of chemicals in agriculture might usher in a new era of industry. Biofertilizers may be useful in giving nutrients to agricultural plants while preserving the natural environment (Tallou et al. 2020).

Biofertilizers are organic in nature and contain secondary metabolites derived by microorganisms (Chang and Yang 2009). Microorganisms are extracted from soil, water, air, or the rhizosphere for the creation of biofertilizers, which are then concentrated for use in the field. Microorganisms begin creating agriculturally important metabolites in response to particular environmental circumstances, and these metabolites may be used by plants to support various biochemical processes (Haouas et al. 2021a). Microorganisms and microbial metabolites facilitate the breakdown of complicated soil minerals into a simpler form that functions as a growth stimulant for a specific crop. Indeed, biofertilizers might be utilized for a variety of purposes and could have a significant positive influence on climate change, chemical pollution, and natural resource degradation (Huang et al. 2022).

10.2.3.4 Resilience

The idea of resilience has lately been used to multiple studies dealing with the behavior of agricultural systems subjected to various types of disruption (El Chami et al. 2020). Reflecting the varying interpretations of this concept, agricultural system resilience has been defined in several ways: sometimes as the degree of resistance to shock in the face of economic or climatic perturbations, but more frequently as the ability of agricultural systems to preserve their intrinsic functions through flexibility (Venkatramanan and Prasad 2020). In these circumstances, resilience is mostly utilized to explore the elements that contribute to agricultural systems' ability to adjust to changes, restructure their structure, foresee future changes, and capitalize on new possibilities (El Chami et al. 2020; Venkatramanan and Prasad 2020).

10.3 Agroecological Approaches

The question of whether new research concepts are required or if suitable concepts and methodologies already exist that can be employed immediately is extremely important and highly controversial in regard to the transition to an agroecological framework of the food system. The usage of the agroecology concept has, in fact, greatly expanded in scope and dimensions, beginning with the application of ecological principles to agriculture (Ferguson and Lovell 2014). Agroecological transformation must be viewed for these purposes as a complicated endeavor requiring a variety of inputs from sources outside of scientific organizations. Therefore, to develop the agroecological transition, agroecologists engage with both traditional and contemporary producers while taking into account "performance" metrics, the dynamics of farmers' networks, or fusing them with complex adaptive systems and transition theories (Mottet et al. 2020; Tittonell et al. 2012). Many resulted agroecological approaches have been adopted in worldwide including conservation agriculture, organic agriculture, integrated farming, agroforestry, and permaculture. These approaches are at the same time a scientific field, a set of agricultural practices and a social movement (Figure 10.2).

10.3.1 Conservation Agriculture

With the goal of increasing crop productivity while lowering production costs, conservation agriculture is a farming system that has been promoted as a substitute for agricultural practices that deplete the organic matter and nutrient content of the soil and cause soil degradation (Kassam et al. 2009). Conservation agriculture revolves around three fundamental pillars: no-till (or minimal soil disturbance), soil cover (more than 30% of the ground area must be covered immediately after the direct seeding operation), and crop rotation (in time and space such as crop rotation and crop association), respectively; at least three different crops according to FAO. The benefits could be the ease of crop management, energy/cost/time savings, and soil and water conservation (Cordeau 2022; Giller et al. 2015; Tittonell et al. 2012).

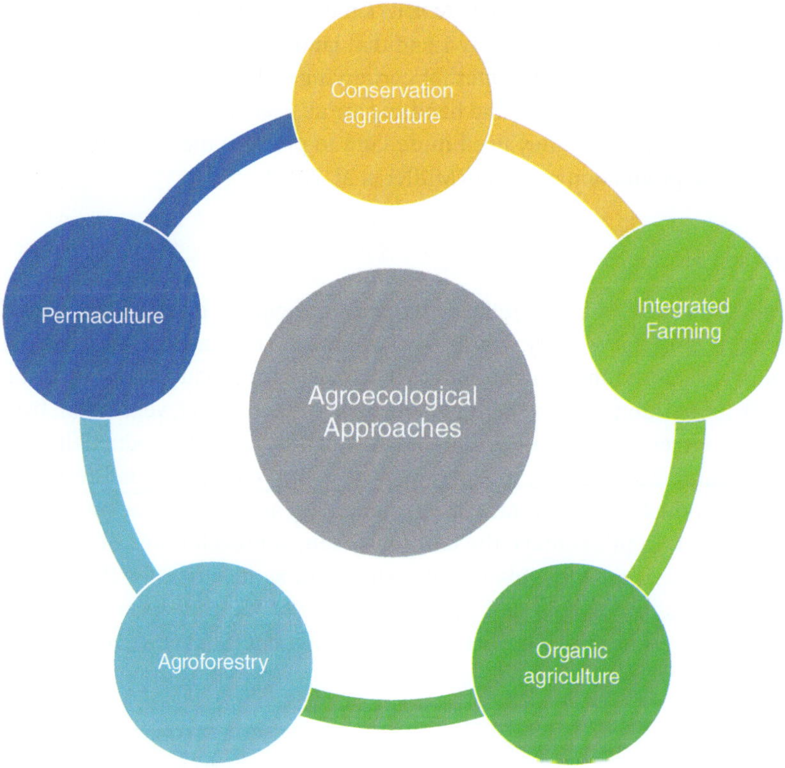

Figure 10.2 Agroecological approaches.

This approach could be high profitable in areas of large farms, where farmers harness the tools of modern science: highly-sophisticated machines, potent agrochemicals, and biotechnology (Giller et al. 2015). The rationale for developing countries, CA may represent a low-investment strategy to increase water productivity and mitigate risks, by breaking the vicious cycle of low rainfall, poor yields, low investment, and soil degradation (Tittonell et al. 2012). The past 10 years have seen a massive wave of enthusiasm for CA among scientists, with strong support from the Food and Agriculture Organization of the United Nations (FAO) (Giller et al. 2015).

10.3.2 Organic Agriculture

Organic agriculture is one of the production systems that tries to promote sustainability. It mixes traditional farming methods with cutting-edge agricultural technology (Reganold and Wachter 2016). By enhancing the soil with compost additions, animal and green manures, and closed nutrient cycles, it avoids utilizing synthetic fertilizers and closed nutrient cycles. It emphasizes using natural methods to manage pests, encourages crop rotation, and concentrates. It has been widely reported that organic farming performs well when compared to a variety of environmental indicators (Amundson et al. 2015; Reganold and Wachter 2016). Due to its many sustainability advantages, organic agriculture can provide a greater portion of the world's food (FAOSTAT n.d.). A little more than 38% of the land on Earth is used for agriculture, but only 1% of that land is used for organic farming. Modern tools, better crop varieties, techniques for conserving soil and water, and the most recent advancements in livestock feeding and handling are all used by organic farmers. But because organic systems yield less, they need more land to produce the same amount of product as conventional production methods (Muller et al. 2017).

10.3.3 Integrated Farming

Small and marginal farmers can find relief from their difficulties through the use of integrated farming, a multidisciplinary, whole-farm approach (Paramesh et al. 2022). It takes innovation in agriculture into account in order to maximize production by utilizing an optimal strategy for harmonizing the joint management of land, water, vegetation, livestock, and human resources. It can result in a quantum leap in sustainable productivity and improve people's livelihood security. By integrating diverse farm enterprises and recycling crop residues and by-products within the farm itself, the strategy aims to increase small-farm income and employment. In this context, the farming system approach is one of the most important solutions for dealing with this peculiar situation, as it allows for the careful implementation of various enterprises and the development of location-specific systems based on available resources, resulting in sustainable development (Soni et al. 2014).

10.3.4 Agroforestry

In 1977, the term "agroforestry" was coined to describe a set of farming techniques in which trees and other woody species are raised alongside other plants and animals (Rosati et al. 2021). In actuality, agroforestry is regarded as an agroecological approach based on the diversification of the production components of an agroecosystem and the intensification of the agroecological relationships between these components in space and time. Potential for the diversification of crop production exists in the vast array of underutilized native foods found in forests and wooded lands, which are frequently richer in micronutrients, fiber, and protein than staple crops (Leakey 1999; Malézieux 2013). Although these foods have traditionally been harvested from forests and woodlands, deforestation and forest degradation are reducing access to these resources (Paudel et al. 2022). In this context, agriculture offers an alternative resource. Moreover, the yield and quality of production can be enhanced during cultivation if genetic improvement and efficient farm management practices are implemented, making planting an attractive option: Genetic selection can improve the yield of many native trees, including fruits, by a factor of two or more (Ickowitz et al. 2022). Because of this, the incorporation of agroforestry into organic farming holds the promise of radical change; it could pave the way for the industry to mature and become more sustainable.

10.3.5 Permaculture

Permanent agriculture is the source of the term "permaculture," which refers to landscapes that are purposefully designed to yield an abundance of food, fiber, and energy to meet local needs by emulating the patterns and relationships found in nature (Ferguson and Lovell 2014). Permaculture is one of these agroecological movements, with a global reach and a distinctive approach to system design.

In order to foster collaboration with diverse groups of researchers, stakeholders, and land users, permaculture can serve as a framework for integrating knowledge and practice from various disciplines. By offering a popular and approachable synthesis of difficult socioecological concepts, permaculture contributes to an applied form of ecological literacy. The permaculture design philosophy offers a unique viewpoint that suggests lines of inquiry for agroecosystem research. Finally, the fact that these factors are embodied in a global movement that largely operates independently of and without the help of powerful institutions suggests opportunities for participatory action research and the mobilization of public interest and support (Ferguson and Lovell 2014; Méndez et al. 2013).

10.4 Limits

Despite experimental evidence demonstrating increased food productivity and crop yields with agroecology approaches, its adoption by smallholder farmers appears to be constrained by a number of factors. The knowledge gap in agroecology concepts, both in terms of a lack of scholarly research and contemporary scientific perspectives, is of the utmost importance. In practical terms, agroecological practices are also hindered by a number of issues, such as initial yield decreases with agroecology, lack of sufficient biomass for effective mulching due to poor crop productivity or competing uses for crop residues in crop-livestock systems, and increased labor requirements when herbicides are not used. A fundamental problem with the adoption of agroecology systems is their promotion as an indivisible package that farmers find difficult to adopt in its entirety, ignoring farmers' participation in the design/selection of agroecological alternatives, and overlooking the fact that the process through which innovation emerges is complex and non-linear (Tittonell et al. 2012).

Furthermore, there is no systematic multisite assessment of agroecological practices, and the claims made by the movement's supporters do not overreach or oversimplify the concept. Further, the challenge of defining agroecology precisely can lead to muddled thinking and prevent methodical discussion (Ferguson and Lovell 2014).

10.5 Prospects

Solving the problems of food and nutritional security requires among other interventions a range of interconnected agricultural approaches, and no one farming system alone will safely feed the planet. Rather, a blend of organic and other innovative farming systems, including agroforestry, integrated farming, conservation agriculture, mixed crop and livestock, and still undiscovered systems, will be needed for future global food and ecosystem security. For example, integrated farming systems that blend mostly organic with some conventional practices have been shown to be more sustainable than conventional farming systems and are likely to play a central role. Achieving global food and ecosystem security requires more than just achieving sustainable farming systems worldwide. We need to reduce food waste, improve food distribution and access, stabilize the human population, eliminate the conversion of food into fuel, and change consumption patterns toward a more plant-based diet (Reganold and Wachter 2016).

It has become evident that agroecological approaches have to be tailored to local conditions to make it more suitable to resource constrained smallholder farmers. This is best done by collaborating with local smallholders, improving their innovation capacity, sharing technological knowledge in co-innovation platforms, and overcoming adoption barriers through gender-sensitive technological solutions and supportive policy measures (Tittonell et al. 2012).

10.6 Conclusion

Farmers absolutely must have a stable source of income in order to keep their families from falling into poverty. To meet the challenges of the current economic, political, and technological environment, progress in production or stable growth in production is required. In order to guarantee production and safeguard the interests of small farmers, an agroecology approach to maintaining the supply chain of a food production system has proven to be both promising and effective. It is recommended that each new agricultural ecosystem should be studied in depth with the input of all relevant stakeholders, including researchers, farmers, policymakers, and financiers.

References

Ali, S., Aqueel, M.A., Shakeel, Q. et al. (2022). Integrated insect pest management under changing climate. In: *Advances in Integrated Pest Management Technology: Innovative and Applied Aspects* (ed. A.S. Tanda), 21–35. Springer International Publishing. https://doi.org/10.1007/978-3-030-94949-5_2.

Amundson, R., Berhe, A.A., Hopmans, J.W. et al. (2015). Soil and human security in the 21st century. *Science* 348 (6235): 1261071. https://doi.org/10.1126/science.1261071.

Anderson, C.R., Bruil, J., Chappell, M.J. et al. (2021a). Conceptualizing processes of agroecological transformations: from scaling to transition to transformation. In: *Agroecology Now!: Transformations Towards More Just and Sustainable Food Systems* (ed. C.R. Anderson, J. Bruil, M.J. Chappell, et al.), 29–46. Springer International Publishing. https://doi.org/10.1007/978-3-030-61315-0_3.

Anderson, C.R., Bruil, J., Chappell, M.J. et al. (2021b). Introduction. In: *Agroecology Now!: Transformations Towards More Just and Sustainable Food Systems* (ed. C.R. Anderson, J. Bruil, M.J. Chappell, et al.), 1–8. Springer International Publishing. https://doi.org/10.1007/978-3-030-61315-0_1.

Barrios, E., Gemmill-Herren, B., Bicksler, A. et al. (2020). The 10 elements of agroecology: enabling transitions towards sustainable agriculture and food systems through visual narratives. *Ecosyst. People* 16 (1): 230–247. https://doi.org/10.1080/26395916.2020.1808705.

Batáry, P., Dicks, L.V., Kleijn, D., and Sutherland, W.J. (2015). The role of agri-environment schemes in conservation and environmental management. *Conserv. Biol.* 29 (4): 1006–1016. https://doi.org/10.1111/cobi.12536.

Béné, C., Oosterveer, P., Lamotte, L. et al. (2019). When food systems meet sustainability – current narratives and implications for actions. *World Dev.* 113: 116–130. https://doi.org/10.1016/j.worlddev.2018.08.011.

Bezner Kerr, R., Madsen, S., Stüber, M. et al. (2021). Can agroecology improve food security and nutrition? A review. *Glob. Food Sec.* 29: 100540. https://doi.org/10.1016/j.gfs.2021.100540.

Calabi-Floody, M., Medina, J., Rumpel, C. et al. (2018). Chapter three—smart fertilizers as a strategy for sustainable agriculture. In: *Advances in Agronomy*, vol. 147 (ed. D.L. Sparks), 119–157. Academic Press. https://doi.org/10.1016/bs.agron.2017.10.003.

Chabert, A. and Sarthou, J.-P. (2020). Conservation agriculture as a promising trade-off between conventional and organic agriculture in bundling ecosystem services. *Agric Ecosyst. Environ.* 292: 106815. https://doi.org/10.1016/j.agee.2019.106815.

Chang, C.-H. and Yang, S.-S. (2009). Thermo-tolerant phosphate-solubilizing microbes for multi-functional biofertilizer preparation. *Bioresour. Technol.* 100 (4): 1648–1658. https://doi.org/10.1016/j.biortech.2008.09.009.

Cordeau, S. (2022). Conservation agriculture and agroecological weed management. *Agronomy* 12 (4): 867. https://doi.org/10.3390/agronomy12040867.

Corson, M.S., Mondière, A., Morel, L., and van der Werf, H.M.G. (2022). Beyond agroecology: agricultural rewilding, a prospect for livestock systems. *Agric. Syst.* 199: 103410. https://doi.org/10.1016/j.agsy.2022.103410.

Dariush, M., Ahad, M., and Meysam, O. (2006). Assessing the land equivalent ratio (LER) of two corn [Zea mays L.] varieties intercropping at various nitrogen levels in Karaj, Iran. *J. Central Eur. Agric.* 7 (2): 359–364.

Duchene, O., Vian, J.-F., and Celette, F. (2017). Intercropping with legume for agroecological cropping systems: complementarity and facilitation processes and the importance of soil microorganisms. A review. *Agric. Ecosyst. Environ.* 240: 148–161. https://doi.org/10.1016/j.agee.2017.02.019.

Duru, M., Therond, O., Martin, G. et al. (2015). How to implement biodiversity-based agriculture to enhance ecosystem services: a review. *Agron. Sustain. Dev.* 35 (4): 1259–1281. https://doi.org/10.1007/s13593-015-0306-1.

El Chami, D., Daccache, A., and El Moujabber, M. (2020). How can sustainable agriculture increase climate resilience? A systematic review. *Sustainability* 12 (8): 3119. https://doi.org/10.3390/su12083119.

FAO (2014a). *Building a Common Vision for Sustainable Food and Agriculture. Principles and Approaches*. Food and Agriculture Organization of the United Nations. http://www.fao.org/agroecology/database/detail/en/c/1187152 (accessed 17 March 2023).

FAO (2014b). Conservation Agriculture Principles. https://www.fao.org/conservation-agriculture/overview/principles-of-ca/en (accessed 13 July 2022).

FAO (2019). *2019—The State of Food Security and Nutrition in the World (SOFI): Safeguarding against Economic Slowdowns and Downturns*. Rome: FAO.

FAOSTAT (n.d.). https://www.fao.org/faostat/en/#home (accessed 14 August 2022).

Ferguson, R.S. and Lovell, S.T. (2014). Permaculture for agroecology: design, movement, practice, and worldview. A review. *Agron. Sustain. Dev.* 34 (2): 251–274. https://doi.org/10.1007/s13593-013-0181-6.

Gerber, J.-D. and Gerber, J.-F. (2017). Decommodification as a foundation for ecological economics. *Ecol. Econ.* 131: 551–556. https://doi.org/10.1016/j.ecolecon.2016.08.030.

Giller, K.E., Andersson, J.A., Corbeels, M. et al. (2015). Beyond conservation agriculture. *Front. Plant Sci.* 6. https://www.frontiersin.org/articles/10.3389/fpls.2015.00870.

Gurr, G.M. and You, M. (2016). Conservation biological control of pests in the molecular era: new opportunities to address old constraints. *Front. Plant Sci.* 6. https://www.frontiersin.org/articles/10.3389/fpls.2015.01255.

Haouas, A., El Modafar, C., Douira, A. et al. (2021a). Alcaligenes aquatilis GTE53: phosphate solubilising and bioremediation bacterium isolated from new biotope "phosphate sludge enriched-compost". *Saudi J. Biol. Sci.* 28 (1): 371–379. https://doi.org/10.1016/j.sjbs.2020.10.015.

Haouas, A., Modafar, C.E., Douira, A. et al. (2021b). Phosphate sludge: opportunities for use as a fertilizer in deficient soils. *Detritus* 0. https://doi.org/10.31025/2611-4135/2021.15112.

Huang, D., Gao, L., Cheng, M. et al. (2022). Carbon and N conservation during composting: a review. *Sci. Total Environ.* 840: 156355. https://doi.org/10.1016/j.scitotenv.2022.156355.

Ickowitz, A., McMullin, S., Rosenstock, T. et al. (2022). Transforming food systems with trees and forests. *Lancet Planet. Health* 6 (7): e632–e639. https://doi.org/10.1016/S2542-5196(22)00091-2.

International Panel of Experts on Sustainable Food Systems (IPES-Food) (2019). Towards a Common Food Policy for the European Union. The Policy Reform and Realignment That Is Required 1e112. http://www.ipes-food.org/_img/upload/files/CFP_FullReport.pdf (accessed 17 March 2023).

Kassam, A., Friedrich, T., Shaxson, F., and Pretty, J. (2009). The spread of conservation agriculture: justification, sustainability and uptake. *Int. J. Agric. Sustain.* 7 (4): 292–320. https://doi.org/10.3763/ijas.2009.0477.

Landis, D.A. and Orr, D.B. (2023). Biological Control: Approaches and Applications | Radcliffe's IPM World Textbook. https://ipmworld.umn.edu/landis (accessed 17 March 2023).

Leakey, R.B. (1999). Potential for novel food products from agroforestry trees. *Food Chem.* 64: 1–14.

Malézieux, E. (2013). *Editorial.* Revue Fruits: Underutilized fruit trees in Africa. Special Issue.

Méndez, V.E., Bacon, C.M., and Cohen, R. (2013). Agroecology as a transdisciplinary, participatory, and action-oriented approach. *Agroecol. Sustain. Food Syst.* 37 (1): 3–18. https://doi.org/10.1080/10440046.2012.736926.

Moragues-Faus, A. and Morgan, K. (2015). Reframing the foodscape: the emergent world of urban food policy. *Environ. Plan. A Econ. Space* 47 (7): 1558–1573. https://doi.org/10.1177/0308518X15595754.

Mottet, A., Bicksler, A., Lucantoni, D. et al. (2020). Assessing transitions to sustainable agricultural and food systems: a tool for agroecology performance evaluation (TAPE). *Front. Sustain. Food Syst.* 4. https://www.frontiersin.org/articles/10.3389/fsufs.2020.579154.

Muller, A., Schader, C., El-Hage Scialabba, N. et al. (2017). Strategies for feeding the world more sustainably with organic agriculture. *Nat. Commun.* 8 (1): 1290. https://doi.org/10.1038/s41467-017-01410-w.

Padró, R., Tello, E., Marco, I. et al. (2020). Modelling the scaling up of sustainable farming into agroecology territories: potentials and bottlenecks at the landscape level in a Mediterranean case study. *J. Clean. Prod.* 275: 124043. https://doi.org/10.1016/j.jclepro.2020.124043.

Paramesh, V., Ravisankar, N., Behera, U. et al. (2022). Integrated farming system approaches to achieve food and nutritional security for enhancing profitability, employment, and climate resilience in India. *Food Energy Sec.* 11 (2): e321. https://doi.org/10.1002/fes3.321.

Paudel, D., Tiwari, K.R., Raut, N. et al. (2022). What affects farmers in choosing better agroforestry practice as a strategy of climate change adaptation? An experience from the mid-hills of Nepal. *Heliyon* 8 (6): e09695. https://doi.org/10.1016/j.heliyon.2022.e09695.

Reganold, J.P. and Wachter, J.M. (2016). Organic agriculture in the twenty-first century. *Nat. Plants* 2 (2): 1–8. https://doi.org/10.1038/nplants.2015.221.

Rosati, A., Borek, R., and Canali, S. (2021). Agroforestry and organic agriculture. *Agrofor. Syst.* 95 (5): 805–821. https://doi.org/10.1007/s10457-020-00559-6.

Sartori, R.B., Severo, I.A., Oliveira, Á.S.d. et al. (2021). Biofertilizers from waste. In: *Biofertilizers* (ed. Inamuddin, I.A. Mohd, B. Rajender, and R. Mashallah), 517–540. Wiley. https://doi.org/10.1002/9781119724995.ch17.

Siebrecht, N. (2020). Sustainable agriculture and its implementation gap—overcoming obstacles to implementation. *Sustainability* 12 (9): 3853. https://doi.org/10.3390/su12093853.

Sinclair, F., Wezel, A., Mbow, C. et al. (2019). The Contribution of Agroecological Approaches to Realizing Climate-Resilient Agriculture. Rotterdam and Washington, DC. www.gca.org (accessed 17 March 2023).

Soni, R.P., Katoch, M., and Sani, R.K. (2014). Integrated farming systems – a review. *IOSR J. Agric. Vet. Sci.* 7 (10): 36–42.

Stanojevic, A.B. and Stanojevic, A.B. (2021). Conservation agriculture and its principles. *Ann. Environ. Sci. Toxicol.* 5 (1): 018–022. https://doi.org/10.17352/aest.000031.

Stenberg, J.A., Sundh, I., Becher, P.G. et al. (2021). When is it biological control? A framework of definitions, mechanisms, and classifications. *J. Pest Sci.* 94 (3): 665–676. https://doi.org/10.1007/s10340-021-01354-7.

Tallou, A., Salcedo, F.P., Haouas, A. et al. (2020). Assessment of biogas and biofertilizer produced from anaerobic co-digestion of olive mill wastewater with municipal wastewater and cow dung. *Environ. Technol. Innov.* 20: 101152. https://doi.org/10.1016/j.eti.2020.101152.

Therond, O., Duru, M., Roger-Estrade, J., and Richard, G. (2017). A new analytical framework of farming system and agriculture model diversities. A review. *Agron. Sustain. Dev.* 37 (3): 21. https://doi.org/10.1007/s13593-017-0429-7.

Tittonell, P., Scopel, E., Andrieu, N. et al. (2012). Agroecology-based aggradation-conservation agriculture (ABACO): targeting innovations to combat soil degradation and food insecurity in semi-arid Africa. *Field Crop Res.* 132: 168–174. https://doi.org/10.1016/j.fcr.2011.12.011.

United Nations (UUNN) (2016). The Sustainable Development Goals. http://www.undp.org/content/undp/es/home/sustainable-developmentgoals.html (accessed 12 September 2019).

Venkatramanan, S. and Prasad, M.N.V. (2020). *Global Climate Change: Resilient and Smart Agriculture*. Springer. https://link.springer.com/book/10.1007/978-981-32-9856-9.

Wezel, A. (2017). *Agroecological Practices for Sustainable Agriculture: Principles, Applications, and Making the Transition*, 485–481. Hackensack, New Jersey, USA: World Scientific Publishing. https://doi.org/10.1142/9781786343062_bmatter.

Willer, H., Yussefi, M., and Sorensen, N. (2010). *The World of Organic Agriculture: Statistics and Emerging Trends 2008*. Earthscan.

11

Alternative Production Systems ("Roof-Top," Vertical, Hydroponic, and Aeroponic Farming)

Ágnes Szepesi

Department of Plant Biology, Institute of Biology, Faculty of Science and Informatics, University of Szeged, Közép fasor 52, Szeged, Hungary

11.1 Introduction

Alternative systems are cost-effective and as a part of urban farming could add value for sustainable environmental systems (Ali et al. 2022). These closed plant production systems give us a really nice opportunity to increase food production in the post-pandemic era independently from the season (Montes Rivera et al. 2022; Shahda and Megahed 2022). These techniques are sensitive to growing conditions as they are affected by resource input (Nemali 2022). It is a trend to achieve zero-emission without using fossil fuels. Hydroponic culture is a growing method with climate-smart solutions (Meena et al. 2022), where the plant roots are immersed in hydroponic solutions. In this case, it is very important to prove the precise water flow and root temperature as well as the carbon dioxide concentration needed for proper growth. In the case of aeroponic culture, plant roots are fixed in the air, and roots are exposed to the nutrient solution by spraying. Other significant and usual technique is the aquaponic system where in a closed system fish and plants are in a common place and the nutrient-rich solution is used for plant growth. The biggest advantage of alternative growing systems is their soilless features providing a controlled and highly regulated system without using pesticides and chemical fertilizers. The used growing substrates are variable, depending on the plant used for growing, e.g., peat, vermiculite, zeolite. If we adjust the optimal growth cycle during the whole life stages, the continuous production of controlled and high-quality food could be evolved. If we mention soilless crops used in alternative farming, it is also a factor to evaluate the preference of the consumer acceptance (Ercilla-Montserrat et al. 2019). There are some concerns, e.g., food safety or the social impact of RA which could be solved by better communication between the consumers and stakeholders (He and Qin 2022). The common problem in all cases is the question of infections by microbes. Different sources of microbes in CEA could enhance the microbial food safety risk decreasing the potential of these systems (Dong and Feng 2022). A nice idea to add some antioxidant compounds to the plants in order to enhance not only the production but also to eliminate microbes, e.g., by adding compost tea (Lu et al. 2022).

Agroecological Approaches for Sustainable Soil Management, First Edition.
Edited by Majeti Narasimha Vara Prasad and Chitranjan Kumar.

11.2 Rooftop Farming/Agriculture (RA) and Vertical Farming

The rooftop farming and vertical farming could reduce the costs of food production and transport to consumers as it could happen very close to each other sometimes at the same places (Al-Kodmany 2018). Rooftop farming has a great advantage as it could be efficient to decrease the number of unused places for society (Table 11.1).

Table 11.1 Advantages of high-tech vertical farming systems.

1. Reliable harvests	Controlled indoor environments are independent of outside weather conditions and would provide consistent and reliable growing cycles to meet delivery schedules and supply contracts.
2. Minimum overheads	Production overheads would decrease by 30%.
Low energy usage	The use of high efficiency LED lighting technology ensures minimum power use for maximum plant growth. Computer management of photosynthetic wavelengths, in harmony with phase of crop growth, further minimizes energy use while ensuring optimized crop yields.
Low labor costs	Fully automated growing systems with automatic SMS text messaging would require manual labor only for on-site planting, harvesting, and packaging.
Low water usage	Vertical farms would use around 10% of the water required for traditional open field farming.
Reduced washing and processing	Positioning of facilities close to the point of sale would dramatically decrease travel times, reducing refrigeration, storage, and transport costs in the process.
	Vertical farms would employ strict bio-security procedures to eliminate pests and diseases.
3. Increased growing areas	Vertical farms would supply nearly 10 times more growing area than traditional farms.
4. Maximum crop yield	Irrespective of external conditions, vertical farms can provide more crop rotations per year than open field agriculture and other farming practices. Crop cycles are also faster due to controlled temperature, humidity, light, etc.
5. Wide range of crops	The vertical farm would provide a wide range of crops.
6. Fully integrated technology	The vertical farm would be fully monitored, controlled, and automated.
Optimum air quality	The temperature, CO_2, and humidity levels of the vertical farm would be optimized at all times.
Optimum nutrient and mineral quality	The vertical farm would use specially formulated, biologically active nutrients in all crop cycles, providing organic minerals and enzymes to ensure healthy plant growth.
Optimum water quality	All fresh water's contaminants would be removed before entering the vertical farm.
Optimum light quality	High-intensity low-energy LED lighting would be specifically developed and used for maximum growth rates, high reliability, and cost-effective operations.

Source: Al-Kodmany (2018) / MDPI / CC BY 4.0.

As cities are rapidly growing, it needs to find a way how to optimize resource consumption optimizing the food-energy-water nexus (Toboso-Chavero et al. 2018). To apply systems like photovoltaic or solar thermal combined with other sustainable options like rainwater harvest, rooftop farming could be an option at the local scale contributing to the reduced CO_2-eq emissions. Vertical farming provides a cooling option to the building surface and an opportunity to people to get super healthy food as microgreens from very close places (Reddy et al. 2022). Herbs, berries, or some spicy plants could be used for rooftop farming adding health value to diets [(Reddy et al. 2022), Table 11.2].

Table 11.2 Summary of the benefits of rooftop farms.

	Environmental benefits
Energy conservation	The use of green roofs compared to conventional roofing surfaces significantly affects the energy balance within a building. Studies have revealed that green roofs have the potential to reduce a building's energy use by as much as 30%.
Stormwater management	A green roof absorbs rainwater and helps to prevent sewer system back-ups and contaminated stormwater overflow. Green roofs can also help to prevent catastrophic environmental events, such as the Ala Wai Canal sewage spill disaster.
Fossil fuel reduction	A rooftop farm can grow hyper-local foods. Growing Low Food Mile organic produce substantially reduces the fossil fuel consumption associated with the traditional food transportation system.
Global warming	Green roofs sequester carbon from the atmosphere, lower the levels of carbon dioxide in the air, eliminate the build up of greenhouse gases, and keep city temperatures cooler by effectively reducing the "Urban Heat Island Effect."
Biodiversity	By replacing inorganic, lifeless roofs with living and thriving green spaces, green roofs support increased biodiversity in urban environments–offering a habitat for a multitude of organisms–from birds to butterflies to countless other beneficial insects.
Environmental stewardship	Organic rooftop farming protects soil and water from toxic pesticides, herbicides, fungicides, and other dangerous chemicals typically used in conventional farming.
	Socioeconomic benefits
Community	A rooftop farm is a beacon of sustainable community building that creates tangible connections between farmers and consumers. Rooftop farms have the power to do this throughout the city, no matter how scarce or valuable the land.
Local food economy	Rooftop farms generate revenue for local farmers and businesses. In this way, rooftop agriculture can also be viewed as an emerging green technology that creates jobs and improves food self-sufficiency by providing organic and hyper-local produce.
Nutrition and health	As an example, the produce of FarmRoof™ is extraordinarily nutritious and healthy. Thanks to a special soil that has been infused with minerals, trace elements, omegas, proteins and microorganisms, all of their crops are packed with enzymes, antioxidants, nutrients, and minerals.
Esthetics and Beauty	Supplanting inorganic, lifeless roofs with vibrant greenery, green roofs can beautify cityscapes and balance an otherwise bleak horizon of concrete and tar.

Source: Al-Kodmany (2018) / MDPI / CC BY 4.0.

In case of microgreens, we have to pay attention to selecting and treating plants not to produce toxic compounds, e.g., lettuce could produce nitrates (Guffanti et al. 2022). During the postharvest period, a flushing method could reduce the level of nitrates in lettuce plants. Indoor vertical farming needs an energy innovation in order to be productive in a sustainable manner (Avgoustaki and Xydis 2020). The photovoltaic system for irrigation could be effective to use solar energy (Prasad et al. 2021). To date, only some specific crops were involved into this type of farming, but there are some efforts to find the potential combinations of crops to the continuous and diverse food supply (Rufí-Salís et al. 2020). This study showed that the biggest impact was in case of spinach and arugula. Paris is a hotspot of innovative agriculture in a study from 2017. The study showed that there are some innovative commercial forms of RA, e.g., on-the-ground farms, "low-tech off-soil" systems added to open-air cropping systems on rooftops, and "high-tech" greenhouse systems (Aubry and Daniel 2017). It is also a question that how to manage the optimal growing conditions in a highly polluted urban region. More study is needed to solve this crucial problem in case of RA. The air quality is also a remarkable point in case of RA. As this air could be moved back in order to prove the heat of the building, its quality and safety by microbes should be taken into account. So, the unused urban places are optional places for UA however there are some risks associated with this type of farming. Specht et al. in 2017 (Specht and Sanyé-Mengual 2017) described the five potential risks which are associated with urban integration, production system, food products, and also some economic and environmental risks. This type of farming could be useful for reducing food waste and loss as these products could be used back as biogas or other energy for plant production [(Sanyé-Mengual et al. 2017; Babbitt et al. 2022), Table 11.3].

Table 11.3 Key sustainable benefits of the vertical farm.

# Benefit	Environmental	Social	Economic
1. Reducing food-miles (travel distances)	Reducing air pollution	Improves environmental and people's health.	Reduce energy, packaging, and fuel to transport food
2. Reducing water consumption for food production by using high-tech irrigation systems and recycling methods	Reducing surface water run off of traditional farms	Making potable water available to more people	
3. Recycling organic waste	Save the environment by reducing needed land fills	Improve food quality and subsequently consumers' health	Turn waste into asset
4. Creating local jobs	People do not have to commute to work and hence will decrease ecological footprint	Create a local community of workers and social networks with farmers	Benefit local people economically

Table 11.3 (Continued)

# Benefit	Environmental	Social	Economic
5. Reduced fertilizers, herbicides, and pesticides	Improve the environmental well-being	Improve food quality and subsequently consumers' health	Decrease costs
6. Improve productivity	Needs less space	Reduce redundant, repetitive work, and save time to do productive and socially rewarding activities	Offer greater yields
7. Avoid crop losses due to floods, droughts, hurricane, over exposure to sun, and seasonal changes	Decrease environmental damage and cleanups of farms after damage	Improve food security	Avoiding economic loss
8. Control product/ produce regardless seasons	Produce regarding season	Increase accessibility year-round and improve respond to population demand	Fuel economic activities
11. Promoting high-tech and green industry	"Green technology" reduce harm and improve environmental performance	Encourage higher education and generate skilled workers	Provide new jobs in engineering, biochemistry, biotechnology, construction and maintenance, and research and development
12 Reducing the activities of traditional farming	Preserving natural ecological system	Improve health of citizens	Saving money required to correct environmental damage
13 Repurposing dilapidated buildings	Enhance the environment. Remove eye sores and stigma from neighborhoods	Create opportunities for social interaction	Revive economy

Source: Al-Kodmany (2018) / MDPI / CC BY 4.0.

The energy flows and energy consumption performance should be in balance in case of RA in order to be sustainable with lower ecological footprint by a raised urban food security (Piezer et al. 2019). Vertical farm cultivation could improve not only fresh weight but also some compounds with health benefits like phytonutrients as it was studied in watercress (Qian et al. 2022). New sensors and monitoring systems are needed to improve conditions of growing plants at vertical farming.

Of vertical farm by Plantagon can be seen in Figures 11.1 and 11.2. The results of growing like speed of growing, flowering, or crop yield (Popkova 2022). A proposed model of these factors could contribute the cities to be self-managed in food production (Lejeune et al. 2022). Other sample is the Sky Greens vertical farm project, the environmental and socioeconomic benefits are summarized in the Table 11.4.

Figure 11.1 A proposed vertical farm by Plantagon. It comprises a helix structure placed in the center of a sphere-shape building and stretches vertically. The helix structure contains a robot belt that instead of moving thousands of plants over the whole belt, it moves each row of plants one by one. Seeds are placed in pots on the belt at the ground floor level. Then, plants are elevated to the top by an elevator and placed on the moving belt which takes them back to the ground, and by the time they reach the ground floor, plants are ready to harvest. *Source:* Al-Kodmany (2018) / from MDPI.

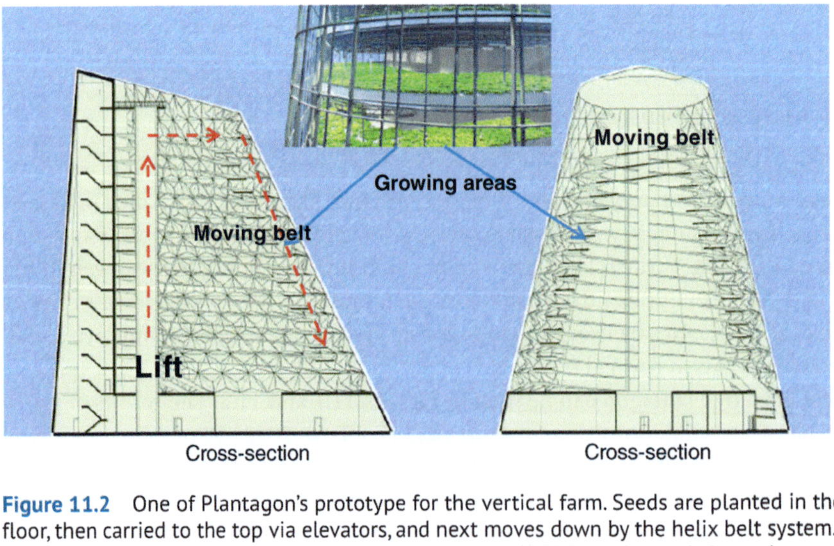

Figure 11.2 One of Plantagon's prototype for the vertical farm. Seeds are planted in the first floor, then carried to the top via elevators, and next moves down by the helix belt system. *Source:* Al-Kodmany (2018), p. 36. http://plantagon.com as per MDPI Open Access Information and Policy.

Table 11.4 The environmental and socioeconomic benefits of the Sky Greens vertical farm project.

	Environmental benefits
Environmentally friendly and high-tech	Sky Greens observes, learns, and works with nature to achieve sustainability for the good of the environment to grow safe, high-quality vegetables using green technologies.
Low energy usage	Outdoor green houses have abundant sunlight in the tropics. The A-Go-Gro system uses patented low carbon hydraulic green technology to power the rotation of the tower at very low energy costs, while still allowing the plants to receive abundant sunlight.
Low water usage	As the troughs of plants rotate, irrigation occurs using an innovative flooding method, using very little water. Water is also recycled and reused.
Good waste and water management	Sustainable water management practices are utilized with all organic wastes being composted at the farm to ensure the use of safe, high-quality fertilizers.
Green technologies	Green technologies have been stringently implemented at the farm to achieve the three Rs (reduce, reuse, and recycle).
	Socioeconomic benefits
Increased productivity	The production yield of Sky Greens Farm is 5–10 times greater per unit of area than traditional Singaporean farms that growing leafy vegetables in a conventional fashion.
Tasty vegetables	Tropical leafy vegetables are grown in special soil-based media, which contribute to good tasting vegetables, suitable for stir-fry and soups. The vegetables are harvested every day and delivered almost immediately to retail outlets for consumers.
Year-round production	As the vertical farm structures are in protected-outdoor green houses, the vegetables are grown in a controlled environment, protected from pests, winds and floods.
Consistent and reliable harvest	A steady supply of fresh leafy vegetables is assured as growing takes place in a controlled environment.
Easy to install and easy to maintain	The modular A-frame rotary system allows quick installation and easy maintenance.
Better ergonomics and automation	The rotary system allows the troughs to be immediately adjusted for easy harvesting. Automation increases the productivity of workers per ton of vegetables grown.
Space savings	The footprint of the vertical system is small but can produce significantly more per unit area than traditional farms. It can also be customized to suit different crop requirements and varying environments.

Source: Al-Kodmany (2018) / MDPI / CC BY 4.0.

11.3 Hydroponic Farming

Soil-based development is currently confronting troubles because of various man-made reasons, for example, industrialization and urbanization. Additionally, unexpected cataclysmic events, environmental change, and unlimited use of synthetic compounds for farming purposes cause the exhaustion of soil fruitfulness and quality. That is the reason, researchers have fostered another elective methodology for development framework specifically soil-less development or tank-farming. Tank-farming is a strategy for developing plants in a water based, supplement-rich arrangement. Through aquaculture, an enormous number of plants and harvests or vegetables can be developed. The nature of yield, taste, and nutritive worth of finished results created through hydroponics is by and large higher than the normal soil-based development. Along these lines, tank-farming would be a superior method to create various types of natural products, vegetables what's more, grain just as fulfill the worldwide nourishment need with making advanced future. In the future, tank-farming could be arising method for providing food to the worldwide populace. In case of hydroponic systems, it is obvious that we have to control the nutrient supply and prove and fine adjust the nutrient balance for optimal growth of plants (Figure 11.3). In a closed system, the used nutrient solution should be cleaned and controlled and recirculated to the roots of plants (Sanjuan-Delmás et al. 2020). There are various techniques for hydroponic cultivation depending upon the space availability, plant growth, and resources. These are (i) Nutrient film technique (NFT), (ii) Deep water culture (DWC), Wick system, (iii) Ebb and Flow, (iv) Drip system, etc. NFT is space-efficient with plentiful oxygen flow, and excess nutrient solution recirculation. It is incredible for lightweight plants that needn't bother with much help, in the same way as other mixed greens. This is on the grounds that

Figure 11.3 *Talinum triangulare* (Jacq.) Wild–Hydroponics.

most NFT clients decide not to utilize a developing mode for extra help. Furthermore, in light of the fact that the water is streaming reliably, NFT frameworks once in a while experience the ill effects of salt formation. In DWC hydroponics system, DWC frameworks are well known for a wide range of reasons, the essential one being that they're perhaps the easiest kind of frameworks, to begin with. The solitary framework that is less difficult is a wicking framework. Here are a couple of different advantages to filling in a DWC framework. DWC works especially for plants with bigger roots and bears fruits in abundance since the roots are sitting in nutrients and get unlimited oxygen.

Wick systems can be easily set up as they are passive, which means they have no moving parts. This makes them simpler and less expensive to keep up than dynamic frameworks like ebb and flow, however, they additionally have the downside of being less proficient and not exceptional for high-support plants, or enormous plants that devour a great deal of water. The best plants to use in this framework are quickly developing lettuces or herbs like rosemary, thyme, etc. plants are put in absorbing developing medium with a nylon "wick" running from plants into a repository of nutrients.

Ebb and flow, otherwise called flood and drain, is quite possibly the most generally perceived aquaculture framework out there. It is middle-level in trouble, moderately minimal expense to set up, and incredibly adaptable. This technique permits us to effortlessly change our nursery, adding or eliminating plants as we wish without influencing any of the encompassing harvests.

Drip hydroponics is a functioning tank-farming framework. This implies that it utilizes a siphon to take care of your plants with supplements and water consistently. It is additionally called a stream, or miniature water system framework. As the name proposes, the framework utilizes little producers to trickle the supplement arrangement straightforwardly onto your plants. A drip hydroponics framework isn't remarkable to aquaculture. Such an arrangement is likewise generally utilized in outside nurseries to convey water and supplements to individual plants.

Hydroponic CEA could use fewer sources than field-based systems and could alleviate water stress and soil degradation in arid regions (Casey et al. 2022). CEA lettuce produced relatively small environmental footprints using renewable electricity compared to the field-based systems. It is important to connect the building air handling to the CEA requirement inputs like cooling, CO_2, and water (Casey et al. 2022). Light is one of the most crucial points in alternative farming systems. Gaining control in light management in controlled systems affects not only the better production but also the synthesis and accumulation of some important bioactive compounds in plants (Neo et al. 2022). Vertical farming adapted to natural lightning could be also a cost-effective manner to manage investments in these systems (Song et al. 2022). Adjusting light gradients showed species-dependent responses in terms of shape and color investigated in seven different species (Banerjee et al. 2022). The horticultural crops like leafy vegetables and berries or cruciferous species indoor hydroponic systems were successful (Mishra et al. 2022) but it needs to be automatized and controlled. The circadian rhythm of crops is crucial in effectively regulating the growing systems (Marie et al. 2022). LED lighting should be adjusted to the natural rhythm of crops and synchronize the environmental conditions (Marie et al. 2022). It is also a critical point to find the best crop cultivars suitable for CEA (Marie et al. 2022). Recycling nutrients in hydroponic systems is a beneficial part of this system, however it could be a threat as a continuous flow is a place for microbes. This challenge could be minimized by using cleaning techniques (Asaduzzaman et al. 2022). Using algae in hydroponic systems could be a new technique in order to treat wastewater streams (Cowan et al. 2022b).

While hydroponic farming may appear to be achievable and advantageous, it is marginally tedious too. The water should be supplanted at standard spans, and you need to consistently beware of the plants to check whether they are doing fine and dandy. So, extra care and regulation of hydroponics are essential. Hydroponic cultivation relies upon a scope of hardware that requires legitimate skill. Two main considerations in hydroponics are power and water. Dissimilar to ordinary soil-based cultivating, hydroponic farming includes costly hardware (at any rate for the primary establishment). Despite the sort of framework, you intend to fabricate, you will require holders, great lighting, an exact clock, and quality supplements. After at long last setting up the framework, the repetitive expense might be restricted to the supplements, water, and power.

In the course of the most recent few years, there has been huge discussion and consultation on whether Hydroponic Farming is without a doubt a confirmed technique for natural cultivation. Numerous ranchers have addressed whether plants become through hydroponic cultivation can be undoubtedly guaranteed as organic as they don't get the microbiomes as accessible in the soil. Notwithstanding, individuals across the globe have created hydroponic plants like tomatoes, lettuces, and other verdant vegetables throughout the last decade. Nations like Australia, the US, and the Netherlands have effectively performed and prevailed in this method of cultivation. Certain natural cultivating techniques are proposed for hydroponic ranchers too. For example, certain producers give the necessary microbiomes to the plants by utilizing natural developing ancient rarities like coco coir or worm projecting. On the other hand, common supplements like fish, cottonseeds, bones, or even hay might be utilized in this interaction.

11.4 Aeroponic Farming

Aeroponic farming faces with microclimate challenges, e.g., in case of roots. It should be taken that controlling the root zone temperature could add the production of subtropical conditions in some cases (He and Qin 2022). The growers have to decide and plan the crops in case of these types of farming adjusting the supply chain and requirements of tastes, preferences, and food ethics (Howitt and Rausser 2022).

11.5 Future Perspectives

It is important to mention that urban farming could provide ecosystem services and additionally some social cohesion people to cooperate together (Petit-Boix and Apul 2018). Urban agriculture could contribute to tremendous societal challenges of urbanization, e.g., climate change, food security, and economic growth, etc. (Artmann and Sartison 2018). The precise crop planning also a key point to enhance the efficiency of vertical farming (Santini et al. 2021). It is basic that UA should meet the sustainability goals and operating maximal efficiency, so growing food with minimized water consumption without using pesticide (Delorme and Santini 2022). Vertical farms with clean source of food produced in a way positive for biosecurity free from pests and fossil fuels could bring a chance for a sustainable and eco-friendly at the same time highly productive food production era (Benke and Tomkins 2017). The integrated energy-food nexus systems in controlled environment agriculture could guarantee that the demand for food and energy could be optimized in urban regions (Vatistas et al. 2022). The technological advancements could efficiently grow the economic and environmental performance of CEA systems (Cowan et al. 2022a). Urban farming has a role in the most problematic challenges of this century. COVID-19 pandemic proved that the local food production in high quality, but scientists paid attention that this system should be adjustable for the local region in different scales and capacities (Shahda and Megahed 2022; Shema and Abdulmalik 2022). It is crucial to plan how to convert the growing space, maintain equipment, selecting suitable plant species (Lubna et al. 2022). Public visibility is very important in order to vertical farming to be accepted and this could promote new findings in this field (Mitchell 2022). High-tech systems could minimize the cost of maintenance in order to maximize yield (Mir et al. 2022; Velazquez-Gonzalez et al. 2022) (Table 11.5).

The role of biostimulants could be a nice perspective in boosting the yield in controlled growth systems. A new study found that phycocyanin-rich microalgae-derived Spirulina extract (PRSE) could be a good candidate for improving the yield and quality of vertical farmed lettuce (Varia et al. 2022). Findings of new natural compounds from plants also are a future perspective to limit the chemical fertilizer and pesticides but enhance the yield in a sustainable manner (Bakacsy et al. 2022). There is an enormous usage of the experiences came from controlled farming systems. These systems are very close to those used in

Table 11.5 High-tech Indoor farming.

Farming method	Key characteristics	Major benefits	Common/Applicable technologies
Hydroponics	Soilless based, uses water as the growing medium	Fosters rapid plant growth; Reduces, even eliminates soil-related cultivation problems; decreases the use of fertilizers or pesticides	Computerized and monitoring systems; cell phones, laptops, and tablets; food growing apps; remote control systems and software (farming-from-afar systems); automated racking, stacking systems, moving belts, and tall towers; programmable LED lighting systems; renewable energy applications (solar panels, wind turbines, geothermal, etc.); closed-loop systems, anaerobic digesters; programmable nutrient systems; climate control, HVAC systems; water recirculating and recycling systems; rainwater collectors; insect-killing systems; robots
Aeroponics	A variant of hydroponics; it involves spraying the roots of plants with mist or nutrient solutions.	In addition to benefits mentioned above, aeroponics requires less water.	
Aquaponics	It integrates aquaculture (fish farming) with hydroponics.	Creates symbiotic relationships between the plants and the fish; it uses the nutrient-rich waste from fish tanks to "fertigate" hydroponics production beds; and hydroponic bed cleans water for fish habitat.	

Source: Al-Kodmany (2018) / MDPI / CC BY 4.0.

Table 11.5, high-tech indoor farming spaceships. These results could contribute to the research how we could manage the long-term space habitation (Mortimer and Gilliham 2022). New analytic techniques also contribute to the rapid development of CEA systems, like screening of lettuce to predict biomass production (Kim and van Iersel 2022).

Acknowledgments

This work was supported by the Hungarian Scientific Research Fund FK129061 project and by the Hungarian government and the European Union within the frames of the Széchenyi Terv 2020 Plusz grant GINOP_PLUSZ-2.1.1-21-2022-00080. Author thanks to László Bakacsy for his valuable comments.

References

Ali, M.S., Zaman, N.B.K., and Othman, N.M.I. (2022). Urban farming. *Proc. Sci. Ethics Civil.* 1: 42–48.

Al-Kodmany, K. (2018). The vertical farm: a review of developments and implications for the Vertical City. *Buildings* 8 (2): 24. https://doi.org/10.3390/buildings8020024.

Artmann, M. and Sartison, K. (2018). The role of urban agriculture as a nature-based solution: a review for developing a systemic assessment framework. *Sustainability* 10 (6): 1937. https://doi.org/10.3390/su10061937.

Asaduzzaman, M., Niu, G., and Asao, T. (2022). Nutrients recycling in hydroponics: opportunities and challenges toward sustainable crop production under controlled environment agriculture. *Front. Plant Sci.* 13: 845472, 403.

Aubry, C. and Daniel, A.C. (2017). Innovative commercial urban agriculture in the Paris metropolitan area. In: *Toward Sustainable Relations Between Agriculture and the City* (ed. C.-T. Soulard, C. Perrin, and E. Valette), 147–162. Cham: Springer International Publishing. https://doi.org/10.1007/978-3-319-71037-2_9.

Avgoustaki, D.D. and Xydis, G. (2020). How energy innovation in indoor vertical farming can improve food security, sustainability, and food safety? *Adv. Food Secur. Sustain.* 5: 1–51. https://doi.org/10.1016/bs.af2s.2020.08.002.

Babbitt, C.W., Neff, R.A., Roe, B.E. et al. (2022). Transforming wasted food will require systemic and sustainable infrastructure innovations. *Curr. Opin. Environ. Sustain.* 54: 101151.

Bakacsy, L., Sípos, L., Barta, A. et al. (2022). Concentration-dependent effects of effusol and juncusol from Juncus compressus on seedling development of Arabidopsis thaliana. *Sci. Rep.* 12: 13870. https://doi.org/10.1038/s41598-022-18063-5.

Banerjee, A., Paul, K., Varshney, A. et al. (2022). Soilless indoor smart agriculture as an emerging enabler technology for food and nutrition security amidst climate change. In: *Plant Nutrition and Food Security in the Era of Climate Change* (ed. V. Kumar, A.K. Srivastava, and P. Suprasanna), 179–225. Academic Press.

Benke, K. and Tomkins, B. (2017). Future food-production systems: vertical farming and controlled-environment agriculture. *Sustain. Sci. Pract. Policy* 13 (1): 13–26.

Casey, L., Freeman, B., Francis, K. et al. (2022). Comparative environmental footprints of lettuce supplied by hydroponic controlled-environment agriculture and field-based supply chains. *J. Clean. Prod.* 369: 133214.

Cowan, N., Ferrier, L., Spears, B. et al. (2022a). CEA systems: the means to achieve future food security and environmental sustainability? *Front. Sustain. Food Syst.* 6: 891256. https://doi.org/10.3389/fsufs.

Cowan, N., White, S., Olszewska, J. et al. (2022b). Integration of algae treatment with hydroponic crop waste to reduce impact of nutrient waste streams. *J. Sustain. Agric. Environ.* 1: 203–215.

Delorme, M. and Santini, A. (2022). Energy-efficient automated vertical farms. *Omega* 109: 102611.

Dong, M. and Feng, H. (2022). Microbial community analysis and food safety practice survey-based hazard identification and risk assessment for controlled environment hydroponic/aquaponic farming systems. *Front. Microbiol.* 13: 879260. https://doi.org/10.3389/fmicb.2022.879260.

Ercilla-Montserrat, M., Sanjuan-Delmás, D., Sanyé-Mengual, E. et al. (2019). Analysis of the consumer's perception of urban food products from a soilless system in rooftop greenhouses: a case study from the Mediterranean area of Barcelona (Spain). *Agric. Hum. Values* 36: 375–393. https://doi.org/10.1007/s10460-019-09920-7.

Guffanti, D., Cocetta, G., Franchetti, B.M., and Ferrante, A. (2022). The effect of flushing on the nitrate content and postharvest quality of lettuce (Lactuca sativa L. Var. Acephala) and rocket (Eruca sativa Mill.) grown in a vertical farm. *Horticulturae* 8 (7): 604.

He, J. and Qin, L. (2022). Microclimate control to increase productivity and nutritional quality of leafy vegetables in a cost-effective manner. *Int. J. Agric. Biol. Eng.* 15 (3): 55–61.

Howitt, R.E. and Rausser, G. (2022). Developments in agricultural crop innovations. *Annu. Rev. Resour. Econ.* 14 (1): 91–108.

van Kim, C. and Iersel, M.W. (2022). Morphological and physiological screening to predict lettuce biomass production in controlled environment agriculture. *Remote Sens. (Basel)* 14 (2): 316.

Lejeune, P., Fratamico, A., Bouché, F. et al. (2022). LED color gradient as a new screening tool for rapid phenotyping of plant responses to light quality. *GigaScience* 11.

Lu, S., Pentico, D., Castro, R. et al. (2022). Effect of ultraviolet light exposure and compost tea supplementation on growth, antioxidant activities, and microbiome of hydroponically grown mustard greens. *ACS Agric. Sci. Technol.* 2 (3): 521–533.

Lubna, F.A., Lewus, D.C., Shelford, T.J., and Both, A.J. (2022). What you may not realize about vertical farming. *Horticulture* 8 (4): 322.

Marie, T.R.J.G., Leonardos, E.D., Lanoue, J. et al. (2022). A perspective emphasizing circadian rhythm entrainment to ensure sustainable crop production in controlled environment agriculture: dynamic use of LED cues. *Front. Sustain. Food Syst.* 6: 856162. https://doi.org/10.3389/fsufs.2022.856162.

Meena, V.D., Dotaniya, M.L., Meena, M.D. et al. (2022). Vertical farming: can solve the challenge of growing food demand? *Biotica Res. Today* 4 (1): 018–022.

Mir, M.S., Naikoo, N.B., Kanth, R.H. et al. (2022). Vertical farming: the future of agriculture: a review. *Pharma Innov. J.* 11 (21): 1175–1195.

Mishra, S., Karetha, K.M., Yau, Y.Y., and Easterling, M. (2022). Vertical cultivation: moving towards a sustainable and eco-friendly farming. In: *Biotechnological Innovations for Environmental Bioremediation* (ed. S. Arora, A. Kumar, S. Ogita, and Y.-Y. Yau), 487–507. Singapore: Springer.

Mitchell, C.A. (2022). History of controlled environment horticulture: indoor farming and its key technologies. *HortScience* 57 (2): 247–256.

Montes Rivera, M., Escalante-Garcia, N., Dena-Aguilar, J.A. et al. (2022). Feature selection to predict LED light energy consumption with specific light recipes in closed plant production systems. *Appl. Sci.* 12 (12): 5901.

Mortimer, J.C. and Gilliham, M. (2022). SpaceHort: redesigning plants to support space exploration and on-earth sustainability. *Curr. Opin. Biotechnol.* 73: 246–252.

Nemali, K. (2022). History of controlled environment horticulture: greenhouses. *HortScience* 57 (2): 239–246.

Neo, D.C.J., Ong, M.M.X., Lee, Y.Y. et al. (2022). Shaping and tuning lighting conditions in controlled environment agriculture: a review. *ACS Agric. Sci. Technol.* 2 (1): 3–16.

Petit-Boix, A. and Apul, D. (2018). From cascade to bottom-up ecosystem services model: how does social cohesion emerge from urban agriculture? *Sustainability* 10 (4): 998. https://doi.org/10.3390/su10040998.

Piezer, K., Petit-Boix, A., Sanjuan-Delmás, D. et al. (2019). Ecological network analysis of growing tomatoes in an urban rooftop greenhouse. *Sci. Total Environ.* https://doi.org/10.1016/j.scitotenv.2018.09.293 651: 1495–1504.

Popkova, E.G. (2022). Vertical farms as a promising direction for the development of sustainable agriculture. In: *Sustainable Agriculture* (ed. E.G. Popkova and B.S. Sergi), 273–278. Singapore: Springer.

Prasad, A.R., Shankar, R., Patil, C.K. et al. (2021). Performance enhancement of solar photovoltaic system for roof top garden. *Environ. Sci. Pollut. Res. Int.* 28 (36): 50017–50027. https://doi.org/10.1007/s11356-021-14191-z.

Qian, Y., Hibbert, L.E., Milner, S. et al. (2022). Improved yield and health benefits of watercress grown in an indoor vertical farm. *Sci. Hortic.* 300: 111068.

Reddy, R.V.S.K., Omprasad, J., and Janakiram, T. (2022). Technological innovations in commercial high tech horticulture, vertical farming and landscaping. *Int. J. Innov. Hortic.* 11 (1): 78–91.

Rufí-Salís, M., Petit-Boix, A., Villalba, G. et al. (2020). Identifying eco-efficient year-round crop combinations for rooftop greenhouse agriculture. *Int. J. Life Cycle Assess.* 25: 564–576. https://doi.org/10.1007/s11367-019-01724-5.

Sanjuan-Delmás, D., Josa, A., Muñoz, P. et al. (2020). Applying nutrient dynamics to adjust the nutrient-water balance in hydroponic crops. A case study with open hydroponic tomato crops from Barcelona. *Sci. Hortic.* 261: 108908. https://doi.org/10.1016/j.scienta.2019.108908.

Santini, A., Bartolini, E., Schneider, M., and de Lemos, V.G. (2021). The crop growth planning problem in vertical farming. *Eur. J. Oper. Res.* 294 (1): 377–390.

Sanyé-Mengual, E., Rieradevall, J., and Montero, J.I. (2017). *Resource Efficiency and Waste Avoidance*, Rooftop Urban Agriculture, 263–276. Springer. https://doi.org/10.1007/978-3-319-57720-3_16.

Shahda, M.M. and Megahed, N.A. (2022). Post-pandemic architecture: a critical review of the expected feasibility of skyscraper-integrated vertical farming (SIVF). *Archit. Eng. Des. Manage.* 1–22. https://doi.org/10.1080/17452007.2022.2109123.

Shema, A.I. and Abdulmalik, H. (2022). Urban vertical farming as a path to healthy and sustainable urban built environment. *A+ Arch. Des.* 8 (1): 67–88.

Song, S., Hou, Y., Lim, R.B. et al. (2022). Comparison of vegetable production, resource-use efficiency and environmental performance of high-technology and conventional farming systems for urban agriculture in the tropical city of Singapore. *Sci. Total Environ.* 807: 150621.

Specht, K. and Sanyé-Mengual, E. (2017). Risks in urban rooftop agriculture: assessing stakeholders' perceptions to ensure efficient policymaking. *Environ. Sci. Policy*. https://doi.org/10.1016/j.envsci.2016.12.001 69: 13–21.

Toboso-Chavero, S., Nadal, A., Petit-Boix, A. et al. (2018). Towards productive cities: environmental assessment of the food-energy-water nexus of the urban roof mosaic. *J. Ind. Ecol.* 23 (4): 767–780.

Varia, J., Kamaleson, C., and Lerer, L. (2022). Biostimulation with phycocyanin-rich Spirulina extract in hydroponic vertical farming. *Sci. Hortic.* 299: 111042.

Vatistas, C., Avgoustaki, D.D., and Bartzanas, T. (2022). A systematic literature review on controlled-environment agriculture: how vertical farms and greenhouses can influence the sustainability and footprint of urban microclimate with local food production. *Atmos.* 13 (8): 1258.

Velazquez-Gonzalez, R.S., Garcia-Garcia, A.L., Ventura-Zapata, E. et al. (2022). A review on hydroponics and the technologies associated for medium-and small-scale operations. *Agriculture* 12 (5): 646.

12

Regaining the Essential Ecosystem Services in Degraded Lands

V. Girijaveni[1], K. Sammi Reddy[1], J.V.N.S. Prasad[1], V.K. Singh[1], and Chitranjan Kumar[2]

[1] *Central Research Institute for Dryland Agriculture, Santoshnagar, Saidabad P.O., Hyderabad, 500 059, India*
[2] *Amity Institute of Organic Agriculture, Amity University, Uttar Pradesh, Noida, India*

12.1 Introduction

Soil is the most complex and diverse ecosystem on the earth providing shelter to a vast number of different species. Taking only soil-based activity into consideration, it accounts for the gross production value of world agriculture, which is around 7.1 trillion USD in 2016. According to the FAO-led Global Soil Partnership, 75 billion ton (Pg) of soil is eroded annually from agricultural areas throughout the world, resulting in a financial loss of 400 billion USD each year (FAO 2011; GSP 2017). Due to poor land use decisions, overuse, and exploitation, the soil is degraded extensively. This is not only seen in our country but happening worldwide. Recently, considerable attention is given to "ecosystem" services of soils. Globally soil ecosystems are detrimentally affected mainly due to population increase and changes in land use patterns. As a result of affected soil ecosystems, they lose their capacity to function properly and fail to sustain crop production. It is well known that soil forms a major component of the terrestrial ecosystem. Thus, soil degradation is an alteration in the status of soil health that reduces the inherent capacity of the soil ecosystem to furnish goods and services to the recipient.

One of the important concepts to be considered for soil policymaking is ecosystem services. Recently, the EO4 cultivar project is providing information for sustainable ecological services and supporting soil sustainability (Bell et al. 2020). In India, soil degradation has reached the alarming stage that accounts for runoff and water erosion (94 million ha), soil acidification (16 million ha), surface flooding (14 million ha), wind erosion (9 million ha), soil salinization (96 million ha), and other degraded soil 7 million ha due to combined problematic soils (Bhattacharyya et al. 2015). The major reasons for soil degradation include both natural and human-induced. Due to climate change, the extent of soil degradation due to natural disasters and calamities has increased in the recent period. The major unsustainable and non-agroecological practices responsible for soil degradation are shown in Figure 12.1.

Agroecological Approaches for Sustainable Soil Management, First Edition.
Edited by Majeti Narasimha Vara Prasad and Chitranjan Kumar.

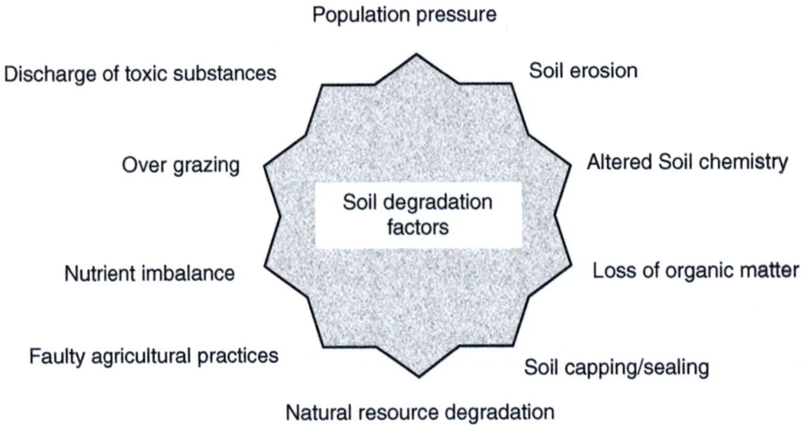

Figure 12.1 The major unsustainable and non-agroecological practices for soil degradation.

Human-induced soil degradation is due to various activities that include poor agricultural practices, deforestation, untreated sewage irrigation, improper management of pasture and forage crops, improper forest regeneration, soil mining, and unplanned urbanization and industrialization. Frequent and deep tillage, imbalanced chemical fertilizers, irrigation with saline and wastewaters, use of pesticides below the threshold limit of pests, and lower organic inputs in soils are common examples of unsustainable agricultural practices. Due to unsustainable management practices, 60–70% of soils are degraded in the European Union and are unable to provide ecological functions (Veerman et al. 2020). Human-induced soil degradation can be addressed through the adoption of proper and scientifically proven technologies. Soil degradation is now threatening crop sustainability, potable and irrigation water quality, soil biodiversity, and overall soil–plant sustainability. Being a responsible member of the United Nations to Combat Desertification (UNCCD), India is committed to minimizing its soil degradation and desertification. The target is to regain a neutral land degradation status by 2030 whereby gains in land reclamation should offset the increase in land degradation.

Soil erosion is the old phenomenon where the topsoil is lifted and eroded, which is further exacerbated within the ecosystem due to inter and intra reactions. Poor land management is one of the major causes of soil erosion. Soil erosion includes soil loosening, transport, and deposition. Reduction in yield potential and surface water quality is the impact of soil erosion. Hence it is the key cause of soil degradation. In China, land eroded through wind and water is about 3 million km^2 covering 32% of the total land area (Li et al. 2016). In the world, ice-free land surface, accelerated erosion by soil accounts for 28% while water accounts for 56% as per the Global Assessment of Human Induced Soil Degradation (GLASOD). Soil loss will be more in fallow land as compared to cultivated land. Soil erosion in China is between 30.9–107.4 $t\,ha^{-1}\,a^{-1}$ and 7.65–49.38 $t\,ha^{-1}\,a^{-1}$ on fallow land and agricultural land under conventional tillage, respectively (Issaka and Ashraf 2017). Guo et al. (2015) reported that permanent cover reduces soil loss, thus the soil loss in forests, shrubs, and grassland is reported as 0–1.89, 0.28–8.06, and 3.98–1.57 $t\,ha^{-1}$ a^{-1}respectively. Now remote sensing technologies are utilized to obtain the accurate distribution of soil erosion. These technologies can estimate wind erosion with over 87%

accuracy using 6 critical factors: number of days covered with snow, soil erodibility, aridity index, vegetation fraction, soil crust index, and wind field intensity (Guo et al. 2017), respectively. The extent of degradation is denoted as severe, moderate, and low. About 25% of the global soil falls under the severely degraded category while nearly half of the global soil (50%) falls under the moderately degraded category (Srivastava et al. 2019). This scenario calls for urgent measures to arrest soil degradation. According to Brady et al. (2008), with a soil formation rate of about 0.04–0.08 mm yr^{-1}, the annual soil accumulation rate is 0.5–1 t ha^{-1}. Upon comparison with annual soil loss rates, it is estimated that presently the rate of soil loss is 10–30 times higher than the rate of soil formation. Globally, the annual average soil loss accounts for up to 5–40 t ha^{-1} via erosion (Pimentel 2006). Soil degradation needs to be arrested to ensure global food demand for future generations. The widely accepted solution to address soil degradation and global climate change is through vegetation, which aids in soil sequestration by improving soil organic carbon status (Srinivasarao et al. 2022). Hence, COP21 has launched the "4 per 1000 Initiative: Soils for Food Security and Climate" to increase soil organic matter (SOM) stock at an annual rate of 0.4% per year in soils (Soussana et al. 2019).

12.2 Soil and Water Conservation Techniques

Soil management, crop management, engineering, and other interventions are widely used to reduce water and soil loss (Figure 12.2). Globally, a severe environmental problem that needs urgent attention is soil degradation. Due to soil degradation, approximately 1.9 billion ha of land worldwide is severely affected and almost 24 billion tons of soil is carried away every year (Borrelli et al. 2017). Soil degradation contributes to annual land depletion of 36–75 billion tons. Thus, soil degradation is a threat to the global food supply. Soil is a basic element by which the ecosystem sustains. Erosion and desertification result in soil degradation where ecological efficiency is seriously impaired. Soil erosion is linked with multiple economic and social issues. The soil degradation on sloppy lands can be prevented by establishing vegetation. Soil bioengineering is an effective option, particularly in developing countries such as India, Nepal, and Bangladesh, to cope with the issues of soil degradation and deep soil erosion (Gobinath et al. 2022). One of the eight major fragile ecosystems of India is the *Shivalik* foothills in the northwest, which is highly susceptible to

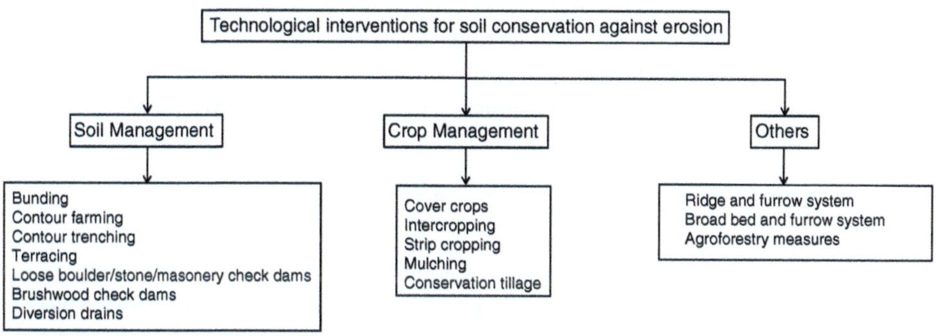

Figure 12.2 Various technological interventions to arrest soil erosion.

accelerated soil erosion. Mainly conservation techniques need to adopt in sloping farmland, barren land, and eroded gullies. However, it is most important to consider the erosion degree and soil type in addition to social and economic benefits for the successful implementation of conservation techniques. Soil conservation is a technique, which prevents soil losses. These conservation measures vary from region to region for example, in hilly regions, vegetation cannot get established as the land topography has a steep slope and is subject to erosion problems. Runoff is the major problem in these areas and thus makes them unsuitable for cultivation. Under the above circumstances, agronomical measures including contouring, strip cropping, and tillage practices can control soil erosion. Thus, the adoption of agronomical measures entirely depends on the soil types, land slope, and rainfall characteristics. Agronomical practices are counted as the first line of defense in soil and water conservation programs, while mechanical or engineering measures are considered the second line of defense. Both control soil erosion effectively. In terms of economics, agronomical measures are found to be more economical, long-lasting, and effective. In mild sloppy lands, vegetation/perennial grasses can control the erosion as dissipates the erosive power of water. In South China, various water and soil conservation techniques has decreased soil erosion by 30–40% since the 1980s (Zheng et al. 2008). Among various conservation techniques, terraces, mulching, and soil management have been widely practiced in tropical countries such as Thailand, India, and Pakistan, etc. (Ali et al. 2018; Donjadee and Tingsanchali 2013; Ghosh et al. 2016; Xiong et al. 2018; Zhou et al. 2020). In the Indo-Gangetic Plains (IGP), India, a study was carried out under conservation tillage (zero tillage with bed planting, zero tillage with flat planting, conventional tillage with bed planting, and cotton/maize residues) on soil aggregates and found that macroaggregates of 0.25–8 mm were highest in zero tillage with bed planting while macroaggregates of 2–8 mm were highest in cotton/maize residues and both maintained greater mean weight diameter (Bhattacharyya et al. 2013).

12.3 Soil Management

Another significant cause of land degradation is the transportation of dissolved nutrients from agricultural fields which in turn is a threat to the capacity of the land to provide ecosystem services. It is a well-known fact that proper ecosystem services are necessary to achieve sustainable development goals (SDGs) (Keesstra et al. 2018; Visser et al. 2019). Agronomical measures play an important role in soil and water conservation techniques. The agronomical measures include the practice of contour cropping, crop rotation, strip cropping, etc., but for the land, which is more erodible, has a steep slope, and prevails with high rainfall intensity, the use of agronomical measures is not justified; only terracing is an alternative measure.

12.3.1 Engineering Measures for Controlling Soil Erosion

12.3.1.1 Bunding

Bunds are constructed in the form of an embankment-like structure across the land slope. It is an engineering technique to conserve soil loss across the land slope. This practice is suitable for slopy lands where the slope ranges from 2% to 10%. Bunding is found successful in shallow and medium-depth soils and is not suitable for black soils due to its swelling and cracking nature. They are of different types such as contour bund (constructed on the contour of the area), graded bund, and stone bund. Stone bunds effectively reduced soil loss and surface runoff even under heavy rainfall events in a field experiment conducted in Ethiopian Highlands (Brenner et al. 2013). The contour bund is effective in low rainfall areas, and the graded bund is effective in medium to high rainfall areas. Contour bunding is widely practiced on Alfisols of the semiarid tropics in India. Conventional contour bunding involves the construction of small bunds across the slope of the land along a contour so that the long slope is reduced to a series of small ones. A modified contour bunding system with gated outlets is more beneficial as compared to conventional contour bunding (Pathak et al. 1989). Contour bunding technology was first introduced in 1993 by the Institut d'Economie Rurale (IER) in Mali and Centre de Coopération Internationale en Recherche Agronomique pour le Développement (CIRAD), a French agricultural research center for international development (Zemadim et al. 2014). In southern Mali, evidence states that the runoff rate ranged from 20% to 40% of total rainfall even on the agricultural field with a gentle slope of 1–3% (Birhanu et al. 2019; Traore et al. 2017). However, with the adoption of contour bunding, the runoff rate was reduced to a larger extent, and farmers benefitted from higher crop productivity (Birhanu et al. 2020). In India, contour bunding is practiced in all types of soil except clay and black cotton soils in Tamil Nadu. It is practiced in land having an average slope > 0.5–6% in well-drained soils and high rainfall areas of Maharashtra while it is practiced in dry areas with slopes ranging up to 6% in Gujarat and Uttar Pradesh. However, it is practiced in fields with slopes up to 15%, with a vertical interval of 1.5–2.0 m in Kerala.

12.3.1.2 Contour Farming

Contour farming is the activity of plowing and planting along the contour. This practice helps to prevent surface runoff and improves infiltration in the soil. Farmers establish crops along the contour lines following the configuration on the ground. Contour farming and balancing fertilization as a part of soil management were studied in coffee grown in Indonesia (Djufry et al. 2022). They practiced tillage, planting, and other farming operations on or near the contour of the field slope to promote positive row drainage and reduce ponding and found that contour farming along with balanced fertilization was found very effective in coffee cultivation in Indonesia. In Zimbabwe, the practice of contour rainwater-harvesting and integrated nutrient management in sorghum could improve soil fertility, reduce soil loss, and improve the grain yield of sorghum grown in semi-arid conditions (Kubiku et al. 2022).

12.3.1.3 Contour Trenching

Contour trenching is mostly practiced in degraded non-arable lands. Contour trenching aids in the in-situ conservation of moisture, soil, nutrients, and energy fluxes (Kurothe et al. 2014). Here without a spillway, microdepression, or tiny reservoir constructed across the slope. It is found effective in semi-arid and arid regions with high rainfall intensities and sandy soils with high permeability. Contour trenching with vegetative barriers is found to be more effective than simple contour trenching (Madhu et al. 2011). Contour trenches break the velocity of runoff and improve the soil moisture regimes. Continuous contour trenches are better for moisture enhancement in small catchment areas (Patode et al. 2015). Contour trenches generate runoff from the upper part of the catchment, thus, protecting the bund against breaching. To lay contour trenches, the cross-sectional area should not exceed $0.3 \times 0.3\,m^2$ and the side slopes should be in the range of 1:1 to 0.5:1, depending on soil type, and need to be excavated at suitable intervals.

12.3.1.4 Terracing

Earthen embankment across a slope to minimize soil runoff is known as Terracing. A study was carried out to study the effectiveness of terraced structures to conserve soil fertility in Pakisthan (Rashid et al. 2016). They found that terraced structures could greatly minimize water erosion and gully development. In highly sloping lands (where the land slope is greater than 10%) especially in hilly and mountainous regions, terracing is followed as an engineering or mechanical measure to control runoff. In terracing, embankment or ridge and steps-like structures are constructed across the land slope (Cao et al. 2013). Deng et al. (2021) have described comprehensively the effect of terraces by describing the mechanisms behind terraced systems. In Southeast Asia, terraced fields were seen 5000 years ago. Terraced farmland is mostly noticed in Portugal, Switzerland, Italy, Nepal, Indonesia, the Philippines, Peru, China, Japan, and Ethiopia (Baryła and Pierzgalski 2008). Terracing can reduce soil erosion by 43–70% and 20–40% on mountainy slopes lands (Shi et al. 2012; Hu et al. 2005). The effectiveness of terracing depends on the design and its maintenance (Table 12.1). Runoff and sediment loss, on average, can be reduced by over 41.9% and 52%,

Table 12.1 Impact of terracing on soil moisture, runoff, and soil erosion.

Technique adopted	Impact	References
Terracing	It increased moisture contents in soils by 4.2–13%.	Chen et al. (2020); Feng et al. (2017)
Terraces with ridges	Retained 18% more runoff water than the terracing alone	Jiao and Wang (1999)
Terraces	Reduced the rate of soil erosion by $1\,t\,ha^{-1}\,yr^{-1}$, when the slope tangent exceeded to 0.1 level	Quine et al. (1999)
Bench terraces	Potential to increase the arable land by 20–40% due to bench terracing	Posthumus and De Graaff (2005)
Terraces	Observed 3–20 times higher soil erosion at non-terraced sites as compared to terraced sites	Hammad et al. (2006)
Terraces	Reduced soil losses over 96% at terraced sites along with proper land management	Wickama et al. (2014)
Half-moon terraces	Water runoff was reduced by 18% and soil/sediment loss by 76%.	Fu et al. (2011)
Sloping terraces	Runoff decreased by 41% and soil/sediment loss by 80%.	Zheng et al. (2008)
Cultivated terraces	Observed 2 times higher runoff coefficients on the abandoned sites than on the cultivated terraced sites.	Arnáez et al. (2015)

respectively due to terracing with proper field management (Deng et al. 2021). Terraces are named after the material with which terracing is done such as grass ridge, grassed bench, broad bed embankment, stone wall, level, etc. which may differ across ecosystems. Terracing increases infiltration, soil moisture, and soil water-holding capacity by increasing soil roughness (Wei et al. 2016). Terraces decrease the connectivity and integrity of overland flow and increase the infiltration (Molina et al. 2014). There are three layers in a terraced profile, the fast-changing layer (0–40 cm), the activity layer (40–100 cm), and the relatively stable layer (under 100 cm) (Zhang et al. 2005). The fast-changing layer of a terrace holds 7.2% higher water than that in sloping land (Huo and Zhu 2013). Terracing in trapping the splashed soil particles due to a reduction in the slope length. However, terracing is not feasible where soil depth is not sufficient.

12.4 Loose Boulder/Stone/Masonry Check Dams/Brushwood Check Dams

Brushwood made of bamboo with a stone check dam was studied in the Benishangul-Gumuz region of Asosa Zone to arrest gully erosion. Due to the implementation of brushwood, gully depth was reduced to 0.58 m from 0.94. Also, previously without brushwood, the soil loss was 468 t/ha/yr, which was reduced to 204 t/ha/yr due to brushwood (Adugna and Alemu 2017). In the Tigray Highlands of Northern Ethiopia, stone bunds were installed to resist soil erosion. These bunds were evaluated in 202 field parcels. There was an average sediment accumulation rate of 58 t/ha/yr behind 3–21 years old stone bunds. Also, the P factor in the universal soil loss equation was 0.32. This shows the sediment accumulation rate decreased with bund age but increased the grain yield by 53% in the study regions (Nyssen et al. 2007). Chen and Bezak (2022) quantified the different levels of soil erosion after extreme rainfall events in China. This study reported that erosion intensities for gully erosion, rill erosion, landslides, and collapse of implement soil and water conservation measures were 787 ± 486, 2815 ± 529, 4433 ± 657, and $3927 + 1009$ t/km^2, respectively. In Spain, steep terraces and earthen check dams were studied in rainfed agricultural systems. They observed a decrease of 25% in the density of step terraces and check dams between 1956 and 2005. Due to soil and water conservation measures, there was good vegetation cover, enhanced spring discharge, and decrease in rill and sheet erosion by 68% in the Tigray highlands (Nyssen et al. 2008). Also, stone bunds and conservation trenches were studied, and found that conservation trenches were more effective than stone bunds in controlling runoff and soil loss (Taye et al. 2015). It is also reported that soil and water conservation measures (stone bunds and trenches) integrated with fodder species could resist soil erosion to a larger extent in Ethiopia with improved soil aggregate stability and soil organic carbon stocks (127%) (Mesfin et al. 2018). Ecological engineering techniques play an important and significant role in preventing the intense gravitational erosion occurring in mountain hazards near the Xianjiang River Basin of China (He et al. 2017). Stone check dams resisted the sediment loss and the gully development to a larger extent on sloppy land (Xiang et al. 2004). Check dams could decrease the runoff rate caused by rains on sloppy lands (Al-Seekh and Mohammad 2009). Stone bunds could result in 53% increase in crop yield (Vancampenhout et al. 2006). Gully rehabilitation was evaluated with gabion check dams, brushwood check dams, and livestock exclusion in wooded range lands using sequential Digital Elevator models (DEMs) (Alfonso-Torreño et al. 2022). This study revealed that the implementation of these measures could reduce the suspended sediment concentration by 65% at the outlet. There was also a dramatic reduction in runoff thereby changing the gully dynamics and sediment production.

12.5 Crop Management

Cover crops are spontaneous vegetation and sown plant species associated with a perennial crop in the rows or interrows with a high potential to provide ecosystem services (Fernández-Mena et al. 2021). Agri-environmental benefits from the use of cover crops in vineyard interrow include improved soil protection that contributes to reducing soil erosion (Bagagiolo et al. 2018). Cover crops serve important ecosystem services such as arresting soil degradation, mitigating water pollution, and resisting erosion (García-González et al. 2018) In Spain, compared barley (*Hordeum valgare* L.) and vetch (*Vicia sp* L.) cover crops were assessed in degraded lands and found that adoption of cover crops enhanced C sequestration and N retention in addition to the promotion of structural stability, water bolding capacity, infiltration rate, and the saturated hydraulic conductivity (García-González et al. 2018). In Spain, cover crops grown in inter rows of vineyards on a rolling landscape were studied and found that cover crops could reduce the loss of organic matter by 3.6 times compared to control (López-Vicente et al. 2020). With sustainable soil management, the economics of soil ecological services can be improved by enhanced soil organic carbon (SOC) storage (Brar et al. 2013; Portela and Rademacher 2001). Cover crops also perform multiple soil ecosystem services though it varies from region to region. In low precipitation regions (with <500 mm precipitation), cover crops could produce $2.37 \pm 2.30\,\text{Mg}\,\text{ha}^{-1}$ of aboveground biomass and increase soil wet aggregate stability up to 63%. Additionally, growing cover crops has the benefit of weed suppression (Blanco-Canqui et al. 2022). A study carried out in temperate humid climate revealed that cover crops could improve the SOC content by $10\text{–}20\,\text{Mg}\,\text{C}\,\text{ha}^{-1}$ (nearly 6 times over eight years) and the study suggest that there is the long-term potential of cover crops to improve agroecosystem resilience (Chahal et al. 2020). Porwollik et al. (2022) assessed the biophysical and biogeochemical impacts of growing cover crops using a Global Dynamic Vegetation Model: Lund Potsdam Jena managed Land (LPJmL) V. 5.0. Cover crops reduced N-leaching from different farmlands (Abdalla et al. 2019; Thapa et al. 2018; Valkama et al. 2015).

Impact of intercropping in rainfed maize–wheat rotation was studied at Himalaya, India. In this study, maize + cowpea/okra in the ratio of 1 : 1 and 1 : 2 was evaluated and found that the intercropping adopted could reduce soil loss and runoff by 43% and 26%, respectively (Sharma et al. 2017). Winter cover crops–rye (*Secale cereal* L.) and ryegrass (*Lolum multiflorum* Lam.) were studied in a continuous silage maize cropping system (Laloy and Bielders 2010), reduced erosion and runoff by >94% and could generate excess biomass for incorporation that in turn improved soil quality. Intercropping is an efficient source of resource utilization. Intercropping also improved soil chemical properties and enzyme activities and thus helps in sustaining agricultural ecosystems. In subtropical Asia and tropical Africa, intercropping is becoming the most promising technology for improved productivity.

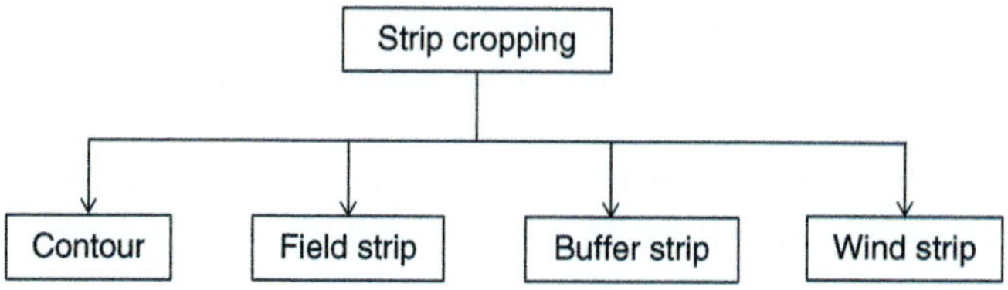

Figure 12.3 Different types of strip cropping practiced to arrest soil erosion.

In strip cropping, alternative strips are made along the slope and on the strip, different crops are grown. This consists of a combination of erosion permitting and erosion-resistant crops. This is only feasible in areas where the slope length is small. There are different types of strip cropping followed in India based on soil types, topography, etc. (Figure 12.3). Corn and Soya or dried beans are practiced in strip cropping in the parts of the United States East and Midwest (Bhat et al. 2019). it is beneficial to cultivate legumes in strip cropping. They aid in the biological fixation of N, resist erosion, improves soil fertility and overall soil quality. In case of a buffer strip, the width to be maintained is in between 2 and 4 m.

Interrcropping results in efficient utilization of resources and reduced soil erosion (Table 12.2). In the Shivalik foothills of India, field experiments were conducted to evaluate the effect of land slopes by adopting maize and cowpea strip-intercropping. The results revealed that on smaller slopes (1%), maize grain yield was significantly higher yields with higher net returns (US$ 428 ha^{-1}), while cowpea recorded higher yields on a 2% slope. Runoff, soil, and nutrient losses were higher on a 3% slope as compared to 1% and 2% slopes. The adoption of 4.8 m maize strip width and 1.2 m cowpea strip width resulted in significantly higher maize equivalent y cxzield than sole maize and other strip-intercropping systems (Khokhar et al. 2021).

Table 12.2 Impact of intercropping on soil conservation and soil health.

Sl. No.	Intercropping	Effect	References
1.	Vetch and barley	Decrease in runoff and soil loss	Sharaiha and Ziadat (2008)
2.	Maize and cowpea (1:1)	Decrease in runoff by 26% and soil loss by 43%	Sharma et al. (2017)
3.	Maize and common bean	Improved resource efficiency	Himmelstein et al. (2017)
4.	Maize–black gram/ cowpea (2:2)	Improved soil organic carbon status	Naresh et al. (2014)
5.	Maize+ legume [cowpea(1:1)/soyabean (1:1)]	Improved soil fertility	Baishya et al. (2021)
6.	Maize + cowpea (1:1)	Improved land use efficiency	Biruk et al. (2021)
7.	Maize + cowpea with plastic mulch	Reduced soil loss by 41%	Ranjan et al. (2021)
8.	Maize and potato	Reduced runoff by 15.65% in 1st year and 46.73% in 2nd year	Zhang et al. (2020)

Mucuna, lablab, and calliandra in sandy clay loam soil with a slope of around 20% could report the lowest soil erosion as compared to groundnut, maize/common bean intercrop in western Kenya (Muoni et al. 2019).

In semi-arid and arid areas in Al-Khalil district, west bank, intercropping with native perennial plants could reduce runoff by 34–89% and soil erosion by 45–95% (Salah 2008).

Mulching is a practice of covering the soil surface with crop/vegetative biomass. It acts as a protective cover. Mulching acts as a protection against raindrop impact and thus resists soil loss. Biomass adds SOM to the soil, improves infiltration rate, decreases evaporation rate, and conserves soil moisture. In Rodrigue's soils with slope of 8%, prone to erosion was mulched with banana (*Musa sp*) leaves, coconut (*Cocos nucifera*) leaves, and vetiver (*Vetiveria zizanoides*) were done at different rates (10 t/ha, 20 t/ha and 40 t/ha) in maize, which is the main crop grown in that region. The results revealed that application of coconut leaves at 40 t/ha and banana leaves at 10 t/ha reduced the soil and nutrient erosion by 57.3 and 28.9%, respectively. The mitigation potential thus depends on leaf area coverage (Lalljee 2013).

12.5.1 Conservation Tillage

The term "conservation tillage" is also known as minimum tillage, reduced tillage, zero tillage, mulch tillage, stubble-mulching, etc. In several countries, particularly in USA and Australia, the use of conservation tillage has been recommended to follow on cropland, strictly. However, due to its significant benefits, this technology is now in widespread use in several other countries. As far as its use regarding soil conservation, it has been found well suitable, mainly for highly mechanized farming; and also for controlling wind erosion. Conservation tillage is also found very effective to develop a good cover of crop residues by using vigorous vegetative growth-producing crops in the combination of conservation tillage operations. This technique also offers advantages regarding the elimination of the construction of terraces, bunds, or other permanent structures to check the soil erosion from highly sloping lands, however, there are a few disadvantages, too, which normally make this technique not favorable for use. Especially, for semi-arid conditions, a few main disadvantages of conservation tillage are outlined as under:

i) A higher planting density is incompatible with the situation of low moisture availability in the soil.
ii) Animals may graze the crop residues, allowed to remain over the soil surface for erosion control.
iii) Planting under mulched conditions is not easy for animal-drawn planters.

Minimum tillage is a general term, that covers several methods. One method is to reduce the number of operations by directly smoothing after plowing which is usually carried out for making a fine seedbed. It also involves the killing of weeds present in the field, using chemicals followed by secondary tillage practices. Minimum tillage causes greater erosion control and also reduces the tillage cost very much, as compared to the conventional methods.

12.5.2 Objectives of Minimum Tillage

The feasibility and success of minimum tillage practices depend on the following two main objectives:

1) The first objective is to facilitate a conducive environment for seed germination, and plant seedlings. It is basically done by performing tillage activities around the seeds or adjacent to the crop row areas.
2) The second objective refers to improving water use efficiency and thereby controlling soil erosion. This is applicable relatively for larger areas; the smaller areas are not suitable. To achieve fast absorption of water, reduction in erosion, and discourage weed growth, minimum tillage has been proved to be very effective, because it creates favorable conditions for water intake, reduction in erosion, and removal of the weeds.

The tillage operations such as plowing, secondary cultivation with one or more disc harrowing, and planking as conventional tillage have been found suitable for a wide range of soils to control erosion. The plowing at inappropriate moisture content produces a rough cloddy surface with variations in the soil surface height ranging from 120 to 160 mm. The secondary cultivation produces a surface roughness of 30–40 mm in height, which forms a favorable condition to control wind erosion, especially. The tillage roughness gets reduced over time by raindrop impacts, water flow, and wind erosion. The soil loss due to water erosion gets decreased with increasing surface roughness, according to the following relationship, given by Cogo et al. (1984).

$$SL \ \alpha \ e^{-0.5R}$$

In which, SL is the soil loss by water erosion and R is the surface roughness caused by tillage practices. This equation demonstrates how a tillage operation is useful for erosion control. The use of this equation can be extended for studies of the effect of different conventional tillage operations on wind erosion. The types of equipment such as moldboard plow and disc plow for plowing operations and disc cultivator for secondary cultivations are conventional tillage implements. The most common type of cultivator is the chisel type, which can cut the soil up to a width ranging from 50 to 75 mm. Sweep or blade type cultivators are also suitable implements, which consist of a "V"-shaped blade of 50–200 cm in width.

12.5.2.1 Listing

Listing is also a tillage technique used for controlling soil erosion. It is preferred in the semi-arid regions where unexpected heavy rainfall is received due to intense storms, and the field permits the use of contour farming along with erosion control measures. Listing is done with the help of a lister/planter (2-row or 4-row). Sometimes, the planters are modified by using furrow openers, to act as mouldboard plows, called mouldboard listers.

Hard Ground Listing: When the listing is done without plowing, it is called hard listing.

Loose-Ground Listing: The practice in which soil is plowed first and then listing is done is called loose-ground listing.

12.5.2.2 Crop Rotation

The protection against soil loss can be made more effective with the adoption of suitable crop rotation. In general, crop rotations are found more effective in which a large number of crop stubbles (plant residue left after harvest) are left on agricultural farms. The stubble covers or mulching over the soil surface is very effective to minimize soil erosion from water by reducing the raindrop impact on soil particle detachment and lowering overland flow velocity. In other words, it reduces the ability of rainwater/wind to detach the soil particles and transport the eroded soil mass or sediments. For example–in case of wheat crop harvesting a part of its plant stem is left uncut, which is called stubble. These stubbles remain over the entire field and provide a very conducive effect on soil erosion control from the turbulent wind blowing. Crops producing very little stubble mass should not be preferred for growing, because they are incapable to hinder soil erosion. For example, the sunflower typically produces less than 40% soil cover after harvest in the form of stubble, as result, there is a significant percentage of the soil surface is left for soil erosion. But if at the place of sunflower, the wheat crop is taken, then the soil surface gets covered with a large percentage of stubbles, and as result, there is less soil erosion.

A corn-soybean crop rotation in a no-till system results in good control of soil erosion. The corn plants produce substantial residue mass after harvest, while soybean is a less residue producing plant, but when it is planted after corn, then a sufficient cover from the previous crops (corn) is found available on the soil surface, which is very helpful in checking soil erosion/soil loss. For reducing soil loss, the mono-cropping low-stubble producing crop should not be grown. The crop residues added through a particular crop rotation with substantial biomass also improve the soil structure. The stubble cover prevents the disruption and detachment of soil aggregates, as a result, the macro-pores remained unblocked; thus, there is a fast infiltration rate and reduced sheet/overland flow. A significant reduction in sheet/overland flow causes an appreciable reduction in soil erosion. The effect of crop rotation on erosion control also varies with the climatic condition of the area. In regions of relatively consistent climate, a particular crop rotation can produce sufficient plant growth, soil cover, and thus reduction in soil erosion. On the other hand, in the regions where climate condition is less predictable, and there is an occurrence of unexpected rain/drought, a flexible crop rotation is very promising producing good soil cover.

In erratic climatic conditions, the opportunity cropping system provides adequate soil cover to check soil erosion. In the opportunity cropping system, the crops are grown when the adequate moisture content is available in the soil. This type of cropping system produces better soil cover, as compared to rigid crop rotation. Apart from the above, crop rotation also affects the timing and length of ground cover. When a field is under fallow conditions, and there takes place intense rainfall, then a cropping system that produces a long-lasting soil cover is very effective to protect the soil from erosion. The effectiveness of crop rotation for soil erosion control can be made better by using soil management practices such as zero tillage, etc. In zero tillage, the crop stubbles remain for a longer duration on the soil surface, or in other words, the crop stubbles remain till the period when crops cannot be grown. This ultimately causes protection of soil from erosion.

12.5.2.3 Grassed Waterways

Growing thick grasses to resist erosion is known as grassed waterways. This is practiced mainly in different countries to drain/divert the runoff from the catchment. In this way, the

Table 12.3 Depth-wase distribution of site selection criteria.

Sl. No.	Soil groups	Soil depth (cm)	Remarks
1.	Shallow	Less than 25	When clay content in the soil is more than 30% then the soil is known as deep clay soil.
2.	Medium	25–40	
3.	Medium-deep	40–90	
4.	Deep-soil	More than 90	

use of grasses in the waterway controls the problem of soil erosion caused by turbulent runoff flow. In general, the grassed waterways are planned and laid down along the slope of the area. It is the oldest practice to control erosion in agriculture fields of North America (Chow et al. 1999). Apart from disposing of the runoff from the area, they also act as an outlet for the terraces or graded bunds. Waterways are a very important means for removing surplus water from the terraced field, and for erosion control. In addition, they are also used for other purposes such as handling natural runoff or carrying the discharge from general fields, contour furrows, diversion channels, or used as emergency spillways in farm ponds or other water storage structures. The grassed waterways should be fully established with grasses before water is turned into them. In other words, the waterways should be ready to hold the water before the bunds, terraces, diversions, etc. are constructed.

12.5.2.4 Site Selection Criteria

The important criteria followed for selecting the sites of waterways or grassed waterways construction are outlined in Table 12.3:

i) The land slope should be less than 20%.
ii) Where flow velocity does not exceed 1.8 m/s.
iii) The other types of waterways can also be constructed on steeper slopes, and also where flow velocities are very high.

12.6 Soil Erosion Models for Quantification

Successful erosion assessment is done by assessing the soil loss risk and its spatial distribution. Also, it is predicted that soil erosion in terms of the loss of soil fertility and reservoir sedimentation accounts for around US $10.8 billion (Emadodin et al. 2012). Data related to soil erosion helps in erosion control, conservation planning, and management of the environment. Quantitative estimation of soil loss rates forms the basis for the adaptation of appropriate conservation approaches, for which various models have been developed across various countries. Some popular models are the Universal Soil Loss Equation (USLE), Revised Universal Soil Loss Equation (RUSLE), and its improved version (Hao et al. 2017; Gao and Wang 2019; Phinzi and Ngetar 2019), Water Erosion Prediction Project (WEPP) (Flanagan et al. 2007; Zhang et al. 2005), European Soil Erosion Model (EUROSEM) (Sogon et al. 1999) and Limburg Soil Erosion Model (LISEM) (de Barros et al. 2014). Different models are used in different countries for example in Asia, LISEM, RUSLE, WEPP v2010.1, EPIC, SWAT, KINEROS, and AGNPS models are widely used.

In Europe, EPIC, WATEM-SEDEM, MEFIDIS, AGNPS, RUSLE, SWAT, WEPP v2010.1, and AnnAGNPS models are widely used. While in America, the widely used models include AnnAGNPS, SWAT, WEPP v2010.1, KINEROS, EPIC, and AGNPS (Guo et al. 2019). Modeling can predict soil erosion rates in future also.

12.7 Integrated Nutrient Management to Address the Soil Degradation

Integrated Nutrient management is a practice of combined use of organic and inorganic sources in appropriate proportions. It is a very established, proven, and viable technology to mitigate soil degradation along with improvement in soil health and C stocks. The use of organic sources aids in nutrient cycling, increases nutrient solubility, increases soil microbial biomass, and moderates soil temperature. Integration of organics with inorganics will sustain higher yields and improve overall soil health than the sole use of inorganic chemical fertilizers (Baishya et al. 2015; Liang et al. 2011; Nayak et al. 2012). There are a huge number of organic sources for use in agriculture. The organic sources are of two categories- bulky organic manures and concentrated organic manures. Concentrated organic manures contain higher N content (Figure 12.4) than bulky organic manures. The addition of organic along with inorganic fertilizers affects the soil microbe's population, function, and composition thereby increasing the agronomic efficiency.

Integrated use of organics and inorganics results in reduced global warming potential (GWP) in croplands. Recently, nutrient management in long-term field experiments that were assessed through CERES–Rice (Crop Environment Resource Synthesis) and DNDC

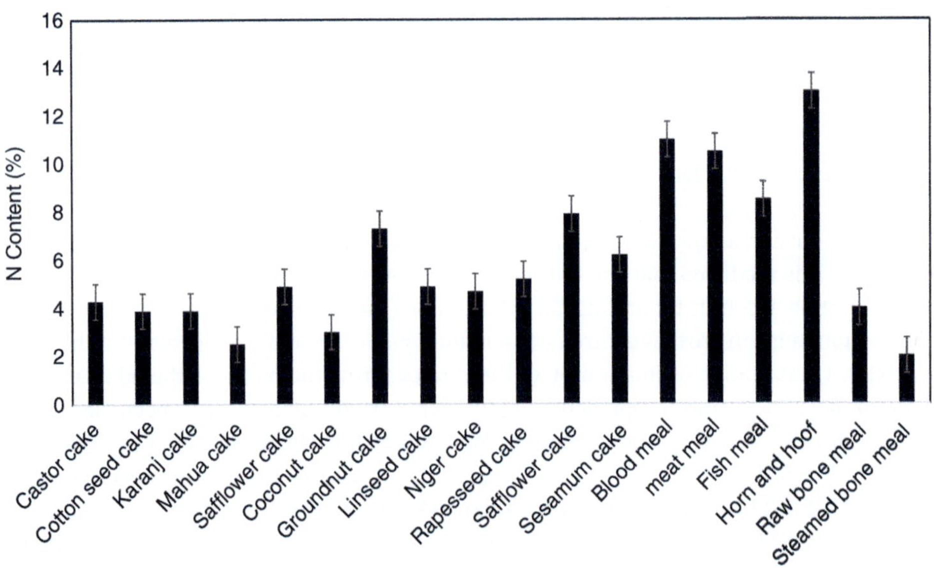

Figure 12.4 N content (%) in different concentrated organic sources.

(De-nitrification and Decomposition model) models revealed that integrated nutrient management using organic fertilizers reduced the GWP in rice fields. In that study, the highest rate of increase in GWP (0.014 and 0.021 kg CO_2eq kg-grain season^{-1}) was observed in RCP 4.5 and 8.5 scenarios, respectively, due to the application of recommended dose of fertilizers (RDF) through chemical fertilizer while integrating organic fertilizers with chemical fertilizers gave almost similar yield in later decades (Abbhishek et al. 2022).

In India, the lesson learned from long-term fertilizer experiments (LTFE), reveals that there was soil P build-up due to continuous application of P, although this practice has increased the productivity of all the cropping systems across the country, yet the P applied was not getting transformed/utilized by the crop. Here the phosphorus management strategy plays a role as it targets to re-utilize the build-up P by a reduction in P dose to half or skipping of P in an alternate year from the fertilizer schedule of the Kharif crops. In case of K application, especially in the mica-rich alluvial soils, the K application did not contribute much toward an increase in the crop yields. However, the application of K is very much essential in red soils (Alfisols), where the crop cannot be grown without K addition (Singh et al. 2017).

Conservation agriculture is practiced in water and nutrient-limiting areas across the globe. Soil hydraulics, water holding capacity, pore spaces, and dimension in a long term (>10 years) are practiced under corn and millet crops in the African continent. Studies (Bell et al. 2018; Mloza-Banda et al. 2014; NAIP 2018) have shown a significant increase in soil total porosity (5–15%), soil moisture (0.06–0.22 cm/min in K_{sat}), water retention (as fine porosity 3–7%), and plant available water capacity (PAWC by 3–6%).

Integrated nutrient management was practiced on a long-term basis (32 years) in sandy clay loam acid soil (Gogoi et al. 2021). In this study, Azolla, FYM, and rice stubbles were studied as organic sources and found that Azolla contributed to soil bacteria (8.95 log CFU g^{-1} soil) while FYM contributed to soil fungi (7.47 log CFU g^{-1} soil). However, the long-term addition of rice stubbles resulted in higher yields as compared to Azolla and FYM. However, Azolla contributed to the highest soil microbial biomass carbon (SMBC) followed by FYM and rice stubble. Regarding enzymatic activities, Azolla in combination with RDF produced maximum dehydrogenase activity up to 223.6 μg TTF g^{-1} soil 24 h^{-1} and urease activity up to 4.1 μg NH_4-N g^{-1} soil h^{-1}. FYM in combination with RDF produced the maximum phospho-monoesterase up to 337.4 μg *p*-nitrophenol g^{-1} soil h^{-1} and fluorescein-diacetate up to 10.0 μg fluorescein g^{-1} soil h^{-1}. In acidic sandy loam soils of Assam, the combined application of RDF and vermicompost at 5 t/ha along with mulching in potato fields resulted in 41.80% more net income due to an increase in tuber yield that ranged from 15.47% to 23.38% (Gogoi et al. 2017). The INM practices had higher values (0.72–0.75) for sustainable yield index (SYI) than that of RDF (0.67) after 10 years of rice mono-cropping (Patra et al. 2022).

Soil ecosystem services are found to be poor in drylands. In Nallabelli and Regonda Mandals of Andhra Pradesh, studies on recycling of tank silt in rainfed areas revealed that soil moisture retention has gone up by four to seven days, reduced cost of cultivation with a reduction in chemical fertilizers that ranged from Rs. 2500/– to Rs. 3750/– per ha. In crops like maize and cotton, tank silt application resulted in an increase in yield up to 700 kg/ha and 1000 kg/ha, respectively. In the United States, an increase in aggregate stability and increased rainwater penetration (25 mm per hour) was observed in an experiment in Texas due to the retention of wheat straw residues in Pullman clay loam. No tillage plots measured less runoff and cumulative infiltration compared with conventional disk tillage (Baumhardt and Lascano 1999). A Long-Term Experiment (LTE) on INM has shown a significant restoration of organic carbon in soils (Table 12.4).

Degraded lands and their ecosystem services can be regenerated through agricultural land management practices and the adoption of low-cost-effective technologies (Rosegrant et al. 2014). Conservation agriculture can improve ecosystem services by improving soil fertility. The concept is to minimize soil disturbances, maximize crop diversification, and enhance crop residue retention (Thierfelder et al. 2018).

12.8 Improving Soil Ecosystem Services Through Soil Microorganisms

Soils provide various ecosystem services that majorly include food, fodder, and fiber. Thus, it is well-known fact that the proper function of soil can contribute to essential ecosystem services. Hence, healthy soils are critical for assured ecosystem services. This is possible by planning and implementing correct and proper management thereby in addition to sustaining ecosystem services, also aiding in climate change mitigation. Soil ecosystem services depend largely on SMBC, SOC, and soil enzyme activity (Jackson et al. 2019). Microbial communities perform multiple roles by acting as natural recyclers, regulators, and engineers. According to Pereira et al. (2018), soil ecosystem services largely depend on environmental parameters that in turn determine the types of ecological services and soil properties. Arbuscular mycorrhizal Fungi (AMF), and plant growth promoting rhizobacteria (PGPR) are capable of improving soil ecological services. In temperate grasslands, bacterial biomass constitutes a large volume (1–2 t ha^{-1}) (Orgiazzi et al. 2016). As such, the estimated microbial genome sequences per gram of soil are 1000 Gbp (Vogel et al. 2009). The diversity and abundance of microbes in the root zone are three-orders higher than the bulk soils (Reinhold-Hurek et al. 2015). The rhizosphere effect denotes the increase in microbial activity, that in turn regulates nutrients, exudates, organic acids, and mucilages released by root hairs (Philippot et al. 2013). Research showed that rhizosphere microbial composition plays a major role in plant nutrition (Hartmann et al. 2008). However, belowground microbial communities help in soil formation, organic matter transformation, nutrient cycling, plant growth, decontamination, bioremediation, disease and pest regulation, etc. which are regulating and supporting services of soil ecosystems. AMF also helps in mitigating soil degradation through the production of glomalin which plays a major role in addressing several problems of soil degradation (Singh et al. 2022). Glomalin improves aggregation by acting as a gluing agent in addition to acting as a substrate for microbes in soils; thus, it helps in ecological restoration even in diverse conditions.

Table 12.4 Impact of long-term experiments on soil health.

S. No.	Type of soil	Place	Cropping System	Impact	References
1.	Rainfed inceptisols	LTE in Rakhdhiansar, India	Maize-black gram system	Observed higher SOC when nitrogen (25 kg N) was applied through compost or compost was applied in combination with N-fertilizer for 8 yr.	Sharma et al. (2019)
2.	Typic ustochrept	LTE in Bihar, India	Rice-wheat system	Increase in SOC (from 0.46–0.76%), available-N (194–226 $kg\,ha^{-1}$), available-P (23.6–49.5 $kg\,ha^{-1}$), and available-K (from 155 to 190 $kg\,ha^{-1}$) due to partial substitution of inorganics with organic sources.	Kumari et al. (2017)
3.	–	Pantnagar, Uttarakhand, India	Rice-wheat system	Observed maximum CEC up to 35.30 Cmol (p+) kg-1, SOC up to 1.20%, available-N up to 282.75 $kg\,ha^{-1}$), and available-P up to 28.3 $kg\,ha^{-1}$) by the combined application of 180 kg N, 80 kg P, 40 kg K, 30 kg Zn, and 5 tons FYM per hectare. This treatment also recorded the highest yield.	Nagar (2017)
4.	Typic haplustept	Pantnagar, Uttarakhand, India	Rice-wheat system	The combined application of FYM and NPK (100%) produced the maximum yield of grain up to 46.7 $q\,ha^{-1}$.	Pant and Ram (2016)

LTE, long-term experiment.

References

Abbhishek, K., Swain, D.K., Dey, S. et al. (2022). Nutrient management may reduce global warming potential of rice cultivation in subtropical India. *Curr. Res. Environ. Sustain.* 4: 100169.

Abdalla, M., Hastings, A., Cheng, K. et al. (2019). A critical review of the impacts of cover crops on nitrogen leaching, net greenhouse gas balance and crop productivity. *Glob. Chang. Biol.* 25: 2530–2543. https://doi.org/10.1111/gcb.14644.

Adugna, O. and Alemu, D. (2017). Evaluation of brush wood with stone check dam on gully rehabilitation. *J. CleanWAS* 1 (2): 10–13.

Alfonso-Torreño, A., Schnabel, S., Gómez-Gutiérrez, Á. et al. (2022). Effects of gully control measures on sediment yield and connectivity in wooded rangelands. *Catena* 214: 106259.

Ali, M., Khan, F., Khan, I. et al. (2018). Soil and water conservation practices in district Swabi, KP, Pakistan. *Adv. Crop Sci. Technol.* 6 (366): 2.

Al-Seekh, S.H. and Mohammad, A.G. (2009). The effect of water harvesting techniques on runoff, sedimentation, and soil properties. *Environ. Manag.* 44 (1): 37–45.

Arnáez, J., Lana-Renault, N., Lasanta, T. et al. (2015). Effects of farming terraces on hydrological and geomorphological processes. A review. *Catena* 128: 122–134.

Bagagiolo, G., Biddoccu, M., Rabino, D., and Cavallo, E. (2018). Effects of rows arrangement, soil management, and rainfall characteristics on water and soil losses in Italian sloping vineyards. *Environ. Res.* 166: 690–704.

Baishya, A., Gogoi, B., Hazarika, J. et al. (2015). Effect of continuous cropping and integrated nutrient management practices on soil properties and yield of rice (Oryza sativa)-rice cropping system in acid soil. *Indian J. Agron.* 60 (4): 493–501.

Baishya, L.K., Jamir, T., Walling, N., and Rajkhowa, D.J. (2021). Evaluation of maize (Zea mays L.)+ legume intercropping system for productivity, profitability, energy budgeting and soil health in hill terraces of eastern Himalayan region. *Legume Res.-Int. J.* 44 (11): 1343–1347.

de Barros, C.A.P., Minella, J.P.G., Dalbianco, L., and Ramon, R. (2014). Description of hydrological and erosion processes determined by applying the LISEM model in a rural catchment in southern Brazil. *J. Soil. Sediment.* 14 (7): 1298–1310.

Baryła, A. and Pierzgalski, E. (2008). Ridged terraces-functions, construction and use. *J. Environ. Eng. Landscape Manage.* 16 (2): 1–6.

Baumhardt, R.L. and Lascano, R.J. (1999). Water budget and yield of dryland cotton intercropped with terminated winter wheat. *Agron. J.* 91: 922–927. https://doi.org/10.2134/agronj1999.916922x.

Bell, P., Namoi, N., Lamanna, C. et al. (2018). *A Practical Guide to Climate-Smart Agriculture Technologies in Africa.*

Bell, G., Parker, N., Parker, J.A., and Smith, M.A.E. (2020). *EO4cultivar Sustainable Livelihoods Case Studies: Ecosystems Services Mapping in the Magdalena Region.* Colombia.

Bhat, S.A., Dar, M.U.D., and Meena, R.S. (2019). Soil erosion and management strategies. In: *Sustainable Management of Soil and Environment*, 73–122. Singapore: Springer.

Bhattacharyya, R., Das, T.K., Pramanik, P. et al. (2013). Impacts of conservation agriculture on soil aggregation and aggregate-associated N under an irrigated agroecosystem of the Indo-Gangetic Plains. *Nutr. Cycl. Agroecosyst.* 96 (2): 185–202.

Bhattacharyya, R., Ghosh, B.N., Mishra, P.K. et al. (2015). Soil degradation in India: challenges and potential solutions. *Sustainability* 7 (4): 3528–3570.

Birhanu, B.Z., Traoré, K., Gumma, M.K. et al. (2019). A watershed approach to managing rainfed agriculture in the semiarid region of southern Mali: integrated research on water and land use. *Environ. Dev. Sustain.* 21 (5): 2459–2485.

Birhanu, B.Z., Traoré, K., Sanogo, K. et al. (2020). Contour bunding technology-evidence and experience in the semiarid region of southern Mali. *Renew. Agric. Food Syst.* 1: 1–9.

Biruk, G., Awoke, T., and Anteneh, T. (2021). Effect of intercropping of maize and cowpea on the yield, land productivity and profitability of components crops in Bena-Tsemay Woreda, Southern Ethiopia. *Int. J. Agric. Res. Innov. Technol.* 11 (2): 147–150.

Blanco-Canqui, H., Ruis, S.J., Holman, J.D. et al. (2022). Can cover crops improve soil ecosystem services in water-limited environments? A review. *Soil Sci. Soc. Am. J.* 86 (1): 1–18.

Borrelli, P., Robinson, D.A., Fleischer, L.R. et al. (2017). An assessment of the global impact of 21st century land use change on soil erosion. *Nat. Commun.* 8 (1): 1–13.

Brady, N.C., Weil, R.R., and Weil, R.R. (2008). *The Nature and Properties of Soils*, vol. 13. Prentice Hall Upper.

Brar, B.S., Singh, K., Dheri, G.S., and Kumar, B. (2013). Carbon sequestration and soil carbon pools in a rice-wheat cropping system: effect of long-term use of inorganic fertilizer and organic manure. *Soil Tillage Res.* 128: 30–36.

Brenner, C., Strohmeier, S., Ziadat, F. and Klik, A. (2013). Soil conservation measures in the Ethiopian Highlands: the effectiveness of stone bunds on soil erosion processes. *EGU General Assembly Conference Abstracts (EGU2013–5392)* (April).

Cao, Y., Wu, Y., Zhang, Y., and Tian, J. (2013). Landscape pattern and sustainability of a 1300-year-old agricultural landscape in subtropical mountain areas, Southwestern China. *Int. J. Sustainable Dev. World Ecol.* 20 (4): 349–357.

Chahal, I., Vyn, R.J., Mayers, D., and Van Eerd, L.L. (2020). Cumulative impact of cover crops on soil carbon sequestration and profitability in a temperate humid climate. *Sci. Rep.* 10 (1): 1–11.

Chen, W. and Bezak, N. (2022). Comment on "Towards improved USLE-based soil erosion modelling in India: a review of prevalent pitfalls and implementation of exemplar methods" by Majhi et al. (2021), Earth-Science Reviews 221, 103786. *Earth-Sci. Rev.* 231: 104095.

Chen, D., Wei, W., Daryanto, S., and Tarolli, P. (2020). Does terracing enhance soil organic carbon sequestration? A national-scale data analysis in China. *Sci. Total Environ.* 721: 137751.

Chow, T.L., Rees, H.W., and Daigle, J.L. (1999). Effectiveness of terraces/grassed waterway systems for soil and water conservation: a field evaluation. *J. Soil Water Conserv.* 54 (3): 577–583.

Cogo, N.P., Moldenhauer, W.C., and Foster, G.R. (1984). Soil loss reductions from conservation tillage practices. *Soil Sci. Soc. Am. J.* 48 (2): 368–373.

Deng, C., Zhang, G., Liu, Y. et al. (2021). Advantages and disadvantages of terracing: a comprehensive review. *Int. Soil Water Conserv. Res.* 9 (3): 344–359.

Djufry, F., Wulandari, S., and Villano, R. (2022). Climate smart agriculture implementation on coffee smallholders in Indonesia and strategy to accelerate. *Land* 11 (7): 1112.

Donjadee, S. and Tingsanchali, T. (2013). Reduction of runoff and soil loss over steep slopes by using vetiver hedgerow systems. *Paddy Water Environ.* 11 (1): 573–581.

Emadodin, I., Narita, D., and Bork, H.R. (2012). Soil degradation and agricultural sustainability: an overview from Iran. *Environ. Dev. Sustain.* 14 (5): 611–625.

FAO (The Food and Agriculture Organization of the United Nations) (2011). *The State of the World's Land and Water Resources for Food and Agriculture (SOLAW)—Managing Systems at Risk; Food and Agriculture Organization of the United Nations*, Rome, Italy. London, UK: Earthscan. http://www.fao.org/docrep/017/i1688e/i1688e.pdf.

Feng, T.J., Wei, W., Chen, L.D. et al. (2017). Comparison of soil hydraulic characteristics under the conditions of long-term land preparation and natural slope in Longtan catchment of the loess hilly region. *Huan Jing ke Xue = Huanjing Kexue* 38 (9): 3860–3870.

Fernández-Mena, H., Frey, H., Celette, F. et al. (2021). Spatial and temporal diversity of service plant management strategies across vineyards in the south of France. Analysis through the coverage index. *Eur. J. Agron.* 123: 126191.

Flanagan, D.C., Gilley, J.E., and Franti, T.G. (2007). Water erosion prediction project (WEPP): development history, model capabilities, and future enhancements. *Trans. ASABE* 50 (5): 1603–1612.

Fu, B., Liu, Y., Lü, Y. et al. (2011). Assessing the soil erosion control service of ecosystems change in the loess plateau of China. *Ecol. Complexity* 8 (4): 284–293.

Gao, J. and Wang, H. (2019). Temporal analysis on quantitative attribution of karst soil erosion: a case study of a peak-cluster depression basin in Southwest China. *Catena* 172: 369–377.

García-González, I., Hontoria, C., Gabriel, J.L. et al. (2018). Cover crops to mitigate soil degradation and enhance soil functionality in irrigated land. *Geoderma* 322: 81–88.

Ghosh, B.N., Meena, V.S., Alam, N.M. et al. (2016). Impact of conservation practices on soil aggregation and the carbon management index after seven years of maize–wheat cropping system in the Indian Himalayas. *Agric. Ecosyst. Environ.* 216: 247–257.

Gobinath, R., Ganapathy, G.P., Gayathiri, E. et al. (2022). Ecoengineering practices for soil degradation protection of vulnerable hill slopes. In: *Computers in Earth and Environmental Sciences*, 255–270. Elsevier.

Gogoi, B., Baishya, A., Deori, B.B. et al. (2017). Soil productivity and potato yield improvement by mulching and integrated supply of nutrients in acid soils of Assam. In: *Natural Resource Management for Climate Smart Sustainable Agriculture* (ed. S. Arora, S. Swami, and S. Bhan), 34–43. New Delhi: SCSI.

Gogoi, B., Borah, N., Baishya, A. et al. (2021). Enhancing soil ecosystem services through sustainable integrated nutrient management in double rice-cropping system of North-East India. *Ecol. Indic.* 132: 108262.

GSP (2017). Global Soil Partnership Endorses Guidelines on Sustainable Soil Management 2017. http://www.fao.org/global-soilpartnership/resources/highlights/detail/en/c/416516 (accessed 15 October 2021).

Guo, Q., Hao, Y., and Liu, B. (2015). Rates of soil erosion in China: a study based on runoff plot data. *Catena* 124: 68–76.

Guo, B., Zhang, F.F., Yang, G. et al. (2017). Improved estimation method of soil wind erosion based on remote sensing and geographic information system in the Xinjiang Uygur autonomous region, China. *Geomat. Nat. Haz. Risk* 8 (2): 1752–1767.

Guo, Y., Peng, C., Zhu, Q. et al. (2019). Modelling the impacts of climate and land use changes on soil water erosion: model applications, limitations and future challenges. *J. Environ. Manage.* 250: 109403.

Hammad, A.H.A., Børresen, T., and Haugen, L.E. (2006). Effects of rain characteristics and terracing on runoff and erosion under the Mediterranean. *Soil Tillage Res.* 87 (1): 39–47.

Hao, C., Oguchi, T., and Pan, W.U. (2017). Assessment for soil loss by using a scheme of alterative sub-models based on the RUSLE in a Karst Basin of Southwest China. *J. Integr. Agric.* 16 (2): 377–388.

Hartmann, A., Rothballer, M., and Schmid, M. (2008). Lorenz Hiltner, a pioneer in rhizosphere microbial ecology and soil bacteriology research. *Plant Soil* 312: 7–14.

He, S., Wang, D., Lan, H., and Fang, Y. (2017). Guidelines for integrating ecological and biological engineering technologies for control of severe erosion in mountainous areas – a case study of the Xiaojing River Basin, China. International Soil and Water Conservation Research 2017. https://agris.fao.org/agris-search/search.do?recordID=US201700090575.

Himmelstein, J., Ares, A., Gallagher, D., and Myers, J. (2017). A meta-analysis of intercropping in Africa: impacts on crop yield, farmer income, and integrated pest management effects. *Int. J. Agric. Sustain.* 15 (1): 1–10.

Hu, J.M., Hu, X., and Zuo, C.Q. (2005). Analysis on soil and water conservation benefit of terracing on red-soil slope land. *Res. Soil Water Conserv.* 12 (4): 271–273.

Huo, Y. and Zhu, B. (2013). Analysis on the benefits of level terrace on soil and water conservation in loess hilly areas. *Res. Soil Water Conserv.* 20 (5): 24–28.

Issaka, S. and Ashraf, M.A. (2017). Impact of soil erosion and degradation on water quality: a review. *Geol. Ecol. Landscapes* 1 (1): 1–11.

Jackson, L.E., Bowles, T.M., Ferris, H. et al. (2019). Plant and soil microfaunal biodiversity across the borders between arable and forest ecosystems in a Mediterranean landscape. *Appl. Soil Ecol.* 136: 122–138.

Jiao, J.Y. and Wang, W.Z. (1999). Quality and soil water conservation effectiveness of level terrace on the Loess Plateau. *Trans. CSAE* 15: 59–63.

Keesstra, S., Mol, G., de Leeuw, J. et al. (2018). Soil-related sustainable development goals: four concepts to make land degradation neutrality and restoration work. *Land* 7: 133.

Khokhar, A., Yousuf, A., Singh, M. et al. (2021). Impact of land configuration and strip-intercropping on runoff, soil loss and crop yields under rainfed conditions in the Shivalik foothills of north-west, India. *Sustainability* 13 (11): 6282.

Kubiku, F.N., Nyamadzawo, G., Nyamangara, J., and Mandumbu, R. (2022). Effect of contour rainwater-harvesting and integrated nutrient management on sorghum grain yield in semi-arid farming environments of Zimbabwe. *Acta Agric. Scand. Sec.B—Soil Plant Sci.* 72 (1): 364–374.

Kumari, R., Kumar, S., Kumar, R. et al. (2017). Effect of long-term integrated nutrient management on crop yield, nutrition and soil fertility under rice-wheat system. *J. Appl. Nat. Sci.* 9 (3): 1801–1807.

Kurothe, R.S., Vishwakarma, A.K., Sena, D.R. et al. (2014). Decision support system for contour trenching. *Indian J. Soil Conserv.* 42 (2): 143–153.

Lalljee, B. (2013). Mulching as a mitigation agricultural technology against land degradation in the wake of climate change. *Int. Soil Water Conserv. Res.* 1 (3): 68–74.

Laloy, E. and Bielders, C.L. (2010). Effect of intercropping period management on runoff and erosion in a maize cropping system. *J. Environ. Qual.* 39 (3): 1001–1008.

Li, X., Niu, X., Wang, B. et al. (2016). Driving forces of dynamic changes in soil erosion in the Dahei Mountain ecological restoration area of northern China based on GIS and RS. *PLoS One* 11 (3): e0142331.

Liang, Y.L., Chen, C., Xue, Q. et al. (2011). Long-term soil organic carbon and crop yield dynamics on cropland in hilly and gully areas of Loess Plateau. *J. Agron.* 10 (2): 40–47.

López-Vicente, M., Calvo-Seas, E., Álvarez, S., and Cerdà, A. (2020). Effectiveness of cover crops to reduce loss of soil organic matter in a rainfed vineyard. *Land* 9 (7): 230.

Madhu, M., Raghunath, B., Tripathi, K.P. et al. (2011). Vegetative barrier with contour staggered trenches for resource conservation in new tea plantations of the Nilgiris, India. *Indian J. Soil Conserv.* 39 (1): 33–36.

Mesfin, S., Taye, G., and Hailemariam, M. (2018). Effects of integrated soil and water conservation measures on soil aggregate stability, soil organic matter and soil organic carbon stock of smallholder farmlands in semi-arid Northern Ethiopia. *Carbon Manage.* 9 (2): 155–164.

Mloza-Banda, M.L., Cornelis, W.M., Mloza-Banda, H.R. et al. (2014). Soil properties after change to conservation agriculture from ridge tillage in sandy clay loams of mid-altitude Central Malawi. *Soil Use Manage.* 30 (4): 569–578.

Molina, A.J., Latron, J., Rubio, C.M. et al. (2014). Spatio-temporal variability of soil water content on the local scale in a Mediterranean mountain area (Vallcebre, 21 North Eastern Spain). How different spatio-temporal scales reflect mean soil water content. *J. Hydrol.* 516: 182–192. https://doi.org/10.1016/j.jhydrol.2014.01.040.

Muoni, T., Barnes, A.P., Öborn, I. et al. (2019). Farmer perceptions of legumes and their functions in smallholder farming systems in east Africa. *Int. J. Agric. Sustain.* 17 (3): 205–218.

Nagar, U.S. (2017). Long-term effects of inorganic fertilizers and FYM on soil chemical properties and yield of wheat under rice-wheat cropping system. *Himalayan Ecol.* 25: 28.

NAIP (2018). Prioritised and Coordinated Agricultural Transformation Plan for Malawi. FY 2017/18–2022/23. https://www.scotland-malawipartnership.org/files/9815/3113/0121/National_Agicultural_Investment_Plan_2018_Final_Signed.pdf (accessed 20 March 2023).

Naresh, R.K., Singh, S.P., Kumar, P. et al. (2014). Sustainability of maize (Zea mays)-wheat (Triticum aestivum) cropping system under legumes intercropping and effect of nitrogen level on light distribution, soil temperature and crop productivity. *Indian J. Agric. Sci.* 84 (6): 695–701.

Nayak, A.K., Gangwar, B., Shukla, A.K. et al. (2012). Long-term effect of different integrated nutrient management on soil organic carbon and its fractions and sustainability of rice–wheat system in Indo Gangetic Plains of India. *Field Crop Res.* 127: 129–139.

Nyssen, J., Poesen, J., Gebremichael, D. et al. (2007). Interdisciplinary on-site evaluation of stone bunds to control soil erosion on cropland in Northern Ethiopia. *Soil Tillage Res.* 94 (1): 151–163.

Nyssen, J., Poesen, J., Moeyersons, J. et al. (2008). Dynamics of soil erosion rates and controlling factors in the Northern Ethiopian Highlands–towards a sediment budget. *Earth Surf. Process. Landf.* 33 (5): 695–711.

Orgiazzi, A., Bardgett, R.D., Barrios, E. et al. (2016). *Global Soil Biodiversity Atlas*. Luxembourg: Publications Office of the European Union. http://esdac.jrc.ec.europa.eu/content/global-soil-biodiversity-atlas.

Pant, P.K. and Ram, S. (2016). Long term manuring and fertilization effects on yields and nutrients uptake under Rice-wheat sequence in a Indian Mollisols. *Prog. Res. – An Int. J.* 11 (4): 583–589.

Pathak, P., Laryea, K.B., and Singh, S. (1989). A modified contour bunding system for Alfisols of the semi-arid tropics. *Agric. Water Manag.* 16 (3): 187–199.

Patode, R.S., Nagdeve, M.B., and Reddy, K.R. (2015). Effect of continuous contour trenches on soil moisture and leaf area index of perennial plantation in small catchment. *IOSR J. Agric. Vet. Sci. (IOSR-JAVS)* 8 (2): 50–54.

Patra, A., Sharma, V.K., Nath, D.J. et al. (2022). Long-term impact of integrated nutrient management on sustainable yield index of rice and soil quality under acidic inceptisol. *Arch. Agron. Soil Sci.* 1–18.

Pereira, P., Bogunovic, I., Muñoz-Rojas, M., and Brevik, E.C. (2018). Soil ecosystem services, sustainability, valuation and management. *Curr. Opin. Environ. Sci. Health* 5: 7–13.

Philippot, L., Raaijmakers, J.M., Lemanceau, P., and van der Putten, W.H. (2013). Going back to the roots: the microbial ecology of the rhizosphere. *Nat. Rev. Microbiol.* 11: 789–799.

Phinzi, K. and Ngetar, N.S. (2019). The assessment of water-borne erosion at catchment level using GIS-based RUSLE and remote sensing: a review. *Int. Soil Water Conserv. Res.* 7 (1): 27–46.

Pimentel, D. (2006). Soil erosion: a food and environmental threat. *Environ. Dev. Sustain.* 8 (1): 119–137. https://doi.org/10.1007/s10668-005-1262-8.

Portela, R. and Rademacher, I. (2001). A dynamic model of patterns of deforestation and their effect on the ability of the Brazilian Amazonia to provide ecosystem services. *Ecol. Model.* 143: 115–146.

Porwollik, V., Rolinski, S., Heinke, J. et al. (2022). The role of cover crops for cropland soil carbon, nitrogen leaching, and agricultural yields – a global simulation study with LPJmL (V. 5.0-tillage-cc). *Biogeosciences* 19: 957–977. https://doi.org/10.5194/bg-19-957-2022.

Posthumus, H. and De Graaff, J. (2005). Cost-benefit analysis of bench terraces, a case study in Peru. *Land Degrad. Dev.* 16 (1): 1–11.

Quine, T.A., Walling, D.E., Chakela, Q.K. et al. (1999). Rates and patterns of tillage and water erosion on terraces and contour strips: evidence from caesium-137 measurements. *Catena* 36 (1–2): 115–142.

Ranjan, R., Sharma, N.K., Kumar, A. et al. (2021). Evaluation of maize-based intercropping on runoff, soil loss, and yield in foothills of the Indian sub-Himalayas. *Exp. Agric.* 57 (2): 69–84.

Rashid, M., Alvi, S., Kausar, R., and Akram, M.I. (2016). The effectiveness of soil and water conservation terrace structures for improvement of crops and soil productivity in rainfed terraced system. *Pakistan J. Agric. Sci.* 53 (1): 241–248.

Reinhold-Hurek, B., Bünger, W., Burbano, C.S. et al. (2015). Roots shaping their microbiome: global hotspots for microbial activity. *Annu. Rev. Phytopathol.* 53: 403–424.

Rosegrant, M.W., Koo, J., Cenacchi, N. et al. (2014). *Food Security in a World of Natural Resource Scarcity: The Role of Agricultural Technologies*. Washington, D.C.: International Food Policy Research Institute (IFPRI). https://doi.org/10.2499/9780896298477.

Salah, A.M.D.A. (2008). Intercroppin annual fields with perennial plants: a strategy to reduce land degradation in semi-arid regions. Doctoral dissertation. Gottfried Wilhelm Leibniz Universität Hannover, Hannover.

Sharaiha, R.K. and Ziadat, F.M. (2008). Alternative cropping systems to control soil erosion in the arid to semi-arid areas of Jordan. *Arid Land Res. Manag.* 22 (1): 16–28.

Sharma, N.K., Singh, R.J., Mandal, D. et al. (2017). Increasing farmer's income and reducing soil erosion using intercropping in rainfed maize-wheat rotation of Himalaya, India. *Agr. Ecosyst. Environ.* 247: 43–53.

Sharma, K.L., Ravindra Chary, G., Sammi Reddy, K. et al. (2019). Effect of integrated nutrient management on soil quality indicators and soil quality indices in hill and mountainous inceptisol soils in northern India under maize (Zea mays) – black gram (*Vigna mungo*) system. *Ind. J. Dryl. Agric. Res. Dev.* 34 (1): 1–8. https://doi.org/10.5958/2231-6701.2019.00001.0.

Shi, Z.H., Ai, L., Fang, N.F., and Zhu, H.D. (2012). Modeling the impacts of integrated small watershed management on soil erosion and sediment delivery: a case study in the Three Gorges Area, China. *J. Hydrol.* 438: 156–167.

Singh, M., Wanjari, R.H., and Jatav, R.C. (2017). Phosphorus and potassium management under long-term manuring and fertilisation. *Ind. J. Fertilisers* 13 (4): 98–109.

Singh, A.K., Zhu, X., Chen, C. et al. (2022). The role of glomalin in mitigation of multiple soil degradation problems. *Crit. Rev. Environ. Sci. Technol.* 52 (9): 1604–1638.

Sogon, S., Penven, M.J., Bonte, P., and Muxart, T. (1999). Estimation of sediment yield and soil loss using suspended sediment load and 137 Cs measurements on agricultural land, Brie Plateau, France. *Hydrobiologia* 410: 251–261.

Soussana, J.F., Lutfalla, S., Ehrhardt, F. et al. (2019). Matching policy and science: rationale for the '4 per 1000 - soils for food security and climate' initiative. *Soil Tillage Res.* 188: 3–15. https://doi.org/10.1016/j.still.2017.12.002.

Srinivasarao, C., Jasti, V.N.S.P., Kondru, V.R. et al. (2022). Land and water conservation technologies for building carbon positive villages in India. *Land Degrad. Dev.* 33 (3): 395–412.

Srivastava, P., Giri, N., and Mandal, D. (2019). 137 Cs technology for soil erosion and soil carbon redistribution. *Curr. Sci.* 116 (6): 888–889.

Taye, G., Poesen, J., Vanmaercke, M. et al. (2015). Evolution of the effectiveness of stone bunds and trenches in reducing runoff and soil loss in the semi-arid Ethiopian highlands. *Z. Geomorphol.* 59 (4): 477–493.

Thapa, R., Mirsky, S.B., and Tully, K.L. (2018). Cover crops reduce nitrate leaching in agroecosystems: a global meta-analysis. *J. Environ. Qual.* 47: 1400–1411. https://doi.org/10.2134/jeq2018.03.0107.

Thierfelder, C., Baudron, F., Setimela, P. et al. (2018). Complementary practices supporting conservation agriculture in southern Africa. A review. *Agron. Sustain. Dev.* 38: 16. https://doi.org/10.1007/s13593-018-0492-8.

Traore, K., Sidibe, D.K., Coulibaly, H., and Bayala, J. (2017). Optimizing yield of improved varieties of millet and sorghum under highly variable rainfall conditions using contour ridges in Cinzana, Mali. *Agric. Food Secur.* 6 (1): 1–13.

Valkama, E., Lemola, R., Känkänen, H., and Turtola, E. (2015). Meta-analysis of the effects of undersown catch crops on nitrogen leaching loss and grain yields in the Nordic countries. *Agr. Ecosyst. Environ.* 203: 93–101. https://doi.org/10.1016/j.agee.2015.01.023.

Vancampenhout, K., Nyssen, J., Gebremichael, D. et al. (2006). Stone bunds for soil conservation in the northern Ethiopian highlands: impacts on soil fertility and crop yield. *Soil Tillage Res.* 90 (1–2): 1–15.

Veerman, C., Correia, T.P., Bastioli, C. et al. (2020). Caring for soil is caring for life: ensure 75% of soils are healthy by 2030 for healthy food, people, nature and climate: interim report of the mission board for soil health and food: study. *European Commission*. https://doi.org/10.2777/918775.

Visser, S., Keesstra, S., Maas, G., and De Cleen, M. (2019). Soil as a basis to create enabling conditions for transitions towards sustainable land management as a key to achieve the SDGs by 2030. *Sustainability* 11: 6792.

Vogel, T.M., Simonet, P., Jansson, J.K. et al. (2009). TerraGenome: a consortium for the sequencing of a soil metagenome. *Nat. Rev. Microbiol.* 7: 252.

Wei, W., Chen, D., Wang, L. et al. (2016). Global synthesis of the classifications, distributions, benefits and issues of terracing. *Earth-Sci. Rev.* 159: 388–403.

Wickama, J., Okoba, B., and Sterk, G. (2014). Effectiveness of sustainable land management measures in West Usambara highlands, Tanzania. *Catena* 118: 91–102.

Xiang, X., Hong-Wu, Z., and Ouyang, Z. (2004). Development of check-dam systems in gullies on the Loess Plateau, China. *Environ. Sci. Policy* 7: 79–86.

Xiong, M., Sun, R., and Chen, L. (2018). Effects of soil conservation techniques on water erosion control: a global analysis. *Sci. Total Environ.* 645: 753–760.

Zemadim, B., Tabo, R., Sogoba, B. et al. (2014). Assessment of contour bunding technology for improved land and water management in Mali (No. 63, p. 34). Resilient Dryland Systems Research Report.

Zhang, Y., Cao, N., Wu, M., and Wu, F. (2005). Analysis on soil moisture character of level terrace on the south Loess Plateau. *Chin. Agric. Sci. Bull.* 21 (08): 215–220.

Zhang, X.Y., Wu, K.X., Fullen, M.A., and Wu, B.Z. (2020). Synergistic effects of vegetation layers of maize and potato intercropping on soil erosion on sloping land in Yunnan Province, China. *J. Mountain Sci.* 17 (2): 423–434.

Zheng, H., Chen, F., Ouyang, Z. et al. (2008). Impacts of reforestation approaches on runoff control in the hilly red soil region of Southern China. *J. Hydrol.* 356 (1–2): 174–184.

Zhou, L., Wang, X., Wang, Z. et al. (2020). The challenge of soil loss control and vegetation restoration in the karst area of southwestern China. *Int. Soil Water Conserv. Res.* 8 (1): 26–34.

13

Phytochemicals as an Eco-Friendly Source for Sustainable Management of Soil-Borne Plant Pathogens in Soil Ecosystem

Shikha Tiwari[1], *Nawal K. Dubey*[2], *and Chitranjan Kumar*[3]

[1] *Department of Botany, S.S.S.V. S. Government Post Graduate College, Chunar, Mirzapur, 231304, India*
[2] *Laboratory of Herbal Pesticides, Centre of Advanced Study (CAS) in Botany, Institute of Science, Banaras Hindu University, Varanasi, 221005, India*
[3] *Amity Institute of Organic Agriculture, Amity University, Noida, India*

13.1 Introduction

A soil ecosystem is a dynamic living environment, comprising thousands of microbial and animal species and plant roots. The rhizospheric zone of plants is considered as a microbial hotspot that secretes carbon substances (rhizodeposits) responsible for the growth and activity of several microorganisms. The diversity and number of harmful and beneficial rhizospheric microbes relied on microbial interaction as well as qualitative and quantitative aspects of rhizodeposits. Soil microbial communities exhibit a wide range of diversity and play a significant role in ecosystem services. However, several rhizospheric microorganisms comprising bacteria, fungi, oomycetes, and nematodes negatively affect the growth and development of plants causing soil-borne diseases *viz.* root rot, damping off, wilt, and root knot leading to huge loss of agricultural productivity. Plant diseases caused by soil-borne plant pathogens are of great importance because most of the strategies applied for their control affect the soil ecosystem and are a major risk for the soil ecosystem. Oerke (2006) and Mavrodi et al. (2017) reported approximately 7–15% harvest loss of major crops such as potato, wheat, maize, rice, and soybean due to bacteria and fungi. In addition, Raaijmakers et al. (2009) reported a 10–20% loss of agriculture products due to nematodes while it is estimated to be nearly 50% of crop loss occurs due to soil-borne pathogens. Most soil-borne diseases are difficult to diagnose as they mostly exhibited below-ground symptoms and above-ground symptoms are identical to abiotic stress symptoms like water and nutrient unavailability. Hence, soil-borne pathogens are a major challenge to agriculture causing nutrient and water deficiency thereby reducing plant growth and development. Looking toward the highly challenging role of soil-borne pathogens, different crop protection methods such as chemical pesticides, resistant variety and crop rotation are frequently adopted. Although, chemical pesticides are strongly relied on all over the world to control soil pests, however, due to their toxic effects on the soil ecosystem through infiltration into

Agroecological Approaches for Sustainable Soil Management, First Edition.
Edited by Majeti Narasimha Vara Prasad and Chitranjan Kumar.

groundwater and nonbiodegradable nature European Protection Agency restricted or banned most synthetic pesticides. Synthetic pesticides are reported to cause destructive effects on natural soil habitat by imposing deleterious effects on soil mycobiota and soil health by affecting soil structure and functionality (Sattler et al. 2007; Yang et al. 2021). In addition, synthetic pesticides are also reported to have chronic and acute toxicity effects that make them harmful to non-targeted organisms in the soil ecosystem. Such chemicals enter the food chain by contaminating soil and water causing adverse effects on human health and other organisms as well as promote the development of resistance in pest varieties (Tripathi et al. 2020). Therefore, considering the vast adverse effects of synthetic pesticides different plant-based products have currently attracted the attention of scientists as a sustainable approach to combat different soil-borne pests because of their renewable, biodegradable nature, potent bio-efficacy, safer nature for soil health and environment, diverse mode of action and easy availability that support sustainable agriculture (Shabana et al. 2017). Soil is an important fragile ecosystem having a mutualistic and parasitic dynamic interaction between plant root and soil microbes that depend on chemical host signaling. Pathogen *Phytophthora sojae* zoospores are attracted by soybean due to the secretion of two chemicals daidzein and genistein (Hirsch et al. 2003). Besides, secondary chemicals like phytoalexins and phytoanticipins are also secreted by plants as defense molecules. In addition, nematodes attack their host through chemosensory perception by recognizing nonprotein signals released by the host. In addition, nematodes also release cytokinin that enhances cell division and creates a nutrient-rich feeding site in host roots (Badri et al. 2009; Wu et al. 2021). Traditionally, different plants were used as a green strategy to improve soil nutrients and also to combat soil-borne diseases. Cover crops planted prior to cash crops are used to inhibit pest infection through two different mechanisms either by enhancing competition (space, nutrition, water) between soil microbes through improving microbial activity and diversity or by producing volatile or non-volatile fungitoxic compounds (Baker and Cook 1974; Larkin 2015). *Crambe abyssinica* contains two potent nematicidal compounds, *i.e.*, allyl isothiocyanate and allyl sulphide, which are effective in the control of soil-borne nematodes (Donkin et al. 1995). Similarly, biocidal isothiocyanate compounds produced from the hydrolysis of glucosinolates containing plants have been reported to inhibit soil-borne diseases without affecting the beneficial rhizospheric microbes such as *Trichoderma* (Baysal-Gurel et al. 2018). There are reports on plant-based volatile and nonvolatile chemicals in the control of soil-borne pests as green pesticides. Hence, there are enormous prospects for plant-based chemicals in the soil ecosystem. Different plant secondary metabolites especially essential oils (EOs) are also reported to control soil-borne pathogens. EOs are basically defense molecule of plants that is volatile, lipophilic, and aromatic compounds. EOs are obtained from different plant parts such as flowers, stems, leaves, roots, fruits, twigs, and bark. Different low molecular weight volatile bioactive components such as terpenes (monoterpene, diterpene, sesquiterpene), alcohol, ester, sulphides, ketones, aldehydes, amines, and epoxides constitute the EOs that are also responsible to various biological activity of EOs (Bakkali et al. 2008). EOs are reported to be green alternatives for agriculture due to their potent bio-efficacy as antifungal, antimicrobial, nematocidal, antioxidant, antiviral, and insecticidal properties (Turek and Stintzing 2013). Due to minimum adverse effects, EOs are categorized by the U.S. Food and Drug Administration under the Generally Recognized as Safe Category (GRAS) (Ali et al. 2015). Despite having tremendous bio-efficacy against pathogenic microbes EOs and plant-based

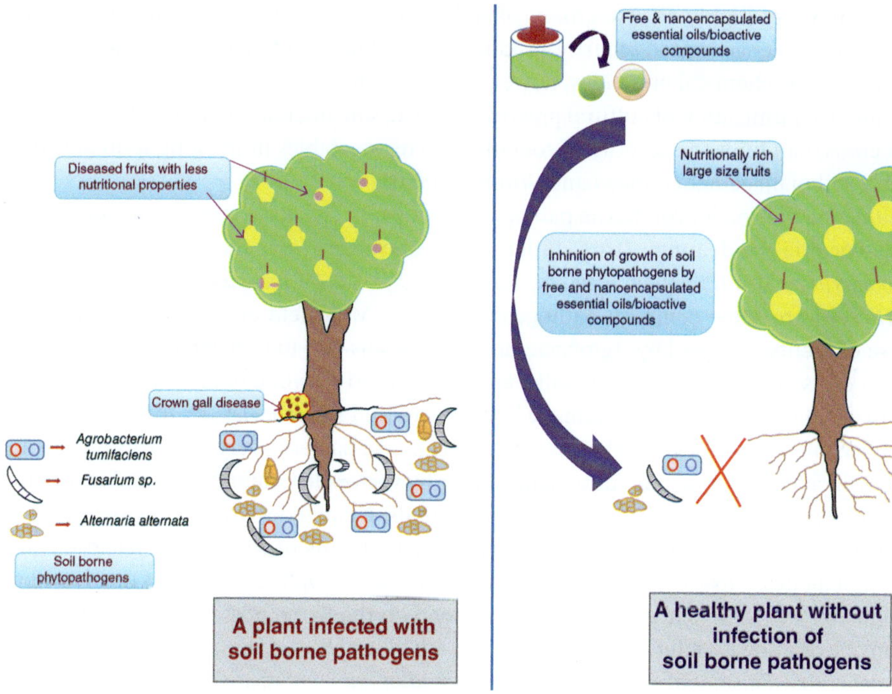

Figure 13.1 Showing effects of plant products on plant disease management.

chemicals have some limitations in terms of their volatility, low water solubility, and high sensitivity to oxidation (Liao et al. 2021). Inorganic nanoparticles have shown size-dependent properties and the particle size larger than 30 nm require regulatory scrutiny from environmental, health, and safety perspectives (Auffan et al. 2009). These limitations on plant-based chemicals can be overcome through the incorporation of nanoencapsulation technology (Figure 13.1). Therefore, the article focuses on the deleterious effect of soil-borne pathogens with special reference to plant-based green pesticides as a sustainable approach to managing soil-borne pests. In addition, the role of nanoencapsulation strategy as an innovative booster approach in the practical application of plant-based chemicals as green pesticides has been also highlighted in the article.

13.2 Soil-Borne Pathogens: Major Threat to Agroecosystem

Soil-borne pathogens exhibit a devastating role in most soil ecosystems by causing plant diseases. It is reported that nearly 10–15% of the global food crops are destroyed due to plant diseases (Strange and Scott 2005; Peng et al. 2021). The agricultural economy mostly depends upon soil fertility so there must be the incorporation of an awareness program to maintain soil health, quality, and productivity sustainably. Therefore, to enhance soil quality, soil-borne pathogens have to be managed that ultimately not only improve soil fertility by increasing the organic matter content of soil but also prevent soil-borne plant diseases and enhance the productivity of plants. It is also reported that warmer condition supports the occurrence of highly diverse groups of soil-borne fungal pathogens. These pathogens are a major challenge to food security, productivity, and health of plants, which are tried to be controlled by chemical pesticides, however these chemical pesticides ineffective against these fungi. Optimization of cultural practices (tillage, soil amendments, crop rotation, and cover crops) can limit the severity of root diseases of vegetables in the long term and sustainable soil management (Abawi and Widmer 2000).

Among the four major soil-borne pathogenic groups, *i.e.*, bacteria, virus, nematodes, and fungi, bacteria, and fungi (true fungi and oomycetes) are the most devastating. *Agrobacterium, Xanthomonas, Erwinia, Ralstonia, Xylella, Pectobacterium, Pseudomonas,* and *Dickeya* are some common pathogenic bacteria (Mansfield et al. 2012). Crown gall disease of plants is caused by *Agrobacterium tumifaciens* due to transferring and integrating part of its DNA, i.e., T-DNA (Gohlke and Deeken 2014); while potato (main food crop after rice and wheat globally) is damaged by *Dickeya* and *Plectobacterium* through backleg and tuber soft rot diseases (Czajkowski et al. 2011). Like bacteria, very few viruses are soil-borne, which can be transmitted to plants either by vectors like nematodes, and zoosporic fungi *i.e.,Olpidium* and *Polymyxa* or through wounds. Nematodes are the phytopathogens that live in soil and obtain food either by parasitizing plants or from fungi, bacteria, and other nematodes. Most of the soil-borne fungi are necrotrophic, *i.e.*, they kill their host through enzymes and toxins as well as not depend on the living host for their food. Fungi responsible for downy mildew and rust disease are biotrophic while some lower fungi and oomycetes like *Plasmopara halstedii* and *Plasmodiophora brassicae* are semi-biotrophic. Nectrotrophic fungi are having wide host range while biotrophic fungi are having narrow host range that suggests less chance of single gene resistance in plants. Plant root pathogens are considered as micro-herbivores that release different enzymes and develop different structures (haustorium, appressorium) or stylets by nematodes to enter the roots. Delgado-Baquerizo et al. (2020) reported *Venturia, Alternaria, Phoma,* and *Fusarium* as major soil fungal pathogens that destroyed the growth and productivity of the medicinal plants such as *Aloe vera, Hibiscus,* and food source plants such as potato, sunflower, tomato, cabbage, and wheat. *Fusarium* is a notorious fungus for stem rot, root rot, and ear rot diseases causing huge economic loss all over the world by declining about 10–40% yield loss (Bottalico and Perrone 2002). In addition, *Fusarium oxysporum* f. sp. *lycopersici* damages *Lycopersicum esculantum* (tomato) causing wilt disease while *F. oxysporum* f. sp. *racini* infects castor plants leading to a significant economic threat (Shaw et al. 2022). Tomato, a nutritionally and economically important crop is also damaged by soil-borne pathogens such as *Sclerotium rolfsii, Fusarium solani,* and *Rhizoctonia solani* that cause root rot and wilt diseases of tomato (El-Mohamedy et al. 2014). Among different species of *Alternaria*, *A. brassicae, A. helianthi,* and *A. citri* cause black spot disease of rapeseed-mustard, blight of sunflower, and brown spot of citrus, respectively (Kadam et al. 2009).

13.3 Green Chemicals as Better Alternatives to Synthetic Pesticides to Combat Soil-Borne Pests

Agricultural practices can be modified to enhance agricultural productivity and soil suppressiveness. Different chemical pesticides are in use to combat soil-borne pathogens but the use of most of them is cautioned due to adverse effects on health, and the environment, and also cause detrimental effects on non-targeted organisms. Traditionally farmers used to incorporate suppressive soils into conducive soil because it contains microbes inhibitory to soil-borne pathogens. This gave rise to the concept of biological control of soil-borne pathogens. Biological control agents can protect crops from diseases via directly hyper-parasitizing the mycelia, resting structures, and spores of fungal and bacterial pathogens or by secreting some antimicrobial substances. In addition, the indirect mechanism of bio-control agents, *i.e.*, competition for nutrients and space as well as induction of resistance in plants has been also reported as a plant disease management strategy (Kohl et al. 2019). There are 2–5% of bacterial communities reported to produce plant growth-promoting factors that support plant growth as well as it can produce antagonistic chemicals such as ammonia, cyanide, antibiotics, and siderophore that help to combat soil-borne pathogens (Bumunang et al. 2013). Although biological control measure is effective, at a practical level, it is highly challenging, because in nature pathogens can develop a resistance against the controlling microbe or due to frequent changes in environmental conditions. The current global trend is to use utilize food produced by applying plant-based safer products. Hence, researchers and scientists are looking toward some novel green alternatives. In this regard, some botanical products of higher plant origin including plant extract, plant essential oils, and bioactive components have shown their fruitfulness as a safer strategy in the control of soil-borne plant pests. Botanical pesticides are highly reliable, acceptable, and effective sustainable crop protection approach due to their less expensive, less toxic, lack of persistence or bioaccumulation in the environment, and easy availability in the environment. Plant extracts of *Azadirachta indica, Nicotiana tobacum, Vinca rosea,* and *Tagetes erecta* were found to be efficacious against *F. oxysporum Schl* f.sp. *phaseoli*, which are soil-borne pathogens of *Phaseolus vulgaris*. Among all four plant extracts, *A. indica* was highly efficacious and inhibited the growth of *Fusarium* and controlled the wilt disease (Obongoya et al. 2010). Tej et al. (2018) reported that the methanolic extract of *Lycium europaeum* having rutin as a major phenolic compound exhibited detrimental effects on soil-borne pathogens such as *Sclerotinia sclerotiorum, Verticillium dahlia,* and *Harpophora maydis,* and reduced late wilt of maize disease occurrence suggesting its significance in crop protection. Further, *Flourensia cernua, Flourensia retinophylla,* and *Flourensia microphylla* ethanolic extracts were inhibitory toward the mycelial growth of pathogens like *R. solani, Alternaria* sp., and *F. oxysporum*, recommending botanicals as a novel sustainable approach in pest management (De Rodriguez et al. 2007). Svecova et al. (2013) investigated that *Vitex agnus castus* methanolic extract was effective against the pathogen *Pythium ultimum* in tomatoes under both *in vitro* and *in vivo* conditions. The defense system in tomato plants was enhanced by plant extracts through activating PR genes expression. These allelopathic effects of phytoproducts are mainly due to their bioactive components that act directly on phytopathogen, enhance systemic resistance in plants and reduce disease occurrence. Among different plant products plant volatiles (EOs) are of interest as a viable substitute for synthetic pesticides due to their environmentally sound and safety status. Different volatile oils and other groups of bioactive components are registered efficacious natural biopesticides such as Rosemary EO (Sporan and EcoTrol), *A. indica* EO (Azatrol EC, Agroneem, Trilogy), clove EO (Matron EC, Burnout), Eugenol (EcoSMART, EcoPCOR), Menthol and Thymol (Apilife VATRAM) (Saroj et al. 2020). Pradhanang et al. (2003) reported that plant EOs (*Cymbopogon citratus, Cymbopogon martini*, and thymol) as soil fumigants effectively control bacterial wilt in tomatoes caused by *Rhizoctonia solanacearum*. Lee et al. (2007) reported *Origanum vulgare, Cuminum cyminum Eucalyptus citriodora* EO to be effective against three soil-borne fungi, *i.e., P. ultimum, F. oxysporum*, and *R. solani* while *Thymus vulgaris EO* inhibitory to *F. oxysporum, R. solani*, and *C. citratus* EO against *F. oxysporum. Ocimum basilicum*, and *C. citratus* are reported as eco-friendly alternatives to synthetic pesticide mancozeb and metalaxyl-mancozeb against soil-borne phytopathogens *Pythophthora drechsleri, Pythophthora capsici, Pythophthora melonis*, and protect cucumber, melon, and pepper plant from the disease caused by them such as damping off, rot of crown and root (Amini et al. 2016). *O. vulgare EO* was also reported to be a potential biopesticide against four soil-borne pathogenic fungi, *i.e., V. dahliae, F. oxysporum* f. sp. *lycopersici, Pythium*, and *S. sclerotiorum*, based on its phenolic bioactive compounds carvacrol and thymol (Wogiatzi et al. 2009). *Salvia sclarea* (Clary sage) EO as well as its bioactive components, *i.e.*, Linalool and linanyl acetate were found efficacious against three phytopathogenic fungi such as *F. oxysporum, Sclerotium cepivorum*, and *S. sclerotiorum* (Pitarokili et al. 2002). In the study, it was found that the inhibitory action of *S. sclarea EO* is due to the synergistic action between its compounds. The linalool-rich *Lippia alba* essential oil was found as an efficacious botanical pesticide to control damping off disease caused by *R. solani* in two commercial plants, i.e., *Plantago ovata* and *O. basilicum* (Saroj et al. 2019). The EOs pose a broad range of antimicrobial actions because they are comprised of various bioactive components, and each has its own function. The major action of EOs is due to its lipophilic attribute that assists easy interaction of botanicals to the lipids of the cell membrane and mitochondria that disrupt the membrane integrity ultimately leading to the leakage of vital cellular ions and cellular content leading ultimately to cell death (Lambert et al. 2001). Perina et al. (2015) reported that *T. vulgaris* EO and its bioactive component thymol effectively suppress the occurrence of brown spot disease caused by *Alternaria alternata via* disrupting cell wall and cell membrane and attacking mitochondria and other cellular organelles. Further, *Cinnamomum verum* EO and its bioactive components, *i.e., trans*-Cinnamaldehyde and salicylaldehyde depicted strong anti-microbial activity against *A. tumifaciens*. The mode of action involved to inhibit *A. tumifaciens* growth was the disruption of cell membrane integrity, expression of genes, and ROS generation which is linked with the Fenton reaction that causes down-regulation of ATP synthesis related gene cluster, damage of ROS defense system, and corruption of iron ion homeostasis (Lee et al. 2020). *Thymus zygis* EO was reported to significantly inhibit different soil-borne pathogens such as *F. oxysporum, S. sclerotiorum, R. solani, Pythium irregulare*, and *Colletotrichum acutatum* due to their bioactive compounds, *i.e., α-terpinene* and *3-octanol* (Perez-Sanchez et al. 2007). El-Zemity et al. (2008) also reported caraway, clove, chenopodium, thyme, and rosemary EO and thymol, carvacrol, chlorothymol as the effective inexpensive natural bactericidal substance to control the growth of plant pathogenic bacteria *Erwinia caratovora var. caratovora* and *Agrobacterium tumefaciens* in agriculture. Excess ROS generation ultimately leads to cell death. *Monarda fistulosa* and *Monarda didyma* EO and their major bioactive component carvacrol are also reported as potent sustainable products as they inhibited the growth of *Meloidogyne incognita* nematodes and prevent gall formation in tomato roots (Laquale et al. 2018). Thymol and Eugenol main bioactive components of thyme and clove EO respectively reported to have antibacterial efficacy against *Ralstonia solanacearum*, the causal organism of the bacterial wilt of potato, and are considered as a potent natural pesticide (George et al. 2021). According to Mukherjee and

SinhaBabu (2014), plant bioactive compounds citral and menthol are reported as efficacious products to manage root-knot disease causing nematode *M. incognita* without any phytotoxic effect. Further, defense chemicals like phenolic compounds and phenylalanine ammonia-lyase activity have also been boosted in *Abelmoschus esculentus* plant after the treatment of soil with menthol and citral to combat against *Megalaima incognita*. Ojaghian et al. (2015) also reported a major bioactive component of *Cinnamomum cassia*, *i.e.*, E-cinnamaldehyde as an efficacious natural product to control the causal organism of potato white mold disease, *S. sclerotiorum* under both *in vivo* and *in vitro* conditions. Table 13.1 represents a list of some green chemicals as potent biopesticides.

13.4 Nanoencapsulation as a Booster to Green Pesticides

The encapsulation technique is basically the entrapment of bioactive components within the wall matrix that cause a sustainable release of bioactive molecules and thereby enhancing their bio-efficacy. The nanoencapsulation of green pesticides is basically the entrapment of plant products of pesticidal efficacy inside the biopolymer of nanometric size. Nanoencapsulation technique has emerged as a novel innovative technique that has several advantages such as acceleration of bioactivity of encapsulated substance at low concentration, prevention of its degradation driven by environmental factors as well as also controlled negative sensorial effects through control release of molecules (Abreu et al. 2012). Nanoencapsulation of green pesticides enhances their pest control duration due to the controlled release of the entrapped molecules. The nanostructured system that carries botanical products is basically of two types, *i.e.*, polymer-based nanostructured system and lipid-based nanostructured system (Figure 13.2). Lipid-based nanostructures have the potential to involve both hydrophilic and hydrophobic compounds that enhances their targeted release, improve solubility, and decrease volatility. The lipid-based nanostructure is basically comprised of four different types of nanocarriers such as solid lipid nanocarrier (SLN), nanoemulsion, liposomes, and surfactant-based nanocarrier (Blanco-Padilla et al. 2014; Assadpour and Jafari 2019). In SLN, the suspension of solid-lipid particles is dispersed in the nanometric range in the aqueous media. Nanoemulsion is comprised of two immiscible solvents (aqueous and organic phases) where one phase is dispersed into another as nanometric droplets. Liposomes are lipid vesicles where lipids are organized in bilayers having an outer hydrophilic region while an inner hydrophobic region where the bioactive components are entrapped. Surfactant-based nanocarriers are systems formed due to the self-assembly of different surfactants in liquid media contributing to different nanodelivery systems like niosome, hexosome, cubosome, etc. Similarly, polymer-based nanostructured systems (EOs are encapsulated in food-grade biopolymers) are comprised of nanocapsules or nanosphere, and nanogel. Nanocapsule basically deals with the presence of a hydrophobic oil cavity surrounded by the polymer in which bioactive substances are entrapped inside the oil cavity. Nanocapsules are produced through the nanoprecipitation strategy. While in the nanosphere, there is the presence of a continuous polymeric network where either the drug is inside, or it is adsorbed on the surface. Different encapsulation strategies such as nanoprecipitation complex coacervation, ionic gelation, spray drying, and freeze drying are used for its production.

Table 13.1 Some green chemicals as potent biopesticides.

S. No.	Botanical pesticide	Targeted phytopathogen	Remark	References
1.	*Cymbopogon citratus, Cymbopogon martinii*, and *Pelargonium graveolens*	*Fusarium oxysporum* f.sp. *niveum*	EOs completely inhibited the growth of pathogenic fungus under *in-vitro* conditions and thus recommended to be green pesticides to control wilt of watermelon disease, effectively.	Naveenkumar et al. (2017)
2.	*Mentha piperita*	*Agrobacterium tumefaciens*	Peppermint EO exhibited both *in-vivo* and *in-vitro* efficacy against *A. tumefaciens* leading to protect tomatoes from tumor formation.	Hsouna et al. (2019)
3.	Garlic EO and diallyl disulfide	*Phytophthora nicotianae*	Both EO and its bioactive compound show efficacy to control tobacco black shank disease through inhibiting the growth of *Plagithmysus nicotianae* by damaging its cell membrane.	Wang et al. (2019)
4.	*Zingiber officinale* EO	*Phytophthora colocasiae*	EO effectively inhibited the growth of phytopathogen under *in-vitro* conditions and are thus can be a promising candidate to control the leaf blight of taro.	Kalhoro et al. (2022)
5.	Cassia EO and Thyme EO	*Alternaria alternata*	Both Cassia and Thyme EO effectively inhibited the growth of fungus responsible to cause disease in tomatoes. Thus, both acts as promising botanical fungicides.	Feng and Zheng (2007)
6.	*Tetradium glabrifolium* EO	*Phytophthora capsici*	EO showed *in-vivo* and *in-vitro* efficacy against pathogen and can be a promising candidate to control *Phytophthora* blight of *Capsicum annum*.	Wang et al. (2022)
7.	*Syzygium aromaticum* and *Zanthoxylum xanthoxyloides* EO	*Phytophthora megakarya*	Growth and sporulation of pathogen were effectively suppressed on the fumigation of EOs which suggests EOs as a natural alternative to control cocoa black pod diseases.	Nana et al. (2015)
8.	*Tagetes erecta* EO	*Pythium aphanidermatum*	EO was effective over synthetic fungicides, *i.e.*, Agrosan G.N., Captan, and Dithane Z-78, and effectively manage damping off seedlings of tomato disease.	Kishore and Dwivedi (1991)

9.	Citronella EO	*A. alternata*	EO showed promising action under both *in vivo* and *in vitro* conditions in controlling black rot disease caused by *Lycopersicon esculentum* through inhibiting the growth *A. alternata*.	Chen et al. (2014)
10.	*Cinnamomum zeylanicum* (bark), *C. zeylanicum* (leaf), *Cinnamomum cassia* (cassia), *S. aromaticum* (clove) EO	*F. oxysporum* and *Alternaria porri*	EOs effectively inhibited the growth of pathogenic *F. oxysporum* and *A. porri* and act as prosing alternative of synthetic fungicide.	Pawar and Thaker (2007)
11.	*Lavandula officinalis, Eucalyptus citriodora, Artemisia absinthium, Mentha arvensis, Z. officinale, Hypericum perforatum, Citrus bergamia, Ocimum basilicum* EO	*Meloidogyne incognita*	EOs effectively suppressed the root gall formation in tomato plants through inhibiting the pathogenic nematode.	Ozdemir and Gozel (2018)
12.	*Dysphania ambrosioides* EO	*M. incognita*	EO inhibited several galls and eggs of *Megalaima incognita* on tomato plants thus providing agriculture an exciting botanical pesticide.	Barros et al. (2019)
13.	*Satureja montana, Ruta graveolens, D. ambrosioides, Thymbra capitata, Filipendula ulmaria EO*	*Meloidogyne chitwoodi*	EOs exhibited strong anti-hatching potency due to their bioactive components. Hence, EOs can be an efficacious natural alternative to cope with root-knot nematodes.	Faria et al. (2016)
14.	*Eugenol*	*Rhizoctonia solani*	Eugenol major bioactive component of *S. aromaticum* inhibited the growth of pathogens through disrupting their membrane integrity, inhibiting ergosterol synthesis, enhancing ROS production, and downregulating tricarboxylic acid cycle and oxidative phosphorylation. This can be a novel and safer approach to controlling rice sheath blight disease.	Zhao et al. (2021)
15.	*Origanum onites, O. vulgare, Thymus vulgaris, Origanum minutiflorum, Origanum syriacum EO*	*Verticillium dahliae*	EOs effectively act as bio-fumigant and inhibited the growth of phytopathogen leading to the protection of plant from it.	Arslan and Dervis (2010)

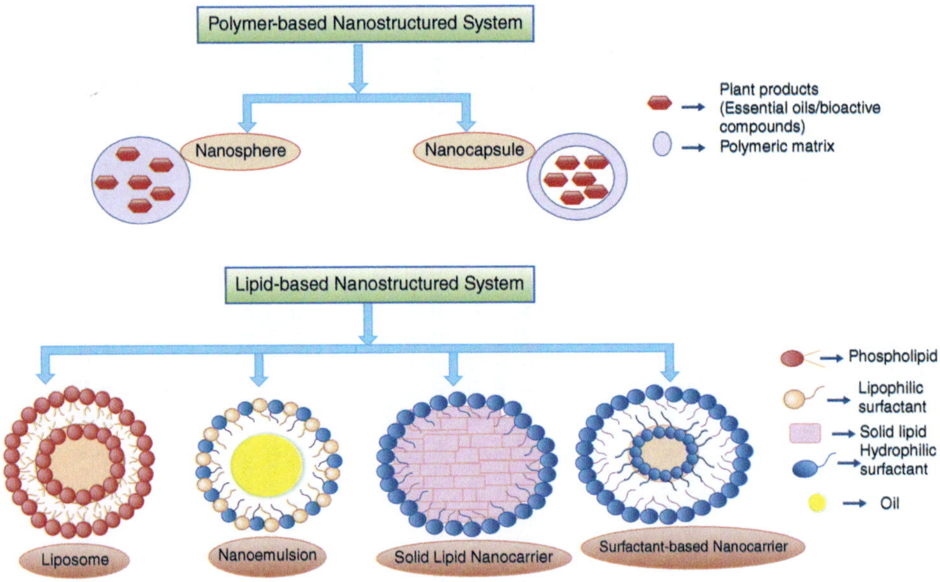

Figure 13.2 Diagram showing different nanostructured system.

Nanoencapsulated bioactive plant products have shown promising efficacy against different soil-borne pathogens. Co-nanoencapsulated *Carum copticum* and *Peganum harmala* EO showed enhanced antifungal activity and stability of substances that lead to suppression of the growth of the soil-borne pathogen *A. alternata*, responsible to cause the early blight disease of tomatoes in tropical and subtropical countries (Izadi et al. 2021). Oluoch et al. (2021) reported improved bio-efficacy of nanoencapsulated bioactive molecules, *i.e.*, thymol and eugenol in comparison to their free form due to controlled release and improved stability of compounds. Further, the encapsulated compounds exhibited great bactericidal efficacy against wilt-causing bacteria in potatoes, *i.e.*, *R. solanacearum*. EO-loaded lipid nanoemulsion, which is a ternary complex of encapsulated chili oil, cinnamon oil, and garlic oil, exhibited strong antifungal activity against the leaf spot causing pathogen *A. alternata* without causing any negative effect on plant growth (Nguyen et al. 2022). Nanoformulation of sweet basil (*O. basilicum*) and thyme (*T. vulgaris*) exhibited remarkable antifungal potential against *F. oxysporum f. sp. Cumini* because it protects entrapped bioactive molecules and reduced the dose required (Hassanin et al. 2017). Ali et al. (2017) reported that *A. indica* EO and *Cymbopogon nardus* EO nanoemulsion is having noticeable antifungal activity against *S. rolfsii* and *R. solani*. In addition, nanoemulsion of monoterpenes such as citral, ®-carvone, cinnamaldehyde, and geraniol pulegone was also found to be more efficacious in comparison to their free forms against soil-borne pathogenic fungi like *F. solani*, *R. solani*, and *Pythium digitatum* as well as also inducing biochemical defense mechanisms in host plant tomato by enhancing the activity of enzymes like peroxidase, polyphenoloxidase, and total phenolic contents leading to protect against damping off disease of tomato (Abdelrasoul and Amer 2018). Nanoformulations have significant impact on active ingredients and their solubility (Kah et al. 2013). Hence, the botanical-based nanoformulated pesticides are safer, efficacious, biodegradable, target-specific, and eco-friendly, strongly suggesting their role in a future agricultural approach that should be also linked with an integrated pest management system in the soil ecosystem.

Nanoparticles are very small, so they are easily inhaled by workers and accumulated inside the liver, lymph system, nervous system, and other body parts causing health problems. Hence, considering the environmental, as well as health problems associated with nanoparticles, the selected encapsulating material should be biodegradable (corn oil, bees wax, cashew gum, and lecithin) in nature to load green pesticides, and workers also need to bear protective covering to avoid exposure with nanoparticles. The cost–benefit ratio of nanoencapsulated materials must be properly analyzed. In addition, the fate of nanomaterials in the environment also provides a base to frame regulatory measures for its commercialization. For preparing commercial pesticides organic solvents are utilized to make them stable while nanoencapsulated pesticides provide organic solvent-free safe application. In addition, nanoencapsulated green pesticides are also eco-protective in nature, because the amount of pesticides needed is reduced and due to sustainable release, pesticides runoff in very less amounts that prevent water pollution (Nuruzzaman et al. 2016). Further, it is also reported that nanoencapsulation alter the character of pesticidal compounds, allow their controlled release, and make them non-toxic toward non-targeted organism (Grillo et al. 2012). Overall, nanotechnology is a bio-safe, ecofriendly, and efficacious novel technology. However, for its long-term and safer applicability, more research related to its fate in the environment and toxicity is needed to be explored.

13.5 Conclusion

Soil-borne pathogens damage the quality and yield of crops to a large extent. Conventional synthetic pesticides are frequently used to manage soil pathogens, however, their side effects as environmental pollutants and health hazards limit their reliability and applicability to the agriculture sector. Hence, there is an urgent need to adopt efficacious environment friendly control practices against soil-borne plant diseases to maintain agricultural quality and productivity. In this context, green chemistry has revolutionized the incorporation of a sustainable approach to the management of phytopathogens. Plant bioactive components play a tremendous role to combat soil-borne pathogens effectively without any harmful effects. Despite having huge merits such bioactive components are also having some practical challenges that can be effectively curbed through the implementation of green nanotechnology. Green nanopesticides through their proper safety analysis can be a sustainable revolutionary approach to manage soil phytopathogens and improving agricultural yield that also supports a nation's economy and development.

References

Abawi, G.S. and Widmer, T.L. (2000). Impact of soil health management practices on soilborne pathogens, nematodes and root diseases of vegetable crops. *Appl. Soil Ecol.* 15 (1): 37–47.

Abdelrasoul, M.A. and Amer, G.M. (2018). Comparative fungicidal activity of some pure and nanoemulsion monoterpenes effects on soil-borne plant pathogenic fungi. *J. Plant Prot. Pathol.* 9 (10): 707–713.

Abreu, F.O., Oliveira, E.F., Paula, H.C., and de Paula, R.C. (2012). Chitosan/cashew gum nanogels for essential oil encapsulation. *Carbohydr. Polym.* 89 (4): 1277–1282.

Ali, B., Al-Wabel, N.A., Shams, S. et al. (2015). Essential oils used in aromatherapy: a systemic review. *Asian Pac. J. Trop. Biomed.* 5 (8): 601–611.

Ali, E.O.M., Shakil, N.A., Rana, V.S. et al. (2017). Antifungal activity of nano emulsions of neem and citronella oils against phytopathogenic fungi, *Rhizoctonia solani* and *Sclerotium rolfsii*. *Ind. Crop Prod.* 108: 379–387.

Amini, J., Farhang, V., Javadi, T., and Nazemi, J. (2016). Antifungal effect of plant essential oils on controlling Phytophthora species. *Plant Pathol. J.* 32 (1): 16.

Arslan, M. and Dervis, S. (2010). Antifungal activity of essential oils against three vegetative-compatibility groups of *Verticillium dahliae*. *World J. Microbiol. Biotechnol.* 26 (10): 1813–1821.

Assadpour, E. and Jafari, S.M. (2019). An overview of lipid-based nanostructures for encapsulation of food ingredients. In: *Lipid-Based Nanostructures for Food Encapsulation Purposes*, vol. 2 (ed. S. Jafari), 1–34. Elsevier.

Auffan, M., Rose, J., Bottero, J.Y. et al. (2009). Towards a definition of inorganic nanoparticles from an environmental, health and safety perspective. *Nat. Nanotechnol.* 4 (10): 634–641.

Badri, D.V., Weir, T.L., van der Lelie, D., and Vivanco, J.M. (2009). Rhizosphere chemical dialogues: plant–microbe interactions. *Curr. Opin. Biotechnol.* 20 (6): 642–650.

Baker, K.F. and Cook, R.J. (1974). *Biological Control of Plant Pathogens*, 433. WH Freeman and Company.

Bakkali, F., Averbeck, S., Averbeck, D., and Idaomar, M. (2008). Biological effects of essential oils – a review. *Food Chem. Toxicol.* 46 (2): 446–475.

Barros, A.F., Campos, V.P., de Paula, L.L. et al. (2019). Nematicidal screening of essential oils and potent toxicity of *Dysphania ambrosioides* essential oil against *Meloidogyne incognita* in vitro and in vivo. *J. Phytopathol.* 167 (7–8): 380–389.

Baysal-Gurel, F., Liyanapathiranage, P., and Mullican, J. (2018). Biofumigation: opportunities and challenges for control of soilborne diseases in nursery production. *Plant Health Prog.* 19 (4): 332–337.

Blanco-Padilla, A., Soto, K.M., Hernández Iturriaga, M., and Mendoza, S. (2014). Food antimicrobials nanocarriers. *Sci. World J.* 2014. https://doi.org/10.1155/2014/837215.

Bottalico, A. and Perrone, G. (2002). Toxigenic *Fusarium* species and mycotoxins associated with head blight in small-grain cereals in Europe. In: *Mycotoxins in Plant Disease* (ed. A. Logrieco, J.A. Bailey, L. Corazza, and B.M. Cooke), 611–624. Dordrecht: Springer. https://doi.org/10.1007/978-94-010-0001-7_2.

Bumunang, E.W., Babalola, O.O., and Barros, E. (2013). Bacterial community profiling in the rhizosphere of field grown GM and non-GM maize. *J. Pure Appl. Microbiol.* 7 (1): 157–164.

Chen, Q., Xu, S., Wu, T. et al. (2014). Effect of citronella essential oil on the inhibition of postharvest *Alternaria alternata* in cherry tomato. *J. Sci. Food Agric.* 94 (12): 2441–2447.

Czajkowski, R., Perombelon, M.C., van Veen, J.A., and van der Wolf, J.M. (2011). Control of blackleg and tuber soft rot of potato caused by Pectobacterium and Dickeya species: a review. *Plant Pathol.* 60 (6): 999–1013.

De Rodriguez, D.J., Hernández-Castillo, D., Angulo-Sánchez, J.L. et al. (2007). Antifungal activity in vitro of Flourensia spp. extracts on *Alternaria sp., Rhizoctonia solani*, and *Fusarium oxysporum*. *Ind. Crop Prod.* 25 (2): 111–116.

Delgado-Baquerizo, M., Guerra, C.A., Cano-Díaz, C. et al. (2020). The proportion of soil-borne pathogens increases with warming at the global scale. *Nat. Clim. Change* 10 (6): 550–554.

Donkin, S.G., Eiteman, M.A., and Williams, P.L. (1995). Toxicity of glucosinolates and their enzymatic decomposition products to *Caenorhabditis elegans*. *J. Nematol.* 27 (3): 258.

El-Mohamedy, R.S., Jabnoun-Khiareddine, H., and Daami-Remadi, M. (2014). Control of root rot diseases of tomato plants caused by *Fusarium solani, Rhizoctonia solani* and *Sclerotium rolfsii* using different chemical plant resistance inducers. *Tunisian J. Plant Prot.* 9 (1): 56.

El-Zemity, S.R., Radwan, M.A., El-Monam Mohamed, S.A., and Sherby, S.M. (2008). Antibacterial screening of some essential oils, monoterpenoids and novel N-methyl carbamates based on monoterpenoids against *Agrobacterium tumefaciens* and *Erwinia carotovora*. *Arch. Phytopathol. Plant Prot.* 41 (6): 451–461.

Faria, J.M.S., Sena, I., Ribeiro, B. et al. (2016). First report on Meloidogyne chitwoodi hatching inhibition activity of essential oils and essential oils fractions. *J. Pest Sci.* 89 (1): 207–217.

Feng, W. and Zheng, X. (2007). Essential oils to control *Alternaria alternata* in vitro and in vivo. *Food Control* 18 (9): 1126–1130.

George, O., Edward, G.M., Viviene, M., and Moses, N. (2021). Efficacy of thymol and eugenol against bacterial wilt bacterium *Ralstonia solanacearum*. *Afr. J. Biotechnol.* 20 (6): 256–265.

Gohlke, J. and Deeken, R. (2014). Plant responses to *Agrobacterium tumefaciens* and crown gall development. *Front. Plant Sci.* 5: 155.

Grillo, R., dos Santos, N.Z.P., Maruyama, C.R. et al. (2012). Poly (ϵ-caprolactone) nanocapsules as carrier systems for herbicides: physico-chemical characterization and genotoxicity evaluation. *J. Hazard. Mater.* 231: 1–9.

Hassanin, M.M.H., Abd-El-Sayed, M.A., and Abdallah, M.A. (2017). Antifungal activity of some essential oil emulsions and nanoemulsions against *Fusarium oxysporum* pathogen affecting cumin and geranium plants. *Sci. J. Flowers Ornamental Plants* 4 (3): 245–258.

Hirsch, A.M., Bauer, W.D., Bird, D.M. et al. (2003). Molecular signals and receptors: controlling rhizosphere interactions between plants and other organisms. *Ecology* 84 (4): 858–868.

Hsouna, A.B., Touj, N., Hammami, I. et al. (2019). Chemical composition and in vivo efficacy of the essential oil of *Mentha piperita* L. in the suppression of crown gall disease on tomato plants. *J. Oleo Sci.* 68 (5): 419–426. https://doi.org/10.5650/jos.ess18261.

Izadi, M., Jorf, S.A.M., Nikkhah, M., and Moradi, S. (2021). Antifungal activity of hydrocolloid nano encapsulated *Carum copticum* essential oil and *Peganum harmala* extract on the pathogenic fungi *Alternaria alternata*. *Physiol. Mol. Plant Pathol.* 116: 101714.

Kadam, B.P., Chavhan, R.L., Chakrabarty, P.K., and Patil, F.S. (2009). Characterization of variability in some pathogenic species of Alternaria based on the nucleotide sequence of ribosomal DNA. *J. Plant Biochem. Biotechnol.* 18 (1): 59–64.

Kah, M., Beulke, S., Tiede, K., and Hofmann, T. (2013). Nanopesticides: state of knowledge, environmental fate, and exposure modeling. *Crit. Rev. Environ. Sci. Technol.* 43 (16): 1823–1867.

Kalhoro, M.T., Zhang, H., Kalhoro, G.M. et al. (2022). Fungicidal properties of ginger (Zingiber officinale) essential oils against Phytophthora colocasiae. *Sci. Rep.* 12 (1): 1–10.

Kishore, N. and Dwivedi, R.S. (1991). Fungitoxicity of the essential oil of *Tagetes erecta* L. against *Pythium aphanidermatum* Fitz. The damping-off pathogen. *Flavour Fragr. J.* 6 (4): 291–294.

Kohl, J., Kolnaar, R., and Ravensberg, W.J. (2019). Mode of action of microbial biological control agents against plant diseases: relevance beyond efficacy. *Front. Plant Sci.* 10: 845. https://doi.org/10.3389/fpls.2019.00845.

Lambert, R.J.W., Skandamis, P.N., Coote, P.J., and Nychas, G.J. (2001). A study of the minimum inhibitory concentration and mode of action of oregano essential oil, thymol and carvacrol. *J. Appl. Microbiol.* 91 (3): 453–462.

Laquale, S., Avato, P., Argentieri, M.P. et al. (2018). Nematotoxic activity of essential oils from Monarda species. *J. Pest Sci.* 91 (3): 1115–1125.

Larkin, R.P. (2015). Soil health paradigms and implications for disease management. *Annu. Rev. Phytopathol.* 53 (1): 199–221.

Lee, S.O., Choi, G.J., Jang, K.S. et al. (2007). Antifungal activity of five plant essential oils as fumigant against postharvest and soilborne plant pathogenic fungi. *Plant Pathol. J.* 23 (2): 97–102.

Lee, J.E., Jung, M., Lee, S.C. et al. (2020). Antibacterial mode of action of trans-cinnamaldehyde derived from cinnamon bark (*Cinnamomum verum*) essential oil against *Agrobacterium tumefaciens*. *Pestic. Biochem. Physiol.* 165: 104546.

Liao, W., Badri, W., Dumas, E. et al. (2021). Nanoencapsulation of essential oils as natural food antimicrobial agents: an overview. *Appl. Sci.* 11 (13): 5778.

Mansfield, J., Genin, S., Magori, S. et al. (2012). Top 10 plant pathogenic bacteria in molecular plant pathology. *Mol. Plant Pathol.* 13 (6): 614–629.

Mavrodi, D.V., Yang, M., Mavrodi, O.V., and Wen, S. (2017). Management of soilborne plant pathogens with beneficial root-colonizing *Pseudomonas*. *Adv. PGPR Res.* 147: 147–164.

Mukherjee, A. and SinhaBabu, S.P. (2014). Potential of citral and menthol for suppression of *Meloidogyne incognita* infection of okra plants. *J. Essent. Oil Bear. Plants* 17 (3): 359–365.

Nana, W.L., Eke, P., Fokom, R. et al. (2015). Antimicrobial activity of *Syzygium aromaticum* and *Zanthoxylum xanthoxyloides* essential oils against *Phytophthora megakarya*. *J. Phytopathol.* 163 (7–8): 632–641.

Naveenkumar, R., Muthukumar, A., and Mohanapriya, R. (2017). Occurrence, virulence and evaluation of essential oils against *Fusarium oxysporum* f. sp. niveum causing Wilt of Watermelon. *Vegetos* 30 (4): 4473.

Nguyen, M.H., Tran, T.N.M., and Vu, N.B.D. (2022). Antifungal activity of essential oil-encapsulated lipid nanoemulsions formulations against leaf spot disease on tomato caused by *Alternaria alternata*. *Arch. Phytopathol. Plant Prot.* 55 (2): 235–257.

Nuruzzaman, M.D., Rahman, M.M., Liu, Y., and Naidu, R. (2016). Nanoencapsulation, nano-guard for pesticides: a new window for safe application. *J. Agric. Food Chem.* 64 (7): 1447–1483.

Obongoya, B.O., Wagai, S.O., and Odhiambo, G. (2010). Phytotoxic effect of selected crude plant extracts on soil-borne fungi of common bean. *Afr. Crop Sci. J.* 18 (1): 15–22.

Oerke, E.C. (2006). Crop losses to pests. *J. Agric. Sci.* 144 (1): 31–43.

Ojaghian, M.R., Sun, X., Zhang, L. et al. (2015). Effect of E-cinnamaldehyde against *Sclerotinia sclerotiorum* on potato and induction of glutathione S-transferase genes. *Physiol. Mol. Plant Pathol.* 91: 66–71.

Oluoch, G., Matiru, V., Mamati, E.G., and Nyongesa, M. (2021). Nanoencapsulation of thymol and eugenol with chitosan nanoparticles and the effect against *Ralstonia solanacearum*. *Adv. Microbiol.* 11 (12): 723–739.

Ozdemir, E. and Gozel, U. (2018). Nematicidal activities of essential oils against *Meloidogyne incognita* on tomato plant. *Fresen. Environ. Bull.* 27 (6): 4511–4517.

Pawar, V.C. and Thaker, V.S. (2007). Evaluation of the anti-*Fusarium oxysporum* f. sp cicer and anti-*Alternaria porri* effects of some essential oils. *World J. Microbiol. Biotechnol.* 23 (8): 1099–1106.

Peng, Y., Li, S.J., Yan, J. et al. (2021). Research progress on phytopathogenic fungi and their role as biocontrol agents. *Front. Microbiol.* 12: 1209, 670135. https://doi.org/10.3389/fmicb.2021.670135.

Perez-Sanchez, R., Infante, F., Galvez, C., and Ubera, J.L. (2007). Fungitoxic activity against phytopathogenic fungi and the chemical composition of *Thymus zygis* essential oils. *Food Sci. Technol. Int.* 13 (5): 341–347.

Perina, F.J., Amaral, D.C., Fernandes, R.S. et al. (2015). *Thymus vulgaris* essential oil and thymol against *Alternaria alternata* (Fr.) Keissler: effects on growth, viability, early infection and cellular mode of action. *Pest Manag. Sci.* 71 (10): 1371–1378.

Pitarokili, D., Couladis, M., Petsikos-Panayotarou, N., and Tzakou, O. (2002). Composition and antifungal activity on soil-borne pathogens of the essential oil of *Salvia sclarea* from Greece. *J. Agric. Food Chem.* 50 (23): 6688–6691.

Pradhanang, P.M., Momol, M.T., Olson, S.M., and Jones, J.B. (2003). Effects of plant essential oils on *Ralstonia solanacearum* population density and bacterial wilt incidence in tomato. *Plant Dis.* 87 (4): 423–427.

Raaijmakers, J.M., Paulitz, T.C., Steinberg, C. et al. (2009). The rhizosphere: a playground and battlefield for soilborne pathogens and beneficial microorganisms. *Plant and Soil* 321 (1): 341–361.

Saroj, A., Chanotiya, C.S., Maurya, R. et al. (2019). Antifungal action of *Lippia alba* essential oil in *Rhizoctonia solani* disease management. *SN Appl. Sci.* 1 (10): 1–12.

Saroj, A., Oriyomi, O.V., Nayak, A.K., and Haider, S.Z. (2020). Phytochemicals of plant-derived essential oils: a novel green approach against pests. In: *Natural Remedies for Pest, Disease and Weed Control* (ed. C. Egbuna and B. Sawicka), 65–79. Academic Press.

Sattler, C., Kächele, H., and Verch, G. (2007). Assessing the intensity of pesticide use in agriculture. *Agr. Ecosyst. Environ.* 119 (3–4): 299–304.

Shabana, Y.M., Abdalla, M.E., Shahin, A.A. et al. (2017). Efficacy of plant extracts in controlling wheat leaf rust disease caused by *Puccinia triticina*. *Egypt. J. Basic Appl. Sci.* 4 (1): 67–73.

Shaw, R.K., Shaik, M., Prasad, M.S.L. et al. (2022). Genomic regions associated with resistance to Fusarium wilt in castor identified through linkage and association mapping approaches. *Genome* 65 (3): 123–136.

Strange, R.N. and Scott, P.R. (2005). Plant disease: a threat to global food security. *Annu. Rev. Phytopathol.* 43 (1): 83–116.

Svecova, E., Proietti, S., Caruso, C. et al. (2013). Antifungal activity of *Vitex agnus-castus* extract against *Pythium ultimum* in tomato. *Crop Prot.* 43: 223–230.

Tej, R., Rodríguez-Mallol, C., Rodríguez-Arcos, R. et al. (2018). Inhibitory effect of *Lycium europaeum* extracts on phytopathogenic soil-borne fungi and the reduction of late wilt in maize. *Eur. J. Plant Pathol.* 152 (1): 249–265.

Tripathi, S., Srivastava, P., Devi, R.S., and Bhadouria, R. (2020). Influence of synthetic fertilizers and pesticides on soil health and soil microbiology. In: *Agrochemicals Detection, Treatment and Remediation.* (ed. M.N.V. Prasad), 25–54. Butterworth-Heinemann.

Turek, C. and Stintzing, F.C. (2013). Stability of essential oils: a review. *Compr. Rev. Food Sci. Food Saf.* 12 (1): 40–53.

Wang, Y., Wei, K., Han, X. et al. (2019). The antifungal effect of garlic essential oil on Phytophthora nicotianae and the inhibitory component involved. *Biomolecules* 9 (10): 632.

Wang, B., Li, P., Yang, J. et al. (2022). Inhibition efficacy of *Tetradium glabrifolium* fruit essential oil against *Phytophthora capsici* and potential mechanism. *Ind. Crop Prod.* 176: 114310.

Wogiatzi, E., Gougoulias, N., Papachatzis, A. et al. (2009). Chemical composition and antimicrobial effects of Greek origanum species essential oil. *Biotechnol. Biotechnol. Equip.* 23 (3): 1322–1324.

Wu, D., Wang, L., Zhang, Y. et al. (2021). Emerging roles of pathogen-secreted host mimics in plant disease development. *Trends Parasitol.* 37 (12): 1082–1095.

Yang, T., Lupwayi, N., Marc, S.A. et al. (2021). Anthropogenic drivers of soil microbial communities and impacts on soil biological functions in agroecosystems. *Global Ecol. Conserv.* 27: e01521.

Zhao, Y., Wang, Q., Wu, X. et al. (2021). Unraveling the polypharmacology of a natural antifungal product, eugenol, against *Rhizoctonia solani*. *Pest Manag. Sci.* 77 (7): 3469–3483.

14

Restoration of Saline Soils for Sustainable Crop Production

Bülent OKUR[1], Nesrin ÖRÇEN[2], and Nur OKUR[1]

[1] *Faculty of Agriculture, Soil Science and Plant Nutrition Department, Ege University, 35100, Bornova, İzmir, Turkey*
[2] *Faculty of Agriculture, Field Crop Science, Ege University, 35100, Bornova, İzmir, Turkey*

14.1 Introduction

According to the current information from 118 countries covering 85% of global land area, more than 424 million ha of topsoil (0–30 cm) and 833 million ha of subsoil (30–100 cm) are salt-affected (FAO 2020). Meanwhile, degraded soils stretch up to 6 billion ha worldwide (Ayub et al. 2020). The countries affected by salinization are in the arid and semi-arid regions of North Africa, the Near East, the former Soviet Union, and Central Asia. They are also widespread in India, Iran, Iraq, Australia, and the Americas. Arid and semi-arid areas cover approximately 46% of the terrestrial area in the world. In these climatic regions, approximately 50% of the irrigated areas have salinity problems at different levels. In the next 75 years, it is estimated that agricultural land will increase by only 10%, while the world population will double, and most of this increase will be in semi-arid and arid regions of the world where salinity is very common. In response to the increasing population in the world, the area of agricultural lands is decreasing every year due to urbanization and industrialization. In addition, the increase in irrigated areas also increases the risk of salinization of soils in arid and semi-arid areas. Salinization develops without feeling the salinity in over-irrigated lands and is understood by the decrease in plant yield after certain stages. For this reason, the World Food and Agriculture Organization (FAO) has defined salinization as the "Silent Enemy."

There are three main sources of salts in soils. These are the salts produced by the disintegration of the minerals in the soil, the salts transferred from the atmosphere to the soil by precipitation, and the salts entering the soil from the ancient fossil-based salt deposits in salty seas, lakes, and similar places. Soil salinity problems are further complicated where the groundwater is highly saline, especially near coastal region. The coastal region is likely to face severe challenges in the future due to rise in sea level resulting from global warming. Irrigation in arid and semi-arid regions also causes salinization. With irrigation, the salt in the lower layers of the soil is carried up by capillarity during evaporation and accumulates in the root zone of the plant. Excessive irrigation, use of poor-quality irrigation

Agroecological Approaches for Sustainable Soil Management, First Edition.
Edited by Majeti Narasimha Vara Prasad and Chitranjan Kumar.

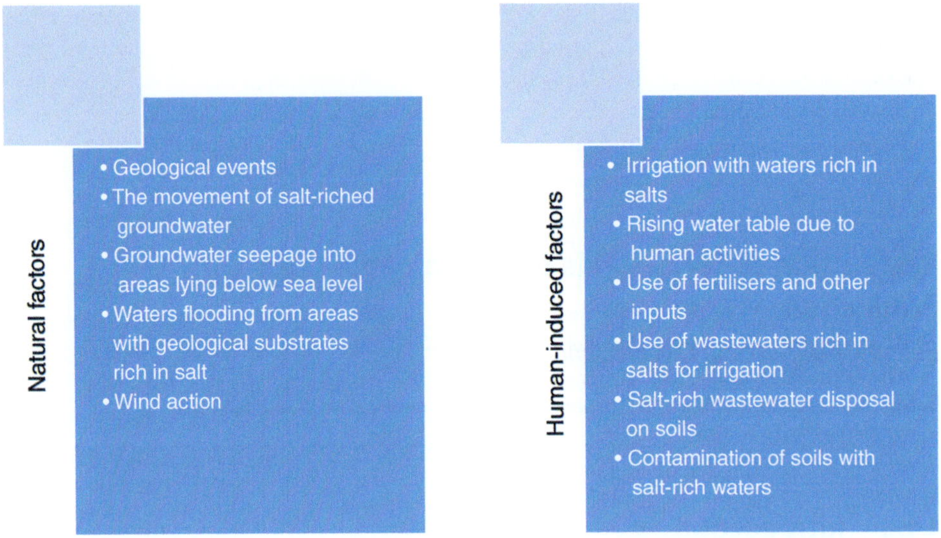

Figure 14.1 Salinization factors.

water, inadequate drainage systems, accumulation of soluble salts in soil water and especially in the upper soil layers in contact with the sun are among the other causes of salinization (Epstein et al. 1980). Natural or anthropogenic factors leading to salinity in soil are given in Figure 14.1.

The most influential human-induced factors are land use, farming systems, land management, and land degradation. Inappropriate irrigation practices (such as the use of salt-rich irrigation water) and insufficient drainage both cause salinization. Due to the close relationship between climate and agriculture, climate change also has an impact on the salinization of soils. Both precipitation and temperature will change as a result of climate change and affect agriculture soils. Coastal agriculture will also face added impacts from sea-level rise, inundation, seawater intrusion, rising salinity, storm surges, tropical cyclones, and flooding (Gopalakrishnan et al. 2019). Sustainable management of natural resources in coastal and the other salt effected regions is, therefore, of utmost importance to ensure the stability of this system of production. Today, the improvement of salt-affected soils with traditional reclamation methods has lost its currency due to its negative effects on both the economy and the environment. In this section, we review new and innovative ways of restoring salt-affected soils alongside classical methods for agricultural sustainability and sustainable food production.

14.2 Characteristics of Saline Soils

Saline soils are soils that have a saline horizon within 125 cm of the surface (125 cm in coarse textured, 90 cm in medium textured and 75 cm in fine textured soils) or have an electrical conductivity of more than 2 dS m^{-1} in a 25 cm layer from the surface. Salt-affected soils include soils containing soluble salts or their ions in at least one of their horizons in quantities that are above the threshold of toxicity–the maximal permissible concentration of salts that does not suppress plant growth. Saline soils have an electrical conductivity of the saturation soil extract (ECe) of more than 4 dS m^{-1} at 25 °C. This value is generally used the world over although the terminology committee of the Soil Science Society of America has lowered the boundary between saline and non-saline soils to 2 dS m^{-1}in the saturation extract (FAO 2018). Soluble salts most commonly present are the chlorides and sulphates of sodium, calcium, and magnesium. The most dominants ions are sodium and chloride. Calcium and magnesium are generally in sufficient concentrations for plant need. Soluble carbonates are always absent. The pH value of the saturated soil paste is always less than 8.2 and more often near neutrality depending on the ions present in the composition of the salt. Soils affected by salts could be categorized as saline, sodic, and saline-sodic soils (Table 14.1).

Although saline soils are the ones that contain soluble salts, mostly sodium, magnesium, chlorides, sulphates, etc. that affect the growth of plants, sodic soils specifically contain sodium salts (mainly Na_2CO_3) that are capable of alkaline hydrolysis. Chhabra (2004) proposed additional criteria for defining saline and alkali soils taking into account pH, EC, ESP, or SAR, nature of the soluble and insoluble salts (Table 14.2).

Salinity and sodicity affect flocculation, dispersion, infiltration, hydraulic conductivity, runoff, and seal formation of soil. The high salt concentration in the soil solution promotes

Table 14.1 Classification of salt-affected soils.

Soil class	ECe (dS m^{-1})	ESP[a] (%)	pH
Saline	≥4	<15	<8, 5
Saline-sodics	≥4	≥15	≥8, 5
Sodic	<4	≥15	>8, 5

[a] Exchangeable Sodium Percent.
Source: Adapted from USSL Staff (1954).

Table 14.2 Additional criteria for classification of salt-affected soils proposed.

Soil properties	Saline soils (salic horizon)	Alkali soils (natric horizon)
pH	<8.2	>8.2
ESP	<15 SAR < 13	>15; SAR > 13 >6 for vertisolsv
ECe	>4 dS m^{-1}	Variable, mostly < 4 dS m^{-1} Not to be considered for classification
Nature of soluble salts	Neutral mostly Cl^- and SO_4^{2-} of Na^+, Ca^{2+}, and Mg^{2+}. HCO_3^- may be present but mostly CO_3^{2-} are absent.	Capable of alkaline hydrolysis, preponderance of HCO_3^-,–CO_3^{2-},–, and silicates of Na^+
Nature of sparingly soluble salts	May contain gypsum	Always contain free calcium carbonate

Source: Chabbra (2004) / Permission of Taylor & Francis.

flocculation of the clay particles. As a result of this formation, the spaces between soil aggregates are wider than disperse soils, the soil has a higher permeability value and the volume of water-filled pores in the soil increases. This increase in aggregation benefits root development and root penetration rather than aeration (Öztürk 2004). Flocculation generally increases when soil salinity exceeds approximately 1.5dS m^{-1} or when the salinity value of irrigation water exceeds 0.5 dS m^{-1}. The proportionally high salt concentration in the soil solution pushes the adsorbed cations to the surface of the soil particles and increases the uptake of cations by the soil particles (Bauder and Brock 2001). While the increase in salinity in the soil solution has a positive effect on soil aggregation or stabilization, it negatively affects plant production. The decrease in plant yield varies according to the amount of salt and plant variety.

When evaluating the salinity in the soil, it is necessary to consider the soil texture. Clay soils have higher water holding capacity and slow drainage than other soils. On the other hand, sandy soils hold less water and drain faster. For this reason, sandy soils can resist irrigation with higher salinity waters. Since sandy soils have higher leaching fraction (LF) than fine-textured soils, salts can be washed down to the root zone in these soils. Clay soils can be easily dispersed because they have the capacity to bind excessive amounts of Na^+ ions. For this reason, many salts, including Na^+ ions, accumulate in clay soils with relatively low LF and high surface area values, as opposed to sandy soils. The permeability reducing effect of Na^+ ion is greater in soils dominated by montmorillonite group clays than in soils dominated by illite and kaolinite group clays. The high Na^+ ion concentration in the soil also affects the infiltration of water. Clay particles dispersed due to sodium can clog soil pores. The clogged pores and the resulting structure limit the growth of plants, prevent water infiltration and water flow. This change in soil hydraulic properties is important in two respects; (i) The decrease in the amount of water infiltrating the soil, and (ii) The surface flow losses and the increase in erosion accordingly (Warrence et al. 2002). The dispersion of soil particles often results in crust formation. Crust formation is a common in arid and semiarid region soils under the influence of salt and sodium.

14.3 Impact of Soil Salinization on Plant Growth

The impact of salinity on plant growth is a result of complex interactions between the physiological, morphological, and biological processes of plant growth, seed germination, and nutrient and water uptake (Akbarimoghaddam et al. 2011; Singh and Chatrath 2001). Salt accumulation inhibits plant growth and reduces the ability to uptake water and nutrients, leading to osmotic or water-deficit stress. Salt is also causing injury of the young photosynthetic leaves and acceleration of their senescence, as the Na^+ cation is toxic when accumulating in cell cytosol resulting in ionic imbalance and toxicity of transpiring leaves (Hanin et al. 2016). Salinity composes stress by damaging ionic and osmotic balances in plants. Osmotic stress caused by increasing the amount of salt in soil, decreases the amount of water that plant use and as a result physiological drought occurs. After these conditions, ionic stress occurs in the plant with deterioration of plant ion balance. Na and Cl ions, which increases in medium with ionic stress, get in competition with essential nutrients such as K^+, Ca^{2+}, and Mg^{2+} lead to nutrient deficiency in plant. While the direct effect of salinity is osmotic and ionic stresses, deteriorations in structure and synthesis of toxic components composes secondary effect (Botella et al. 2007). The plants growing in saline soils show some specific features, such as stunted growth, smaller leaves with dark blue-green coloration and increased succulence (Talat 2020).

As the amount of salt in the soil increases, significant decreases in plant yield occur. For most species, a 10% yield decrease is observed as the salinity increases over the 4–8 dS m^{-1} range. On the other hand, the growth of many halophytic species is stimulated within a salinity range of 15–25 dS m^{-1} (Rozema and Schat 2013). Plants show different abilities and coping mechanisms against salinity, depending on their variety. There are differences in salt tolerance among families, genera, and species, even among genotypes of the same species. The majority of the crop and forage species used in modern agriculture are salt sensitive (glycophytes). The effect of the amount of salt in the soil on crop growth and relative loss in yield of some crop plants due to the increase in salinity are given in Tables 14.3 and 14.4, respectively.

Although chlorine, which is the most abundant ion besides sodium in saline soils, is a vital element for all plants, sodium ion is only important for halophyte plants and some C4 plants (Tang et al. 2015). Plants usually contain 0.01–10% sodium under normal conditions. At salinity concentrations that will cause salt stress, there are sodium and chlorine ions far above the amount needed by plants. Sodium is an element that can move in both phloem and xylem in plants (Elmacı and Yagmur 2021). In salt stress, the ion balance in the root region of the plant is disturbed and the increased sodium intake competes with the uptake of other mineral substances, causing nutritional deficiency. Ion imbalance and disruption of cell membrane permeability in the roots affect the plant's nutritional regime, preventing the uptake of some basic elements used in metabolic events, and thus physiological problems occur in the plant (Villora et al. 1997). In the presence of high sodium ions, the potassium uptake of the plant decreases (Wakeel 2013). The reaction of plants to

Table 14.3 Soil salinity classes and crop growth.

Soil salinity class	Conductivity of the saturation extract (dS m^{-1})	Plant yield
Non saline	0–2	There are no visible effects of salts on the growth of crops. No yield loss
Slightly saline	2–4	Yields of very sensitive crops may be restricted. Cereals are generally unaffected.
Moderately saline	4–8	Yields of many crops may be restricted. Wheat yields may be reduced by 30%.
Strongly saline	8–16	Only tolerant crops yield satisfactorily. Wheat yields may be reduced by 60%.
Very strongly saline	>16	Only a few very tolerant crops yield satisfactorily. Wheat yields may be reduced by 80–100%.

Note: Yield loss estimates are based on recent research and only apply to the saline soils, not to the entire delineated area.
Source: Adapted from Miyamoto et al. (2004).

Table 14.4 Relative loss in yield of some crop plants due to the increase in salinity.

Plant species	ECe ($dS\,m^{-1}$) (Thereshold value)	Relative loss in yield (%) (Loss per $dS\,m^{-1}$)
Phaseolus vulgaris L. (Beans)	1.0	19.0
Solanum melongena L. (Eggplant)	1.1	6.9
Allium cepa L. (Onion)	1.2	16.0
Capsicum annuum L. (Pepper)	1.5	14.0
Zea mays L. (Maize)	1.7	12.0
Solanum tuberosum L. (Potato)	1.7	12.0
Saccharum officinarum L. (Sugarcane)	1.7	5.9
Brassica oleracea L. (Cabbage)	1.8	9.7
Lycopersicon esculentum Mill. (Tomato)	2.5	9.9
Oryza sativa L. (Rice)	3.0	12.0
Arachis hypogaea L. (Peanut)	3.2	29.0
Soja hispida Moench. (Soya bean)	5.0	20.0
Triticum sp. L. (Wheat)	6.0	7.1
Gossypium sp. L. (Cotton)	7.7	5.2
Hordeum vulgare L. (Barley)	8.0	5.0

Source: Chinnusamy (2005) / John Wiley & Sons.

salt may vary depending on the development period of the plant, the concentration of salt, the time the salt affects the plant, and the climate and soil characteristics.

Amongst plants there is a continuum of tolerance to Cl^- and Na^+ in their environment, from the very sensitive (e.g., chickpea; Flowers et al. 2010) to the tolerant (e.g., some *Tecticornia* species; English and Colmer 2013). Most of the plant families contain a few salt-tolerant species (Halophytes), about 100 of which were identified by Aronson (1989). Crops differ in their abilities to deal with soil, climate, and saline conditions. This differential ability is not a fixed characteristic of a particular species. Rather, it changes depending on the stage of the plant growth, climatic conditions, and different varieties within the same species (Talat 2020). Plants have developed various biochemical and molecular mechanisms against osmotic and ion stress caused by salinity. These biochemical mechanisms are selective accumulation or removal of ions, control of ion uptake from the root and transmission to the shoot, accumulation of ions in certain parts in the whole plant and cells, and formation of antioxidant systems by synthesis of osmotic regulators (Parida and Das 2005). Thresholds for both soil salinity and irrigation water salinity for some fruit, nut crops, vegetables, herbaceous crops, field grasses, forages, and field crops are given in Table 14.5.

Table 14.5 Salinity threshold values of some fruit, nut crops, vegetables, herbaceous crops, field grasses, forages, and field crops.

	Common name	Botanical name	Salinity threshold, EC (dS m^{-1})	
			Irrigation water	Soil
Very sensitive	Green bean	*Phaseolus vulgaris*	1	1
	Kidney bean	*Phaseolus vulgaris*	1	0–1
	Broad bean	*Vicia faba*	1	1–2
	Cabbage	*Brassica oleracea var. capitata*	1	1–2
	Sweet corn	*Zea mays*	1–2	1–2
	Eggplant	*Solanum melongena var. esculentum*	1	1–2
	Kale	*Brassica oleracea var. acephala*	1	1–2
	Lettuce	*Lactuca sativa*	1–2	1–2
	Onion	*Allium cepa*	0–1	1–2
	Sweet potato	*Ipomoea batatas*	1	1–2
	Pumpkin	*Cucurbita pepo var. pepo*	1	1–2
	Radish	*Raphanus sativus*	0–1	1–2
	Cherry tomato	*Solanum lycopersicum var. cerasiforme*	1	1–2
	Turnip	*Brassica rapa*	1	0–1
	Almond	*Pranus dulcis*	1–2	1–2
	Apple and crabapple	*Malus species and cultivars*	1–2	1–2
	Apricot	*Prunus armeniaca*	1–2	1–2
	Lemon	*Citrus limon*	1–2	1–2
	Orange	*Citrsus sinensis*	1	1–2
	Pear	*Pyrus communis*	1–2	1–2
	Plum	*Prunus domestica*	1	1–2
	Strawberry	*Fragaria sp.*	1	1–2
	Redclover	*Trifolium pratense*	1	1–2
	Cowpea	*Vigna unguiculata*	1	0–1
	Lupine	*Lupinus hardwegii*	1	0–1
	Sugarcane	*Saccharum spp.*	1	1–2

(Continued)

Table 14.5 (Continued)

			Salinity threshold, EC (dS m^{-1})	
	Common name	**Botanical name**	**Irrigation water**	**Soil**
Sensitive	Broccoli	*Brassica oleracea var. botrytis*	1–2	2–3
	Canola	*Brassica napus*	1–1.5	2–3
	Cucumber	*Cucumis sativus*	1–2	2–3
	Garlic	*Allium sativum*	1.5–2	3–4
	Muskmelon	*Cucumis melo*	1–2	2–4
	Pea	*Pisum sativum*	1.5–2	3–4
	Spinach	*Spinaca oleracea*	1–2	2–3
	Tomato	*Solanum lycopersicum var. lycoprsicum*	1–2	2–3
	Watermelon	*Citrullus lanatus*	1	2
	Cherry	*Prunus avium*	1.5	<3
	Fig	*Ficus carica*	1–2	2–3
	Alfalfa	*Medicago sativa*	1–2	2–3
	Oat	*Avena sativa*	1–2	2–4
	Rice (paddy)	*Oryza sativa*	1–1.5	2–3
	Tobacco	*Nicotiana tabacum*	1–2	2–4
	Durum wheat	*Triticum durum*	1–1.5	2–3
Moderately tol	Asparagus	*Asparagus officinalis*	2–3	4–5
	Sugar beet	*Beta vulgaris*	2–3	4–6
	Date palm	*Phoenix dactylifera*	1–2	4–6
	Parsley	*Petroselinium crispum*	2–3	4–6
	Soyabean	*Gycline max*	2–3	4–6
	Sunflower	*Helianthus annuus*	2–3	4–6
	Wheat	*Triticum aestivum*	4	4–5
Tolerant	Jojoba	*Simmondsia chinensis*	5–6	6–8
	European olive	*Olea europaea*	3–4	6–8
	Barley	*Hordeum vulgare*	5	8
	Coastal bermudagrass	*Cynodon dactylon*	4.6	6–8
	Common bermudagrass	*Cynodon dactylon*	4.6	6–8
	Castor bean	*Ricinus communis*	3–3.5	6–7
	Cotton	*Gossypium hirsutum*	3–5	6–10
	Red fescue	*Festuca rubra*	3–4	6–8
	Rye (Cereal)	*Secale cereale*	3–4	6–8
	Sorghum	*Sorghum bicolor*	3–5	6–10
	Triticale	*xTriticosecale*	3–3.5	6–7

Table 14.5 (Continued)

	Common name	Botanical name	Salinity threshold, EC (dS m^{-1}) Irrigation water	Soil
Very tolerant	Fults alkaligrass	*Puccinellia distans*	5	>10
	Gulf cordgrass	*Spartina spartinae*	5	>10
	Alkali muhyl	*Muhlenbergia asperifolia*	5	>10
	Alkali sacaton	*Sporobolus airoides*	5	>10
	Desert /inland saltgrass	*Distichlis spicata*	5	>10
	Tall wheatgrass	*Thinopyrum ponticum*	5	>10
	Western wheatgrass	*Agropyron smithii*	5	>10

Source: Adapted from McFarland et al. (2014).

14.4 Restoration of Saline Soils

The main purpose in the restoration of salt-affected soils is to reduce the amount of soluble salt and exchangeable sodium in the soil to levels that will not damage plant growth. Restoration of saline soils can be realized using the different methods. Selection of an appropriate site-specific restoration method depends on a number of soil properties and other considerations such as the depth of soil to be restored, presence of compacted layer(s) in the subsoil, the type and content of salts present, the quality and quantity of water available for leaching, the quality and depth of groundwater, the topographic features of land, the nature of the crop(s) to be grown after amelioration, the climatic conditions and the time available for amelioration (Qadir et al. 2000). For a successful restoration, there must be an adequate drainage system (natural or artificial) in the field. Otherwise, the soil restoration will not be sustainable.

A number of methods have been used for the restoration of saline soils. These methods are leaching of excess salt with good quality water, surface flushing of salts, physical remediation (land leveling, deep plowing, subsoiling, sanding, and profile inversion), electroreclamation (treatment with electric current), growing halophytes, and salt-tolerant plants which have economic value, and inoculation of microorganisms which have capacity to tolerate and improve plant salinity tolerance (Qadir et al. 2000; Raychev et al. 2001; Cho et al. 2009; Lakhdar et al. 2009; Wang et al. 2018; Mokrani et al. 2020). Soil amendments such as elemental sulfur, gypsum, other calcium materials, and other soil amendments do not help reclaim saline soils, despite claims to the contrary. Instead, these materials add salts and compound the problem (Horneck et al. 2007).

14.4.1 Leaching of Excess Salt along Soil Profile

Leaching of salts from the topsoil to lower depths down below the root zone is a prevalent method of saline soil restoration. But it is applicable to those areas where sufficient and quality water is available to enforce a net downward movement of soil solution through the root zone (Qadir et al. 2000). The success rate of leaching method was very low and effective in only small areas due to the high cost, high quality water requirement for washing and inability to remove drainage water (Hanay et al. 2004). In heavy textured soils, water infiltration takes place over a long period of time. Moreover, a significant part of the applied water evaporates, and the leaching efficiency decreases. Therefore, leaching processes during plant growth can cause problems in terms of water supply. Another important issue is the correct determination of the amount of leaching water and the restoration time required for improvement. A lot of experimental and theoretical studies have been made to determine a functional formula in accordance with the laws of water movement in the soil to calculate the amount of leaching water (Bahçeci et al. 2021). In the field unsaturated conditions during leaching were obtained by adopting intermittent ponding or by intermittent sprinkling at rates less than the infiltration rate of the soil. A number of leaching methods have been used for the restoration of saline soils. These include continuous ponding, intermittent ponding and sprinkling. Coarser textured (sandy) soils require less management to leach salts because infiltration rates are high, and more water can be applied over a shorter period of time. Fine-textured soils (clays) require more management to move salts out of the soil because the infiltration rate is lower and less water can be added at one time, potentially creating problems with ponding and runoff (Horneck et al. 2007).

14.4.2 Surface Flushing of Salts

Some saline soils may contain visible crusts of salts at the surface, a shallow water table or a highly impermeable layer near to the surface. Under these conditions, surface flushing method can be used. Water is passed over the soil surface and wasted the runoff water collected at a sloppy end of the field. Because this improvement method is not economical and restoration efficiency is low, physical methods are recommended for the removal of salts on the soil surface (Qadir et al. 2000).

14.4.3 Physical Remediation

Mechanical treatments of the soil, commonly adopted to overcome the problems of soil alkalinity are deep plowing, sub-soiling, sanding, and profile inversion. The purpose of the first three treatments is to increase soil permeability directly by mixing fine and coarse textured layers and to obtain a more uniform soil (deep plowing), by breaking up impermeable layers (subsoiling), and by incorporating sand into a fine textured soil (sanding) ultimately for enhancing the infiltration and transportation of water-soluble salts in the deep soil layer.

Proper land leveling allows for a more uniform water application for better leaching and salinity control. Since land leveling with heavy machinery causes significant soil compaction, this process should be followed by mechanical applications such as deep plowing, subsoil tillage, sanding, and profile inversion to increase soil infiltration and permeability. Tillage can improve soil permeability by breaking the surface crusts. However, if it is not done well, it can cause the plow layer or salt layer to form closer to the surface. In this application, heavy machinery should be avoided, and wet soil conditions should be avoided. Deep tillage is the type of tillage that provides the most beneficial effect on soils with impermeable layers lying between permeable layers. With deep plowing, the soils affected by sodium are brought under the soils containing gypsum. While this process provides soluble calcium, it also increases root growth and water movement. Subsoil tillage provides an increase in soil permeability through gaps or cracks formed as a result of the breaking of the hard layer in the soil. If the hard layer is formed at a depth of 25–30 cm from the soil surface, it can be broken with a chisel if it is at a depth of 30–50 cm with a bottom digger if it is at a depth of 50–90 cm with a ripper. A fine-textured surface soil can be made more permeable by mixing sand. A successful sanding improves penetration and provides better air and water permeability under saline conditions. The addition of approximately 10–20 cm of sand to the soil surface can provide a better environment for plant growth. Profile inversion can be applied if the surface soil has good physical properties, but the underlying soil has poor physical properties (Sönmez and Beyazgül 2008).

14.4.4 Electro-Kinetic Remediation

Electro-kinetic (EK) processing of soil involves the application of a low-density direct current through a wet-soil mass, which results in the development of electrical, hydraulic, and chemical gradients. Thus, an electric field is created by inserting electrodes into the contaminated site and passing the current through it, which makes the contaminant particles mobile in the soil media (Acar and Alshawabkeh 1993). The main removal mechanisms of contaminants in the EK process are electromigration and electroosmosis (Altaee et al. 2008). Many researchers suggested that the EK process is suitable for the removal of salts from the soil since the salts exist as charged ions in the soil (Jia et al. 2005; Jia et al. 2006; Manokararajah and Ranjan 2005). In addition, current flow is very efficient because saline soil has very high electrical conductivity (EC). Cho et al. (2009) effectively removed salts such as nitrate, chloride, potassium, and sodium salts from saline greenhouse soil by using electrokinetic remediation. Lageman and Pool (2011) concluded that electrokinetic remediation can be used to remove sodium and chloride from saline water, which thus allows water to meet the water-quality standards for irrigation and drinking purposes.

The use of electrokinetic remediation to remove salts in contaminated soils does have challenges, for example the formation of complexes with magnesium and calcium. These complexes have positive charges and move toward the cathode by electromigration in the direction opposite to that of chloride migration. In addition, some chlorides form a complex with sodium as dissolved NaCl. There is a need to gain more insight into electrokinetic applications in remediation of saline salts (Hamdan et al. 2014).

14.4.5 Salt-Tolerant Plants, Halophytes, and Organic Matter Applications

When good-quality water and/or adequate drainage are not available, another remediation way of saline soils may be to growth crops tolerant to saline soil conditions also known as a biological remediation method. Barley, sugar beet, cotton, millet, wheat, date palm, spinach, red beet, pumpkin, and asparagus can tolerate salinity levels of ECe 5–10 $dS\,m^{-1}$ (Table 14.5). Crop growth and harvest will aid in salt removal, improve water penetration, and supply organic matter. Therefore, vegetation management is an important part of reclamation (Horneck et al. 2007). Reclamation of salt affected soils with different degree of severity can be done through a plant assisted approach generically termed phytoremediation. It is thought that the use of halophyte plants is more effective in improving saline soils with the phytoremediation method (Nouri et al. 2017). Phytoremediation occurs through several mechanisms in salt-affected soils: (ii) Plant roots increase the dissolution rate of inherent or precipitated calcite ($CaCO_3$) or dolomite ($CaCO_3.MgCO_3$) as the source of Ca^{2+} and/or Mg^{2+}, thereby the enhanced levels of Ca^{2+} and/or Mg^{2+} in soil solution effectively replace Na^+ from the cation exchange complex, (ii) the generation of protons (H^+) released by roots of certain plant species enhances Na^+ uptake by plants and its subsequent removal from the field at harvest, and (ii) the pores formed by roots and root penetration improve drainage and salt leaching (Qadir et al. 2007).

Most of the plants cannot tolerate high salt concentrations of the soil and cannot be grown on a salt affected land. Such plants are known as non-salt tolerating plants, nonhalophytes, or glycophytes (Xiong and Zhu 2002). Some of the plants have the ability to grow under salinity due to the presence of different mechanisms in them for salt tolerance such plants are known as salt resisting plants, salt tolerating plants or halophytes (Flowers and Colmer 2008). The growth of halophytes is highly dependent on salt and complete their life cycle in the environment at a salt concentration of at least 200 mM NaCl (Shabala and Mackay 2011). The largest number of halophytes is included in *Chenopodiaceae* and it consists of about 550 halophyte species (Aronson 1989). Only a few plant species can survive only in saline conditions and not in low salt conditions (*Salicornia herbecea* and *Atriplex vericaria*). Another group of halophyte plants that are not obligate halophytes (such as *Aster atripalium* and *Plantago vericaria*) can continue to grow normally at low salt levels. Plants growing in saline conditions have a low growth rate and show a stunted structure. Halophytes respond to salt stress at three levels i.e., cellular, tissue, and whole plant level. Most of the halophytes utilize the basic mechanism of controlled accumulation and sequestration of inorganic ions which they utilize to adjust their internal osmotic balance to external salinity (Flowers and Yeo 1986). Salt-tolerant crops (cotton, maize, sorghum, *triangularis, Kostelezkya virginica*), salt-tolerant herbage (alfalfa, Caudan grass, *Amaranthus* spp.), and salt-tolerant trees (cv. Hongyechun, Chinese Ash, willow, *Tamarix chinensis*, thread jujube, *Lycium barbarum*, purple leaf cherry plum, hibiscus, *Prunus maritima*) can improve moderate and severe saline alkali soils in the Huang-Huai-Hai Plain of China (Li et al. 2006). After cultivation of forage grasses in an saline-alkali soil of the Yellow River Delta, desalination rate in decreasing order was: alfalfa (41.81%) > Sesbania (35.93%) > white sweet clover (30.82%) > chicory (27.52%) > sweet sorghum (Hou et al. 2014). The effect of *Leptochloa fusca* on the reclamation of saline-sodic soils was investigated for five years and it was determined that the initial EC values of soils at 0–20 cm, 40–60 cm and 80–100 cm depths decreased from 22.0to 2.0 $dS\,m^{-1}$, from 22.2to 2.1 $dS\,m^{-1}$, and from 12.5to 3.2 $dS\,m^{-1}$, respectively (Akhter 2003). In a 120-day growing period of *Sueda maritama* and *Sesuvium portulacastrum* halophyte plants in a saline soil, the soil EC value was decreased from 4.9to 1.4 $dS\,m^{-1}$ by *S. maritama* and from 4.9to 2.5 $dS\,m^{-1}$ by *S. portulacastrum* (Ravindran et al. 2007).

Many mechanisms that provide resistance to salinity also create tolerance to other abiotic stress factors. The salt tolerance feature ensures that the transgenic plants obtained by transferring through gene transfer are also resistant to drought and frost damage.

A high K/Na or Ca/Na ratio in plants increases salt tolerance. The high K/Na ratio indicates that the plant prefers the K^+ ion instead of the Na^+ ion and either takes it from outside or carries it from the old leaves to the young leaves. The fact that the plants prefer to take K^+ or Ca^{+2} ions instead of Na^+ ions in salty conditions and the high K/Na and Ca/Na ratios measured accordingly are reliable parameters that can be used in salt tolerant genotype selections (Maathuis and Amtmann 1999).

There are differences between plant species and even genotypes within the same species in terms of salt tolerance. Studies are carried out all over the world to develop plants that are tolerant to salt stress. Although there are differences in salt tolerance among genetic sources, genotypes that are truly salt tolerant are quite limited. For this reason, it is important to evaluate the current genetic potential and to determine tolerant genotypes from it. In particular, plants with hyperaccumulator mechanism should be identified. Hyperaccumulators are plants that can accumulate high amounts of pollutants in their above-ground tissues and can grow, thrive, and tolerate soils containing high amounts of pollutants. Different researchers have defined hyperaccumulators as ""plants that can accumulate pollutants in their above-ground tissues in much higher amounts than those found in soil or adjacent plants."" It is possible to define hyperaccumulators as a special and extreme case of a broader perspective category. They are extremely tolerant of the pollutants they accumulate on the shoots. The criteria for hyperaccumulators have not been defined in any empirical way. However, hyperaccumulators have been described by certain authors or research groups. To characterize plants for Ni concentrations >1000 mg/kg dry weight, the term "hyperaccumulator" was first used. Scientists have sought to define Ni hyperaccumulation more precisely because a Ni hyperaccumulator is a plant in which at least 1000 mg/kg Ni concentration has been reported in at least one sample of any aboveground tissue growing in dry matter in its natural habitat. Aboveground tissue can only be viewed as plant leaves for determination of hyperaccumulatory state. The phrase ""grow in their natural habitat"" suggests that excess pollutants accumulation must be accomplished by hyperaccumulators while maintaining adequate protection to maintain a self-sustaining population.

The application of dead organic matter (crops, stems, straw, green manure, barnyard manure, compost, sewage sludge) to saline soils is also another biological amelioration method. The addition of organic matter can accelerate the leaching of Na, decrease the ESP and electrical conductivity (EC), and increase water infiltration, water-holding capacity, and aggregate stability (Qadir et al. 2000). Amending saline soil with a chemically stable organic material having a high humification degree increase amelioration effects. Such material of high cation change capacity can adsorb part of soluble salts, decrease pH, and promote aggregation (Mahdy 2011). Not only the physical and chemical properties of the soil, but also its biological properties can be affected by organic material addition. It is determined a significant stimulation of soil respiration, urease, and alkaline phosphatase activity after incorporation of organic manure to a soil derived from alluvial and marine deposits with $3.3\,g\,kg^{-1}$ total salts (Liang et al. 2003).

14.4.6 Inoculation of Microorganisms

Soil microorganisms play an important role in the sustainability of all ecosystems with functions such as nutrient cycling, regulation of soil organic matter, and soil carbon sequestration, improvement of soil physical properties and water regimes, enhancing the amount and efficiency of nutrient acquisition by the vegetation and enhancing plant health, suppression of pathogen organisms and detoxification of harmful chemicals (Okur 2018). Soil microorganisms belong to various groups of fungi, archaea, bacteria, and protozoa. Some of these microorganisms are recognized for their capacity to tolerate and improve plant salinity tolerance (Egamberdieva et al. 2019). Several bacterial genera, like *Streptomyces, Azospirillum, Clostridium, Alcaligenes, Bacillus, Rhizobium, Pseudomonas, Thiobacillus, Serratia*, and *Klebsiella* are recognized as PGPR (Plant Growth Promotion Rhizobacter) working under saline conditions (Whipps 2001). PGPR are rhizosphereic/endophytic bacteria that colonises root interior or exterior. PGPR represent a possible solution to mitigate salinity-induced plant stress (Yao et al. 2010). The impact of five plant growth stimulating halotolerant bacteria on wheat growth revealed that bio-formulation of these halotolerant bacterial strains can improve salinity stress (80, 160, and 320 mM) in wheat seedlings, resulting in a 70% improvement of root length comparatively to negative controls (Ramadoss et al. 2013). The ability of PGPR to improve crop yields during salt stress includes many direct and indirect pathways such as ferrous iron minerals and inorganic phosphate solubilization, exopolysaccharides, and biofilms synthesis (Kasim et al. 2016), production of phytohormones (Naz et al. 2009), increased ACC (1-aminocyclo propane-1-carboxylate) deaminase activity (Bal et al. 2013), and nitrogen fixation (Sadiki and Rabih 2001).

The effects of PGPR on yield and growth of different plants/crops under salt stress are given in Table 14.6.

14.5 Conclusion

Soil salinization is an important soil degradation because salt- affected soils increase due to the climate change while the agricultural lands required for food production decreases. Sustainable agriculture management practices in salt-affected areas can be either preventive or curative. Since the reclamation, improvement, and management of salt-affected soils require complex and expensive technologies, first of all, it is necessary to take measures to prevent the salinization of the soils. Adequate soil and water conservation practices, based on a comprehensive soil or land degradation assessment, can provide an "early warning system" that provides possibilities for efficient salinity (or alkalinity) control, the prevention of these environmental stresses and their undesirable ecological, economic and social consequences. Determination of salt-tolerant plant species and varieties is one of the economical methods that can be applied in the improvement of salt-affected soils. With the breeding of salinity-resistant plants, agriculture can be made with these plants in areas with salinity problems. Also, the phytoremediation of a salt-affected land with halophytes can also provide an easy, economical, and permanent solution for the reclamation of saline soils. Recent studies showed that halo-tolerant PGPR play a vital role in the amelioration of physiological abnormalities induced by salts in plants. These microbial communities are also helpful to trigger the plant stress responsive genes and plants show enhanced growth, yield, and development under stress conditions. In recent years, while a method suitable for the region, climate and soil characteristics is selected for restoration of saline soils for

Table 14.6 Various roles of PGPR on yield and growth of different plants/crops under salt stress.

Plants/cropsc	PGPR	Beneficial effects	Reference
Maize	*Pseudomonas putida*	An increase in plant height, leaf area, 1000 grain weight and grain yield	Ullah and Bano 2019
Oat	PGPR and AMF	An increase in dry weight and stem height	Xun et al. 2015
Lettuce	*Azotobacter chroococcum*	Enhanced the radicle lengths and plumule of germinated seeds	Hussein and Joo 2018
Sunflowers	*Pseudomonas fluorescens* CECT 378T	Fresh weight significantly improved by more than 10%; $K+/Na+$ proportion	Shilev et al. 2012
Strawberry	*Bacilus* sp.	Ameliorated fruit productivity, leaf water amount, ionic constitution, and membrane permeability	Karlidag et al. 2010
Wheat	*Serratia* sp. SL-12	Improvement in plant increase, as determined by factors like root/shoot length, dry/fresh weight, and augmentation of photosynthetic pigment	Singh and Jha 2016
Maize	*Trichoderma harzianum* Th-6	Improved water composition and stomatal conductance, elevated pigment amounts, and increased photosynthetic performance	Yasmeen and Siddiqui 2017
Elaeagnus angustifolia L., (halophyte)	AMF (*Glomus mosseae*)	Promoted biomass accumulation in aboveground organs, increased the numbers of leaves and branches, and improved the leaf areas, stem diameters and plant height	Pan et al. 2020

sustainable crop production, the using of halo-tolerant microorganisms and plants of the soil have started to gain more importance.

By adopting these restoration strategies, degraded soils can be restored, and economic losses due to land degradation can be minimized. However, the point that should not be forgotten here is that the measures we need to take in order not to make our soils salty are more important and economically cheaper.

References

Acar, Y.B. and Alshawabkeh, A.N. (1993). Principles of electrokinetic remediation. *Environ. Sci. Technol.* 27 (13): 2638–2647.

Akbarimoghaddam, H., Galavi, M., Ghanbari, A., and Panjehkeh, N. (2011). Salinity effects on seed germination and seedling growth of bread wheat cultivars. *Trakia J. Sci.* 9 (1): 43–50.

Akhter, J., Mahmood, K., Malik, K.A. et al. (2003). Amelioration of a saline sodic soil through cultivation of a salt-tolerant grass Leptochloa fusca. *Environ. Conserv.* 30 (2): 168–174.

Altaee, A., Smith, R., and Mikhalovsky, S. (2008). The feasibility of decontamination of reduced saline sediments from copper using the electrokinetic process. *J. Environ. Manag.* 88 (4): 1611–1618.

Aronson, J. (1989). *Haloph. A Database of Salt Tolerant Plants of the World.* Arizona: Office of Arid Lands Studies, University of Arizona Tucson.

Ayub, M.A., Usman, M., Faiz, T. et al. (2020). Restoration of degraded soil for sustainable agriculture. In: *Soil health restoration and management* (ed. R.M. Meena), 31–81. Singapore: Springer.

Bahçeci, B., Tarı, A.F., and Bahçeci, İ. (2021). Burdur Gölü Çevresindeki Tuzlu ve Borlu Toprakların Islahı İçin Gerekli Yıkama Suyu Miktarı ve Islah Süresi. *Çukurova Tarım ve Gıda Bilimleri Dergisi* 36 (2): 289–300.

Bal, H.B., Nayak, L., Das, S., and Adhya, T.K. (2013). Isolation of ACC deaminase producing PGPR from rice rhizosphere and evaluating their plant growth promoting activity under salt stress. *Plant Soil* 366 (1): 93–105.

Bauder, J.W. and Brock, T.A. (2001). Irrigation water quality, soil amendment and crop effects on sodium leaching. *Arid Land Res. Manag.* 15: 101–113.

Botella, M.A., Rosado, A., Bressan, R., and Hasegawa, P.M. (2007). Plant adaptive responses to salinity stress. In: *Plant Abiotic Stress* (ed. M.A. Jenks and P.M. Hasegawa), 37–70. USA, Iowa: Blackwell Publishing. https://doi.org/10.1002/9780470988503.ch3.

Chhabra, R. (2004). Classification of salt-affected soils. *Arid Land Res. Manag.* 19 (1): 61–79.

Chinnusamy, V., Jagendorf, A., and Zhu, J.K. (2005). Understanding and improving salt tolerance in plants. *Crop Sci.* 45 (2): 437–448.

Cho, J.M., Kim, K.J., Chung, K.Y. et al. (2009). Restoration of saline soil in cultivated land using electrokinetic process. *Sep. Sci. Technol.* 44 (10): 2371–2384.

Egamberdieva, D., Wirth, S., Bellingrath-Kimura, S.D. et al. (2019). Salt-tolerant plant growth promoting rhizobacteria for enhancing crop productivity of saline soils. *Front. Microbiol.* 2791.

Elmacı, Ö.L. and Yagmur, B. (2021). Plant Nutrients. In: *Soil Science and Plant Nutrition* (ed. N. Okur). Ankara, Turkey: Nobel Academic Publishing. ISBN: 978-625-439-788-2.

English, J.P. and Colmer, T.D. (2013). Tolerance of extreme salinity in two stem-succulent halophytes (*Tecticornia* species). *Funct. Plant Biol.* 40 (9): 897–912.

Epstein, E., Norlyn, J.D., Rush, D.W. et al. (1980). Saline culture of crops: a genetic approach. *Science* 210: 399–404.

FAO (2018). Handbook for saline soil management. In: *Food and Agriculture Organization of the United Nations and Lomonosov Moscow State University* (ed. R. Vargas, E.I. Pankova, S.A. Balyuk, et al.). FAO. ISBN: 978-92-5-130141-8.

FAO (2020). *Mapping of Salt-Affected Soils–Technical Specifications and Country Guidelines.* Rome: FAO Publications.

Flowers, T.J. and Colmer, T.D. (2008). Salinity tolerance in halophytes. *New Phytol.* 179: 945–963.

Flowers, T.J. and Yeo, A.R. (1986). Ion relations of plants under drought and salinity. Austr. *J. Plant Physiol.* 13: 75–91.

Flowers, T.J., Galal, H.K., and Bromham, L. (2010). Evolution of halophytes: multiple origins of salt tolerance in land plants. *Funct. Plant Biol.* 37 (7): 604–612.

Gopalakrishnan, T., Hasan, M.K., Haque, A.T.M. et al. (2019). Sustainability of coastal agriculture under climate change. *Sustainability* 11 (24): 7200.

Hamdan, S.H., Molelekwa, G.F., and Van der Bruggen, B. (2014). Electrokinetic remediation technique: an integrated approach to finding new strategies for restoration of saline soil and to control seawater intrusion. *ChemElectroChem* 1 (7): 1104–1117.

Hanay, A., Büyüksönmez, F., Kiziloglu, F.M., and Canbolat, M.Y. (2004). Reclamation of saline-sodic soils with gypsum and MSW compost. *Compost Sci. Util.* 12 (2): 175–179.

Hanin, M., Ebel, C., Ngom, M. et al. (2016). New insights on plant salt tolerance mechanisms and their potential use for breeding. *Front. Plant Sci.* 7: 1787.

Horneck, D.A., Ellsworth, J.W., Hopkins, B.G. et al. (2007). *Managing Salt-Affected Soils for Crop Production*. PNW 601-E. Oregon State University, University of Idaho, Washington State University. SASCP.

Hou, H.H., Wang, C.T., Wang, X.D., and Gao, F.H. (2014). Study of improvement effects by biological measures in the Yellow River delta saline-alkali soil. *China Rural Water Hydropower* 7: 1–6.

Hussein, K.A. and Joo, J.H. (2018). Plant growth-promoting rhizobacteria improved salinity tolerance of Lactuca sativa and Raphanus sativus. *J. Microbiol. Biotechnol.* 28: 938–945.

Jia, X., Larson, D., Slack, D., and Walworth, J. (2005). Electrokinetic control of nitrate movement in soil. *Eng. Geol.* 77 (3–4): 273–283.

Jia, X., Larson, D.L., and Zimmt, W.S. (2006). Effective nitrate control with electrokinetics in sandy soil. *Trans. ASABE* 49 (3): 803–809.

Karlidag, H., Esitken, A., Yildirim, E. et al. (2010). Effects of plant growth promoting bacteria on yield, growth, leaf water content, membrane permeability, and ionic composition of strawberry under saline conditions. *J. Plant Nutr.* 34 (1): 34–45.

Kasim, W.A., Gaafar, R.M., Abou-Ali, R.M. et al. (2016). Effect of biofilm forming plant growth promoting rhizobacteria on salinity tolerance in barley. *Ann. Agric. Sci.* 61 (2): 217–227.

Lageman, R. and Pool, W. (2011). Elektrokinetische schermen tegen intrusie zout water. *H twee O* 44 (21): 45.

Lakhdar, A., Rabhi, M., Ghnaya, T. et al. (2009). Effectiveness of compost use in salt-affected soil. *J. Hazard. Mater.* 171 (1–3): 29–37.

Li, Z.J., Li, J.H., and Ma, W.P. (2006). Study on biological improvement and utilization of saline soil. In: *National Exchange of Soil Pollution Control. Remediation and Salt Soil Improvement*, 352–354. Shenzhen: China Environmental Science.

Liang, Y., Yang, Y., Yang, C. et al. (2003). Soil enzymatic activity and growth of rice and barley as influenced by organic manure in an anthropogenic soil. *Geoderma* 115 (1–2): 149–160.

Maathuis, F.J. and Amtmann, A.N.N.A. (1999). K^+ nutrition and Na^+ toxicity: the basis of cellular K^+/Na^+ ratios. *Ann. Bot.* 84 (2): 123–133.

Mahdy, A.M. (2011). Comparative effects of different soil amendments on amelioration of saline-sodic soils. *Int. Soil Water Conserv. Res.* 6 (4): 205–216.

Manokararajah, K. and Ranjan, R.S. (2005). Electrokinetic retention, migration and remediation of nitrates in silty loam soil under hydraulic gradients. *Eng. Geol.* 77 (3–4): 263–272.

McFarland, M.L., Provin, T.L., Redmon, L.A. et al. (2014). *An Index of Salinity and Boron Tolerance of Common Native and Introduced Plant Species in Texas*, vol. 1, 3–45. Texas Agrilife Extension.

Miyamoto, S., Martinez, I., Padilla, M. et al. (2004). *Landscape Plant Lists for Salt Tolerance Assessment*, 1–13. USDI Bureau of Reclamation.

Mokrani, S., Nabti, E.H., and Cruz, C. (2020). Current advances in plant growth promoting bacteria alleviating salt stress for sustainable agriculture. *Appl. Sci.* 10 (20): 7025.

Naz, I., Bano, A., and Ul-Hassan, T. (2009). Isolation of phytohormones producing plant growth promoting rhizobacteria from weeds growing in Khewra salt range, Pakistan and their implication in providing salt tolerance to Glycine max L. *Afr. J. Biotechnol.* 8 (21).

Nouri, H., Borujeni, S.C., Nirola, R. et al. (2017). Application of green remediation on soil salinity treatment: a review on halophytoremediation. *Process. Saf. Environ. Prot.* 107: 94–107.

Okur, N. (2018). A review: bio-fertilizers- power of beneficial microorganisms in soils. *Biomed J. Sci. Tech. Res.* 4 (4): 1–2.

Öztürk, A. (2004). Tuzluluk ve sodyumluluğun oluşumu, bitki ve toprağa etkileri. In: *Sulanan Alanlarda Tuzluluk Yönetimi Sempozyumu* (20–21 Mayıs) (ed. T.C. Ankara), 1–15. Enerji ve Tabii Kaynaklar Bakanlığı DSİ Gen. Müd. Bildiriler Kitabı.

Pan, J., Huang, C., Peng, F. et al. (2020). Effect of arbuscular mycorrhizal fungi (AMF) and plant growth-promoting bacteria (PGPR) inoculations on Elaeagnus angustifolia L. in saline soil. *Appl. Sci.* 10 (3): 945.

Parida, A.K. and Das, A.B. (2005). Salt tolerance and salinity effects on plants: a review. *Ecotoxicol. Environ. Saf.* 60: 324–349.

Qadir, M., Ghafoor, A., and Murtaza, G. (2000). Amelioration strategies for saline soils: a review. *Land Degrad. Dev.* 11 (6): 501–521.

Qadir, M., Oster, J.D., Schubert, S. et al. (2007). Phytoremediation of sodic and saline-sodic soils. *Adv. Agron.* 96: 197–247.

Ramadoss, D., Lakkineni, V.K., Bose, P. et al. (2013). Mitigation of salt stress in wheat seedlings by halotolerant bacteria isolated from saline habitats. *Springerplus* 2 (1): 1–7.

Ravindran, K.C., Venkatesan, K., Balakrishnan, V. et al. (2007). Restoration of saline land by halophytes for Indian soils. *Soil Biol. Biochem.* 39 (10): 2661–2664.

Raychev, T., Popandova, S., Józefaciuk, G. et al. (2001). Physicochemical reclamation of saline soils using coal powder. *Int. Agrophys.* 15 (1).

Rozema, J. and Schat, H. (2013). Salt tolerance of halophytes, research questions reviewed in the perspective of saline agriculture. *Environ. Exp. Bot.* 92: 83–95.

Sadiki, M. and Rabih, K. (2001). Selection of chickpea (Cicer arietinum) for yield and symbiotic nitrogen fixation ability under salt stress. *Agronomie* 21 (6–7): 659–666.

Shabala, S. and Mackay, A. (2011). Ion transport in halophytes. *Adv. Bot. Res.* 57: 151–199.

Shilev, S., Sancho, E.D., and Benlloch-González, M. (2012). Rhizospheric bacteria alleviate salt-produced stress in sunflower. *J. Environ. Manag.* 95: S37–S41.

Singh, K.N. and Chatrath, R. (2001). Breeding for adaptation to environmental factors. In: *Chapter 8. Salinity Tolerance*, Application of physiology in wheat breeding (ed. M.P. Reynolds, J.J. Ortiz-Monasterio, and A. McNab), 101–110. Mexico: CIMMYT.

Singh, R.P. and Jha, P.N. (2016). Alleviation of salinity-induced damage on wheat plant by an ACC deaminase-producing halophilic bacterium Serratia sp. SL-12 isolated from a salt lake. *Symbiosis* 69 (2): 101–111.

Sönmez, B. and Beyazgül, M. (2008). 'Türkiye'de Tuzlu ve Sodyumlu Toprakların Islahı ve Yönetimi. *Sulama ve Tuzlanma Konferansı*, Şanlı Urfa, Turkey (12–13 June 2008).

Talat, N. (2020). Alleviation of soil salinization and the management of saline soils, climate change, and soil interactions. In: *Climate Change C and Soil Interactions SI* (ed. M.N.V. Prasad and M. Pietrzykowski). Elsevier.

Tang, X., Mu, X., Shao, H. et al. (2015). Global plant-responding mechanisms to salt stress: physiological and molecular levels and implications in biotechnology. *Crit. Rev. Biotechnol.* 35 (4): 425–437.

Ullah, A. and Bano, A. (2019). Role of PGPR in the reclamation and revegetation of saline land. *Pak. J. Bot.* 51 (1): 27–35.

USSL Staff (1954). *Diagnosis and Improvement of Saline and Alkali Soils.* USDA Handbook No 60, 160. Washington DC, USA: USDA.

Villora, G., Pulgar, G., Moreno, D.A., and Romero, L. (1997). Effect of salinity treatments on nutrient concentration in zucchini plants (*Cucurbita pepo* L. var. Moschata). *Aust. J. Exp. Agric.* 37 (5): 605–608.

Wakeel, A. (2013). Potassium–sodium interactions in soil and plant under saline-sodic conditions. *J. Plant Nutr. Soil Sci.* 176 (3): 344–354.

Wang, X., Xing, X., Zhang, F., and Xin, K. (2018). Biological improvement of saline alkali soil reference system: a review. *Sci. Cold Arid. Reg.* 10 (6): 516–521.

Warrence, N.J., Bauder, J.W., and Pearson, K.E. (2002). *Basics of Salinity and Sodicity Effects SE on Soil Physical Properties SPP*, vol. 129, 1–29. MT: Departement of Land Resources and Environmental Sciences, Montana State University-Bozeman.

Whipps, J.M. (2001). Microbial interactions and biocontrol in the rhizosphere. *J. Exp. Bot.* 52 (suppl_1): 487–511.

Xiong, L. and Zhu, J.K. (2002). Molecular and genetic aspects of plant responses to osmotic stress. *Plant Cell Environ.* 25: 131–139.

Xun, F., Xie, B., Liu, S., and Guo, C. (2015). Effect of plant growth-promoting bacteria (PGPR) and arbuscular mycorrhizal fungi (AMF) inoculation on oats in saline-alkali soil contaminated by petroleum to enhance phytoremediation. *Environ. Sci. Pollut. Res.* 22 (1): 598–608.

Yao, L., Wu, Z., Zheng, Y. et al. (2010). Growth promotion and protection against salt stress by Pseudomonas putida Rs-198 on cotton. *Eur. J. Soil Biol.* 46 (1): 49–54.

Yasmeen, R. and Shaheed Siddiqui, Z. (2017). Physiological responses of crop plants against Trichoderma harzianum in saline environment. *Acta Bot. Croat.* 76 (2): 154–162.

15

Conservation Agriculture as Sustainable and Smart Soil Management: When Food Systems Meet Sustainability

Rachid Mrabet[1], *Akashdeep Singh*[2], *and Tarun Sharma*[2]

[1] *Institut National de la Recherche Agronomique (INRA), Avenue de la Victoire P.O. Box 415, Rabat, 10000, Morocco*
[2] *Department of Agronomy, CSK Himachal Pradesh Agricultural University, Palampur, India*

15.1 Introduction: Challenging A "Global Syndemic"

Resource-intensive agriculture or so-called modern agriculture was promoted by public policies first in the US and Europe and then, with the Green Revolution, in low- and middle-income countries (Dawson et al. 2016; Sharma, 2022). These public policies supported the intensive use of seeds, fertilizers, pesticides, and machinery in order to increase productivity and farm profits. However, high dependency on energy and with rising price of external inputs limited farmer's opportunities of investment. Wide-spread negative impacts on environment associated with lowering land productivity and exhausting the productive capacity of local ecosystems were the results of such inappropriate policies (Kopittke et al. 2019).

Under current soil management systems, climate change, and associated droughts and extreme events are expected to further lessen the productivity of key crops in numerous regions of the world while exceeding planetary boundaries (Campbell et al. 2017; Kopittke et al. 2021). Tillage-based systems and practices are undermining the ecological foundation of the global food system through mining and overuse of soils and their polluting effects, thereby enhancing soil organic matter decomposition, degradation, reducing ecosystem capacity to generate sustainable yields and threatening livelihood sustainability, food security, and poverty (Kassam et al. 2017; Das et al. 2018; Sharma, 2022; Sykes et al. 2020).

Feeding the world today and in the future and hence closing the growing yield gap requires a significant shift in ways of managing soils and agroecosystems and a systemic thinking of critical threats to food system sustainability (Godfray and Garnett 2014). In other terms, there is an urgent need to change the current trajectory of production to avoid additional failures (Schreefel et al. 2020).

There is a growing concern for new initiatives in order to simultaneously tackle the many dimensions of this complex problem. In fact, healthy growth and stable productivity of crops and livestock require ground-breaking models of food production for

Agroecological Approaches for Sustainable Soil Management, First Edition.
Edited by Majeti Narasimha Vara Prasad and Chitranjan Kumar.

resource-saving, regenerative, environmentally friendly agriculture (Nabuurs et al. 2022; HLPE 2019; FAO et al. 2020; Blay-Palmer et al. 2019). This treatise has boosted conservation agriculture (CA) as well as agro-ecology, climate smart agriculture, and organic farming as inclusive and sustainable approaches, and they now appear in the lexicon of many organizations, academia, and research institutions. CA is supported by a global network of farmers and NGOs.

Worldwide researchers have studied and discussed diverse aspects and facets of CA. Hence, this paper explores, through an in-depth review of the recent literature, the impacts and benefits of well-managed CA systems in improving soil functions and sustainability of food systems. Hence, the scope of the paper is on linking soil functioning to long-term sustainability of agroecosystems through CA adoption and mainstreaming. Specifically, the paper aims at answering the following question: are CA systems viable and durable options for re-carbonizing and regenerating soils, sustaining crop productivity, and ensuring well-being of farmers in a changing and uncertain climate?

15.2 Conservation Agriculture: Exploring Concept, Objectives, and Ambitions

The concept of Conservation Agriculture goes back to early 1930s. During the 1960s and 1970s, a range of terms (i.e., no-tillage [NT], zero tillage, direct seeding, conservation tillage, minimum tillage, mulch tillage, and strip tillage) were used (Lal 2015). CA was fully defined by FAO in early 2000 and has gained distinction in recent years amongst both farmers and policy-makers. Conservation agriculture (CA) consists of three principles: preserving a permanent soil cover either as mulch or cover crops, minimizing soil disturbance through direct seeding, and diversifying crop species in rotations or associations (Derpsch et al. 2014; Kassam et al. 2020). In other terms, CA systems ignore the three components of conventional production systems that are the cause of soil and land degradation–tillage and seed-bed preparation, excessive removal of crop residues, and mono-cropping.

According to Wall et al. (2020), CA attempts to bring agriculture as close as possible to natural systems (i.e., forests, pastures) that are inhabited by diverse and rich plant species, are not tilled, and where dead leaves and plant residues remain on the soil surface. Reicosky (2015) expressed concerns with the use of the term “conservation tillage” as there is no “soil conservation” with any type of tillage.

Lal (2015) added a fourth principle that is integrated nutrient management (INM) and defined CA as a farming system that comprised crop residue mulch, cover cropping, INM, and NT techniques in a rotation cycle for effective soil and water conservation, soil organic carbon (SOC) sequestration, sustainable intensification, and climate change adaptation and mitigation. However, still different views and interpretations prevail amongst professionals in applying the principles of CA systems in order to contribute to sustainability.

Although several definitions of sustainability and resilience have been discussed in the literature, these concepts can be generally understood as the intersection of three dimensions or pillars: social development and equality, inclusive economic growth and environmental stewardship and protection (Sinclair et al. 2019; Sinclair and Coe 2019). A new appreciation of the relevance of CA for the climate change agenda is emerging as part of the recent IPCC land-use report (IPCC 2019; Smith et al. 2019).

Multiple concerns ranging from environmental to social and economic issues have been the founding adoption and scale-up of CA worldwide. According to Kassam et al. (2022), CA acreage has reached 205 million ha in 2018 which represent 15% of agricultural lands. This represents an increase of 93% in global CA cropland area since 2008/2009 and represents an annual increase of about 10 Mha since 2018. The leading but contrasting countries in terms of agricultural policies and farming features that practice CA in their agricultural lands are USA (44.0 Mha), Brazil (43.9 Mha), Argentina (32.9 Mha), Australia (22.9 Mha), and Canada (21.7 Mha). In other terms, CA systems are being practiced in most climatic regimes and altitudes. They are experienced in both rainfed and irrigated systems, annual, perennial and mixed cropping systems, orchards and plantation systems, agroforestry systems, pasture and livestock-based systems, organic and rice-based systems. Scenario analysis by Prestele et al. (2018) suggests a future potential of CA in the range of 533–1130 Mha (38%–81% of global arable land).

The development, implementation and scaling up of CA systems require an appropriate enabling environment and long-term and sustained availability of resources. In fact, favoring CA often necessitates disabling structural constraints that lock in conventional models of agricultural improvement, necessitating fundamental shifts in the ways food systems function and are organized (HLPE 2019). One of the principal bottlenecks constraining CA transitions beyond the availability of context fitting technical solutions is whether the right policies can be put in place to enable their adoption at scale and in a transformational way (Sinclair et al. 2019). In other terms, policies in terms of instruments and designs for transition to CA systems are dependent on context specificities, target visions and starting points. In here, persuasion and sensitization for a wide range of stakeholders including policymakers and opinion leaders are extremely important and mainly through information sharing platforms and educational campaigns (e.g., enhancing CA messaging in extension and learning systems). However, actionable programs, characterized by instruments, rules, budget, human resources, social participation are needed and should be adapted to contexts. In addition, availability of incentives, brokers, and social skills for implementing collectively CA at the local level often pose a critical challenge.

15.3 Harnessing Soil Functioning under Conservation Agriculture

In order to boost production, modern agriculture has replaced ecosystem services with exogenous inputs such as fertilizers, irrigation, tillage, and pesticides. However, such human-modified practices have affected the well-being of soils which is crucial for securing food and feed worldwide (Rinot et al. 2019). Soil dynamic properties are used to monitor and assess the consequences of restoration activities on soil and widely ecosystem functioning and services. A substantive body of research has been carried to understand and quantify soil functioning under CA systems in comparison to traditional and conventional systems of soil management across climates and farming systems.

Most of these studies confirmed that CA support implementing soil-based strategies and long-term soil fertility dynamics that restore soil functions and health (Powlson et al. 2014). Most land degradation processes that are sensitive to climate change pressures (e.g., erosion, decline in soil organic matter, compaction, waterlogging, salinization, drought, etc.) benefit from CA systems.

Although much of the attention has gone to carbon sequestration in forests as central to global mitigation efforts, there is space for agriculture and food systems at large to contribute via carbon sequestration in agricultural soils (Nabuurs et al. 2022). IPCC scenarios showed that limiting global temperature increases to 1.5 or 2 °C by the end of the century will not be possible without significant CO_2 removals from the atmosphere, and net soil carbon sequestration can make an important contribution to this process (IPCC 2019). According to 4 per 1000 initiative (https://www.4p1000.org), if an annual growth rate of 0.4% in the soil carbon stocks in the first 30–40 cm is applied only to agricultural soils (not to all soils), it would effectively offset 20–35% of global anthropogenic greenhouse gas emissions (Minasny et al. 2017).

CA systems are among the most widely advocated options for sequestering C on croplands throughout the world (Bruce et al. 1999; Minasny et al. 2017; Nicoloso et al. 2020; Pacala and Socolow 2004; Post et al. 2004; Smith et al. 2007; Sperow 2016; Thomson et al. 2008; World Bank 2012; Wang et al. 2020; Zhang et al. 2014; Zomer et al. 2017; Karbin et al. 2021). Lal et al. (2021) showed evidence that CA-based systems allow re-carbonization of depleted soils. However, the amount of additional SOC potentially stored or sequestered in the soil is the subject of much debate in the scientific community (Nabuurs et al. 2022; Chenu et al. 2019). Rates of carbon accumulation in soils under CA systems reported in the literature have varied widely (Table 15.1). Leaving crop residues (often a biowaste) was found most effective in increasing soil carbon content under CA systems (Schmer et al. 2020).

Many authors have reported that CO_2 emissions are higher in conventional tillage agriculture compared to CA systems (Gonzalez-Sanchez et al. 2020; Reicosky 2021). Wang et al. (2020) found that CA lowered CO_2 fluxes and increased SOC stocks at 0–60 cm soil depth compared with conventional tillage. Such results were confirmed by Tellez-Rio et al. (2017) and concluded that no-tillage systems decreased the net global warming potential

Table 15.1 Carbon sequestration due to CA compared to CT under contrasting climatic regions and locations.

Climate regime or Location	Depth (cm)	Carbon Storage[a] (kg C m^{-2})	Rate of change in SOC (kg C ha^{-1} y^{-1})	Reference
Temperate	0–60	0.15 ± 0.22		Meurer et al. (2018)
Temperate	0–150	0.15 ± 0.34		Haddaway et al. (2017)
Various	0–30	0.34		Virto et al. (2012)
Global (24 studies)	0–100	0.49		Angers and Eriksen-Hamel (2008)
Mediterranean Drylands	0–117	0.3		Alvaro-Fuentes et al. (2008)
Various	0–22	0.62 ± 0.16		West and Post (2002)
Southeastern Australia			+180 to 315	Grace et al. (2010)
Global (67 studies)			+300	Janzen et al. (1998)
Global (52 Studies)			+47 to 620	Puget and Lal (2005)

[a] SOC storage difference between CA and CT.

due to the enhancement of C stock. Karbin et al. (2021) reported that most soil health indicators were improved in early year of adoption of CA systems in India and that CA systems mitigated CO_2 emissions.

Long-term research by Olson et al. (2013) also showed that crop residue-returned farming system such as CA can significantly reduce SOC losses from erosion when compared to conventional tillage. Several studies indicate that conservation agriculture has a positive effect on range of soil biota and biological properties. Rao et al. (2022) concluded that CA can potentially stop deterioration of soil fertility and improve crop production in the middle Indo-Gangetic Plains of India (a region producing staple food for 1 billion people).

The soil organic matter stock is easier to deplete but difficult to restore and its restoration is a slow process. For CA, rates of soil C gain are highly variable, reflecting complex interactions of land history, soil properties, climatic conditions, and complementary agronomic practices (Virto et al. 2012; Yang et al. 2013; VandenBygaart 2016). Notably, soil C gain is not a function of the current practice alone, but rather of the change in practice and therefore depends on land use history (Powlson et al. 2022). Thus, for example, land under no-till practices might be gaining C if recently converted from a degradative system (Nicoloso et al. 2020) but might be losing C if recently converted from grassland (DuPont et al. 2010). A global review of CA systems with a management duration of 5–29 years by Govaerts et al. (2009) found that out of 78 cases, SOC stock in CA was higher in 40, not significantly different in 31, and lower in 7 compared with conventionally tilled soils.

In Brazil, the second country with the largest CA area in the world, CA when compared to CT systems or pastures, predominantly resulted in SOC gains varying between 9% and 25%, and represents an alternative to promote sustainable agriculture, which can contribute to the mitigation of GHG emissions and food security (Maia et al. 2022).

Removal of plant C as biomass for livestock feeding or biofuel will certainly not allow carbon storage to reach levels to mitigate climate change or recuperate soil resilience (Bruni et al. 2021; Riggers et al. 2021; Ruis and Blanco-Canqui 2017). Worldwide studies showed that CA with residue management practices over extended periods of time have increased SOC concentrations, mainly near the surface (Reicosky 2021).

Fujisaki et al. (2018) concluded that the SOC accumulation rates increased linearly with C inputs, and the conversion rate of C inputs to SOC was $8.2 \pm 0.8\%$." According to Bohoussou et al. (2022), CA practices increased SOC by 21.39% and total nitrogen stock by 54.34%. The authors also showed that it is necessary to ensure that legumes are included in rotations for greater residue retention, and manure-based fertilizer mixes are applied to obtain maximum increases in SOC and TN. Legume residues provide a rapid available source of nitrogen for microbes (Virk et al. 2022), which strengthen and increase the organo-mineral associations in the soil and stabilize soil organic matter (Kopittke et al. 2020). Other studies suggest that retention of root C in soil organic matter is greater than that of above-ground residues and hence confirmed the need not to disturb or till soils. However, substantial amounts of soil carbon are lost when farmers interrupt continuous use of CA systems with conventional tillage (VandenBygaart 2016).

CA systems can enhance several key environment-related SDGs. Page et al. (2020a) showed that CA systems are climate proof and water conserving schemes for both dry and irrigated regimes. CA systems were initially adopted for their capacity to mitigate soil erosion and reduce runoff and sediment loads (Page et al. 2020a,b; Sapkota et al. 2015, 2017) but also for improving soil resilience, structure, and aggregation (Su et al. 2021c). SOC and its lability are highly recommended as an indicator of soil sustainability as it is associated with short-term nutrient cycling (Bowhmik et al. 2017). Fertilizer costs may be partially offset by nutrients from retained residues under CA systems (Mrabet et al. 2021).

Ecosystem services provided by soil under CA systems are numerous (Lal 2013; Palm et al. 2013; Sapkota et al. 2017; Reicosky 2020; Nyanga et al. 2020; Jayaraman et al. 2021). CA practices have the potential to reduce the erosion of soil by wind and water due to the permanent soil cover using either the crop residues or cover crops (Wilson et al. 2004; West and Post 2002; Ghosh et al. 2019). Regardless of farm size, Nyanga et al. (2020) reported lower risks of downstream sedimentation and flood damage to infrastructure due to slower runoff processes.

Surface cover also restricts evaporation from the soil surface through reduction of the solar radiation reaching the soil. This allows greater water availability to the crops and mainly in situations where water is scare or in periods of moisture stress. In other terms, improved soil water balance under CA systems is especially central under water-limited conditions (drier areas and regions with seasonal droughts) in order to increase water use efficiency and hence stabilize and surge crop yields (Corbeels et al. 2020). In a meta-analysis by Corbeels et al. (2020) in sub-Saharan Africa, soil moisture conservation as a result of crop residue mulching lead to improved rainfall-use efficiency of crops and hence to higher yields under CA compared to CT. From this study, it was also clearly shown that under seasonal rainfall below 800 mm, CA outperformed CT on all soil types, with no significant differences between soil types.

Irrigated farms also benefit from CA principles through savings in energy and water amounts and reduction of number of applications without affecting crop productivity (Wall et al. 2020). As a result of the water conservation effects, CA has increasingly been recognized as a "climate-smart agriculture" system.

The immediate effect of the retention of crop residues is a reduction in the availability of nutrients due to an increase in the carbon to nitrogen (C : N) ratio (nutrient immobilization). Over the period, there is an increment in nutrient availability along with organic matter generation which promotes crop production (Garai et al. 2020).

Soil aggregation and porosity have fundamental roles in root penetration and development, mass and energy transport processes within the soil, and between the soil and the environment. Based upon a recent meta-analysis, compared to conventional tillage, CA increased mean weight diameter of aggregates, water stable aggregates, and macro-aggregates by averages (0–30 cm) of 25, 10, and 22%, respectively, although predominantly in 0–10 and/or 10–20 cm layers, with an accompanying reduction in micro-aggregates (Mondal and Chakraborty 2022). From the same study, aggregate formation allowed that the adoption of CA reduced the total porosity by only 2.5–2.9% over conventional tillage up to a depth of 20 cm and that bulk density was marginally affected by CA. Such trend seems to disappear with time and associated with increased macro-porosity. It was also reported that longer period would likely to make CA effects visible in deeper soil layers through increased root system. Blanco-Canqui and Ruis (2018) reported that no-tillage impact in reducing the susceptibility of soil to compaction through synthesis of published global data, while Li et al. (2019) concluded a 1.4% increase in soil bulk density in their global meta-analysis. Hence, while controlling erosion, CA systems aid improving porosity, structural stability, infiltration, and water retention due to higher microporosity (Sapkota et al. 2015, 2017). In other terms, CA systems obviate the moisture loss occasioned by tillage practices.

15.4 Food Security Under Conservation Agriculture: From Farm to Fork

Agriculture was the pillar of most economies and a significant reason that led to the growth of ancient civilizations. In the present-day scenario, there is enormous pressure on the existing land for increasing food production to feed the burgeoning population. With the advancement in tools and techniques for crop production, the nations can keep a steady supply of food grains but at the expense of environmental degradation. Agriculture in itself has a substantial ecological footprint and is a core driver of land-use change globally (Searchinger 2019). As per the FAO (2021) reports, agri-food systems produce a total of 16.5 billion tons of greenhouse gas emissions of which 7.2 billion tones came from within the farm gate, 3.5 from land-use change, and another 5.8 billion tones from supply-chain processes. Food processing, packaging, transport, consumption, and waste disposal are pushing the food supply chain to top the list of greenhouse gas emitters (Tubiello et al. 2021). Amidst the perplexities, there are a group of researchers who are eyeing a sustainable way to produce food.

To achieve global food security, we need to approximately double food production over the coming decades. In fact, intensive conventional agriculture has more than tripled yield in the last century. Moreover, to maintain stable and durable food production while mitigating environmental degradation across contrasting regions and policies will be difficult challenges under a changing and less predictable climate.

With rising world population, the demand for crops will increase to satisfy the caloric base of diets worldwide (FAO 2021). With livestock, cereals (especially maize, wheat, and rice) continue to play an important role in agri-food systems around the globe but are hotspots to climate change and degrading resources. These staple cereals provide the foundation for food security for most of the world's population. However, considering the importance of balanced diets to combat malnutrition and disease, there is a call to diversify the production systems to include non-cereal crops, such as food legumes and vegetables, which reduce land devoted to cereals. Climate change is expected to have a severe impact on most crops by altering the patterns of floods, droughts, and heat and increased spreading of pests and diseases.

Reversing food insecurity spiral while bridging yield gaps and simultaneously enhancing social and well-being goals as committed to in the SDGs are other genuine aims of CA systems (Richards et al. 2014; Mkomwa and Kassam 2021; Su et al. 2021a,b; Jat et al. 2019, 2020). Globally and dependent on crop species, CA either produced equivalent or greater yields than conventional tillage (Hobbs et al. 2008; Pittelkow et al. 2015; Tellez-Rio et al. 2017). With proper implementation, there are numerous examples of better yield in CA than in CT, especially in well-drained soils prone to water runoff and accelerated erosion (Liao et al. 2015; Moraru and Rusu 2013). Several authors showed that after two to three years, crop yields under CA systems are consistently better than under conventional systems (Brouder and Gomez-Macpherson 2014; Thierfelder et al. 2015; Mrabet et al. 2022).

The meta-analysis by Corbeels et al. (2020) revealed that yields were significantly higher under CA than under CT for maize (4.0%) and when all three CA principles are implemented, the CA effect more than doubled (8.4%). The authors did not find differences for the other crops (cotton, cowpea, rice, sorghum, and soybean) among CA and CT systems. The results of this analysis also support the claim that the crop yield performance of CA systems improves over time which has been attributed to gradually improved soil quality under CA compared to CT.

CA systems were also reported to increase resilience of crops and farm livelihoods, especially in dry areas and among the most vulnerable food producers (Knapp and van der Heijden 2018). It is well recognized that in dry lands or years, CA enhances water infiltration and storage while reducing evaporation and temperature extremes. Furthermore, the yield improvements can be highly sustainable and secure because CA systems maintain soil fertility and can even restore degraded lands.

According to Lal (2015) the objective from CA systems is not to produce the highest yield, but a sustainable optimum yield, especially during bad growing seasons. Thus, sustainability of a CA systems is judged not by the highest yield in a good season, but a respectable/stable yield in a poor (dry) season. In this regard, CA are climate-resilient management system (Mrabet et al. 2021; Corbeels et al. 2020).

According to Knapp and van der Heijden (2018), yield stability did not vary between CA and conventional tillage indicating that a transition to CA generally does not affect yield stability. CA systems had reduced yield and yield stability in humid climate compared to dry climate. These differences are probably due to better soil water retention and higher yields of CA soils in dry climate. However, reduced yields were mostly due to the negative effects of delayed soil warming, nutrient mineralization, and reduced soil aeration in no-tilled, wet, and heavy soils. I was also noted that breeding of crops and selection of varieties for conservation agriculture are not yet widespread. Hence, adapted cultivars to CA systems may enhance yield and yield stability. The inclusion of leguminous crops would regenerate and enhance the natural biological processes and hence sustain the crop production and quality (Montgomery 2020).

15.5 CA Systems as Drivers for Social Development and Economic Growth

Worldwide farmers and especially small ones, even though the most vulnerable to climate and land challenges, are important for the rural economy, labor market, food and nutrition security, biodiversity, etc. Hence, countries should seek to improve the well-being of farmers. In other terms, the search for improved social and economic attributes of farming systems should lead to more sustainable and healthier food systems. There has been a growing body of scientific and grey literature showing that CA systems are cost-effective, energy efficient, and allow farmers higher and more stable incomes (Das et al. 2021; Gonzalez-Sanchez et al. 2016; Jat et al. 2014). With the reduced expenses in terms of labor, energy, and monetary inputs, CA systems reduce the cost of cultivation and hence increased farm profits (Chatterjee et al. 2020; Khan et al. 2020; Ogle et al. 2019). Improvement in soil quality leads to higher input use efficiencies, which add to the economics of the adoption of CA practices.

In mechanized farming systems, conversion of plow-based tillage to no-tillage based CA systems lead to immediate machinery and maintenance costs, energy savings, and reduced labor (Wall et al. 2020). Corsi et al. (2012) found 35–80% reduction of fuel use under CA systems. Sharma and Biswas (2022) found a 20–25% reduction in energy consumption when shifting to CA systems in India. In the global south, CA generally reduces labor, drudgery, operation time, human and animal energy, and production costs and generally leads to increased returns to family labor (Aryal et al. 2015). These improvements are keys to increasing total farm productivity and enhancing farm family livelihoods. Such profit-driven advantages should consent farmers to gain confidence in CA systems. In addition, when farmers shift to CA systems, the lessening of energy per unit area, GHG emissions from cultivation are reduced and hence farmers are contributing to mitigating climate change.

Rewarding farmers for implementation of CA practices could be whether through direct subsidy and payments for public goods or through the development of private offset markets. Yet, the later mechanism requires in-depth technical, economic, and social viability of CA systems (Sykes et al. 2020).

15.6 Challenges and Socio-Economic Barriers for CA Adoption

CA benefits cannot be established overnight (Corbeels et al. 2020). Brown et al. (2017) found that institutional and community constraints hindered the use of financial, physical, human, and informational resources to implement CA systems. In other words, in order to take a fuller advantage of the potential of CA systems in addressing the three dimensions of sustainability, end-to-end strategies based upon systemic and comprehensive inclusion of stakeholders, actors, and partnerships at different scales and levels are necessary if not mandatory.

It is of prime importance that CA systems do not induce any marginalization process and that the introduction of innovation do not create any polarizing social effects. In other terms, CA systems should pay more attention to dynamics beyond agricultural production. Researchers have identified a number of scale-related issues which shape the likelihood of adoption of conservation agriculture. These include farmer's education, poverty, social status, profitability, total farm assets, farm characteristics (type and size, farm location), diversification strategy and access to subsidies, access to support services and advice, market integration (Amundson and Biardeau 2018; Gerber et al. 2013). In African studies, the low productivity and low profitability of CA farming is the primary explanation of reduced adoption and uptake (Giller et al. 2015). In addition, agriculture is perceived as the fallback position by young generations and the changes in perceptions, motivations, and attitudes vis-à-vis farming opportunity space. Social norms as well as psychological and behavioral factors must be considered for widespread and durable adoption of CA systems (Wreford et al. 2017).

CA is considered to be accurately practiced only when its three pillars are applied all together (Reicosky 2015). According to several authors, it becomes evident that when CA is not well managed, crop yields are compromised and extra-costs are generated than expected (Rosenstock et al. 2014; Brown et al. 2017; Su et al. 2021b,c; Corbeels et al. 2020; Yigezu et al. 2021). In other terms, while each principle can be used in separate, the fluid integration of all three principles is key for successfully delivering full on and off benefits of CA systems.

Long-term social, financial, and economic benefits of CA are still difficult to assume by farmers and ex-ante assessments are desirable. In addition, to overcoming past mindsets, farmers, and policymakers often find it challenging to make precise estimates about long-term returns on investments in CA systems, due to system intricacy and lengthened planning horizons.

15.7 Conclusion: Bridging and Bonding CA Science and Policy

Farming will no doubt remain a central component of rural livelihoods both for provision of food and feed and as a source of income and well-being. But out-of-the-box thinking is required to shift in paradigms and address needed structural changes in food systems. For perpetuating such development dynamic, CA systems can offer high-yielding and sustainable food systems with multiple SDG benefits. As a regenerative agriculture, CA is a soil-centered approach that works on the foundation of the health of soil, plant, animals, humans, and the environment is one (Lal 2020; Kassam 2022; Mugandani et al. 2021). Due to higher crop yields under CA systems in moisture limited areas, the highest immediate return on investment in CA research and development is to be expected when crops are experiencing recurrent and prolonged dry spells.

The diversification of cropping systems required by CA systems can be an engine of economic growth and provide welfare benefits and food baskets that meet the household's nutritional essential needs. A bold, broad and orchestrated set of visions should be practiced linking diversification with intensification in order to improve soils, crops, animals and human assets, and lead to higher ecosystem functionality (Dawson et al. 2019).

Resisting disruptive change due to necessity of implement simultaneously the three CA pillars has reduced large scale uptake and mainstreaming (Corbeels et al. 2020). In here, providing in continuous basis state of the art and relevant knowledge and advice on CA systems and related or associated natural resource and crop-livestock management information is of paramount importance to succeed in the transition and adoption. This requires strengthening research and advisory in CA systems and significant commitment and flexibility of institutions through re-allocation of funds, the development of quality scientists and strong advisors and extension specialists.

Translating objectives into working frameworks and instruments requires a systems approach for leveraging synergistic effects and managing co-benefits while mitigating and anticipating trade-offs. Cross-country sharing CA experiences and success stories may accelerate the learning and implementation process and unlock key constraints to CA transitions while creating space for robust decisions, evidence, and opportunities for wider use and adoption. Consequently, for durable impacts, funds should increasingly come from foundations and industry to support traditional publicly-funded bilateral and multilateral donors and development organizations.

It is imperative for CA farmers to get adequate recognition for the off-farm benefits of CA systems, principally those related to ecosystem services and soil functionalities and security. Payments of these services require development of metrics or rigorous indicators to allocate cost/benefit, validate impacts, and the ability to monitor participation. The principles, additionality, permanence, and leakage, that are successfully used for other payment program should be applied to CA systems (Lal 2019, 2020; Newton et al. 2020).

From this analysis of literature, CA systems can be promoted as a triple win situation by addressing GHG mitigation goals, enhancing resilience to meet adaptation goals and for food security. For addition effectiveness, there is great momentum in merging principles of CA with those of regenerative types of farming and especially those related to tillage, synthetic fertilizers, and pesticide use (Newton et al. 2020). Also, Kassam and Kassam (2020) proposed an inclusive ethical and responsible system to assimilate CA systems with plant-based diets and organic farming practices in order to move from corporate agriculture.

References

Álvaro-Fuentes, J., López, M.V., Cantero-Martínez, C. and Arrúe, J.L. (2008). Tillage effects on soil organic carbon fractions in Mediterranean dryland agroecosystems. *Soil Sci. Soc Am. J.* 72: 541–547.

Amundson, R. and Biardeau, L. (2018). Opinion: soil carbon sequestration is an elusive climate mitigation tool. *PNAS* 115 (46): 11652–11656. https://doi.org/10.1073/pnas.1815901115.

Angers, D.A. and Eriksen-Hamel, N.S. (2008). Full-inversion tillage and organic carbon distribution in soil profiles: a meta-analysis. *Soil Sci. Soc. Am. J.* 72: 1370–1374.

Aryal, J.P., Sapkota, T.B., Jat, M.L., and Bishnoi, D.K. (2015). On-farm economic and environmental impact of zero-tillage wheat: a case of north-West India. *Exp. Agric.* 51: 1–16. https://doi.org/10.1017/S001447971400012X.

Bhowmik, A., Fortuna, A., Cihacek, L.J. et al. (2017). Potential carbon sequestration and nitrogen cycling in long-term organic management systems. *Renew Agr. Food Syst.* 1–13.

Blanco-Canqui, H. and Ruis, S.J. (2018). No-tillage and soil physical environment. *Geoderma* 326: 164–200.

Blay-Palmer, A., Conare, D., Meter, K. et al. (2019). *Sustainable Food System Assessment. Lessons from Global Practice*, 1e. London: Routledge.

Bohoussou, Y.N.D., Kou, Y.H., Yu, W.B. et al. (2022). Impacts of the components of conservation agriculture on soil organic carbon and total nitrogen storage: a global meta-analysis. *Sci. Total Environ.* 842: 156822. https://doi.org/10.1016/j.scitotenv.2022.156822.

Brouder, S.M. and Gomez-Macpherson, H. (2014). The impact of conservation agriculture on smallholder agricultural yields: a scoping review of the evidence. *Agric. Ecosyst. Environ.* 187: 11–32.

Brown, B., Nuberg, I., and Llewellyn, R. (2017). Negative evaluation of conservation agriculture: perspectives from African smallholder farmers. *Int. J. Agric. Sustain.* 15: 467–481. https://doi.org/10.1080/14735903.2017.1336051.

Bruce, J.P., Frome, M., Haites, E. et al. (1999). Carbon sequestration in soils. *J. Soil Water Conserv.* 54 (1): 382–389.

Bruni, E., Guenet, B., Huang, Y.Y. et al. (2021). Additional carbon inputs to reach a 4 per 1000 objective in Europe: feasibility and projected impacts of climate change based on century simulations of long-term arable experiments. *Biogeosciences* 18 (13): 3981–4004. https://doi.org/10.5194/bg-18-3981-2021.

Campbell, B.M., Beare, D.J., Bennett, E.M. et al. (2017). Agriculture production as a major driver of the earth system exceeding planetary boundaries. *Ecol. Soc.* 22. https://doi.org/10.5751/ES-09595-220408.

Chatterjee, S., Ghosh, S., and Pal, P. (2020). *Soil Carbon Restoration through Conservation Agriculture, Natural Resources Management and Biological Sciences, Edward R.* Rhodes and Humood Naser, IntechOpen. http://dx.doi.org/10.5772/intechopen.93006. https://www.intechopen.com/chapters/72731.

Chenu, C., Angers, D.A., Barré, P. et al. (2019). Increasing organic stocks in agricultural soils: knowledge gaps and potential innovations. *Soil Tillage Res.* 118: 42–51.

Corbeels, M., Naudin, K., Whitbread, A.M. et al. (2020). Limits of conservation agriculture to overcome low crop yields in sub-Saharan Africa. *Nat. Food* 1: 447–454.

Corsi, S., Friedrich, T., Kassam, A. et al. (2012). *Soil Organic Carbon Accumulation and Greenhouse Gas Emission Reduction from Conservation Agriculture: A Literature Review. Plant Protection Division Integrated Crop Management*, vol. 16, 89. Rome: FAO.

Das, A., Mohapatra, K.P., Ngachan, S.V. et al. (ed.) (2018). *Conservation Agriculture for Advancing Food Security in Changing Climate*. (2 volumes), 867. Today & Tomorrow's Printers and Publishers. ISBN 1-55528-439-6.

Das, T.K., Ghosh, S., Das, A. et al. (2021). Conservation agriculture impacts on productivity, resource-use efficiency and environmental sustainability: a holistic review. *Indian J. Agron.* 66: 111–127.

Dawson, N., Martin, A., and Sikor, T. (2016). Green revolution in sub-Saharan Africa: implications of imposed innovation for the wellbeing of rural smallholders. *World Dev.* 78: 204–218. https://doi.org/10.1016/j.worlddev.2015.10.008.

Dawson, I.K., Park, S.E., Attwood, S.J. et al. (2019). Contributions of biodiversity to the sustainable intensification of food production. *Glob. Food Sec.* 21: 23–37. https://doi.org/10.1016/j.gfs.2019.07.002.

Derpsch, R., Franzluebbers, A.J., Duiker, S.W. et al. (2014). Why do we need to standardize no-tillage research? *Soil Tillage Res.* 137: 16–22. https://doi.org/10.1016/j.still.2013.10.002.

DuPont, S.T., Culman, S.W., Ferris, H. et al. (2010). No-tillage conversion of harvested perennial grassland to annual cropland reduces root biomass, decreases active carbon stocks, and impacts soil biota. *Agric. Ecosyst. Environ.* 137 (1–2): 25–32. https://doi.org/10.1016/j.agee.2009.12.021.

FAO, IFAD, UNICEF, WFP, and WHO (2020). *The State of Food Security and Nutrition in the World 2020. Transforming Food Systems for Affordable Healthy Diets*. Rome, Italy: FAO. https://doi.org/10.4060/ca9692en.

Food and Agriculture Organization (2021). Supply Chain Joins Deforestation and Farming Practices as Main Source of Emissions in Agri-food Sector. https://www.fao.org/north-america/news/detail/en/c/1455485 (accessed 20 March 2023).

Fujisaki, K., Chevallier, T., Chapuis-Lardy, L. et al. (2018). Soil carbon stock changes in tropical croplands are mainly driven by carbon inputs: a synthesis. *Agric. Ecosyst. Environ.* 259: 147–158. https://doi.org/10.1016/j.agee.2017.12.008.

Garai, S., Mondal, M., and Mukherjee, S. (2020). Smart practices and adaptive technologies for climate resilient agriculture. In: (ed. S. Maitra and B. Pramanick), 327–358. Kolkata, India: New Delhi Publishers.

Gerber, P. et al. (2013). *Tackling Climate Change through Livestock: A Global Assessment of Emissions and Mitigation Opportunities*. Rome: Food and Agriculture Organisation of the United Nations (FAO). http://www.fao.org/3/a-i3437e.pdf.

Ghosh, S., Das, T.K., Sharma, D.K., and Gupta, K. (2019). Potential of conservation agriculture for ecosystem services: a review. *Indian J. Agric. Sci.* 89 (10): 1572–1579.

Giller, K. et al. (2015). Beyond conservation agriculture. *Front. Plant Sci.* 6: 870. https://doi.org/10.3389/fpls.2015.00870.

Godfray, H.C.J. and Garnett, T. (2014). Food security and sustainable intensification. *Philos. Trans. R. Soc. Lond. Ser. B Biol. Sci.* 369: 20120273.

Gonzalez-Sanchez, E.J., Kassam, A., Basch, G. et al. (2016). Conservation agriculture and its contribution to the achievement of Agri-environmental and economic challenges in Europe. *Aims Agricultural. Food* 1: 387–408. 10.3934/agrfood.2016.4.387.

Gonzalez-Sanchez, E.J., Veroz-Gonzalez, O., Moreno-Garcia, M. et al. (2020). Chapter 12: climate change adaptability and mitigation with conservation agriculture. In: *Rethinking Food and Agriculture: New Ways Forward* (ed. A. Kassam and L. Kassam), 231–246. ISBN: 978-0-12-816410-5. https://doi.org/10.1016/B978-0-12-816410-5.00012-8.

Govaerts, B., Verhulst, N., Sayre, K.D. et al. (2009). Conservation agriculture and soil carbon sequestration; between myth and farmer reality. *Crit. Rev. Plant Sci.* 28: 97–122.

Grace, P.R., Antle, J., Ogle, S. et al. (2010). Soil carbon sequestration rates and associated economic costs for farming systems of South-Eastern Australia. *Aust. J. Soil Res.* 48 (8): 720–729.

Haddaway, N.R., Hedlund, K., Jackson, L.E. et al. (2017). Potential carbon sequestration and nitrogen cycling in long-term organic management systems. *Renew. Agric. Food Syst.* 32: 498–510. https://doi.org/10.1017/S1742170516000429.

HLPE (2019). Agroecological and other innovative approaches for sustainable agriculture and food systems that enhance food security and nutrition. *HLPE on Food Security and Nutrition*, Report 14. Rome, Italy: HLPE. https://www.fao.org/3/ca5602en/ca5602en.pdf.

Hobbs, P.R., Sayre, K., and Gupta, R. (2008). The role of conservation agriculture in sustainable agriculture. *Philos. Trans. R. Soc. B: Biol. Sci.* 363: 543–555.

IPCC (2019). Climate Change and Land: An IPCC Special Report on Climate Change, Desertification, Land Degradation, Sustainable Land Management, Food Security, and Greenhouse Gas Fluxes in Terrestrial Ecosystems. https://www.ipcc.ch/srccl (accessed 20 March 2023).

Janzen, H.H., Campbell, C.A., Izaurralde, R.C. et al. (1998). Management effects on soil C storage on the Canadian prairies. *Soil Tillage Res.* 47 (3–4): 181–195.

Jat, R.K., Sapkota, T.B., Singh, R.G. et al. (2014). Seven-year-old conservation agriculture in rice-wheat rotation of eastern Gangetic Plains of South Asia: yield trends and economic profitability. *Field Crops Res.* 164: 199–210. 10.1016/j.fcr.2014.04.015.

Jat, H.S., Datta, A., Choudhary, M. et al. (2019). Climate smart agriculture practices improve soil organic carbon pools, biological properties and crop productivity in cereal-based systems of north-West India. *Catena* 181: 104059. https://doi.org/10.1016/j.catena.2019.05.005.

Jat, H.S., Choudhary, K.M., Nandal, D.P. et al. (2020). Conservation agriculture–based sustainable intensification of cereal systems leads to energy conservation, higher productivity and farm profitability. *Environ. Manag.* 65 (6): 774–786.

Jayaraman, S., Dang, Y.P., Naorem, A. et al. (2021). Conservation agriculture as a system to enhance ecosystem services. *Agriculture* 11 (8): 718. https://doi.org/10.3390/agriculture11080718.

Karbin, S., Kassam, A., Oza, A. et al. (2021). Initiating conservation agriculture shows reduced soil CO2 emissions and improved soil aggregate stability in the first season in rainfed cropping in India. *Int. J. Environ. Stud.* https://doi.org/10.1080/00207233.2021.1987050.

Kassam, A. (2022). Conservation agriculture for regenerative and resilient production systems. In: *Biological Approaches to Regenerative Soil Systems* (ed. N. Uphoff and J.E. Thies). CRC Press.

Kassam, A. and Kassam, L. (2020). *Rethinking Food and Agriculture: New Ways Forward.* Elsevier. ISBN 978-0-12-816410-5.

Kassam, A.H., Mkomwa, S., and Friedrich, T. (2017). Weather-proofing agriculture with conservation agriculture. In: *Conservation Agriculture for Africa: Building Resilient Farming Systems in a Changing Climate* (ed. A. Kassam, S. Mkomwa, and T. Friedrich), 10–29. Wallingford, UK: CAB International.

Kassam, A., Derpsh, R., and Friedrich, T. (2020). Development of conservation agriculture systems globally. In: *Advances in Conservation Agriculture Volume 1: Systems and Science* (ed. A. Kassam), 1–56. Cambridge, UK: Burleigh Dodds Science Publishing.

Kassam, A., Friedrich, T., and Derpsh, R. (2022). State of the global adoption and spread of conservation agriculture. In: *Advances in Conservation Agriculture. Volume 3: Adoption and Spread* (ed. A. Kassam). ISBN-13: 9781786764751.

Khan, R., Biswas, S., Kundu, C.K., and Jana, K. (2020). Effect of conservation tillage on yield and economics of fodder crops. *Int. J. Environ. Clim. Chang.* 10 (12): 529–539.

Knapp, S. and van der Heijden, M.G.A. (2018). A global meta-analysis of yield stability in organic and conservation agriculture. *Nat. Commun.* 9: 3632. https://doi.org/10.1038/s41467-018-05956-1.

Kopittke, P.M., Menzies, N.W., Wang, P. et al. (2019). Soil and the intensification of agriculture for global food security. *Environ. Int.* 132: 105078. https://doi.org/10.1016/j.envint.2019.105078.

Kopittke, P.M., Dalal, R.C., Hoeschen, C. et al. (2020). Soil organic matter is stabilized by organo-mineral associations through two key processes: the role of the carbon to nitrogen ratio. *Geoderma* 357: 113974. https://doi.org/10.1016/j.geoderma.2019.113974.

Kopittke, P.M., Menzies, N.W., Dalal, R.C. et al. (2021). The role of soil in defining planetary boundaries and the safe operating space for humanity. *Environ. Int.* 146: 106245. https://doi.org/10.1016/j.envint.2020.106245.

Lal, R. (2013). Enhancing ecosystem services with no-till. *Renew. Agric. Food Syst.* 28: 102–114. https://doi.org/10.1017/S1742170 512000452.

Lal, R. (2015). Sequestering carbon and increasing productivity by conservation agriculture. *J. Soil Water Conserv.* 70 (3): 55A–62A. https://doi.org/10.2489/jswc.70.3.55A.

Lal, R. (2019). Conceptual basis of managing soil carbon: inspired by nature and driven by science. *J. Soil Water Conserv.* 74 (2): 29A–34A. https://doi.org/10.2489/jswc.74.2.29A.

Lal, R. (2020). Regenerative agriculture for food and climate. *J. Soil Water Conserv.* 77 (3): 1a–2a. https://doi.org/10.2489/jswc.2020.0620A.

Lal, R., Monger, C., Nave, L., and Smith, P. (2021). The role of soil in regulation of climate. *Philos. Trans. R. Soc. Lond. Ser. B Biol. Sci.* 376 (1838): 20210420.

Li, Y., Li, Z., Cui, S. et al. (2019). Residue retention and minimum tillage improve physical environment of the soil in croplands: a global metaanalysis. *Soil Tillage Res.* 194: 104292.

Liao, Y., Wu, W.L., Meng, F.Q., and Lal, R. (2015). Increase in soil organic carbon by agricultural intensification in northern China. *Biogeosciences* 12: 1403–1413.

Maia, S.M.F., de Souza Medeiros, A., dos Santos, T.C. et al. (2022). Potential of no-till agriculture as a nature-based solution for climate-change mitigation in Brazil. *Soil Tillage Res.* 220: 105368. https://doi.org/10.1016/j.still.2022.105368.

Meurer, K.H.E., Haddaway, N.R., Bolinder, M.A., and Kätterer, T. (2018). Tillage intensity affects total SOC stocks in boreo-temperate regions only in the topsoil—a systematic review using an ESM approach. *Earth Sci. Rev.* 177: 613–622.

Minasny, B., Malone, B.P., McBratney, A.B. et al. (2017). Soil carbon 4 per mille. *Geoderma* 292: 59–86. https://doi.org/10.1016/j.geoderma.2017.01.002.

Mkomwa, S. and Kassam, A. (ed.) (2021). *Conservation Agriculture in Africa: Climate Smart Agricultural Development*. Wallingford, UK: CAB International.

Mondal, S. and Chakraborty, D. (2022). Global meta-analysis suggests that no-tillage favourably changes soil structure and porosity. *Geoderma* 405: 115443. https://doi.org/10.1016/j.geoderma.2021.115443.

Montgomery, D.R. (2020). Chapter 11: soil health and the revolutionary potential of conservation agriculture. In: *Rethinking Food and Agriculture: New Ways Forward* (ed. A. Kassam and L. Kassam), 219–229. ISBN: 978-0-12-816410-5. https://doi.org/10.1016/B978-0-12-816410-5.00011-6.

Moraru, P.I. and Rusu, T. (2013). Effect of different tillage systems on soil properties and production on wheat, maize, and soybean crop. *Int. J. Sci. Res. Innov.* 7: 679–692.

Mrabet, R., Bahri, H., Zaghouane, O. et al. (2022). Chapter 6: adoption and spread of conservation agriculture in North Africa. In. Kassam, A. (ed). Advances in Conservation Agriculture. Volume 3: Adoption and Spread. Cambridge, UK: Burleigh Dodds. ISBN: 978-1786764751. https://shop.bdspublishing.com/store/bds/detail/workgroup/3-190-106341.

Mugandani, R., Mwadzingeni, L., and Mafongoya, P. (2021). Contribution of conservation agriculture to soil security. *Sustainability* 13: 9857. https://doi.org/10.3390/su13179857.

Nabuurs, G.-J., Mrabet, R., Hatab, A.A. et al. (2022). Agriculture, forestry and other land uses (AFOLU). In: *IPCC, 2022: Climate Change 2022: Mitigation of Climate Change. Contribution of Working Group III to the Sixth Assessment Report of the Intergovernmental Panel on Climate Change* (ed. P.R. Shukla et al.). Cambridge, UK and New York, NY, USA: Cambridge University Press. https://doi.org/10.1017/9781009157926.009.

Newton, P., Civita, N., Frankel-Goldwater, L. et al. (2020). What is regenerative agriculture? A review of scholar and practitioner definitions based on processes and outcomes. *Front. Sustain. Food Syst.* 4: 577723. https://doi.org/10.3389/fsufs.2020.577723.

Nicoloso, R.S., Amado, T.J.C., and Rice, C.W. (2020). Assessing strategies to enhance soil carbon sequestration with the DSSAT-CENTURY model. *Eur. J. Soil Sci.* 71 (6): 1034–1049. https://doi.org/10.1111/ejss.12938.

Nyanga, P.H., Umar, B.B., Chibamba, D. et al. (2020). Chapter 6: reinforcing ecosystem services through conservation agriculture in sustainable food systems. In: *The Role of Ecosystem Services in Sustainable Food Systems* (ed. Rusinamhodzi, L.), 119–133. San Diego, CA: Academic Press.

Ogle, S., Alsaker, C., Baldock, J. et al. (2019). Climate and soil characteristics determine where no-till management can store carbon in soils and mitigate greenhouse gas emissions. *Sci. Rep.* 9: 11665.

Olson, K., Ebelhar, S., and Lang, J. (2013). Effects of 24 years of conservation tillage systems on soil organic carbon and soil productivity. *Appl. Environ. Soil Sci.* https://doi.org/10.1155/2013/617504.

Pacala, S. and Socolow, R. (2004). Stabilization wedges: solving the climate problem for the next 50 years with current technologies. *Science* 305 (5686): 968–972. https://doi.org/10.1126/science.1100103.

Page, K.L., Dang, Y.P., Menzies, N.W., and Dalal, R.C. (2020a). No-till systems to sequester soil carbon: potential and reality. In: *No-till Farming Systems for Sustainable Agriculture: Challenges and Opportunities* (ed. Y.P. Dang, R.C. Dalal, and N.W. Menzies), 301–317. Cham, Switzerland: Springer International Publishing.

Page, K.L., Dang, Y.P., and Dalal, R.C. (2020b). The ability of conservation agriculture to conserve soil organic carbon and the subsequent impact on soil physical, chemical, and biological properties and yield. *Front. Sustain. Food Syst.* 4: 31.

Palm, C., Blanco-canqui, H., Declerck, F. et al. (2013). Conservation agriculture and ecosystem services: an overview. *Agric. Ecosyst. Environ.* 187: 87–105. https://doi.org/10.1016/j.agee.2013.10.010.

Pittelkow, C.M. et al. (2015). Productivity limits and potentials of the principles of conservation agriculture. *Nature* 517: 365–368.

Post, W.M., Izaurralde, R.C., Jastrow, J.D. et al. (2004). Enhancement of carbon sequestration in US soils. *Bioscience* 54 (10): 895–908.

Powlson, D.S., Stirling, C.M., Jat, M.L. et al. (2014). Limited potential of no-till agriculture for climate change mitigation. *Nat. Clim. Chang.* 4 (8): 678–683. https://doi.org/10.1038/nclimate2292.

Powlson, D.S., Poulton, P.R., Glendining, M.J. et al. (2022). Is it possible to attain the same soil organic matter content in arable agricultural soils as under natural vegetation? *Outlook Agric.* https://doi.org/10.1177/00307270221082113.

Prestele, R., Hirsch, A.L., Davin, E.L. et al. (2018). A spatially explicit representation of conservation agriculture for application in global change studies. *Glob. Chang. Biol.* 24 (9): 4038–4053.

Puget, P. and Lal, R. (2005). Soil organic carbon and nitrogen in a Mollisol in Central Ohio as affected by tillage and land use. *Soil Tillage Res.* 80 (1–2): 201–213.

Rao, K.K., Samal, S.K., Poonia, S.P. et al. (2022). Conservation agriculture improves soil physical properties and crop productivity: a long-term study in middle Indo-Gangetic Plains of India. *Soil Res.* 60: 442–454. https://doi.org/10.1071/SR20329.

Reicosky, D.C. (2015). Conservation tillage is not conservation agriculture. *J. Soil Water Conserv.* 70 (5): 103A–108A.

Reicosky, D.C. (2020). Conservation agriculture systems: soil health and landscape management. In: *Advances in Conservation Agriculture Volume: Science and Systems* (ed. A. Kassam), 87–154. Cambridge, UK: Burleigh Dodds.

Reicosky, D.C. (2021). Carbon management in conservation agriculture systems. In: *Regenerative Agriculture* (ed. D. Dent and B. Boincean). Cham: Springer. https://doi.org/10.1007/978-3-030-72224-1_3.

Richards, M., Sapkota, T., Stirling, C. et al. (2014). *Conservation Agriculture. Implementation Guidance for Policymakers and Investors. Practice Brief Climate-Smart Agriculture*. Food and Agriculture Organization of the United Nations (FAO). http://www.fao.org/3/a-i4066e.pdf.

Riggers, C., Poeplau, C., Don, A. et al. (2021). How much carbon input is required to preserve or increase projected soil organic carbon stocks in German croplands under climate change? *Plant Soil* 460 (1–2): 417–433. https://doi.org/10.1007/s11104-020-04806-8.

Rinot, O., Levy, G., Steinberger, Y. et al. (2019). Soil health assessment: a review of existing knowledge and a proposed new approach. *Sci. Total Environ.* 648: 1484–1491.

Rosenstock, T.S., Mpanda, M., Rioux, J. et al. (2014). Targeting conservation agriculture in the context of livelihoods and landscapes. *Agric. Ecosyst. Environ.* 187: 47–51.

Ruis, S.J. and Blanco-Canqui, H. (2017). Cover crops could offset crop residue removal effects on soil carbon and other properties: a review. *Agron. J.* 109 (5): 1785–1180.

Sapkota, T.B., Jat, M.L., Aryal, J.P. et al. (2015). Climate change adaptation, greenhouse gas mitigation and economic profitability of conservation agriculture: some examples from cereal systems of Indo-Gangetic Plains. *J. Integr. Agric.* 14: 1524–1533. https://doi.org/10.1016/S2095-3119(15)61093-0.

Sapkota, T. et al. (2017). Soil organic carbon changes after seven years of conservation agriculture based rice-wheat cropping system in the eastern Indo-Gangetic plain of India. *Soil Use Manag.* 1–9. https://doi.org/10.1111/sum.12331.

Schmer, M.R., Jin, V.L., Wienhold, B.J. et al. (2020). Long-term rotation diversity and nitrogen effects on soil organic carbon and nitrogen stocks. *Agrosyst. Geosci. Environ.* 3: 1–9. https://doi.org/10.1002/agg2.20055.

Schreefel, L., Schulte, R.P.O., de Boer, I.J.M. et al. (2020). Regenerative-agriculture – the-soil-is-the-base. *Glob. Food Sec.* 26: 100404. https://doi.org/10.1016/j.gfs.2020.100404.

Searchinger, T. (2019). *Creating a Sustainable Food Future*. Washington, DC, USA: World Resources Institute.

Sharma, A.R. (ed.) (2022). *Conservation Agriculture in India: A Paradigm Shift for Sustainable Production*, 1e, 370. Routledge. https://doi.org/10.4324/9781003292487.

Sharma, A.R. and Biswas, A.K. (2022). Chapter 2. History and current scenario of conservation agriculture. In: *Conservation Agriculture in India: A Paradigm Shift for Sustainable Production*, 1e (ed. A.R. Sharma). Routledge.

Sinclair, F.L. and Coe, R. (2019). The options by context approach: a paradigm shift in agronomy. *Exp. Agric.* 55 (S1): 1–13.

Sinclair, F., Wezel, A., Mbow, C. et al. (2019). *The Contribution of Agroecological Approaches to Realizing Climate-Resilient Agriculture*. Background Paper. Rotterdam and Washington DC: Global Commission on Adaptation. https://gca.org/reports/the-contributionsof-agroecological-approaches-to-realizing-climate-resilient-agriculture.

Smith, P., Martino, D., Cai, Z. et al. (2007). Agriculture. In: *Climate Change 2007* (ed. B. Metz and O. Davidson), 497–540. Cambridge, UK, Mitigation: Cambridge University Press.

Smith, P., Calvin, K., Nkem, J. et al. (2019). Which practices co-deliver food security, climate change mitigation and adaptation, and combat land-degradation and desertifcation? *Glob. Chang. Biol.* gcb14878. https://doi.org/101111/gcb14878.

Sperow, M. (2016). Estimating carbon sequestration potential on US agricultural topsoils. *Soil Tillage Res.* 155: 390–400. https://doi.org/10.1016/j.still.2015.09.006.

Su, Y., Gabrielle, B., and Makowski, D. (2021a). The impact of climate change on the productivity of conservation agriculture. *Nat. Clim. Chang.* 11: 628–633. https://doi.org/10.1038/s41558-021-01075-w.

Su, Y., Gabrielle, B., Beillouin, D., and Makowski, D. (2021b). High probability of yield gain through conservation agriculture in dry regions for major staple crops. *Sci. Rep.* 11: 3344. https://doi.org/10.1038/s41598-021-82375-1.

Su, Y., Gabrielle, B., and Makowski, D. (2021c). A global dataset for crop production under conventional tillage and no tillage systems. *Sci. Data* 8: 1–18. https://doi.org/10.1038/s41597-021-00817-x.

Sykes, A.J., Macleod, M., Eory, V. et al. (2020). Characterizing the biophysical, economic and social impacts of soil carbon sequestration as a greenhouse gas removal technology. *Glob. Chang. Biol.* 26 (3): 1085–1108. https://onlinelibrary.wiley.com/doi/full/10.1111/gcb.14844.

Tellez-Rio, A., Vallejo, A., García-Marco, S. et al. (2017). Conservation agriculture practices reduce the global warming potential of rainfed low N input semi-arid agriculture. *Eur. J. Agron.* 84: 95–104.

Thierfelder, C., Matemba-Mutasa, R., and Rusinamhodzi, L. (2015). Yield response of maize (Zea mays L.) to conservation agriculture cropping system in southern Africa. *Soil Tillage Res.* 146: 230–242.

Thomson, A.M., Izaurralde, R.C., Smith, S.J., and Clarke, L.E. (2008). Integrated estimates of global terrestrial carbon sequestration. *Glob. Environ. Change-Human Policy Dimens.* 18 (1): 192–203. https://doi.org/10.1016/j.gloenvcha.2007.10.002.

Tubiello, F.N., Rosenzweig, C., Conchedda, G. et al. (2021). Greenhouse gas emissions from food systems: building the evidence base. *Environ. Res. Lett.* 16 (6): 065007. https://doi.org/10.1088/1748-9326/ac018e.

VandenBygaart, A.J. (2016). The myth that no-till can mitigate global climate change. *Agric. Ecosyst. Environ.* 216: 98–99.

Virk, A.L., Lin, B.J., Kan, Z.R. et al. (2022). Simultaneous effects of legume cultivation on carbon and nitrogen accumulation in soil. *Adv. Agron.* https://doi.org/10.1016/bs.agron.2021.08.002.

Virto, I., Barré, P., Burlot, A., and Chenu, C. (2012). Carbon input differences as the main factor explaining the variability in soil organic C storage in no-tilled compared to inversion tilled agrosystems. *Biogeochemistry* 108: 17–26.

Wall, P., Thierfelder, C., Hobbs, P. et al. (2020). Benefits of conservation agriculture. In: *Advances in Conservation Agriculture Volume 2: Practice and Benefits*, vol. 2 (ed. A. Kassam), 40. London, UK: Burleigh Dodds and Publishing Science.

Wang, H., Wang, S., Yu, Q. et al. (2020). No tillage increases soil organic carbon storage and decreases carbon dioxide emission in the crop residue-returned farming system. *J. Environ. Manag.* 261 (1): 110261. https://doi.org/10.1016/j.jenvman.2020.110261.

West, T.O. and Post, W.M. (2002). Soil organic carbon sequestration rates by tillage and crop rotation: a global data analysis. *Soil Sci. Soc. Am. J.* 66: 1930–1946.

Wilson, G.V., Dabney, S.M., McGregor, K.C., and Barkoll, B.D. (2004). Tillage and residue effects on runoff and erosion dynamics. *Trans. ASAE* 47: 119–128. https://doi.org/10.13031/2013.15878.

World Bank (2012). *Carbon Sequestration in Agricultural Soils, 67395-GLB*. Washington, D.C.: World Bank.

Wreford, A., Ignaciuk, A., and Gruère, G. (2017). Overcoming barriers to the adoption of climate-friendly practices in agriculture. In: *OECD Food, Agriculture and Fisheries Papers*, No. 101. Paris: OECD Publishing. https://doi.org/10.1787/97767de8-en.

Yang, X.M., Drury, C.F., and Wander, M.M. (2013). A wide view of no-tillage practices and soil organic carbon sequestration. *Acta Agric. Scand. Sect. B-Soil Plant Sci.* 63 (6): 523–530. https://doi.org/10.1080/09064710.2013.816363.

Yigezu, Y.A., El-Shater, T., Boughlala, M. et al. (2021). Can an incremental approach be a better option in the dissemination of conservation agriculture? Some socioeconomic justifications from the drylands of Morocco. *Soil Tillage Res.* 212: 105067. https://doi.org/10.1016/j.still.2021.105067.

Zhang, H.L., Lal, R., Zhao, X. et al. (2014). Opportunities and challenges of soil carbon sequestration by conservation agriculture in China. In: *Advances in Agronomy*, vol. 124 (ed. D.L. Sparks), 1. San Diego, Advances in Agronomy: Elsevier Academic Press Inc.

Zomer, R.J., Bossio, D.A., Sommer, R., and Verchot, L.V. (2017). Global sequestration potential of increased organic carbon in cropland soils. *Sci. Rep.* 7: 810. https://doi.org/1038/s41598-017-15794-8.

16

The Ecology of Intercropping Systems, Tree-Cover Dynamics of Grazing Lands, and Cover Crops for Soil Management

Chitranjan Kumar[1], *Anil K. Singh*[2], *Deepak R. Joshi*[3], *and David E. Clay*[3,4]

[1] *Amity Institute of Organic Agriculture, Amity University, Uttar Pradesh, Noida, India*
[2] *Shri Murli Manohar Town Post Graduate College, Ballia, India*
[3] *Department of Plant Sciences, South Dakota State University, USA*
[4] *American Society of Agronomy, Madison, USA*

16.1 Introduction

The ecology of cropping systems is a broad concept that means different things to different people. To a governmental official, it might mean food security, whereas to an agronomist, soil, or crop scientist it might suggest that research is required to quantify the effects of cover crops, inter-cropping systems, tree-forage dynamics, conservation tillage, crop rotations, and integrated crop-animal systems on the ecological integrity of the system (Branca et al. 2013; Joshi et al. 2021). Tree and forest dynamics highlight the physical and biological forces responsible for shaping the changing of the forest composition and structure over time. The two basic elements of forest dynamics are (i) disturbance and (ii) succession. An event causing a negative change in forest structure, forest composition, forest resources, and the physical environment is known as forest disturbance; however, a gradual replacement of one community of plants by another is known as succession. Organic matters are the natural sources of food for microorganisms. The low organic matter content of agricultural soils is becoming a global concern. Cover crops are playing beneficial roles in reducing the need for off-field organic matter additions in soils, replenishing the active organic matter lost during cultivation, and improving soil structure and tilth. This chapter will focus on research surrounding the ecology of sustainable systems.

Sustainable systems as defined by a soil health expert often highlight the importance of maintaining a living root system, providing soil cover, minimizing soil disturbance, and using a diverse plant rotation that includes livestock (Dumanski et al. 2006; Clay et al. 2014; Joshi et al. 2021). The implementation of locally adjusted sustainable systems has been shown to prevent land degradation and reduce greenhouse gas emissions (Reicks et al. 2021), whereas the failure to implement such practices can decrease ecosystem stability (Danise et al. 2021; Guo et al. 2021; Scavo et al. 2021; Xu et al. 2021; Liu and

Agroecological Approaches for Sustainable Soil Management, First Edition.
Edited by Majeti Narasimha Vara Prasad and Chitranjan Kumar.

Gong 2022; Ma et al. 2022; Wang et al. 2022). There is a gap between what we require and what we implement for attaining sustainable soil systems. For example, a continuous decline in soil organic matter (SOM) and a continuous increase in greenhouse (GHG) emissions from agriculture and allied disciplines is not sustainable, whereas an increase in SOM and a decrease in GHG emissions may be sustainable. The objectives of the present chapter are to highlight sustainable agroecological systems under (i) intercropping systems, (ii) tree forage dynamics and forest management, (iii) cover crops, and (iv) integrated agroecological and co-culture systems for developing sustainable soil management systems.

16.2 Intercropping Systems

Intercropping is the growing of at least two crops at the same time and location. An intercropping system might be two crops that are planted in alternating strips across a field, or it might be groups of plants that are bunched together across a landscape. Benefits from intercropping range from improved pest resistance and resistance to adverse conditions.

Intercropping provides stability of yields under different seasons, better use of growth resources, erosion control, and better control of weeds, pests, and diseases. One crop can provide physical support to others; for example, mango and coconut can support the black pepper vine. One crop can provide shelter to others; for example, albizzia can provide shade to tea plants. On the other hand, crop competitions, allelopathy, date of planting cover crops, and mechanical operations under cover cropping are limiting its adoption in agriculture. There are three broad categories of intercropping competition: (i) mutual inhibition, (ii) mutual cooperation, and (iii) compensation. When the actual yield of each species is less than expected, it is called mutual inhibition, a rare case of competition. When the yield of each species is greater than expected, it is called mutual cooperation. When the yield of one species is less and the other is more than expected, it is called compensation. Any direct or indirect harmful effects that one plant causes on another through the chemical substances or toxins is referred to as allelopathy. Thus, some crops may be unsuitable as intercrops because they can secrete toxins.

Benefits from intercropping have been observed in many ecosystems. For example, intercropping teak forests with short-duration timber species (*Gmelina arborea, Acacia mangium*, etc.) improved the income of smallholders and brought environmental sustainability to silvicultural practices (Sudomo et al. 2021). In a similar example, modeling research suggested that intercropping a bamboo forest (IBF) with rice and other crops could generate 66.9% higher income, increase carbon storage by 22%, soil conservation by 52.6%, nitrogen retention by 28.5%, phosphorus retention by 19.2%, and habitat quality by 17.9% (Ma et al. 2022). Intercropping of sun hemp and millets during the re-planting of sugarcane increased soil porosity, Ca and K contents, and soil aggregates stability, while reducing the soil bulk density (Farhate et al. 2020). Liu et al. (2021) reported that the sugarcane-soybean intercropping system in Guangxi, China increased the number of N-fixing bacteria, whereas, in Uttar Pradesh, India, Shukla et al. (2022) observed that intercropping autumn sugarcane with vegetables (rajmash, winter maize, and garlic) increased profitability.

The impact of intercropping on enhanced economic returns may be related to changes in the soil microbial community structure (Liu and Gong 2022). A minimum tillage organic fertigation system had higher phospholipid fatty acids, soil organic carbon (SOC), and microbial activity than conventional fertilizer (Morugán-Coronado et al. 2022; Kumar et al. 2022). Cuartero et al. (2022) reported that the melon-cowpea intercropped system had a larger soil microbial community, acid phosphatase, and glucosidase activities than a melon monoculture, whereas, in northern China, winter rapeseed reduced wind erosion and increased the economic benefits (Ma et al. 2019).

Change in management practices across agro-ecological zones from continuous maize monocropping to conservation agriculture with maize-legume intercropping potentially improved SOM and soil hydraulic properties in Malawi (Simwaka et al. 2020). Lopez-Ridaura et al. (2021) showed that maize intercropped with bean, potato, and faba was a promising intercropping system. Mwakilili et al. (2021) reported that a Maize-*desmodium* intercropped system improved soil health by increasing endophytes, mycorrhiza, decomposers, saprophytes, and bioprotective fungi. Maize-soybean intercropping largely benefits maize crops rather than soybean; however, the system benefits productivity, soil conservation, and soil fertility through the control of diseases, pests, and weeds (Blessing et al. 2022). Song et al. (2022) isolated the phosphorus solubilizing bacteria *Bacillus aryabhattai* (B8W22) from the soybean rhizosphere in the maize-soybean intercropping system. These bacteria improved P-solubilization through enhanced acid phosphatase activity and produced Indole 3-acetic acid and siderophores, that improved maize seed germination and seedling growth. Intercropping Chicory (*Cichorium intybus* L.) with Odot grass (*Pennisetum purpureum* cv. Mott) yielded maximum production (cutting at 60 days after sowing) of 7.95 and 15.14 Mg ha^{-1}, respectively (Zaini et al. 2021). The intercropping system improved soil nutrient status and minimized fertilizer requirements for successive crops. Similarly, the pasture phase with livestock management of a grass-legume intercropping system (*Vigna unguiculata* + *Glycine max*) under no-tillage increased soybean productivity along with enriched total organic carbon (SOC), microbial biomass carbon (MBC), and soil basal respiration (SBR) (Silva et al. 2022). An increase in SOC under different intercropping systems is presented in Table 16.1.

Valorization of legume-agrobiodiversity in the Mediterranean region can improve water use, soil fertility, ecological safety, and produce healthier foods (Martinelli et al. 2022). Schröder et al. (2022) relaunched the cropping system by integrating high C-amendments, macro-and microelements, and soil microbiome on marginal soils for land reclamation to improve C-sequestration, soil conservation, and water use efficiency, keeping synthetic fertilizer application at a minimum. The intercropping system is becoming vital to crop productivity and soil sustainability. The fractal dimension of soils improved toward a fine-grained structure having better soil-water holding capacity and increased nutrient cycling under the intercropping of *Salvia miltiorrhiza* with forage, and *S. miltiorrhiza* with tree species (Xu et al. 2021).

New rubber tree plantations in the Asia tropical regions are routinely damaged by storms and extreme climatic conditions. The intercropping of rubber trees (*Hevea brasiliensis*) with native flowering trees Champaka (*Michelia macclurei*) and Laosy (*Mytilaria laosensis*) reduced the maximum wind speed by 8.4% and the average wind speed by 19.6%, respectively, as compared to rubber tree monocropping (Qi et al. 2021). Rice cultivation in Asia is affected by the water crisis, soil degradation, and climate change. The system of rice intensification (SRI) has been proposed as a water-saving and sustainable approach for

Table 16.1 Increase in soil organic carbon due to intercropping with crops/trees under various locations, crop rotations, and soil types.

Location	Soil type	Crop rotation	Soil depth	Duration (years)	Intercropping type	% SOC increase	References
China	Clay loam	Salvia-*Lagerstroemia*	0–20 cm	20	Lagerstroemia	57	Xu et al. (2021)
China	Clay loam	Bamboo-rice	0–30 cm	20	Rice	22	Ma et al. (2022)
Taiwan	Silt loam	Broadleaf forest-sugarcane-vegetables	0–20 cm 10–20 cm	21.1 21.1	Sugarcane-cabbage Sugarcane-vegetables	101 65	Lin et al. (2021) Lin et al. (2021)
India	Sandy loam	Dalbergia-chillies	0–30 cm	2	Chillies	70	Sivakumar et al. (2022)
Poland	Sandy loam	Oat-pea	0–15 cm	3	Pea	51.3	Hajduk et al. (2015)
Italy	Sandy loam	Poplar-walnut	0–15 cm	20	Poplar-walnut-Italian alder	73	Danise et al. (2021)
Southern Africa	Sandy loam	Maize-gliricidia	0–20 cm	14	Gliricidia	12	Beedy et al. (2010)

more than 4 billion people across the globe. Intercropping beans with rice under SRI increased rice yield by 33% and decreased water use by 40% and weed infestation by 65% (Shah et al. 2021). Thus, the intercropping system can establish a buffer zone for the sustainable management of agricultural systems. Gliricidia-maize intercropping over a cycle of 20 years improved soil fertility in the dryland areas of Tanzania, Africa, and was found profitable with a high benefit: cost ratio (BCR = 4.27) for the growers (Swamila et al. 2022).

16.3 Sustainable Forest Management

Global warming, extreme weather, forest fire, and land-use changes are emerging topics of discussion. Forest disturbances vary in frequency, type, spatial distribution, and severity. The different types are wind, flood, ice storms, fire, insects, diseases, grazing, road construction, timber harvest, and conversion to croplands or non-forest lands. It is high time for human civilization to protect and manage natural bioresources. Regeneration of old-grown forests as well as sustainable forest management enhance biodiversity, soil microbial biomass, C-sequestration, and maintains soil health (Martin et al. 2020), and partial cutting offers a promising way to improve forest sustainability. Topography (elevation and convexity) of lands and soil properties affect tree species distribution, structure, and diversity of tree communities (Rodrigues et al. 2021). Litter deposition, rainfall variation, stand structure, and microbial colonization significantly affected SOC and other soil nutrient dynamics in home gardens of Bangladesh (Baul et al. 2022).

We need a better understanding of tree-based intercropping systems on soil fertility and crop production. Replacing native tree species in a forest with plantation trees can also reduce SOC in a converted forest ecosystem. A 10-year conversion of forest lands to oil palm plantations reduced SOC by 10 $Mg C ha^{-1}$ in Indonesia (Handayani et al. 2020). Among different age groups of palm plantations, SOC continuously increased from a 5-year oil palm plantation to successive 10-year, 15-year, and 25-year (Handayani et al. 2020). Similarly, conversion of pine-coffee agroforestry into pine-cabbage agricultural practices reduced SOC by 40% and litter fall by 80% (Prayogo et al. 2021). Thus, we must pay careful maintenance to a natural agroforestry system, and it is not wise to convert agroforestry into croplands.

Forest regeneration or afforestation has become a viable option to store up to 13 $Mg ha^{-1}$ in surface (0–10 cm) and 6.7 $Mg ha^{-1}$ in subsurface (10–20 cm) soils (Lin et al. 2021). Asigbaase et al. (2021) quantified nutrient return to soils through litterfall in organic and conventional cocoa agroforestry systems in Ghana. They observed higher mean tree litterfall (5.0 $Mg ha^{-1} yr^{-1}$) under the shade tree of organic cocoa agroforestry than in conventional systems (3.3 $Mg ha^{-1} yr^{-1}$). The use of high organic-based Integrated Soil Fertility Management (ISFM) information by cocoa farmers is contributing to the development of sustainable systems in Ghana; however, the demand for and access to ISFM information is continuously increasing (Quaye et al. 2021). Baietto et al. (2021) determined the nutrient release dynamics of litterfall from *Eucalyptus grandis* Hill and *Pinus taeda* L. In this study, *P. taeda* had a higher nutrient release and availability than *E. grandis*. These findings suggest that forest tree rotations can be used to reduce nutrient stresses. Liu et al. (2022) observed that a conducive walnut-soybean agroforestry system had higher soybean growth than a soybean monoculture.

The traditional land use is transforming toward an economically attractive tree-based farming system using *Acacia decurrens* plantation in Guder (highland) Ethiopia (Berihun et al. 2019). A mixed tree intercropping system increased soil carbon and nitrogen, and enhanced enzyme activities in Central Italy (Danise et al. 2021). The intercropping system included *Populus alba* and *Juglans regia* (economic trees), and *Alnus cordata* (N-fixing trees) have been used to restore soil fertility. Land-use influences carbon sequestration and changes from *A. decurrens* plantations to croplands reduced SOC by 30.2% and increased CO_2 emissions of >8.0 Mg CO_2 $ha^{-1} yr^{-1}$. These changes also reduced soil N, available-P, and the soil cation exchange capacity (CEC) (Tilahun et al. 2022). If the reverse occurred, then SOC increased by 27.7% (Tilahun et al. 2022). Clearcutting in Japanese cedar agroforestry increased nitrate leaching by 3.3–4.8 $kg N ha^{-1} yr^{-1}$ due to the absence of N-uptake by trees and higher mineralization of SOM; however, higher dissolved organic carbon under cedar agroforestry systems reduced SOC loss and nitrate leaching (<5 $kg N ha^{-1} yr^{-1}$) in soil profiles due to the effect of slash treatment in clearcutting and the slow dieback in stem girdling (Fujii et al. 2021).

Recent documentation on climate change emphasizes the need to protect forest strands and to restore degraded lands to a forest. However, due to the high cost of forest restoration, a subsidy may be needed (Korneeva 2021). Coniferous and broad-leaved mixed forests have higher aggregates of SOC than chestnut forests, hawthorn forests, and apple forests by 1.69%, 3.87%, and 4.5%, respectively (Guo et al. 2021). The different growth stages of forests (initial, optimal, and decay stages) significantly affect C, N, and P contents and C/N and

N/P ratios in the fine roots and humus contents under a temperate Beech-forest (Azaryan et al. 2021). These findings suggest that we need to maintain tree diversity to maintain sustainability. The current rate of production of natural teak plantations is very slow, therefore, a clonal teak pattern has been suggested for fastening teak plantations and sustaining ecological systems. Clonal teak pattern (2.39) was found to be the best agroforestry system followed by the teak pattern (2.08), and the pine pattern (2.06) using multicriteria analysis (Sadono et al. 2019). The selection of crops should be based on root trait syndrome and soil nutrient acquisition in an organic agroecosystem (Isaac et al. 2021). Forms of organic phosphorus, the mechanism of root traits, and rhizo-microbiome have been correlated for better management of P-fertilization in soils (Amadou et al. 2021). In a phosphorus-deficient soil, olive trees manage to satisfy their P-demands through the accumulation of sufficient organic matter and balancing P-mineralization in soils (Chatzistathis et al. 2021). Pan et al. (2021) positively correlated total organic carbon and total nitrogen with the fungal Shannon index which was found beneficial for enhanced enzyme activities in soils. Silva-Olaya et al. (2022) reported effective Silvopastoral Systems (SPS) to recover soil health index (SHI) through positive soil functioning in terms of nutrient dynamics, moisture balance, and microbiological activities. The SHI was 24% higher under SPS compared to pasture lands. Tree Outside Forest (TOF) leads to a spatiotemporal increase in tree vegetation in 1453.9 km^2 in Bangladesh increasing C-storage by 9 Tg C (Islam et al. 2022). A diamond pattern of tree removal (thinning) operated in hybrid poplar for consecutive seven-years after canopy closure showed a rapid growth (diameter at breast height, stem volume, stand volume, and stand biomass) of poplar buffers in Canada (Fortier et al. 2022). Yields of Cassava and chilies were ($P < 0.05$) higher up to 17.8 and 1.6 $t\,ha^{-1}$, respectively than Cassava and chilies along with *Dalbergia sissoo* grown in Karnataka, under the Horti-silvicultural system (Sivakumar et al. 2022). Cheng et al. (2022) emphasized transformative agroecological urbanism involving active interactions among policymakers, researchers, farmers, and local people improving socioeconomic synergy for achieving sustainable goals. Transformative agroecology is becoming equipped with Agricultural Knowledge and Innovative Systems (AKIS). Giagnocavo et al. (2022) reported a case study of four niches (Biological Control, Ecological Restoration, Soil Health, and Ecological Pond Management) under the AKIS greenhouse farm in Spain, indicating the role of 'farmer's interaction with nature and ecological services. Similarly, Mariya et al. (2019) suggested restoration strategies and a task force comprising farmers and other stakeholders for improving ecological water flow in the Indo-Gangetic region, India. The co-culture of crops and aquatic animals is also becoming promising in maintaining high crop yield, increasing microbial stability, and improving soil sustainability (Li et al. 2022).

16.4 Cover Crops for Sustainable Soil Management

Cover crops are non-cash crops that are grown to reduce erosion, improve soil structure, and increase water infiltration and SOC in soils. There are many different types of cover crop systems. For example, in annual cropping systems, cover crops can be planted following the harvest of the cash crop. However, cover crops can also be seeded into a growing crop. Cover crop utilized soil nutrients and water which depending on the location can

Table 16.2 Increase in soil organic carbon (SOC) due to cover crops under various locations, crop rotations, and soil types.

Location	Soil type	Crop rotation	Soil depth	Duration (years)	Cover crop type	% SOC increase	References
India	Sandy clay loam	Corn-groundnut	0–15 cm	5	Green gram	40.3	Ansari et al. (2022)
Brazil	Sandy clay	Corn-corn	0–20 cm	16	Velvet bean	27.1	Bettiol et al. (2022)
Brazil	Sandy clay	Cassava-Brachiaria	0–20 cm	2	Brachiaria	18.9%	Soares et al. (2022)
Kenya	Clay loam	Corn-corn	0–15 cm	2	Mucuna prurier	3.6	Ngome et al. (2011)
USA	Silt loam	Corn-corn	0–10 cm	3	Rye	38.6	Cates et al. (2019)
Italy	Sandy loam	Apricot-clover	0–20 cm	4	Clover	15%	Scavo et al. (2021)
Germany	Sandy loam	Corn-pea	0–30 cm	19	Pea-clover	12%	Seitz et al. (2022)

increase or decrease cash crop yields. The cover crop impact on the cash crop yield may be related to several factors including improvements in soil structure and SOC. An increase in SOC due to cover crops under various locations, crop rotations, and soil types is presented in Table 16.2. Qi et al. (2022) showed that incorporating straw and ryegrass cover crops in spring barley, increased SOC and soil aggregate stability, and decreased clay dispersibility. Scavo et al. (2021) reported that Mediterranean orchard soils increased SOM by 15%, ammonium-N by 194%, nitrate-N by 308%, assimilable P by 5%, exchangeable K by 16%, Fe by 15%, Zn by 36%, and Cu by 24%. The cover crop *Trifolium subterraneum* also suppressed the weed seed bank size by 54% and reduced the need for intensive tillage operations and synthetic fertilizer requirements. The selection of suitable cover crops, their growth duration, and methodologies of termination also impact soil C-sequestration in Mediterranean agroecosystems (Moukanni et al. 2022). The incorporation of cover crops in conventional farming resulted in reduced soil bulk density, improved soil structure, soil nutrients, hydraulic properties, and water infiltration (Koudahe et al. 2022). Rye cover crop significantly increases total microbial biomass ($P = 0.004$), bacteria ($P = 0.001$), and fungi ($P = 0.03$) after two years in the corn-corn cropping system in South Dakota, USA (Figure 16.1) (Joshi et al. 2021).

The emission of greenhouse gases from agriculture is a global concern. Cover crops may also affect greenhouse gas emissions. The study of crop-specific emissions of N_2O in the metropolitan areas of Barcelona reveals that rainfed cereals (oats, barley, wheat, Triticale), and almonds have lower N_2O emissions than the irrigated horticultural crops (tomato and artichoke) (Beltran et al. 2022). Olofsson and Ernfors (2022) measured N_2O emissions after frost dying of three cover crops (oilseed radish, phacelia, and oats) for selecting a suitable cover crop in agriculture. They observed minimum N_2O emissions (in kg N_2O-N year^{-1})

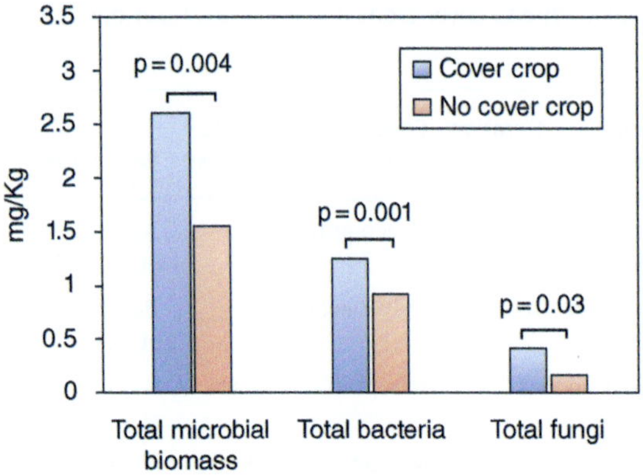

Figure 16.1 Impact of rye cover crop on total microbial biomass, bacteria, and fungi after two years in corn-corn cropping system in South Dakota, USA. *Source:* Adapted from Joshi et al. (2021).

from oats (0.6), followed by phacelia (0.7), and oilseed radish (1.8) but the emission from oats was higher than control (without cover crops). In a two-years study, both sheep grazing and tillage of cover cropped area (oats, beans, and vetch) produced higher emissions of N_2O, CH_4, and CO_2 in a California vineyard. These findings were attributed to higher nitrification rate in the cropped than vineyard soil s (Lazcano et al. 2022). Gou et al. (2021) reported that returning the non-harvested crop residues and using residual film mulching in a maize-wheat intercropping system reduced CO_2 emissions from arid soils. These studies highlight the need for further research in the field of cover crops to minimize greenhouse emissions from soils.

Planting dates of cover crops impact nutrient assimilation and growth of subsequent commercial crops. Different cover crop species (radish, phacelia, etc.) grown in early- (14 August), mid- (7 September), and late-stage (27 September) indicated different dates as their best planting date. Radish sown on 14th August accumulated a maximum of 261 kg N ha^{-1}, and phacelia sown on 27th September outperformed others and accumulated 70 kg N ha^{-1} (Cottney et al. 2022). The study reveals that non-legume crops should be grown from mid-August to early September in the agroecological zones similar to that of Northern Ireland. Five species of cover crops planted in the early season performed 43–53% better crop biomass than mid-season cover crops under ambient climatic conditions in semi-arid agroecosystems, leading to weed suppression, temperature reduction, and soil moisture maintenance in soils (DuPre et al. 2022).

Cover crops are becoming an important source of ecological management and a prime focus for agricultural sustainability (Das et al. 2022). In a seven-year study, rotations with spinach, Chinese cabbage, coriander, and leafy lettuce improved rhizospheric immunity under cucumber plantation by suppressing plant diseases through the fungal microbiome under a plastic shed system in China (Ali et al. 2022). The spinach rotation showed suppression of 13.2% in fusarium wilt in the spring season, and the Chinese cabbage rotation showed suppression of 8.9% in nematode infestation (Ali et al. 2022). Cover crop (vetch) and conventional tillage under long-term fertilization management of cotton production significantly affected functional denitrifying and nitrifying bacterial networks, respectively (Duan et al. 2022). Similarly, vetch-oat cover cropping improved N-cycling and over yielded across soil and management farm gradients as compared to monocropping (Stratton et al. 2022).

Developing sustainable food systems is a global challenge. We need transdisciplinary and action-oriented working models for continuous monitoring of our natural resources such as forest cover, land cover, land use, dynamics of hydraulics, soil nutrient status, especially C-dynamics, and emissions of greenhouse gases from agriculture. We need transformative agroecological urbanism and ruralism with high-income and low-income groups, respectively. Digitization and installation of agricultural knowledge and innovation systems and greenhouse farms are other strategies to sensitize farmers and growers about nature and ecological services (Giagnocavo et al. 2022). Developing a sustainable farm balancing trees with crops and establishing a co-cultural diversity around an agroecological farm will be a milestone in agricultural sustainability (Figure 16.2). Such agroecological farms will be indigenous and community-based experimental laboratories for the adoption of innovative technologies in agriculture, and the generation of employment through sustainable farms.

16.5 Conclusion

Intercropping systems, agroforestry, and cover cropping are becoming eco-friendly, socially acceptable, commercially viable, and resource-generating cropping systems under conservation agriculture to achieve soil sustainability. Intercropping tree plantations with cover crops enrich soil microbial activity, nutrient retention, SOC, and their ability to mitigate adverse climatic conditions. Plant residues, organic wastes, and tree litters are being used as soil conditioners to sustain SOM and improve soil health. Legumes and forage in crop rotations (maize-legume and maize-forage) improve soil flora and fauna, water quality/water use efficiency, and biological N-fixation. They optimize the activities of acid phosphatase, glucosidase, IAA, and siderophores in soil–plant systems. These sustainable farm systems improve nutrition, productivity, and profitability to provide protein-rich foods and reduce the risk of a monoculture system.

The arid and semi-arid regions and the Mediterranean area benefit more through improved water use, soil fertility, ecological safety, and healthier foods under integrated tree-crop intercropping and cover crops. The coniferous and broad-leaved mixed forests and natural agroforestry systems need special attention to preserve their higher aggregates of SOC. Cover crops serve as food sources for vertebrates, invertebrates, and microbes. They attract beneficial insects, stimulate soil life, and increase the diversity of soil organisms. Cover crop foliage shields soils from the rain-splashing and elevates water infiltration rates, establishing a micro-climate in soils. They cause mulching effects, keep the soil cool in summer, and protect organic matter from rapid oxidation.

Agroecological model systems should exist everywhere in rural and urban areas as an essential part of land use management. At the same time, a regular and interactive dialogue among policymakers, researchers, farmers, and local people is a must for improving socioeconomic synergy for achieving sustainable goals. We suggest an integration of tree-crop intercropping with cover crops to maintain a living root system, providing maximum soil cover, minimizing soil disturbances, increasing soil porosity, reducing soil bulk density, and diversifying crop rotations. The integration system should also include co-culturing such as cover crops-livestock and rice-crayfish-turtle. The system benefits agricultural

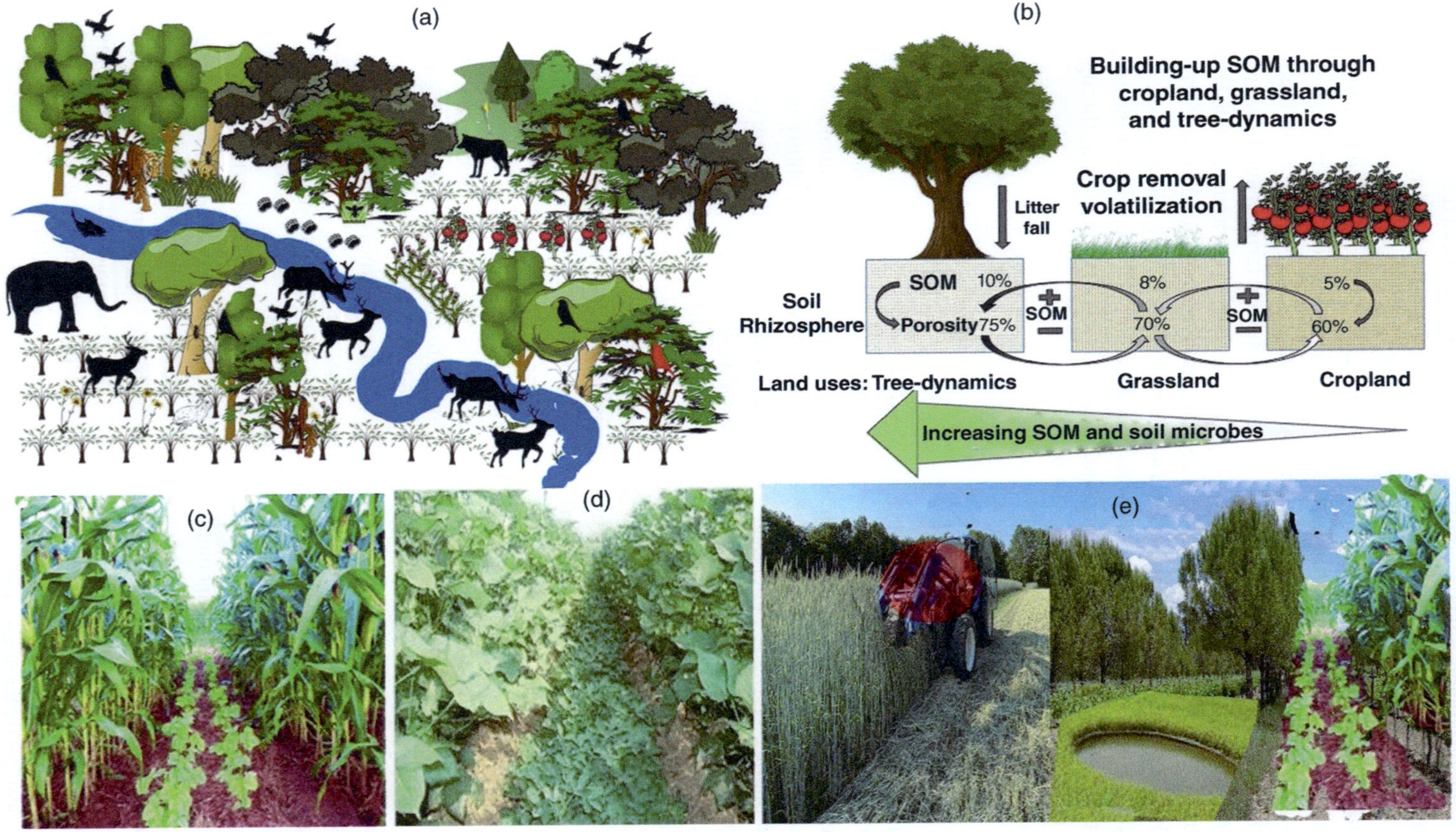

Figure 16.2 A visual presentation of the ecology of forests, intercropping, tree-forage dynamics, and cover crops: (a) forest ecology, (b) tree-forage dynamics, SOM = soil organic matter, (c) maize-soybean inter cropping, (d) cotton-peanuts intercropping, and (e) developing a sustainable farm balancing tree, diversified crops, and co-culture systems.

productivity, soil conservation, soil fertility, and biocontrol of diseases, pests, and weeds in agriculture. In addition, returning the non-harvested crop residues and using residual film mulching in different intercropping systems can reduce emissions of greenhouse gases from arid and semi-arid soils.

References

Ali, A., Elrys, A.S., Liu, L. et al. (2022). Cover plants-mediated suppression of fusarium wilt and root-knot incidence of cucumber is associated with the changes of rhizosphere fungal microbiome structure-under plastic shed system of North China. *Front. Microbiol.* 13: 697815. https://doi.org/10.3389/fmicb.2022.697815.

Amadou, I., Houben, D., and Faucon, M.P. (2021). Unravelling the role of rhizosphere microbiome and root traits in organic phosphorus mobilization for sustainable phosphorus fertilization. A review. *Agronomy* 11: 2267. https://doi.org/10.3390/agronomy11112267.

Ansari, M.A., Choudhury, B.U., Layek, J. et al. (2022). Green manuring and crop residue management: effect on soil organic carbon stock, aggregation, and system productivity in the foothills of eastern Himalaya (India). *Soil Tillage Res.* 218: 105318.

Asigbaase, M., Dawoe, E., Lomax, B.H., and Sjogersten, S. (2021). Temporal changes in litterfall and potential nutrient return in cocoa agroforestry systems under organic and conventional management, Ghana. *Heliyon* 7: e08051.

Azaryan, M., Abrari Vajari, K., and Amanzadeh, B. (2021). Variations in humus and fine root properties related to development stages in a temperate natural Beech forest. *Eur. J. For. Res.* 140 (2): 307–316.

Baietto, A., Hirigoyen, A., Hernández, J., and del Pino, A. (2021). Comparative dynamics of nutrient release through litter decomposition in *Eucalyptus grandis* Hill ex Maiden and *Pinus taeda* L. stands. *Forests* 12: 1227. https://doi.org/10.3390/f12091227.

Baul, T.K., Charkraborty, A., Peuly, T.A. et al. (2022). Effects of varying forest management on soil carbon and nutrients in hill and coastal homegardens in Bangladesh. *J. Soil Sci. Plant Nutr.* 22 (1): 719–731.

Beedy, T.L., Snapp, S.S., Akinnifesi, F.K., and Sileshi, G.W. (2010). Impact of *Gliricidia sepium* intercropping on soil organic matter fractions in a maize-based cropping system. *Agric. Ecosyst. Environ.* 138. https://doi.org/10.1016/j.agee.2010.04.008.

Beltran, A.M., Jepsen, K., Rufí-Salís, M. et al. (2022). Mapping direct N_2O emissions from peri-urban agriculture: the case of the metropolitan area of Barcelona. *Sci. Total Environ.* 822: 153514. https://doi.org/10.1016/j.scitotenv.2022.153514.

Berihun, M.L., Tsunekawa, A., Haregeweyn, N. et al. (2019). Exploring land use/land cover changes, drivers and their implications in contrasting agro-ecological environments of Ethiopia. *Land Use Policy* 87: 104052.

Bettiol, A.C.T., Braos, L.B., Lopes, I.G. et al. (2022). Evaluation of potentially available nitrogen by biological and chemical methods in soil cultivated with maize in succession to cover crops. *J. Plant Nutr.* 45 (13): 1919–1932. https://doi.org/10.1080/01904167.2022.2044488.

Blessing, D.J., Gu, Y., Cao, M. et al. (2022). Overview of the advantages and limitations of maize-soybean intercropping in sustainable agriculture and future prospects: a review. *Chil. J. Agric. Res.* 82 (1). https://doi.org/10.4067/S0718-58392022000100177.

Branca, G., Lipper, L., McCarthy, N., and Jolejole, M.C. (2013). Food security, climate change, and sustainable land management. A review. *Agron. Sustain. Dev.* 33 (4): 635–650.

Cates, A.M., Ruark, M.D., Grandy, A.S., and Jackson, R.D. (2019). Small soil C cycle responses to three years of cover crops in maize cropping systems. *Agric. Ecosyst. Environ.* 286: 106649.

Chatzistathis, T., Monokrousos, N., Psoma, P. et al. (2021). How fully productive olive trees (Olea europaea L., cv. 'Chondrolia Chalkidikis') manage to over-satisfy their P nutritional needs under low Olsen P availability in soils? *Sci. Hortic.* 265: 109251.

Cheng, A., Noor Azmi, N.S., Ng, Y.M. et al. (2022). Appraising agroecological urbanism: a vision for the future of sustainable cities. *Sustainability* 14: 590. https://doi.org/10.3390/su14020590.

Clay, D.E., Clay, S.A., Reitsma, K.D. et al. (2014). Does the conversion of grasslands to row crop production in semi-arid areas threaten global food supplies? *Glob. Food Sec.* 3 (1): 22–30. https://doi.org/10.1016/j.gfs.2013.12.002.

Cottney, P., Black, L., Williams, P., and White, E. (2022). How cover crop sowing date impacts upon their growth, nutrient assimilation and the yield of the subsequent commercial crop. *Agronomy* 12: 369. https://doi.org/10.3390/agronomy12020369.

Cuartero, J., Pascual, J.A., Vivo, J.-M. et al. (2022). A first-year melon/cowpea intercropping system improves soil nutrients and changes the soil microbial community. *Agric. Ecosyst. Environ.* 328: 107856. https://doi.org/10.1016/j.agee.2022.107856.

Danise, T., Andriuzzi, W.S., Battipaglia, G. et al. (2021). Mixed-species plantation effects on soil biological and chemical quality and tree growth of a former agricultural land. *Forests* 12: 842. https://doi.org/10.3390/f12070842.

Das, S., Berns, K., McDonald, M. et al. (2022). Soil health, cover crop, and fertility management: Nebraska producers' perspectives on challenges and adoption. *J. Soil Water Conserv.* 77 (2): 126–134. https://doi.org/10.2489/jswc.2022.00058.

Duan, N., Li, L., Liang, X. et al. (2022). Variation in bacterial community structure under long-term fertilization, tillage, and cover cropping in continuous cotton production. *Front. Microbiol.* 13: 847005. https://doi.org/10.3389/fmicb.2022.847005.

Dumanski, J., Peiretti, R., Benetis, J. et al. (2006). The paradigm of conservation agriculture. *Proc. World Assoc. Soil Water Conserv.* PI: 58–64.

DuPre, M.E., Seipel, T., Bourgault, M. et al. (2022). Predicted climate conditions and cover crop composition modify weed communities in semiarid agroecosystems. *Weed Res.* 62: 38–48. https://doi.org/10.1111/wre.12514.

Farhate, C.V.V., de Souza, Z.M., Cherubin, M.R. et al. (2020). Abiotic soil health indicators that respond to sustainable management practices in sugarcane cultivation. *Sustainability (Switzerland)* 12 (22): 9407.

Fortier, J., Truax, B., Gagnon, D., and Lambert, F. (2022). Thinning and gap harvest effects on soil, tree and stand characteristics in hybrid poplar bioenergy buffers on farmland. *Forests* 13: 194. https://doi.org/10.3390/f13020194.

Fujii, K., Funakawa, S., Hayakawa, C., and Kosaki, T. (2021). Effects of clearcutting and girdling on soil respiration and fluxes of dissolved organic carbon and nitrogen in a Japanese cedar plantation. *For. Ecol. Manage.* 498: 119520.

Giagnocavo, C., de Cara-García, M., González, M. et al. (2022). Reconnecting farmers with nature through agroecological transitions: interacting niches and experimentation and the role of agricultural knowledge and innovation systems. *Agriculture* 12: 137. https://doi.org/10.3390/agriculture12020137.

Gou, Z., Yin, W., and Chai, Q. (2021). Straw and residual film management enhances crop yield and weakens CO_2 emissions in wheat–maize intercropping system. *Sci. Rep.* 11 (1): 14077. https://doi.org/10.1038/s41598-021-93497-x.

Guo, G., Li, X., Zhu, X. et al. (2021). Effect of Forest management operations on aggregate-associated SOC dynamics using a 137Cs tracing method. *Forests* 12 (7): 859.

Hajduk, E., Właśniewski, S., and Szpunar-Krok, E. (2015). Influence of legume crops oncontent of organic carbon in sandy soil. *Soil Sci. Annu.* 66: 52–56.

Handayani, I.P., Widiastuti, H., Coyne, M.S., and Widawati, S. (2020). Soil organic carbon fractions in oil palm management systems. *IOP Conf. Ser. Earth Environ. Sci.* 583: 012006.

Isaac, M.E., Nimmo, V., Gaudin, A.C.M. et al. (2021). Crop domestication, root trait syndromes, and soil nutrient acquisition in organic agroecosystems: a systematic review. *Front. Sustainable Food Syst.* 5: 716480. https://doi.org/10.3389/fsufs.2021.716480.

Islam, I., Cui, S., Hoque, M.Z. et al. (2022). Dynamics of tree outside forest land cover development and ecosystem carbon storage change in eastern coastal zone, Bangladesh. *Land* 11: 76. https://doi.org/10.3390/land11010076.

Joshi, D.R., Ghimire, R., Kharel, T. et al. (2021). Conservation agriculture for food security and climate resilience in Nepal. *Agron. J.* 113: 4484–4493. https://doi.org/10.1002/agj2.20830.

Korneeva, E.A. (2021). Economic evaluation of ecological restoration of degraded lands through protective afforestation in the south of the Russian plain. *Forests* 12 (10): 1317.

Koudahe, K., Allen, S.C., and Djaman, K. (2022). Critical review of the impact of cover crops on soil properties. *Int. Soil Water Conserv. Res.* 10: 343–354. https://doi.org/10.1016/j.iswcr.2022.03.003.

Kumar, C., Ramawat, N., and Verma, A.K. (2022). Organic fertigation system in saline-sodic soils: a new paradigm for the restoration of soil health. *Agron. J.* 114: 317–330. https://doi.org/10.1002/agj2.20957.

Lazcano, C., Gonzalez-Maldonado, N., Yao, E.H. et al. (2022). Sheep grazing as a strategy to manage cover crops in Mediterranean vineyards: short-term effects on soil C, N and greenhouse gas (N_2O, CH_4, CO_2) emissions. *Agric. Ecosyst. Environ.* 327: 107825. https://doi.org/10.1016/j.agee.2021.107825.

Li, P., Wu, G., Li, Y. et al. (2022). Long-term rice-crayfish-turtle co-culture maintains high crop yields by improving soil health and increasing soil microbial community stability. *Geoderma* 413: 115745.

Lin, Y.-H., Lee, P.-C., Menyailo, O.V., and Cheng, C.-H. (2021). Changes in soil organic carbon concentration and stock after forest regeneration of agricultural fields in Taiwan. *Forests* 12: 1222. https://doi.org/10.3390/f12091222.

Liu, C. and Gong, X. (2022). Changes in rhizosphere soil nitrogen fractions associated with enzyme activities are linked to the microbial community following intercropping combined with nitrogen fertilization. *Land Degrad. Dev.* 33 (7): 1101–1113. https://doi.org/10.1002/ldr.4239.

Liu, Y., Ma, W., He, H. et al. (2021). Effects of sugarcane and soybean intercropping on the nitrogen-fixing bacterial community in the Rhizosphere. *Front. Microbiol.* 12: 713349.

Liu, B., Gao, P., and Zhang, S. (2022). Photosynthesis and fine root growth dynamics of soybean in walnut-soybean agroforestry system. *Comput. Intell. Neurosci.* 2246824.

Lopez-Ridaura, S., Barba-Escoto, L., Reyna-Ramirez, C.A. et al. (2021). Maize intercropping in the milpa system. Diversity, extent and importance for nutritional security in the western highlands of Guatemala. *Sci. Rep.* 11: 3696. https://doi.org/10.1038/s41598-021-82784-2.

Ma, L., Wang, X., Pu, Y. et al. (2019). Ecological and economic benefits of planting winter rapeseed (*Brassica rapa* L.) in the wind erosion area of northern China. *Sci. Rep.* 9: 20272. https://doi.org/10.1038/s41598-019-56678-3.

Ma, S., Wang, H.-Y., Zhang, X. et al. (2022). A nature-based solution in forest management to improve ecosystem services and mitigate their trade-offs. *J. Clean. Prod.* 3511: 131557. https://doi.org/10.1016/j.jclepro.2022.131557.

Mariya, A., Kumar, C., Masood, M., and Kumar, N. (2019). The pristine nature of river Ganges: its qualitative deterioration and suggestive restoration strategies. *Environ. Monit. Assess.* 191: 542. https://doi.org/10.1007/s10661-019-7625-7.

Martin, M., Girona, M.M., and Morin, H. (2020). Driving factors of conifer regeneration dynamics in eastern Canadian boreal old-growth forests. *PLoS One* 15 (7): e0230221. https://doi.org/10.1371/journal.pone.0230221.

Martinelli, F., Vollheyde, A.-L., Cebrián-Piqueras, M.A. et al. (2022). LEGU-MED: developing biodiversity-based agriculture with legume cropping systems in the Mediterranean Basin. *Agronomy* 12: 132. https://doi.org/10.3390/agronomy12010132.

Morugán-Coronado, A., Pérez-Rodríguez, P., Insolia, E. et al. (2022). The impact of crop diversification, tillage and fertilization type on soil total microbial, fungal and bacterial abundance: a worldwide meta-analysis of agricultural sites. *Agric. Ecosyst. Environ. Open Access* 3291: 107867. https://doi.org/10.1016/j.agee.2022.107867.

Moukanni, N., Brewer, K.M., Gaudin, A.C.M., and O'Geen, A.T. (2022). Optimizing carbon sequestration through cover cropping in Mediterranean agroecosystems: synthesis of mechanisms and implications for management. *Front. Agron.* 4. https://doi.org/10.3389/fagro.2022.844166.

Mwakilili, A.D., Mwaikono, K.S., Herrera, S.L. et al. (2021, 2021). Long-term maize-*Desmodium* intercropping shifts structure and composition of soil microbiome with stronger impact on fungal communities. *Plant Soil* 467: 437–450. https://doi.org/10.1007/s11104-021-05082-w.

Ngome, A.F.E., Becker, M., and Mtei, K.M. (2011). Leguminous cover crops differentially affect maize yields in three contrasting soil types of Kakamega, Western Kenya. *J. Agric. Rural Dev. Tropics Subtropics (JARTS)* 112 (1): 1–10.

Olofsson, F. and Ernfors, M. (2022). Frost killed cover crops induced high emissions of nitrous oxide. *Sci. Total Environ.* 837: 155634. https://doi.org/10.1016/j.scitotenv.2022.155634.

Pan, J., Guo, Q., Li, H. et al. (2021). Dynamics of soil nutrients, microbial community structure, enzymatic activity, and their relationships along a chronosequence of Pinus massoniana plantations. *Forests* 12: 376. https://doi.org/10.3390/f12030376.

Prayogo, C., Kusumawati, I.A., Qurana, Z. et al. (2021). Does different management and organic inputs in agroforesty system impact the changes on soil respiration and microbial biomass carbon? *IOP Conf. Ser. Earth Environ. Sci.* 743: 012005.

Qi, D., Wu, Z., Yang, C. et al. (2021). Can intercropping with native trees enhance structural stability in young rubber (*Hevea brasiliensis*) agroforestry system? *Eur. J. Agron.* 130: 126353.

Qi, J., Jensen, J.L., Christensen, B.T., and Munkholm, L.J. (2022). Soil structural stability following decades of straw incorporation and use of ryegrass cover crops. *Geoderma* 406: 115463. https://doi.org/10.1016/j.geoderma.2021.115463.

Quaye, A.K., Doe, E.K., Amon-Armah, F. et al. (2021). Predictors of integrated soil fertility management practice among cocoa farmers in Ghana. *J. Agric. Food Res.* 5: 100174. https://doi.org/10.1016/j.jafr.2021.100174.

Reicks, G., Clay, D.E., Clay, S.A. et al. (2021). Winter cereal rye cover crop decreased N_2O-N emissions during early spring. *Agron. J.* 113: (in press). https://doi.org/10.1002/agj2.20658.

Rodrigues, A.C., Villa, P.M., Ferreira-Júnior, W.G. et al. (2021). Effects of topographic variability and forest attributes on fine-scale soil fertility in late-secondary succession of Atlantic Forest. *Ecol. Process.* 10: 62. https://doi.org/10.1186/s13717-021-00333-1.

Sadono, R., Soeprijadi, D., Nikmah, S.F., and Wirabuana, P.Y.A.P. (2019). Determining the best agroforestry system using multicriteria analysis in Banyumas Forest management unit. *IOP Conf. Ser. Earth Environ. Sci.* 449 (2020): 012049. IOP Publishing. https://doi.org/10.1088/1755-1315/449/1/012049.

Scavo, A., Restuccia, A., Abbate, C. et al. (2021). *Trifolium subterraneum* cover cropping enhances soil fertility and weed seedbank dynamics in a Mediterranean apricot orchard. *Agron. Sustain. Dev.* 41: 70. https://doi.org/10.1007/s13593-021-00721-z.

Schröder, P., Mench, M., Povilaitis, V. et al. (2022). Relaunch cropping on marginal soils by incorporating amendments and beneficial trace elements in an interdisciplinary approach. *Sci. Total Environ.* 80310: 149844. https://doi.org/10.1016/j.scitotenv.2021.149844.

Seitz, D., Fischer, L.M., Dechow, R. et al. (2022). The potential of cover crops to increase soil organic carbon storage in German croplands. *Plant Soil.* https://doi.org/10.1007/s11104-022-05438-w.

Shah, T.M., Tasawwar, S., Bhat, M.A., and Otterpohl, R. (2021). Intercropping in rice farming under the system of rice intensification—an agroecological strategy for weed control, better yield, increased returns, and social–ecological sustainability. *Agronomy* 11 (5): 1010.

Shukla, S.K., Sharma, L., Jaiswal, V.P. et al. (2022). Diversification options in sugarcane-based cropping systems for doubling farmers' income in subtropical India. *Sugar Tech* 24: 1212–1229. https://doi.org/10.1007/s12355-022-01127-1.

Silva, L.S., Laroca, J.V.D.S., Coelho, A.P. et al. (2022). Does grass-legume intercropping change soil quality and grain yield in integrated crop-livestock systems? *Appl. Soil Ecol.* 170: 104257. https://doi.org/10.1016/j.apsoil.2021.104257.

Silva-Olaya, A.M., Olaya-Montes, A., Polanía-Hincapié, K.L. et al. (2022). Silvopastoral systems enhance soil health in the Amazon region. *Sustainability* 14: 320. https://doi.org/10.3390/su14010320.

Simwaka, P.L., Tesfamariam, E.H., Ngwira, A.R., and Chirwa, P.W. (2020). Carbon sequestration and selected hydraulic characteristics under conservation agriculture and traditional tillage practices in Malawi. *Soil Res.* 58 (8): 9859. https://www.publish.csiro.au/sr/SR20007.

Sivakumar, K., Ramah, K., Surendran, U. et al. (2022). Assessment of compatibility of intercrops in *Dalbergia sissoo* based Hortisilvicultural system in Mondipatty, Manapparai block of Trichy district, Tamil Nadu. *J. Appl. Nat. Sci.* 14 (1): 94–101. https://doi.org/10.31018/jans.v14i1.3250.

Soares, D.O., Pinto, K.G., Bhowmik, P.C., and Albertino, S.M. (2022). Chemical properties of soil and cassava yield as a function of weed management by cover crops in the Amazon ecosystem. *Sustainability* 14: 1886. https://doi.org/10.3390/su14031886.

Song, C., Wang, W., Gan, Y. et al. (2022). Growth promotion ability of phosphate-solubilizing bacteria from the soybean rhizosphere under maize–soybean intercropping systems. *J. Sci. Food Agric.* 102 (4): 1430–1442. https://doi.org/10.1002/jsfa.11477.

Stratton, A.E., Comin, J.J., Siddique, I. et al. (2022). Assessing cover crop and intercrop performance along a farm management gradient. *Agric. Ecosyst. Environ.* 332: 107925. https://doi.org/10.1016/j.agee.2022.107925.

Sudomo, A., Maharani, D., Swestiani, D. et al. (2021). Intercropping short rotation timber species with teak: enabling smallholder silviculture practices. *Forests* 12 (12): 1761. https://doi.org/10.3390/f12121761.

Swamila, M., Philip, D., Akyoo, A.M. et al. (2022). Profitability of Gliricidia-maize system in selected dryland areas of Dodoma region, Tanzania. *Sustainability* 14: 53. https://doi.org/10.3390/su14010053.

Tilahun, E., Haile, M., Gebresamuel, G., and Zeleke, G. (2022). Spatial and temporal dynamics of soil organic carbon stock and carbon sequestration affected by major land-use conversions in northwestern highlands of Ethiopia. *Geoderma* 406: 115506.

Wang, G., Li, X., Xi, X., and Cong, W.-F. (2022). Crop diversification reinforces soil microbiome functions and soil health. *Plant and Soil* 476: 1–2. https://doi.org/10.1007/s11104-022-05436-y.

Xu, Q., Xiong, K., and Chi, Y. (2021). Effects of intercropping on fractal dimension and physicochemical properties of soil in karst areas. *Forests* 12 (10): 1422.

Zaini, N., Umami, N., Hanim, C. et al. (2021). Growth and biomass production of chicory (*Cichorium intybus* L) planted in intercropping system with *Pennisetum purpureum* cv. Mott and cut at different ages. *IOP Conf. Ser. Earth Environ. Sci.* 667 (2021): 012012. https://doi.org/10.1088/1755-1315/667/1/012012.

17

Strategies for Restoration and Utilization of Degraded Lands for Sustainable Oil Palm Plantation and Industry

Ronny Purwadi[1,2], *Sanggono Adisasmito*[1], *Daniel Pramudita*[1,2], *and Antonius Indarto*[3]

[1] *Department of Chemical Engineering, Institut Teknologi Bandung, Jl. Ganesha 10, Bandung, 40132, Indonesia*
[2] *Department of Food Engineering, Institut Teknologi Bandung, Jl. Let. Jend. Purn. Dr. (HC) Mashudi No.1, Sumedang, 45363, Indonesia*
[3] *Department of Bioenergy Engineering and Chemurgy, Institut Teknologi Bandung, Jl. Let. Jend. Purn. Dr. (HC) Mashudi No.1, Sumedang, 45363, Indonesia*

17.1 Introduction

A large amount of CO_2 is released into the air from decomposition of deforested forest vegetation and accumulated soil organic matter over many years (Mutsaers 2019). The conversion of remaining forests into marginal areas such as peatlands or riparian zones is a setback in the efficiency of agricultural resource use. Because of the large amounts of C stored in these soils (Hooijer et al. 2010), greenhouse gas emissions per site are much higher than conversions of well-drained mineral soils (Carlson et al. 2013). New oil palm plantations contributed to 5% of total deforestation between 1990 and 2000. The proportion then doubled between 2000 and 2010, following the boom of palm oil industry (Corley and Tinker 2015).

Industrial plantations developed shortly after deforestation are likely to be responsible for this deforestation. This is classified as rapid conversion, during which plantations are established around five years after deforestation (Gaveau et al. 2016). The longer the time between deforestation and forest development, the lower the number of plantations usually is. This land is more likely to avoid deforestation directly. Interviews with 14 senior oil palm executives at a meeting of oil palm leaders and NGOs in Singapore in May 2015 indicated that forest conversion to palm plantations industrial oil is usually quick and takes place over two to three years. Interim analyzes have shown that forest area is converted to plantation within five years of clearing, with most being replanted within one to three years (Gaveau et al. 2016). The use of available concession maps to link deforestation to specific companies is often disputed by stakeholders due to inaccurate maps, overlapping claims, and the activities of others; it also cannot account for the removal of external franchise boundaries (Gaveau et al. 2017). In Indonesia, the recent increase in forest

Agroecological Approaches for Sustainable Soil Management, First Edition.
Edited by Majeti Narasimha Vara Prasad and Chitranjan Kumar.

conversion may be related to Indonesia's post-1998 political decentralization. This has led to an increase in local claims to national forest products, creating new opportunities to develop plantations on forest land. Initially, informal land claims can be used to secure local licensing for oil palm plantations, even if national licensing, or land release from forest products, is still ongoing (Gaveau et al. 2016).

Forest restoration in Indonesia mainly takes place on state forest land after degradation due to concessions, illegal logging, and fires (Suhardi et al. 2016). Every day, about 600 ha of Indonesian forests are cleared just by logging. According to data from the World Resources Institute (WRI), only 28% or 40 million hectares of Indonesia's forest reserves remained in 1997. Wildfires are another major cause of forest degradation. Other causes of forest degradation are coal mining and illegal logging. After large-scale logging, especially outside of Java, between 1970 and 1980, the Indonesian government began to reduce logging and issued Presidential Regulation No. 35/1980. Logging franchisees must contribute to the reforestation fund (*Dana Jaminan Reboisasi*), a fee per cubic meter of timber harvested. The National Land and Forest Reclamation Movement (*Gerakan Nasional Restorasi Hutan dan Lahan*) was promoted in 2003 by the Indonesian government during the reign of President Megawati and was planned for at least five years with the participation of several ministries, namely the Ministry of Forestry and the Ministry of Agriculture, Ministry of Social Welfare and Ministry of Education (Suhardi et al. 2016).

(Austin et al. 2019) used high spatial resolution images provided on Google Earth to describe land cover types after random selection of deforestation events in Indonesia, obtained from Global Forest Change dataset. Well known in the area, large-scale oil palm and timber plantations contributed to more than two-fifths of the national deforestation during the study period, with a peak at the end of the 8th month, followed by a significant decline through 2016. The conversion of forests to grasslands, averaging one-fifth of the country's deforestation, increased sharply in the years following periods of severe wildfires. Agriculture and small-scale plantations also contribute to one-fifth of the country's forest loss and are a major cause of damage outside of Indonesia's main islands. Despite their relatively small contribution to total deforestation, logging routes are responsible for the reduction in deforestation rates and the proportion of logging activities increased over the study period (2001–2016) (Austin et al. 2019).

Research by (Gaveau et al. 2016) suggested that the afforestation industry is the main cause of old-growth forest loss in Borneo, Malaysia, as 57–60% of all deforestation over four decades is related to rapid transition in five years of deforestation to industrial plantations (Gaveau et al. 2016). In Indonesian Borneo, only 15–16% of the total deforestation in Kalimantan is related to the rapid conversion to industrial plantations (11–13% is due to oil palm) as most oil palm plantations have been developed on land cleared before 1973 or later and on degraded soil (mainly forest converted to shrubland by drought and continuous burning during El Niño years). Despite this positive assessment, Indonesian Borneo has seen a strong increase in rapid conversion in the five years since 2005 and has become the largest contributor to rapid net forest conversion by area (2.2 million hectares converted from 2005 to 2015 in Borneo, with 1.2 million ha in Indonesian Borneo). Their analysis also showed a higher rate of rapid forest conversion to timber plantations than oil palm in Indonesian Borneo. An important difference between Indonesian pulp wood and oil palm is that the industry was initially dependent on timber from natural forests prior to the

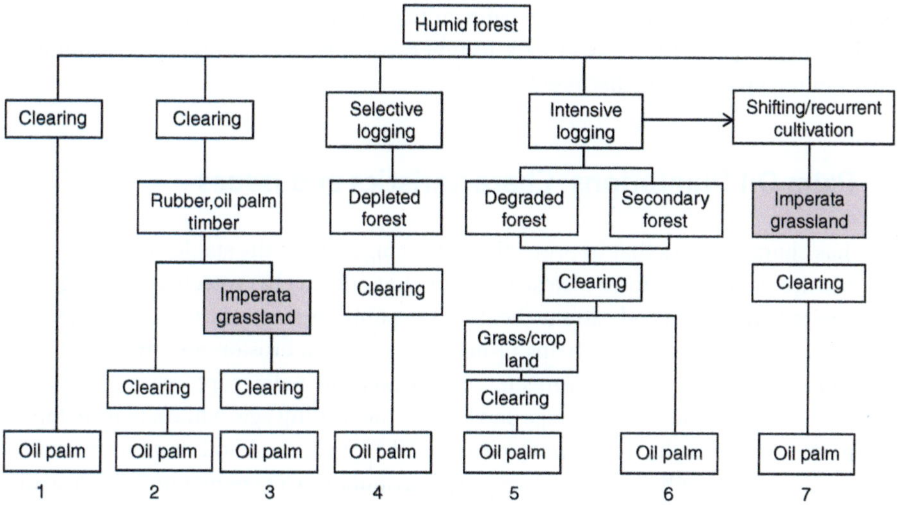

Figure 17.1 Possible land use histories prior to establishment of oil palm plantation (Whitmore 1988). Types 1–4 are particularly common in Indonesia and Malaysia.

establishment of plantations (Pirard and Irland 2007). As a comparison, (Taheripour et al. 2019) found that oil palm share was only 10% (20 Mha) of 200.5 million hectares in Malaysia and Indonesia in 2016, while soybeans covered more than half, equivalent to about 61%. (121 Mha) of the total land. Unlike oil palm plantations, the deforestation is caused by inefficient road construction, logging, and firewood (Etter et al. 2006) (see also Figure 17.1).

The Roundtable on Sustainable Palm Oil's (RSPO) guidance for responding to Indonesia's voluntary sustainability certification requirements states that new plantations in Indonesia must use "cleared and/or formerly degenerated lands." In May 2010, Indonesian President Susilo Bambang Yudhoyono announced a policy of developing oil palm plantations on degraded land instead of forests or peatlands. As part of an ambitious national REDD strategy, this policy has the potential to enable the palm oil industry to continue to grow, generating profits, government revenue, and jobs while reducing greenhouse gas emissions from deforestation and forest degradation (Gingold et al. 2012). In addition to positive economic benefits, the development of degraded land for oil palm cultivation has climate change-related benefits and a reduced impact on biodiversity, especially if key protected areas can be identified and protected. Although these protected areas are difficult to maintain, the development of oil palm concessions on degraded land has a much lower impact on climate and biodiversity than the development of forests, especially peatland forests (Ruysschaert et al. 2011).

In the following sections many aspects of the possibility of expanding oil palm plantation to degraded land are discussed. Discussions start from the issues of palm oil plantations and the opportunities that arise. It is followed by degraded lands and the efforts to rehabilitate such lands using various methods and species. Then enters the center of the discussion, which is the strategies for utilization of degraded lands for oil palm plantations. Only key principle aspects directly related to sustainable business are discussed. Technical details related to common practices of oil palm cultivation and postharvest are not touched and readers are suggested to instead look to other available literatures. Lastly, common challenges in the practices are discussed.

17.2 Palm Oil Plantations: Characteristics and Issues

Oil palm has characteristics that in principle make it an ecologically stable crop: once well planted, mature plantations form a closed canopy that effectively protects the soil against the effects of torrential rain and extreme heat. When palm trees are young, cover crops will often be intercropped, which also helps protect the soil from erosion and, in the case of legumes, also fixes atmospheric nitrogen. In addition, oil palm captures a significant amount of CO_2, both at the juvenile stage and later in the fruiting age, while a mature forest releases more CO_2 from the decomposition of dead biomass and soil organic matter, it binds through new growth and thus does not contribute to CO_2 reduction. The organic matter content in the soil will also reach equilibrium, with more old organic matters decaying as new organic matters forming (Mutsaers 2019). Table 17.1 presents the available area of palm oil plantation in major producing countries. As can be seen, Indonesia and Malaysia are the biggest players, contributing more significantly than other countries to the total. The low percentage of smallholder area in these two countries signify how commercialized the palm oil there.

In its natural and original habitat in West Africa, oil palm forms a stable and self-renewable vegetation composition, from which local people derive many valuable products. The palm tree is also perfectly suited to local periodic farming systems (in repeat farming systems, periods of planting alternate with short [natural] fallow periods) where it creates open upstairs, self-seeding in food crops. Seedlings and young palm trees are protected by farmers along with other perennials, such as timber or fruit trees. The production life cycle continues as long as the fallow vegetation is mainly drought-tolerant shrubs and broadleaf grasses, which are less likely to be affected by wildfires. The system thus

Table 17.1 Estimates of total palm oil plantation area and small ownership in selected producing countries (2010–2012). Adapted from Corley and Tinker (2015); Mutsaers (2019).

Country	Total area (ha)	Percentage of smallholder
Malaysia	4 360 000	40
Indonesia	6 500 000	38
Thailand	645 000	80
Papua New Guinea	134 000	40
Nigeria	457 000	61
Ghana	370 000	89
Cameroon	124 000	71
Ivory Coast	265 000	80
Colombia	300 000	Small estates

constitutes a fairly stable agroforestry system that, as long as the cultivation intensity does not degrade the soil, could lead to encroachment of fallow land by fire-sensitive weeds (Mutsaers 2019).

Originating in Africa, oil palm is a huge success in Asia. The main reason is its innate ability to produce, when grown in favorable climates with high rates of insulation and less frequent periods of drought (Corley and Tinker 2015). Current palm oil yields in Indonesia and Malaysia average between 4 and 4.5 tons/ha and kernel oil about 0.5 tons/ha for mature palm trees, while the average yield in African countries do not exceed 2.5 tons/ha of palm oil. This is not only due to the more favorable physical conditions in Southeast Asia, less water stress than in Africa; Management factors also play a major, and perhaps most important, role (Woittiez et al. 2017). The simulated yield potential of a mature crop under Southeast Asian conditions is more than twice the current average yield and yields above 10 tons/ha of oil are commonly obtained from in-field trials (Corley and Tinker 2015; Hoffmann et al. 2014). Thus, oil palm can be regarded as the most efficient producer of vegetable oils in the world. However, the seedling period is long, with the first fruit clusters appearing after 3–4 years and the crop reaching its peak yield after 8–10 years. Yields begin to decline again at the age of harvest 18–25 years (Mutsaers 2019).

One main strong point of oil palm plantation and processing estates is that they are completely self-sufficient in energy by burning some of the residues (fibers and husks) to generate electricity. In addition to that, the empty fruit bunches (EFBs), often mixed with palm oil mill effluent (POME), and boiler ash, are composted and returned to the fields as fertilizer, preserving organic matter and returning as an important source of phytonutrients (Mutsaers 2019).

More plantations have been established in less pristine, secondary forests, or old rubber or oil palm plantations that were cleared and replanted with oil palms. In Indonesia, and to a lesser extent in Malaysia, the use of peat forest land is increasing due to the drive to further expand oil palm acreage and the lack of suitable land for new plantations. Peat soils, when drained and cleaned, release large amounts of CO_2 through natural oxidation. The tremendous expansion of oil palm plantations over the past half century, primarily on forest soils, has posed a serious environmental threat for some reasons (Mutsaers 2019). The replacement of primary and secondary forests with monoecious oil palm plantations has dramatically reduced plant and animal diversity that is practically irreplaceable, even if oil palm plantations are abandoned and the land "returned to the wild". The low biodiversity also makes the plantations more vulnerable to pests and diseases. Although a well-developed oil palm plantation provides as good a soil protection as the forest it replaces and absorbs a significant amount of CO_2, a large amount of CO_2 is also released during and after deforestation, due to burning or rotting of vegetation and organic matter accumulated in the soil.

There are cases in which oil palm developers are responsible for deforestation without the plantations emerging. Occasionally, new plantations were approved and then the land was cleared, only for the investor to withdraw without establishing the plantation (Casson 2000). However, it may also be the result of profitable scams that generate income from timber from logging before the investor disappears (Casson 2000; Padmanaba and Sheil 2007). Investors and companies that acquire a land to develop plantations hold it for a long time before starting. Acquiring suitable land is known to improve enterprise value and borrowing capacity, and this certainly facilitates land banking (Cramb and McCarthy 2016).

As the predominant palm oil certification scheme, RSPO plays a key role in defining best practices and meeting consumer expectations as the oil palm industry continues to expand globally (Gilroy et al. 2015; Wich et al. 2014). Compliance with current RSPO standards requires clear evidence of previous land clearing prior to plantation development using reference data. According to this data, companies that cause forest loss cannot be certified unless appropriate compensation (financial or land acquisition) is paid. The RSPO unfortunately lacks information on the reclamation routes of many members and evaluations are usually made on a case-by-case basis.

In addition to certification, many palm oil companies are making (net) zero-deforestation commitments to ease consumer concerns and overcome import restrictions, especially in Europe. There is a lack of a clear, realistic, and uniform definition of what this commitment means. The time interval from deforestation to plantation growth can be a useful criterion. Technically, it is not difficult to develop an online system that displays past land cover and current land use and ownership. Such systems will improve transparency and accountability and will greatly help organizations such as RSPO and non-deforestation pledges to build the credibility they desire. The data of the times of deforestation and plantation establishment will be very important (Gaveau et al. 2016).

A land use change analysis (LUCA) has been conducted to estimate the total expansion of palm oil according to the new planting practices (NPPs) 2015 on an annual basis (RSPO 2017). However, the costly process and expert interpretation of farmland imagery (satellite imagery) make it a bad place to start, especially for small palm oil owners. Nonetheless, this disadvantage has not been recognized by NPPs even for smallholder farmers, or they have not changed their LUCA procedures to adapt them to current users, making farmland tracking unreliable and impractical. On the other hand, working with LUCA will help small farmers improve their production, allowing them to abandon the old and instead look to more lucrative growing options in the future (Dislich et al. 2018).

In addition to RSPO, several national organizations in oil palm exporting countries such as Indonesia Sustainable Palm Oil (ISPO), Malaysia Palm Oil Board (MPOB), and Malaysia Biomass Industries Confederation (MBIC) have formulated various plans to create mandatory standards in the industry and support the United Nation's Sustainable Development Goals (SDG) (Halim and Aziz 2019). For example, the usual burning of oil palm residues, e.g., the trunks of oil palm trees after deforestation, has recently been superseded by zero-incineration policies. Malaysia has outright banned open-burn farm liquidation methods since 1987. Indonesia has also moved toward a sustainable approach coupled with the 2014 government law to reduce pollution in the palm oil industry (Kaniapan et al. 2021). Faced with all these governmental and non-governmental bodies, the European Renewable Energy Directive (RED) has filled a gap to address issues not covered by the RSPO. RSPO-RED applies additional conditions to conforming RED specifications required to support the production of bioenergy fuels from palm oil in the European Union (Halim and Aziz 2019).

Aside of the deforestation associated with the establishment of plantations, another possibility of degradation of soil quality is due to the oil palm itself. Guillaume et al. (2016a) conducted an assessment study on the similarity of the soil properties between rubber and oil palm plantations of Muaro Jambi. Heterogeneity of soil properties (C and N content, C stocks, C/N ratio and bulk density) within each plantation was determined by a linear mixed-effects model, with the four operational zones as fixed factors and the plantation as a random effect. Soils under oil palm are, on average, more degraded than under rubber, which is indicated by lower C and N levels and reserves as well as higher bulk density. Under oil palm plantations C content in most soils was <2.2%, although in about 20% of the total plantation area it was more than 9%. This is mainly due to the fact that the plantations were established in scarce land degraded from previous used. The oil palm cultivation led to subsequent soil degradation (decreasing C content and increasing bulk density), especially in interrows. From the obtained non-linear relationship between C content and tree age, it can be deduced that loss of soil organic compounds (SOCs) is higher in the first few years after plantation establishment than in the latter half of plantation maturity. Additionally, younger plantations have more variable C contents than the older ones. The majority of the plantations had C contents about half or less than the topsoil beneath these well-drained forests. Nineteen percent of oil palm plantations and 3% of intensive rubber plantations had C levels higher than the range found in well-drained forests.

Soil erosion is a common yet often overlooked cause of SOC loss in oil palm plantations (Gharibreza et al. 2013; Guillaume et al. 2015). Erosion is more intense in interrows due to lower canopy and higher runoff intensity after heavy precipitation. Additionally, use of trucks or pickups for collecting the fresh fruit bunches by farmers facilitate the soil compaction in interrows. Rubber plantations, on the other hand, do not use large vehicles (Guillaume et al. 2016a).

(Guillaume et al. 2016b) investigated the effects of the three land-use types on microbial activity in well-drained soils in the regency of Batanghari. Microbial biomass and respiration were tolerant to SOC losses up to 2.7% of C but declined sharply below this threshold. Although this threshold may vary with soil type, most of the soil under oil palm was well below this threshold. Moreover, the bulk density of 1.3 gcm^{-3} is close to that found in subsoil clay sediments where water infiltration is greatly reduced (Allen et al. 2015), further increasing the risk of erosion through runoff and groundwater recharge (Bruijnzeel 2004). Significant losses of ecosystem services (C sequestration, water storage and infiltration, nutrient cycling) were observed in the soil under the oil palm. After the disturbances induced by land-use change, the C content should stabilize and reach a new equilibrium corresponding to the new biophysical conditions (de Blécourt et al. 2013; Smith et al. 2012; Van Straaten et al. 2015).

(Smit et al. 2013) argued that in principle oil palm expansion does not necessarily have to be the driver of deforestation and degradation in West Kalimantan. By implementing the right methodology, there are several options for economic and sustainable development of oil palm to co-exist in the region and other tropical areas emerge. Limiting global consumption or use of palm oil is thus not the only rational option or driver for addressing the problem of deforestation. It also opens a potential platform for growing other oil crops in protected areas and woodlands to replace oil palm plantations. The new use of palm oil as a biofuel will come at a "price" if aggregate demand exceeds productivity and exporting countries suffer consequences in the form of higher food prices despite increased export earnings (Kaniapan et al. 2021).

Low risk areas for oil palm are a subset of the land cover categories grass and scrublands, which are qualified as degraded lands (Fargione et al. 2008; Fitzherbert et al. 2008; Phalan et al. 2009). In these areas, oil palm expansion can have a positive impact on the environment by better protecting the soil against erosion and even providing more biological value to forest species than grass and scrublands (Fitzherbert et al. 2008). Replacing the degraded vegetations with products of high commercial value will reduce the risk of large-scale fires, because of the potential financial losses from fires in plantations on degraded soil will ensure the ability to prevent them (Germer and Sauerborn 2008). A study by (Ruysschaert et al. 2011) showed that the development of degraded land on mineral soils provided economic benefits equal to or greater than the development of oil palm plantations on woodlands. Developing oil palm plantations on degraded lands of mineral soils could be more economically viable than doing so on peatlands due to lower capital costs and potentially higher yields under best management practices (BAPPENAS, 2009). Peatlands are often preferred over degraded land simply/mainly due to the fact that land tends to be less claimed by local communities, which can be time-consuming and conflict-prone. With the potential scenario of utilizing degraded lands for oil palm plantation, the targeted sustainability levels of oil palm, such as that set in the recent Amsterdam Declaration for Europe (RSPO 2019), can potentially be reached.

In order to implement the Indonesian Sustainable Palm Oil (ISPO) certification system, the Ministry of Agriculture issued Regulation no. 19/Permentan/OT.140/3/2011 concerning Guidelines for ISPO. It stipulates guidelines that form the basis for encouraging oil palm plantation businesses to fulfill their obligations in accordance with statutory regulations. This regulation is also intended to protect and promote sustainable oil palm plantations in accordance with market demands. The regulation contains the ISPO Principles and Criteria for Sustainable Oil Palm Plantation, which are:

1) plantation licensing and management system.
2) application of technical guidelines for cultivation and processing of oil palm.
3) environmental management and monitoring.
4) responsibilities to workers.
5) social and community responsibility.
6) empowerment of community economic activities.
7) continuous business improvement.

For oil palm plantations to be sustainable, they must meet four main aspects, namely (i) technical aspects; (ii) economic aspects; (iii) social aspects; and (iv) legal aspects. Sustainable palm oil agribusiness must meet technical aspects which include land conditions, climatic conditions, and technology availability, to ensure high productivity can be achieved. Even if it is done on non-ideal lands, plantations must also be economically profitable. The income obtained must also be able to meet the needs of plant maintenance and even develop or expand its business. Oil palm plantations must be accepted by the community as a potential for agribusiness development and the products produced can be accepted by local and foreign communities. With the existence of oil palm plantations, it will open up employment opportunities and provide direct and indirect labor and the community can make it a livelihood. Oil palm cultivation must follow environmentally friendly procedures, do not damage nature, and have a negative impact on the surrounding environment. Lastly, no less important in the palm oil agribusiness business is the legal aspect, which will guarantee business certainty. Examples include the plantation business permits and cultivation rights, which must be met first in the development of oil palm plantations.

17.3 Degraded Land: Definition and Rehabilitation Efforts

Under the Indonesian constitution, forest land is under state control. A forest is described as an ecosystem whose biodiversity is mainly trees (Suhardi et al. 2016). Seventy-five percent of Indonesia's total land area, or 191 million ha, is classified as forest land, and the tropical rainforest component accounts for most of the forest cover. This is especially the case in Kalimantan, Sumatra, and Indonesian Papua. Previous governments have sought to impose state ownership, but Indonesia's forestland management laws are still changing according to national, region, district, and tradition requirements (Colfer and Resosudarmo 2010). This has resulted in a complex system in which forests can be transformed in many ways by multiple actors with contrasting and sometimes overlapping powers. This complexity has contributed to an often-unclear picture of who has what rights over forests and forest lands (Paoli et al. 2013).

Forests in Indonesia are divided into three groups based on their function, namely (i) production forests, (ii) conservation forests, and (iii) protection forests. Production forests are used to produce timber and non-timber forest products. Protection forests, on the other hand, are intended for watershed management, erosion control and wildlife conservation, whereas conservation forests are specifically designated to protect ecosystems, including their biodiversity. In the provincial land use planning of Indonesia, lands are classified into two types, namely forestry land and land used for other purposes, with the country law allowing only establishment of oil palm plantation only on the latter. Taking this and practical limitations into consideration, four categories, from the highest to the lowest potential, can be further defined: land for other uses (*Areal Penggunaan Lain*), conversion forest (*Hutan Produksi Yang Dapat Dikonversi*), production forest (*Hutan Produksi, Hutan Produksi Terbatas*), and protection forest (*Hutan Lindung, Hutan Suaka Alamand Hutan Wisata*) (Smit et al. 2013).

Peatlands or frequently waterlogged soils is often preferred as plantation sites as they have a higher initial SOC than typical forests growing on well-drained soils (Guillaume et al. 2016a). Clearing peatlands in fact causes more destruction than conventional soil reclamation (Kaniapan et al. 2021). The carbon bank beneath the bog forest is a vast carbon bank that releases huge amounts of carbon when cleared, and in the worst case of a shootout releases a significant amount of carbon (Page and Hooijer 2016). Peatlands are also associated with wetlands with critical hydrological functions for soil and water protection. Peatlands are not considered suitable for oil palm cultivation under the National REDD Strategy and Indonesian Presidential Instruction No. 10/2011 on a two-year moratorium on permits for new plantation establishment from primary natural forest and peatland (Gingold et al. 2012).

The scarcity of unused land resources during the oil palm boom not only led farmers to establish new plantations on high-C soils, but also on previously degraded soils (Guillaume et al. 2016a). However, defining "degraded land" is challenging because of the many definitions available (Eswaran et al. 2019; Kent et al. 1980). For example, the WWF definition of degraded land is: "Area where natural vegetation has been altered by human activity, resulting in a reduction in canopy cover, available biomass, or biodiversity that the system cannot recover within a period of time without assistance." In Indonesia alone, there are several definitions of degraded land in law (Fairhurst and McLaughlin 2009), leading to confusion about the size, location, and legal status of these areas (Gingold et al. 2012). In the broadest context, soil that is not in its original state is called degraded or damaged soil. In the context of land use, land degradation means loss of agricultural biological productivity or erosion of the ability of soil to support desired vegetation and maintain effectiveness over many years of use (Kudesia et al. 2009). The term ""degraded land"" can also refer to land cover that was originally a primary forest. In this case, the term "degradation" refers only to the vegetation above ground and does not take into account any deterioration in soil quality. These degraded lands have evolved over time into grasslands, shrub/scrubland, and even secondary forests (Ruysschaert et al. 2011). They are often considered "wasteland" available for development. In practice, however, some people might have made a "degraded land" or "fallow land" from the former forest, and then claimed its ownership (Ruysschaert et al. 2011).

In the context of promoting more sustainable palm oil expansion, degraded land usually refers to areas of low carbon stocks and low biodiversity, rather than characteristics associated with agricultural suitability or legal availability. For example, Indonesia's National REDD Policy Draft suggests that land is considered degraded if it contains less than 35 tons/ha of carbon. The good thing about this definition is that it includes areas that were long ago deforested and now have low levels of low carbon stocks and biodiversity, such as spear grass (*Imperata cylindrica*), called *lalang* or *alang-alang* in Malaysia and Indonesia, while excluding most areas of natural forests and peatlands, which typically contain over 200 tons of carbon per hectare. Vegetation degradation does not necessarily affect soil productivity (Fairhurst and McLaughlin 2009). This means that high agricultural yields can be still achieved in such areas, such as through oil palm cultivation. The downside is, however, that this degraded land definition also includes many areas that are currently cultivated or are of regional importance for social or cultural reasons and may not be suitable for expansion of oil palm cultivation. In addition, this degraded land definition also includes areas that are legally designated as part of a forest estate and therefore not currently available for oil palm cultivation (Gingold et al. 2012). Most degraded forests are still considered forest production areas. However, some of these areas have already been legally or illegally occupied and converted to gardens, oil palms, and rubber plantations (Suhardi et al. 2016).

As the term "degraded land" becomes more appropriate, quantitative assessments will be required. Since the above-ground carbon stocks of oil palm plantations average about 40 tons/ha over time, converting land with high carbon stocks results in a "carbon debt," while converting land below this threshold may increase carbon stocks (Chase and Henson 2010). Carbon stock values may therefore serve as a threshold for what can be termed "degraded" land. Because orangutans live in high-quality lowland rainforest canopies characterized by carbon stocks well in excess of 100 tons/ha, the threshold of 40 tons/ha is far below the orangutan habitat requirements (Ruysschaert et al. 2011).

Indonesia, the current largest oil palm producer, has the potential for the utilization of degraded land for palm oil plantations, which can reduce the need for forest conversion. In Kalimantan (Borneo) island, more degraded lands are available in the Indonesian side than in the Malaysian side (Gaveau et al. 2016) (see Figure 17.2). The high availability is also there in Sumatera Island, which is also known for its oil palm plantations. In Aceh province alone, around 29 000 ha, which is approximately 39% of the total degraded land area, is on mineral soils and is not in the area of state forest, which means it is suitable for oil palm plantations (Ruysschaert et al. 2011).

Establishment of primary forest plantations on degraded agricultural land may facilitate biodiversity conservation or restoration in agricultural landscapes. Plantations also help stabilize the rural population in areas where agricultural transfers are the dominant land use. Plantations are often combined with subsistence and commercial crops (agroforestry) or livestock (agrosilvopastoral system) and used as a tool for rural development projects around

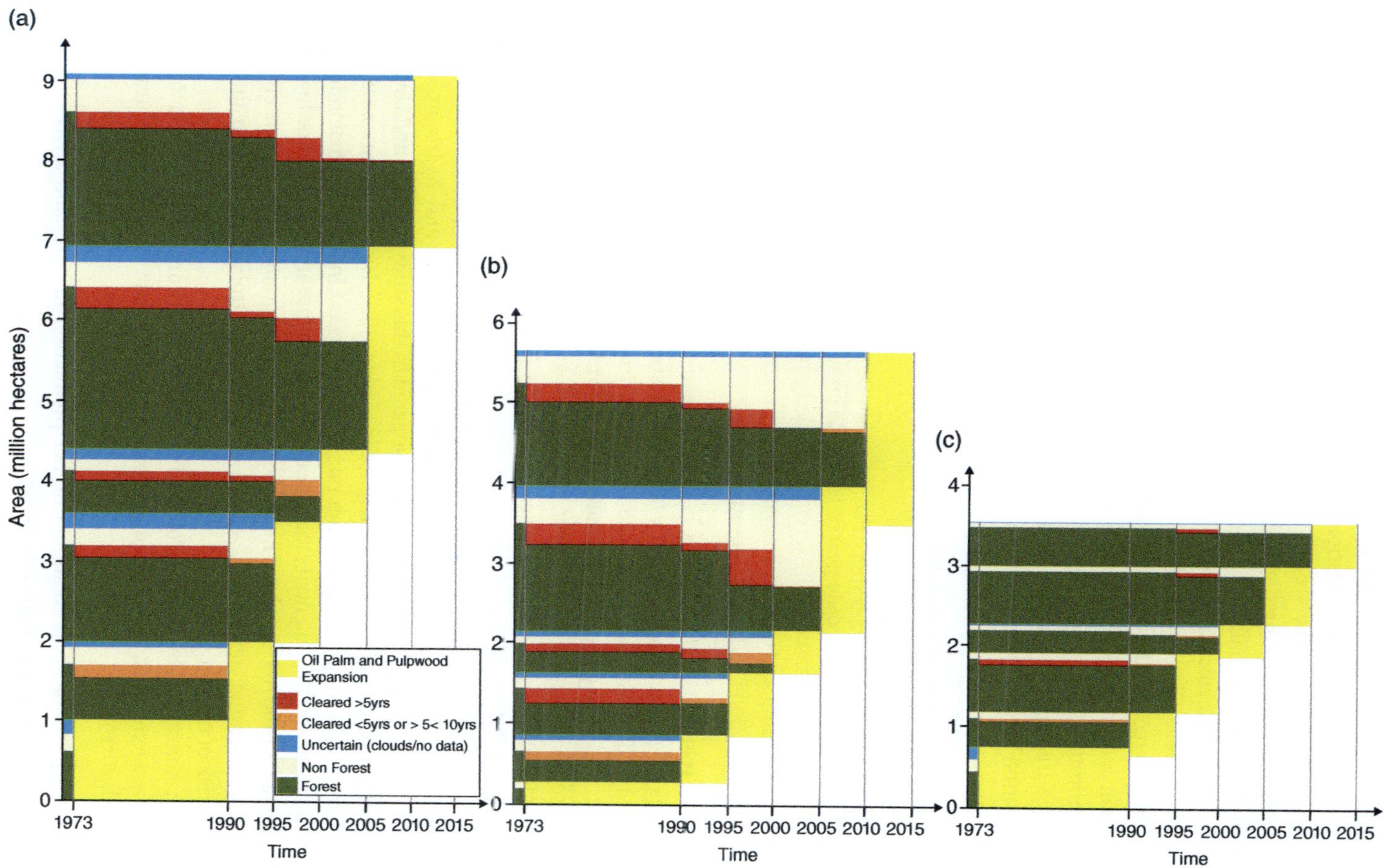

Figure 17.2 Expanding area of oil palm (and pulpwood) plantations in the whole Borneo (Kalimantan) Island (a), the Indonesian Borneo area (b), and the Malaysian Borneo area (c) (Gaveau et al. (2016) / Springer Nature / CC BY 4.0.

the world. They contribute to soil rehabilitation and forest regeneration. Selecting planting species with knowledge of their nutrient utilization efficiency and recycling capacity can lead to higher productivity and assist in ecosystem restoration projects (Montagnini 2001).

In many countries, forest rehabilitation has been carried out by planting various tree species, ideally those native to the area. A rehabilitation program in three regions of Latin America investigated how 29 three species influence soil conditions and how fast they grow (Montagnini 2001). About half of those species showed positive effects, attracting interest of farmers. In India, medicinal plants have been used for reclamation of degraded lands (Kudesia et al. 2009). Similarly in Morocco, Eucalyptus plantations were established in semi-arid lands to recover the degraded soils (Boulmane et al. 2017). It was found that soil organic carbon and soil fertility can be restored by short-rotation forestry based on eucalyptus, although initially the opposite effect on soil fertility was predicted. In Ucayali, Peru, various native tree species have been used to successfully restore degraded forest land under the Campo Verde Project (Rodríguez and Sabogal 2019). They attributed the success to the following factors:

1) involvement of private investors for high-risk activities
2) efficient plant production in nursery
3) continuous improvement of technology and management
4) adoption of a working schedule geared to reduce costs and be sustainable
5) development of skills based on in-service knowledge transfer
6) effective monitoring and evaluation system
7) constructive relations with local communities

Many forest rehabilitation programs have also been done in Indonesia. In Sumatera grassland area are recovered using mostly fast yielding species. The aim was to replant with *Acacia mangium* deforested areas covered by *I. cylindrica*, which resulted from illegal logging and slash-and-burn cultivation (Suhardi et al. 2016). In Manna watershed of Bengkulu province, agroforestry models for timber production were chosen based on the physiographical landscape of the area, which waving and hilly (Barchia et al. 2018).

(Sarminah et al. 2018) introduced agroforestry systems of sengon (*Falcataria moluccana*) and peanuts (*Arachis hypogaea*) to degraded lands on different soil slopes (steep and moderately steep) and analyzedanalyzed the impact on silvicultural and hydro-orological aspects of degraded lands. The parameters investigated in this study were ground cover for peanut growth and stem diameter and height for sengon trees. Hydraulic parameters, on the other hand, included potential erosion rate, erosion hazard index, and erosion hazard level. It was found that the potential erosion rate and overall hazard level become low after the introduction of the tree species. They concluded that restoration and soil conservation with the sengon-peanut agroforestry system effectively controlled the erosion rate and reduced the risk of erosion.

Some more forest rehabilitation projects involving consortiums of local and foreign researchers, academicians, NGOs, and companies in Indonesia have been documented (Suhardi et al. 2016):

1) *Acacia mangium* for short rotational timber production in bare land areas in South Sumatera, covering more than 100,000 ha.

2) Tropical rain forest rehabilitation in Jambi involving KEEC and Universitas Gajah Mada, by testing a rehabilitation model. This resulted in the return of wildlife species and income generation from rubber trees.
3) Tropical rain forest rehabilitation by BIOTROP.
4) *Paraserianthes falcataria* plantings by Perhutani.
5) Forest Education Project emphasizing on ecotourism, conducted in a collaboration between ministries of Indonesia and several organizations.
6) Rehabilitation of coastal area (approx. 450 ha) in south of Yogyakarta by planting wind breaking vegetations, which was to protect nearby agricultural areas.
   (Note: In areas with extensive agricultural land and far from dispersal sources, windbreaks and pasture remnants can be important reservoirs for native species. Windshields may attract birds if they contain native fruit-bearing trees, if there are many tree species and are structurally complex, and if they are placed between patches of forest to facilitate bird migration [Montagnini 2001]).
7) Mangrove rehabilitation in Denpasar, which was aimed at reduction of abrasion of the Ngurah Rai airport area.
8) Plantings of *A. mangium* and *Eucalyptus* in South Kalimantan, in collaboration with Finland.
9) Teak rehabilitation for timber production in Southeast Sulawesi.
10) Rehabilitation of mine areas in Papua via mangrove plantation by PT Freeport Indonesia.

For supporting and enhancing rehabilitation efforts, the following still need to be done (Suhardi et al. 2016):

1) improving and revising existing policies
2) building research and education capacities
3) reconciling global and national policies
4) partnership and collaboration with private sectors
5) creating public awareness and support
6) community involvement
7) monitoring and evaluation for more effective rehabilitation/restoration
8) effective and practical applications
9) financing for forest restoration
10) home garden/indigenous systems
11) enhancement of human well-being and natural environment

Degraded lands usually had the poorest soils in terms of total nitrogen and available forms of phosphorus and calcium. Soils under native forest, on the other hand, have the highest P and Ca contents and up to one order of magnitude higher carbon content. Deforestation for crop production significantly reduced his SOC pool, but not nutrients such as K, P, and Mg. In the aforementioned project in Morocco (Boulmane et al. 2017), the soil under the eucalyptus forest was moderate. In the topsoil, eucalyptus plantations gradually increased soil levels of total nitrogen, available phosphorus, and calcium. Potassium and magnesium patterns were less pronounced, but after decades of eucalyptus (Euca-C3 and Euca-C4) presence, soil nutrient status was comparably only slightly lower than that of natural forest soils. In other words, eucalyptus growth had no appreciable adverse effects on soil fertility, but instead have resulted in high biomass and carbon stocks.

Poorly managed plantations or old ones that have not been replanted or converted to other crops are prone to fire-vulnerable but very tenacious grass species, especially spear grass. Their presence makes restructuring of forest vegetation virtually impossible. The combination of above-ground and underground competition, wildfires, and the allelopathic effects of spear grass on some plant species make it very difficult to grow crops on spear grass-infested lands. As fragments of rhizomes quickly grow into new plants, they are virtually impossible to destroy by mechanical means alone (Mutsaers 2019). The vast expanse of spear grass (*I. cylindrica*) often found in oil palm area is usually resulted from deforestation and poorly managed plantations.

Although spear grass is difficult to remove, many annuals and perennials can suppress grasses and open a window for planting oil palms on degraded land. Established oil palm crops keep grass at bay because it does not tolerate shade. In this way, the public image of the oil palm as an ecological catastrophe can be subverted into a tool for restoring the productive capacity of degraded soils (Mutsaers 2019). There is a considerable amount of scientific literature on the remediation of spear grass-infested soils (e.g., [Murniati 2002]), with the first publications from Indonesia dating well before World War II. They mainly described labor-intensive manual and mechanical control methods, but to control grasses and create favorable conditions for the establishment of perennials such as rubber, creeping (legume), or oil palm. It soon became apparent that a combination, or continuum, of plant species was required. The perennials, once established and with sufficient vigor, will prevent shade-intolerant spear grass from reappearing (Mutsaers 2019). A simpler approach was tested in a study on the potential of agroforestry to rehabilitate spear grass-contaminated land in Kalimantan, Indonesia. We found that a series of tillages, herbicides, and the planting of *Pueraria phaseoroides* as a cover crop created a favorable environment for moderately shaded tree species such as candlenut (*Aleurites moluccana*) with maize as an intercrop (Murniati 2002).

The development of oil palm plantations from degraded soil on mineral soils can have a direct positive impact on climate mitigation. Indeed, oil palm plantations sequestered an average of 40 tons of carbon per hectare over their 25-year planting cycle, with peaks of around 60–80 tons of carbon per hectare in mature plantations. It is much more than the grasslands, shrubs, and bushes that make up most of the degraded land (Ruysschaert et al. 2011). Where degraded land is secondary forest, the conversion to oil palm plantations can often result in net carbon emissions. In most cases, this secondary forest should be preserved as a protected area because of its biodiversity value.

Expanding oil palm on deforested or degraded land is potentially viable from economic point of view (Fairhurst and McLaughlin 2009), promising in helping Indonesia achieve multiple goals of socially and environmentally sustainable economic growth (Daily 1995; Koh et al. 2009). Increasing overall yields and the practices of palm oil in low-yielding pastures would benefit the environment and local economies (Kaniapan et al. 2021). In the next section, the steps (strategy) needed for the sustainable palm oil practices in degraded land are discussed, taking into account economic, environmental, social, and other aspects.

17.4 Operation Strategies

17.4.1 Identification of Initial Constraints

Infertile lands are often thought of as degraded lands with little biodiversity value. At the same time, some have suggested that these degraded or abandoned lands could be a very unique habitat for certain species of flora and fauna (Ruysschaert et al. 2011). Identification of important constraints is often overlooked at the beginning and should arguably be initial step in ecosystem rehabilitation projects (Montagnini 2001). Before conducting studies for selecting suitable lands, there are two key policy obstacles that must be addressed first. They are the amount and type of degraded land available and the willingness of local farmers (small farmers) to participate (Ruysschaert et al. 2011). Local farmer involvement and business model are strongly influence by the definition of the lands degraded by people and their respective rights. Therefore, development of wasteland requires clarification of land rights and prior consent of the owner. Land tenure security is a prerequisite for any serious business investment in this country. Companies looking to develop degraded land will need to rethink their business practices toward improved community-business partnership models or innovative community-based plantation models. As companies realize that their comparative advantage lies primarily in processing rather than primary production, there may be new win-win solutions for farmers and companies. Government intervention will also clarify land tenure, facilitate the process of identifying and certifying individual tenure, and develop a system of customary land rights to own and use land.

Smit et al. (2013) classified the indicators needed to verify the criteria for sustainability initiatives into six criteria for the methodology, divided into three principles: (i) diversity biology and ecosystem services; (ii) provision of services and land (use) rights; and (iii) biophysical and economic suitability (see Table 17.2). Accordingly, the risk levels are defined in Table 17.3. In analyzing the soils, the following properties are considered: cation exchange capacity, base saturation, soil nitrogen, available phosphorus, exchangeable potassium, anion fixation, mineral reserve, soil reaction, aluminum toxicity, salinity, and acid sulphate potential. Lands are therefore distinguished between "active" concessions, i.e., officially operating, and the "inactive" ones, which are not officially operating, but in the process of obtaining permits (Smit et al. 2013). These criteria can be used to evaluate existing plans for plantation development and to assess future oil palm development options in line with sustainability criteria.

Since the indicators selected are based on prominent sustainability initiatives, the results are useful for identifying land that can be used in line with these sustainability initiatives and the implementation of policy regulations. The reasons are (Smit et al. 2013):

1) Violation risks of sustainability initiatives can be mapped.
2) The impacts of sustainability initiatives can be clearly presented in space.
3) Conservation priorities can be identified by mapping threatened areas of high conservation value.
4) Land-use planners can be informed of options to reduce the risks associated with unsustainable development and maximize the trade-off between economic development and conservation goals.

Table 17.2 Methodology of selected principles, criteria, and indicators to identify suitable areas for sustainable oil palm expansion based on RSPO, RSB, and RES-D sustainability criteria.

Principle	Criteria	Indicator
1: Conservation values must be maintained or enhanced	1.1: Valuable biodiversity is protected or enhanced at the population, human, and ecosystem level	1.1.1: Formal protection and conservation areas (HCV 1.1)
		1.1.2: Distribution and habitat of protected and endangered species (Red List, CITES) (HCV1.2 – HCV 1.3 – HCV 1.4)
		1.1.3: Threatened ecosystems intacti landscape and large-scale primary forest (HCV 2 and 3)
	1.2: Ecosystem services are maintained	1.2.1: Hydrological functions (HCV 4.1)
		1.2.2: Erosion risk (HCV 4.2)
		1.2.3: Buffer zones large scale fire (HCV 4.3)
		1.2.4: Carbon stocks
2: Human well-being is ensured and land (use) rights are respected	2.1: Community use is respected	2.1.1: Provisioning services crucial for subsistence (HCV 5) or culturalc sites (HCV 6)
		2.1.2: Customary land rightslr
3: The area is biophysically suitable for oil palm cultivation	3.1: Suitable climate	3.1.1: Rainfall
	3.2: Suitable topography	3.2.1: Slope
		3.2.2: Elevation
	3.3: Suitable soil	3.3.1: Drainage
		3.3.2: Soil texture
		3.3.3: Soil depth
		3.3.4: Soil erosion risk
		3.3.5: Soil chemical properties

Source: Smit et al. (2013) / Smit et al / CC BY 4.0.

Life Cycle Analysis (LCA) is one of the sustainability assessment tools used to assess the environmental sustainability of the palm oil industry. Additionally, LCA is an environmental impact assessment technique associated with all stages of a commercial product, process, or service. Basically, the palm oil industry involves several important processes such as planting, oil collection, oil production, by-products, and waste utilization. The application of LCA is an essential technique to assess the entire lifecycle including the conversion of palm oil fruits to palm oil and residual biowaste in the palm oil industry (Kaniapan et al. 2021).

Table 17.3 Classification of risk levels according to the indicators given in Table 17.2.

Indicator	Low risk	Medium risk	High risk	Very highh risk
1.1.1: Formal protection and conservation areas (HCV 1.1)	No overlap with IUCN areas or conservation and protected areas and buffer zones.	Buffer zones 1 km		IUCN I–IV, IUCN V–VII, protected forest, Ramsar, and national conservation areas
1.1.2: Distribution and habitats protected and endangered species (Redlist, CITES) (HCV1.2 – – HCV 1.3 – HCV 1.4)	No overlap with distribution or habitats of protected and endangered species	Overlap with distribution of protected and endangered species	Overlap with habitat of protected and endangered species (HCV 1.3)	Breeding grounds and nesting places, grazing/browsing for endangered species and temporal habitats for migratory species (HCV 1.3 and 1.4)
1.1.3: Endangered ecosystem Intact landscapes, and large-scale intact forest (HCV 2 and 3)	No overlap endangered ecosystems, important ecotone regions, and large-scale forest	Forest area > 20 000 ha plus buffer 3 km (HCV 2.1)	2 or more eco-tone regions (HCV 2.2) Endangered ecosystems	Rare ecosystems: Karst class 1, peat, freshwater swamp, mangrove, *hutan kerangas* (heath forest), cloud rain forest: peat, mangrove, cloud rain forest (HCV 3)
1.2.1: Hydrological functions (HCV 4.1)	No overlap with water source/riparian zones, mangrove peat, karst, or DAS super priority	DAS Super priority	Mangrove, peat, wetland, and karst forest, cloud forest, *hutan punggung*	Water sources (spring), riparian zones, and buffer zones.
1.2.2: Erosion risk (HCV 4.2)	<15 ton/ha/yr	15–60 ton/ha/yr	60–180 ton/ha/yr	>180 ton/ha/yr
1.2.3: Buffer zones large scale fire (HCV 4.3)	No overlap with barriers for the spread of large-scale fire, the area recently burned more than once in the last 10 yrsrs		Overlap with barriers for the spread of large-scale fire, but (partly) burned in the last 10 yrsr	Area contains barriers for large-scale fire i.e., large forest blocks or peat swap areas and not burned during the last 10 yrs
1.2.4: Carbon stocks	Carbon stock. 0–60 ton/ha	Carbon stock 60–70 ton/ha	Carbon stock. 70–80 ton/ha	Carbon stock >80 ton/ha

(*Continued*)

Table 17.3 (Continued)

Indicator	Low risk	Medium risk	High risk	Very highh risk
2.1.1: Provisioning services crucial for subsistence (HCV 5) or culturalc sites (HCV 6)	Areas providing for <10% for subsistence	Areas providing >10% <25% for subsistence	Areas providing >25% <50% for subsistence, or containing cultural sites	Areas providing >50% for subsistence, or containing cultural sites
2.2.2: Customary land rightslr	No overlap with land rights	Idle land; community interested to change the use	Idle land; *tanah pera*, community not interest to change the use	Active use of land (*tembawang; simpung; limbo*; *tana'jaka*; *tana'ulen*; *gupung*; community protected forest)
3.1.1: Rainfall	1750–5000 mm	1500–1750 mm	1250–1500 mm	<1250 mm; > 5000 mm
3.2.1: Slope	<8%	8–15%	15–30%	>30% (>12°)
3.2.2: Elevation	<200 m	200–500 m	500–1000 m	>1000 m
3.3.1: Drainage	Well to moderately well	imperfect	Extreme; poor	Excessive; very poor; stagnant
3.3.2: Soil texture	Silt loam; sandy clay loam; silty clay loam; clay loam	Clay; silty clay, sandy loam; loam	Sandy clay; silt; loamy sand	Heavy clay; sand
3.3.3: Soil depth	>100 cm	75–100 cm	50–75 cm	< 50 cm
3.3.4: Soil erosion risk	<15 ton/ha/yr	15–59 ton/ha/yr	60–179 ton/ha/yr	> 180 ton/ha/yr
3.3.5: Soil chemical properties#	Well drained and deep mineral soils, such as Nitisols, Alfisols	Weathered and deeply developed mineral soils; Tropudults (Ultisol)	Shallow and infertile mineral soils, e.g., Lithic Dystrudepts (Inceptisol)	Infertile sands (Tropopsamment, Placaquods), soils with acid-sulphate potential (Sulfihemist, Sulfaquept, Sulfaquent)

Source: Smit et al. (2013) / Smit et al / CC BY 4.0.

17.4.2 Selecting Suitable Degraded Land

Investigation on available degraded lands to find the potential ones for plant oil plantations is perhaps the most important step, as it is tied to the most important aspects of the whole project. Clear and spatially accurate access to data is necessary to form the basis of sustainable planning, to ensure the successful formulation, enactment, and enforcement of laws and regulations, and to provide a basis for equity and equality among all stakeholders involved in land planning and management (Smit et al. 2013). More research should be started to reduce deforestation methods to make it a more sustainable approach through smart farming techniques such as artificial intelligence (AI) farming techniques, aerial drone system to visualize the whole farmland tracking, and automatic watering system (Kaniapan et al. 2021).

Several established standards provide guidelines for sustainable palm oil production. Gingold et al. 2012 designed a method in accordance with established standards for sustainable palm oil production, such as those of the RSPO; incorporates relevant Indonesian laws and policies; and is consistent with proposed national REDD+ strategies to support palm oil production on low carbon degraded land. The method has been applied within the framework of Project POTICO (http://www.wri.org/project/potico) by the World Resources Institute (WRI) and Sekala to identify potentially suitable areas for palm oil plantation in the West Kalimantan. The guidelines cover economic, environmental, social, and legal considerations to consider when identifying new areas for oil palm cultivation. The considerations are: (i) carbon and biodiversity; (ii) soil and water protection; (iii) crop productivity; (iv) financial viability; (v) zoning; (vi) rights; (vii) land use; and (viii) local interests (Gingold et al. 2012). Based on a selection of eight considerations grouped into environmental, economic, legal, and social categories, a set of sites potentially suitable for oil palm cultivation can be identified. The method must be able to choose priority areas for expansion according to sustainability standards, forecast potential impacts and issues, and provide guidance on trade-offs between socio-economic development and sustainable land-use planning (Smit et al. 2013). Often found are small lands that are suitable from many aspects but not economically viable mainly due to the extra costs of transport and communications between many small sites the extra costs of transport and connection (communication, etc.) between many small sites. For this reason, a few but large areas are usually more preferred (Ruysschaert et al. 2011).

Smit et al. (2013) have built on a previous approach to define what is known as a Responsible Development Area (RCA). Criteria from three key sustainability initiatives relevant in the context of sustainable oil palm expansion have been translated into spatially explicit indicators that can be used to check compliance with the sustainability initiatives are described in the introduction. The risk of non-compliance with sustainability criteria is visualized with a "risk indicator map." To create options that meet the sustainability criteria in existing land-use plans as well as for future plantation development, the planned afforestation concessions as well as legal expansion options were compared with a map of risk indicators.

The HCV toolkit criteria (Jennings et al. 2003) overlap significantly with those defined by RSB and RES-D (Table 17.4). To develop a holistic methodology for spatially referenced indicators, we combined the most rigorous criteria from various sustainability

Table 17.4 Overlap of requirements between RSPO Principles and Criteria, Renewable Energy Sources-Directive (Article 17) and RSB Principles and Criteria (Smit et al. 2013).

RSPO	RES-D	RSB
HCV 1.1	3b i	P1, P7a, P7c
HCV 1.2	3b ii	P7a
HCV 1.3	3b ii	P7a
HCV 1.4		P7a
HCV 2.1		P7a, P7c
HCV 2.2		P7a
HCV 2.3		P7a
HCV 3	3b ii	P7a
HCV 4.1		P7a, P7b, P9
HCV 4.2		P7a, P7b, P8
HCV 4.3		P7a, P7b
HCV 5		P5, P6, P12a
HCV6		P12a
C7.5:FPIC		P2b, P12b

initiatives. In general, the evaluation criteria defined by the HCV Toolkit and referenced by the RSPO are the most rigorous and are therefore used as the basis of the methodology. These criteria can be supplemented with the RSB or RES-D criteria that are not covered in the HCV toolkit (Smit et al. 2013). To maximize the practical application of the method and comply with all sustainability criteria related to site selection, relevant indicators of the biophysical suitability of oil palm are also included (RSPO 2007). Although oil palm can be grown economically in a variety of tropical environments with specific management interventions, its yield is often reduced when grown under suboptimal conditions, such as areas that are frequently flooded (Corley and Tinker 2015; Fairhurst and Härdter 2003).

The systematic comparison of the criteria of the RSPO, RSB, and RES-D sustainability standards, and the resulting spatially explicit indicators are useful for three purposes (Smit et al. 2013):

1) The identification of candidate areas for sustainable oil palm expansion deemed appropriate under these programs becomes more transparent, spatially visible, and regionally relevant.
2) It visualizes the impact of the sustainability standards for biofuel feedstock cultivation.
3) It provides specific recommendations for collection of field data.

The idea of converting spear grass fields to oil palm plantations is indeed attractive. The lands that are used to be paralyzed and destroyed by fire each year, producing biomass that released all the accumulated CO_2, are now vigorous palm vegetation sequestrating significant amounts of CO_2, turning it into valuable oil and generate electrical energy from the earth.

They also supply the energy needed by the processing plants, power rural areas, and add crop residues to the soil to cycle nutrients and enrich the soil organic matter (Mutsaers 2019).

Oil palm can be an excellent seed for restoring spear grass-infested areas. However, its early growth is relatively slow (although not as slow as most other perennials) (Mutsaers 2019). Several studies have highlighted the feasibility and desirability of exploiting spear grass-covered lands for oil palm, and some companies had experiences in controlling spear grass using a combination of mechanical, chemical, and biological methods (Fairhurst and McLaughlin 2009). In general, the same criteria apply in selecting suitable spear grass-covered lands:

- land suitability, based on land cover, soil type, slope, erosion risk.
- yield potential, based on elevation, rainfall, soil type, drainage, soil pH. (This can me simulated by using available models.)
- cost related to the areas and their distribution, and scope of process operation.
- land ownership and scale of land holders.

Sustainable agricultural practices can be initiated by reducing soil erosion and leaching, which has been shown to be possible with legumes (*Mucuna bracteate*), thereby increasing soil moisture by preventing soil moisture evaporation and improving soil through decomposition of organic matter, even in hilly land (Wawan et al. 2019). It has also been found that primary carbon leaching is attributed to soil drying under trees, which decreases as soil moisture increases, addressing the central question behind the sustainability of land transitions to palm plantations (Manning et al. 2019). Furthermore, the prevention of palm oil cultivation on hilly land is essential to prevent soil erosion and promote sustainable agricultural practices in palm oil production (Kaniapan et al. 2021). In addition, the judicious application of controlled-release fertilizers offers promising results in preventing nutrient leaching, especially in hilly areas where a sustainable palm oil is promoted. In addition, frequent changes in land use must be monitored, which leads to long-term reductions in organic carbon stocks (Dislich et al. 2018).

17.4.3 Species Selection (for Rotation Farming and Interrow Covering)

The selection of suitable tree species for plantation forestry or agroforestry is influenced by knowledge of species performance and their economic and environmental benefits. About 85% of tropical plantations are dominated by three genera: pine, eucalyptus, and tectona, although there may be thousands of native species suitable for similar purposes (Evans, 1999). Native species are preferred because (i) they are better adapted to local environmental conditions, (ii) the availability of seeds and other reproductive organs locally, and (iii) farmers are familiar with them and their uses. Native species of trees may be more suitable. Also, the use of native trees in production systems helps maintain genetic diversity and provides a better balance with local flora and fauna (Montagnini 2001).

(Rodríguez and Sabogal 2019) has proposed a guideline for selecting species based on their field experience. A biophysical assessment characterizing vegetation (herbs, shrubs, trees), soils, and fauna (with an emphasis on entomata) should consider:

a) biophysical diagnosis of the species.
b) socioeconomic assessment of the zone of influence to gain knowledge and understanding of the characteristics and aspirations of the locals.

c) technical proposal design (including species selection, soil preparation, quality of plants to use according to the dominant vegetation and planting design, spacing, management regimes).
d) establishment of a central nursery, organic fertilizer production center, irrigation system, and shading.
e) establishment of the forest plantation.
f) maintenance and silvicultural practices targeting mortality level, high growth, and yield, and mitigating the risk of pests and diseases.
g) research in collaboration of various stakeholders.
h) monitoring of soil quality (especially carbon stocks, leakages, and emissions), plant survival and growth rates, unit costs, and environmental and social impacts of the project.
i) use of geographical information system (GIS) to obtain spatial representation of the area and a database containing species provenance, weed management, site preparation, fertilizer application, etc. This database should be linked to the spatial database obtained using the GIS.
j) social issues.

Questions arise about the definition of "native vegetation," the land threshold used in terms of biodiversity conservation and biomass accumulation, and the timeframe for restoring primary productivity of ecosystems (Smit et al. 2013). Taking a slightly different approach, Tyrie and Gunawan 1998 devised a method to identify areas for sustainable forest management, protected areas and other uses by adopting an inclusive approach to degradation risk, forest regeneration capacity, legal status of the land, and the physiological suitability of the soil. Their method, however, is not easy to be implemented by companies and land users as the compatibility to the international sustainability criteria was not considered. Smit et al. 2013 on the other hand included more detailed environmental, social and biodiversity criteria.

17.4.4 Nursery Practices

Raw materials and their traceability are of paramount importance to ensure high-quality end products from plantations (Rodríguez and Sabogal 2019). High quality seed oil palm can produce a good harvest and high yield for 25–30 years. High quality seeds are usually certified. Poor quality seeds, with unreliable certification or no certification at all, will have low yields for 25 years. In oil palms, the best crops come from seeds known as "hybrids." Hybrid seeds are the result of crossing two different types of parents. Hybrid seeds are produced through a special selection process. Hybrid oil palm seed parents differ in the thickness of the fruit "shell." The female mother of the hybrid palm, which is called the "dura,", has a thick shell. On the other hand, the male progenitor, called "pisifera," has no skin at all. A hybrid between dura and pisifera produced a hybrid derivative called "tenera." Tenera hybrids have thinner skin and produce 30% more oil than the two parents (Woittiez et al. 2016).

Independent plantations with poor planting material usually consist of 50% tenera, 25% dura, and 25% pisifera (Woittiez et al. 2016). In commercial plantations, almost only tenera is found, with maximum percentage of dura not exceeding 1%. Small farmers who work closely with the company also grow only tenera. It is very important not to buy uncertified seeds or to grow seeds from own farms. When the seeds are obtained from own farms, it is possible that the parent combination is wrong and instead of getting top quality hybrid tenera, one will get a mixture of tenera, dura, and pisifera. The resulting oil palm fruits might have thick or thin shell, or no shell at all.

It is very important to accompany planted seeds through their regrowth or natural regeneration (Rodríguez and Sabogal 2019). Regular evaluation of farmer's planting material needs to be carried out to make sure that productive oil palms in the plantation are of tenera type or interspersed with the dura type, identify the number of dura palms, increase overall productivity by cutting down the plants that are not productive, and identify manufacturer-related problems or large price reductions that may occur in the future (Woittiez et al. 2016).

This plant material can be tested using the following steps (Woittiez et al. 2016). First, choose a freshly harvested palm tree and harvest four fruits. Then cut the ripe fruit is in half to find out the exact thickness of the peel and determine its type, whether it is dura or tenera. In addition to the thin skin, tenera fruits usually have brown/black fibers covering the core. If all the fruits are tenera, mark this tree and move on to the next one. If the fruit found is dura, check the other four fruits in the same tree and see if they are dura too. The fruit can have different shapes and sizes. If fruits are confirmed as dura, mark the tree (differently). All the fruits of the same group always have the same type (dura or tenera), and they also apply to all bunches of the same tree. Repeat this verification process for up to 20 trees. If there are one or more solid palms, all oil palms on the farm should be checked at each harvest using the methods described above. Some palms may not produce mature panicles. This is common in dura palm plantations. If some palms do not produce any fruit, keep an eye on them for a year. If they do not produce any bunch during the time, they are classified as "sterile." These palms can be of the pisifera type or simply suffering from other causes. They should be cut down immediately so that they do not feed on nutrients from the soil and fertilizers.

This planting material can be identified by the following steps (Woittiez et al. 2016). First, select a palm tree that has just been harvested and collect four fruits (lodges) from the bunch. Then cut the ripe fruit right in the middle to see the exact thickness of the shell and determine the type, whether dura or tenera. Tenera fruit, apart from having a thin shell, usually also has brown/black fibers covering the core. If all the fruit in the bunch is tenera: mark the tree and then move on to the next tree. If the fruit found belongs to the type of dura, examine the otherfour fruits in the same bunch and determine whether they are also of the type of dura. These fruits can look different in shape and size. If the fruit is confirmed to be dura, mark the tree. All fruits in the same bunch always have the same type (dura or tenera), and also applies to all bunches that come from the same tree. Repeat this checking process for up to 20 palms. If one or more dura palms are found, then all oil palms in the plantation should be inspected using the above methods at each harvest season. Some palms may not produce mature bunches at all. This is very common in plantations with dura palms. If palms that do not bear fruit at all are found, monitor them for a period of one year. If they do not produce any bunches at all during this time, then they are categorized as "sterile." These sterile palms may be of the pisifera type or caused by other reasons. They must be cut down immediately so that they do not get nutrients from the land and fertilizers.

Household farmers with dura and/or pisifera palms in their plantations should be aware of problems that may arise in the future and should think about possible solutions. Having some dura palm trees in the plantation is not good for a number of reasons (Woittiez et al. 2016). The main reason is dura bunches do not produce enough oil, which makes oil mills reluctant to spend money to buy them. In addition, the thick shell of the dura is difficult to cleave and can damage factory machinery. If the mills find some dura palm bunches, they can rightfully refuse cancel the purchase or demand steep discounts to compensate for the low oil content. Moreover, the mills may decide not to accept the product from the farmer anymore, or request proof that the entire yield is just tenera beams.

Farmers should be aware of the types of palm oil available to them and should be prepared to take steps to deal with problems with the mills. If a mill finds a bunch of dura and problems arise from it, the farmers can either (i) cut down all the plants and replant with the new ones, or (ii) replace only the dura and sterile palms (Woittiez et al. 2016). The former is easier to do and has the advantage that the new palms can grow optimally at the same time. The problems are that it is not resource efficient and there will be no produces for three years. For the second option, ideally the palms are replaced with certified tenera. While wasting less resource, it is in more difficult to do and may not give optimal quality due to the difference in growth stages. It is also possible to replace the dura and sterile palms with other plants. This might be a good option when the farmers want to reduce the business risk by having more diverse product. The choice of plants is very important as their presence could either benefit or harm the palms, depending on the height, nutrient requirement, and coverage.

17.4.5 Cultivation and Maintenance

Oil palm requires a balanced and sufficient supply of nutrients in the soil for optimal growth and achieving high production in the long run. Basically, there are already nutrients in the soil that are generally not in the right balance, and generally there are not enough nutrients for the palm to continue growing and producing. Nutrients are also slowly lost during the harvesting and transportation of oil palm fruit and are stored in the stem and roots of the growing oil palm. Fertilizer application is essential to provide nutrients for oil palm growth and fruit production. The main nutrients for all oil palm plants are nitrogen (N), phosphorus (P), potassium (K), and magnesium (Mg). Oil palm plants also need a small amount of boron (B). Fertilization with copper (Cu) and zinc (Zn) is usually required in peat soils. Other nutrients such as calcium (Ca), sulfur (S), manganese (Mn), iron (F), and chlorine (Cl) are very important, but not really necessary when using fertilizers. Things that affect the dose of fertilization on oil palm plants produce, among others (Suprihatin and Waluyo 2015):

1) Nutrient status of plants and soil. If the soil and plant nutrient status is below the optimal value, the fertilizer dose will increase from the standard dose, on the contrary, there will be a decrease in the fertilizer dose to the standard if the soil and plant nutrient status exceeds the optimal nutrient status.
2) Production targets to be achieved. The production target should be based on the age of the oil palm plantation and the previous year's production achievement.

3) There are symptoms of nutrient deficiency in oil palm plants from field observations. If the level of nutrient deficiency symptoms in oil palm plants, the additional dose of fertilizer to be applied will also be higher.

If the amount of nutrients in the soil is too high, these elements can poison the oil palm (Woittiez et al. 2016). Soil organic matter can be increased as a result of good maintenance and is of great importance for soil quality and efficient use of fertilizers.

Oil palm fertilization is very expensive; it can represent almost 60% of the total cost. If fertilization is not done correctly, up to 50% of the nutrients can be lost. This means spending more money on fertilizer, which is not good for the oil palm or to increase production. Fertilizer must be used efficiently, keeping in mind the "4Rs": the right type, the right amount, the right time, and the right place (Woittiez et al. 2016). Oil palms of different ages require different amounts of fertilizer. Fertilizer requirements depend not only on differences in the age of the oil palm, but also on the type of soil (especially mineral or peat soil), soil fertility (the nutrients and organic matter contained in the soil), plant material, and crop yield. Fertilizers can be applied in the form of single and compound fertilizers. The dose of fertilizer given for one year is divided into two applications. The first fertilization was applied in April/May and the second fertilization in September/October, each at 50% of the annual dose of fertilization. The condition of the plantation area is relatively flat, so the single application of fertilizer can be done by sowing (broadcast) (Suprihatin and Waluyo 2015).

Recommendations from extension agents, associated experts, experienced farmers (with the best reputation in the area), and nearby crop companies are needed to determine the amount of fertilizer required. When fertilizer prices increase, companies can provide information on the balance of required nutrients (i.e., the ratio of N, P, K, and Mg). Developing companies can also provide information on the best types of fertilizer and choose the best time for the application. In addition, information from manuals and other available materials, some of which may come from first-hand field/hands-on experience, will also be helpful.

Checking the pH (acidity level) of the soil is very important before choosing which type of fertilizer to use. Measurement must be done both inside and outside the weeded circle, as acidification of the soil can be local and occurs mostly in the circle. If the pH is in the range of 4–5, then the soil can be considered quite acidic, and the use of fertilizers can be reduced slightly. If the acidity level is below 4, then the soil can be considered very acidic. The pH can be increased by applying rock phosphate, dolomite (containing Mg), lime ($CaCO_3$), and ash from burning of EFBs (Woittiez et al. 2016).

Sampling of the midribs is carried out to determine the nutrient content available in the plantation. This helps to identify nutrient deficiencies and determine the exact fertilizers needed by the palm tree. Samples are taken in the form of leaves and leaf bones, which are then taken to the laboratory for analysis of their nutrient content. The sampling process must be carried out by trained workers or field extension workers and must comply with applicable protocols. The leaf sampling process must be carried out once a year, approximately at the same time every year, at least three months after the application of fertilizer and not during the peak of the rainy or dry season. The midrib sampling process is quite difficult and laboratory analysis is expensive (can reach $30 per sample) (Woittiez et al. 2016).

Therefore, the determination of the sampling process should be discussed in advance with plantation extension workers or expert workers.

Pests and diseases can reduce oil palm yields, damage fruit bunches, and even kill the trees themselves. Oil palm pests are generally mammals (such as rats, which damage fruit bunches) and insects (such as rhinoceros beetles and caterpillars, which damage leaves and interfere with the growth of young plants). In the early stages of plantation development, female rats, rhinoceros beetles (*Orycteshinoceros*), and nettle moths (*Parasa lepida*) are often found (Ruysschaert et al. 2011). The best method of pest control in mixed native forest plantations is biological control using entomopathogens (Rodríguez and Sabogal 2019). Establishing biological corridors that provide alternative hosts and refuge for parasites is a good option for maintaining a balance in pest populations. Plant diseases are caused by micro-organisms such as fungi, bacteria, and viruses. A fairly common disease is stem rot, which is caused by a severe fungal infection and can kill the tree.

Good plantation maintenance will help prevent the emergence of pest and disease problems. It is very important to carry out plantation monitoring at least every two months, to find pest and disease problems as soon as possible. There are several ways that can be done so that damage caused by pests and diseases can be minimized (Woittiez et al. 2016): (i) ensuring the growth of cover crops such as soft weeds or legumes that attract natural enemies of pests, (ii) ensuring the cleanliness of the plantation to prevent the emergence of pests, (iii) allowing natural predators to live in the plantation to control the pests, and (iv) ensuring good access to each palm tree so that it is easy to spot pests and diseases in the early stages. The most common predator of insects is birds (Ruysschaert et al. 2011). The use of pesticides should be as little as possible, because can unnecessarily harm insects and other animals that live in and around the plantation area.

Analysis of different land uses shows that the use of the most biodiversity-friendly methods available, i.e., biological methods, offers some potential for pest control, because the rich local biodiversity helps maintain a balanced ecosystem, maximizing the possibility that a natural predator-prey relationship may continue to function. However, maintaining high levels of biodiversity, both in terms of number of species and individuals, can have both positive and negative effects. Some species can become pests, attacking oil palm trees, while others are their natural predators and help control these pests. It is therefore essential to make every effort to maintain ecological balance and, if necessary, to supplement it with targeted actions against specific pests, if absolutely necessary (Ruysschaert et al. 2011).

Imperata cylindrical, a dominant grass that established itself as a dense mulch on degraded soil prior to experimental studies, remains an issue. It spreads both by small, easily carried seeds and by rhizomes. Manual control of *I. cylindrical* is necessary, but difficult because the dense rhizome system prevents the formation and growth of other beneficial crops, especially cover crops planted before the oil palm (Ruysschaert et al. 2011). Interest in purely mechanical and biological methods has slowly diminished as powerful herbicides are available. A combination of mowing and leveling grass with tractor-mounted rollers, repeated applications of glyphosate and imazapyr, and growing legumes is currently considered the most effective method for oil palm plantations (Fairhurst and McLaughlin 2009). A variety of cover crops, rather than a single species, is required to prevent spear grass dominance even when heavy application of herbicides is considered undesirable (Mutsaers 2019).

A ground cover using soft plants tightly in the planting area is necessary to prevent soil erosion, maintain soil moisture, prevent soil nutrient loss, and attract as many pest predators as possible. Additionally, cover crops will provide easy access for weed control, fertilizing, etc. (Woittiez et al. 2016). Good quality cover crops should be planted whole and kept to a maximum of knee height. Ground cover by legumes, such as *Desmodium ovalifolium*, has been shown to improve weed biological activity due to the contributions of low-cost establishment of weeds, persistent and non-aggressive supportive shade in plantations, woody stems, and high biomass (Rodríguez and Sabogal 2019). Leguminous cover crops provide both N to the crop and protection against undesirable weeds and erosion. Legumes can be grown where available, whereas woody weeds must be eliminated. The right time to do so is during planting, or early in the recovery process, after noxious weeds have been removed. In addition, land acquisition should not be carried out at the peak of the rainy season or the peak of the dry season. The provision of legumes as ground cover is done only once during the life of the plantation, with maintenance every six months.

A good pruning process (removing the middle segments of old and ineffective oil palm) is essential for more efficient fertilizer use and easier access to the plantation. Cutting other creeping plants is also important (Rodríguez and Sabogal 2019). It is advisable to maintain an optimal number of leaves per palm tree. Pruning can be done twice a year. When pruning, keep in mind that the more leaves the better for production (as the palms get more sun). Dead or dying leaves retain nutrients that need to be recycled. If the palm tree is tall, it will be difficult to get a good harvest if there are too many leaves. The pruned leaves will decompose faster if they come into contact with the ground and will be an important food source for the roots of the oil palm. By stacking palm leaves in a box shape, nutrients and organic matter will be dispersed, and the leaves will decompose faster because the pile is not too high. Stacking the leaves also allows the reaper to move more freely and avoids soil erosion (Woittiez et al. 2016).

17.4.6 Harvesting and Marketing

Good harvesting practices are methods that lead to the acquisition of large quantities of fresh fruit bunches, high oil extraction rates, and good oil quality. Good harvesting practices include (Woittiez et al. 2016) harvesting with correct procedures (palm leaf pruning, bunch cutting), harvesting only ripe bunches, good and fast bunch transportation system to the mill, and limiting the loss of loose fruit from bunches or during transportation. For good transportation and distribution, road conditions need to be maintained. Although road maintenance can absorb a very large investment, poor road conditions will make the transport process slower and more expensive. To avoid loss of fruit bunches or oil quality during the transportation process to the mill, farmer groups or traders must maintain the condition of the haul trucks, ensure the speed of transportation, and cover all parts of the tailgate with nets. Road maintenance can be the responsibility of (local) authorities or plantation companies. If neither party feels responsible, the maintenance of this road will be determined by the consortium. During the periodic road maintenance, attention should be paid to the construction of the distribution road. Repair potholes and remove weeds from roadsides and drains. Road conditions need to be checked regularly, especially potholes and ruts on both sides of the road. If the condition of the road is too severe, it is possible to consider hiring local construction workers to construct the road works (Woittiez et al. 2016).

Selling produces through cooperations not only result in better prices, but also helpful in distinguishing the production from conventional oil palm development in the area that converts native forests into large-scale plantations. In the case of developing fertile land for oil palm cultivation, it is indeed expected that a premium can be obtained for such a product, since it is directly related to the land of the farmers, local communities, and respect for biodiversity (Ruysschaert et al. 2011).

It is also important to mention that by using organic methods to grow degraded soil, afforestation not only minimizes its impact on the environment, but also gives farmers the opportunity to produce their own crops and organic fertilizers (Ruysschaert et al. 2011). This reduces maintenance costs, while providing new business opportunities. Application of biological methods also makes it possible to go beyond the RSPO requirements which still allow hazardous chemicals. A product that is more environmentally and socially sustainable can reach the niche market.

17.5 Challenges and Opportunities

The main challenges in converting degraded land to oil palm are (Rodríguez and Sabogal 2019; Ruysschaert et al. 2011):

1) All claims must be investigated, and any controversies resolved prior to any development on the land.
2) Each accredited individual owner must agree to participate in all required studies on oil palm.
3) Difficulty in establishing orderly and orderly plantations due to the encroachment of some villagers from the neighboring town
4) Low availability of plant planting equipment of obvious origin (seedlings)
5) Low supply of skilled human resources
6) Low availability of tillage machinery
7) High cost to control invasive weed species

Once these challenges are overcome, the final issue related to land rights is the formal registration of land at the government office in order to grant titles to owners who did not previously own them. Land certificates can be used as collateral to access bank capital. This is important because oil palm cultivation is particularly capital intensive for at least the first fouryears, from the first land preparation work to the first production (Ruysschaert et al. 2011). Inactive concession use on low-risk lands and use of low-risk areas on lands under the responsibility of the forestry department are currently limited by the complex procedures required necessary to adjust the land use plan. The only type where change of status is relatively easy today is transition forest, but the amount of land that is considered transition forest and is considered low risk during this time is relatively small. In the event that additional concessions are still awarded, it is strongly recommended that the concessions be planned on land that has been classified as low-risk in areas designated as other uses. For example, forest expansion plans in other provinces could be redirected to these areas. Only after this potential has been exhausted should forest land, especially heavily exploited or fire-ravaged forest land, be considered for expansion, and then it is best opened on land identified as low-risk in conversion forests (Smit et al. 2013).

At the beginning of the plantation development, it is usually difficult to give an exact picture of what will happen during the 20-years oil palm cycle. However, it is technically relatively easy to develop oil palm plantations on degraded land and is in all respects similar to the development of other small oil palm plantations. Indeed, the limiting factors of oil palm growth appear to be climate (humidity and temperature) and the amount of fertilizer applied. Since these degraded lands are adjacent to areas with the highest recorded palm oil production in the world, it can be assumed that the climate is optimal. As far as fertilizers are concerned, the program strives to provide optimal fertilizer, using organic methods (Ruysschaert et al. 2011).

When establishing a plantation, there is a risk of illegal logging and development of monocultures such as oil palm and *A. mangium* by large private companies in the surrounding area. Illegal logging and land grabbing must be kept to a minimum and adequate funding should be provided for security. In addition, there have been conflicts with local people living in the surrounding areas due to property issues. By pursuing a social approach to meet local needs such as growing vegetables, raising livestock, and other social development projects, these conflicts can be avoided.

Another challenge comes from the fact smallholders. Based on their culture, educational background, and experience in palm oil business, it is sometimes difficult for smallholders to trust and rely on or depend on each other. Therefore, it is not easy for them to create solidarity to achieve common goals. Farmers prefer to borrow money from individual lenders or agricultural agencies at the local village level rather than from other organizations. A major challenge is therefore developing an organization that can cover the technical aspects (i.e., crop access, fertilizer management, equipment rental, etc. to be reasonably priced and quality control guaranteed) (Ruysschaert et al. 2011). There are two types of smallholders (Vermeulen and Goad 2006):

1) *Supported smallholders*: Producers growing palm oil with direct government or private sector support. The basic concept is that a government agency or private farming company provides technical assistance and inputs of seeds, fertilizers, and pesticides, on a loan basis, sometimes subsidized by the government. There may be a verbal or written contract that outlines the agreement and may include guarantees of sale, as well as how the factory price is calculated. Examples of small-scale programs supported are Plasma Nuclear (PIR) in Indonesia and the variety of resettlement and land restoration programs in Malaysia (RISDA, FELCRA, FELDA).
2) *Collective landowner schemes:* Another option for local communities to hold recognized land title or customary land use rights. These are land leases or joint ventures whereby local landowners lease their land use rights to a plantation company or receive a share of the profits based on the net worth of their land. This is not quite a smallholder model, but it could be an attractive alternative for local landowners.

The variety of smallholder approaches reflects the available capital and time they are able to spend on degraded lands, as well as their overall desire to reduce risk. With higher income, smallholders tend to hire relatives and directly participate more in pilot research. These small owners take on greater financial risk but expect higher returns. With lower incomes, smallholders tend to do the bare minimum on their own and actually reduce their participation in the pilot study. They are thus less prone to financial risk. "*Independent smallholders.*" They will likely grow crops without the help of the government or private companies. If not supported, they will sell directly or through traders to local factories. Such a situation at the beginning of the development of degraded land into oil palm plantations can have a very negative impact in the long run. Indeed, independent smallholders tend to earn much lower incomes than smallholders or assisted collective landowners, as independent smallholders tend to be less productive (due to lack of technical support to develop plantations) and tend to sell their products at lower prices (due to lack of negotiating positions) (Ruysschaert et al. 2011).

From the point of view of smallholders, developing oil palm plantations from previously degraded land is not feasible for three main reasons, in addition to the challenge of land ownership mentioned above. They are (Ruysschaert et al. 2011):

1) capital to meet upfront expenses.
2) access to good seedlings and necessary start-up technology.
3) smallholder aversion to risk.

Smallholders face several challenges in maximizing their palm oil production potential while maintaining local choice and autonomy. There are distinct examples of innovations to address these issues and improve smallholder contributions to sustainable palm oil production. Examples of constraints and responses include (Vermeulen and Goad 2006):

1) *Ownership status*: Disagreement and uncertainty about land ownership are common and can be violent. While strong public policy is key to resolving protracted land disputes, initiatives by companies and government agencies to improve fairness and transparency have included granting full set of land ownership rights after repayment of loans, going beyond the legal provisions on land dispute resolution and land management based on sharing rather than individual blocks.
2) *Securing capital to meet upfront expenses*: Smallholders often cannot meet the basic guarantee and minimum loan size requirements for bank financing. Microfinance institutions are the main solution. These may include interest-free loans for specific inputs, negotiable terms, and equity based on other forms of land ownership recognition other than official property rights.
3) *Getting good technical, policy and market information*: Access to reliable information–on pricing and pricing policies, market opportunities, technical aspects of production and site management, and more basic rights and options under national or formal agreements–is also a big difficulty. Measures to address this include access to information services by NGOs and international agencies, but their scope is geographically specific.
4) *Balancing cash crops with food security*: Early palm oil development often left smallholder farmers with insufficient land and time to feed their households. Intercropping between young oilseed crops and agricultural land has become more common. Some companies have adopted flexible work arrangements for smallholder farmers and workers to help them allocate their time optimally.

5) *Coping with market risk*: Independent smallholder farmers are particularly affected by fluctuations in crop prices, and this is the main incentive to join or stay in supported programs. The monopolistic trading of enterprises and the lack of bargaining power among smallholder farmers exacerbate the problem. Innovations have included nationally or internationally indexed price standards and the emergence of small independent mills, but formal insurance for small palm oil farmers remains elusive.

Cooperations and company can help finance the initial work needed (i.e., site cleanup, initial insecticides, ground cover crops). This financial assistance can be incorporated into a loan from a credit union to farmers and helps fill this financial gap (Ruysschaert et al. 2011). The lack of access to good seedlings and the necessary start-up technologies was addressed through the support of cooperation. In fact, it is especially so when smallholder farmers were promised seedlings, they showed a genuine interest in the project and a desire to develop their plantation (Ruysschaert et al. 2011).

Local participation should be encouraged on two levels. One is internal: to maintain a well-trained and motivated workforce, and the other is external: to reach out and raise awareness among neighboring villagers and communities as part of community development programs (Rodríguez and Sabogal 2019). It will also be important to discuss the exact role of the reserve in the years to come, both from the perspective of smallholder farmers, as its economic value can be reinforced, for example, by adding timber species or other trees or plants of economic value. From a biodiversity conservation point of view, the potential its ability to conserve biodiversity can also be enhanced, for example by adding plant species that can especially attract endangered species (Ruysschaert et al. 2011).

It is important to note that an approach that uses existing sustainability standards and regulations to provide guidance on sustainable agricultural expansion may not be sufficient for every product. Whether adequate spatial indicators can be extracted from existing sustainability regulations and standards is highly dependent on the state of industry and national progress on sustainability. In addition, some metrics turn out to be more important than others, depending on the importance of the criteria examined, as well as the availability and quality of verification data. Indicators must be compared and given "weights." Assignment of these weights can be done on the basis of political goals and scientific knowledge but must also take into account the views of the stakeholders. Furthermore, to enhance the practical value of the method, at least one additional indicator needs to be considered, namely proximity to the infrastructure. For oil palm farming, access to infrastructure is critical as the fresh pods must be processed in oil palm refineries within 24 hours of harvest to maintain good quality (Smit et al. 2013).

17.6 Conclusion

Developing degraded land on mineral soils to secure oil palm supplies and protect the remaining biodiversity of lowland rainforests appears very attractive for combining the two goals that seemingly oppose each other: economic development of the oil palm industry and forest protection for mitigation climate change. Developing oil palm plantations on degraded lands can have a positive impact on mitigating climate, as long as it is well calculated and carefully planned. A well-developed methodology can support: (i) non-governmental organizations seeking to identify priority protected areas; (ii) private sector to identify future agricultural expansion in areas that meet sustainability criteria; (iii) local and central governments to align land use plans for agricultural expansion with a commitment to social acceptability and environmental sustainability. (iv) scientists and land-use planners to identify alternatives for expanding agriculture in areas of conservation value, and (v) global policymaker on biofuels and carbon sequestration programs (Smit et al. 2013). By doing so, the link between oil palm expansion and environmental problems can hopefully be broken.

References

Allen, K., Corre, M.D., Tjoa, A., and Veldkamp, E. (2015). Soil nitrogen-cycling responses to conversion of lowland forests to oil palm and rubber plantations in Sumatra, Indonesia. *PLoS One* 10: e0133325. https://doi.org/10.1371/journal.pone.0133325.

Austin, K.G., Schwantes, A., Gu, Y., and Kasibhatla, P.S. (2019). What causes deforestation in Indonesia? *Environ. Res. Lett.* 14: 024007. https://doi.org/10.1088/1748-9326/aaf6db.

BAPPENAS (2009). Reducing carbon missions from Indonesia's peat lands. Interim Report of a Multi-Disciplinary Study.

Barchia, M.F., Amri, K., and Apriantoni, R. (2018). Land degradation and option of practical conservation concepts in Manna mwatershed Bengkulu Indonesia. *TERRA J. L. Restor.* 1: 23–30. https://doi.org/10.31186/terra.1.2.23-30.

de Blécourt, M., Brumme, R., Xu, J. et al. (2013). Soil carbon stocks decrease following conversion of secondary forests to rubber (Hevea brasiliensis) plantations. *PLoS One* 8: e69357. https://doi.org/10.1371/journal.pone.0069357.

Boulmane, M., Oubrahim, H., Halim, M. et al. (2017). The potential of Eucalyptus plantations to restore degraded soils in semi-arid Morocco (NW Africa). *Ann. For. Sci.* 74: 57–66. https://doi.org/10.1007/s13595-017-0652-z.

Bruijnzeel, L.A. (2004). Hydrological functions of tropical forests: notn seeing the soil for the trees? *Agric. Ecosyst. Environ.* 104: 185–228. https://doi.org/10.1016/j.agee.2004.01.015.

Carlson, K.M., Curran, L.M., Asner, G.P. et al. (2013). Carbon emissions from forest conversion by Kalimantan oil palm plantations. *Nat. Clim. Chang.* 3: 283–287. https://doi.org/10.1038/nclimate1702.

Casson, A. (2000). *The Hesitant Boom: Indonesia's Oil Palm Sub-SectorB'OPSS in an EraE of Economic CrisisEC and Political Change, the Hesitant Boom: Indonesia's Oil Palm Sub-Sector in an Era of Economic Crisis and Political Change*. Center for International Forestry Research. https://doi.org/10.17528/cifor/000625.

Chase, L.D.C. and Henson, I.E. (2010). A detailed greenhouse gas budget for palm oil production. *Int. J. Agric. Sustain.* 8: 199–214. https://doi.org/10.3763/ijas.2010.0461.

Colfer, C.J.P. and Resosudarmo, I.A.P. (2010). *WFFP*. Routlege. https://doi.org/10.4324/9781936331772.

Corley, R.H.V. and Tinker, P.B. (2015). *The Oil Palm*, 55e. Wiley. https://doi.org/10.1002/9781118953297.

Cramb, R. and McCarthy, J.F. (2016). *The Oil Palm Complex: Smallholders, Agribusiness and the State in Indonesia and Malaysia*. NUS Press. https://doi.org/10.21315/km2016.34.2.7.

Daily, G.C. (1995). Restoring value to the world's degraded lands. *Science* 269: 199–214. https://doi.org/10.1126/science.269.5222.350.

Dislich, C., Hettig, E., Salecker, J. et al. (2018). Land-use change in oil palm dominated tropical landscapes—an agent-based model to explore ecological and socio-economic trade-offs. *PLoS One* 13: e0190506. https://doi.org/10.1371/journal.pone.0190506.

Eswaran, H., Lal, R., and Reich, P.F. (2019). Land degradation: an overview. Responses to land degradation. In: *Proceedings of the 2nd International Conference on Land Degradation and Desertification*, 20–35. Khon Kaen: Oxford Press. https://doi.org/10.1201/9780429187957-4.

Etter, A., McAlpine, C., Wilson, K. et al. (2006). Regional patterns of agricultural land use and deforestation in Colombia. *Agric. Ecosyst. Environ.* 114: 369–386. https://doi.org/10.1016/j.agee.2005.11.013.

Evans, J. (1999). Planted forests of the wet and dry tropics: their variety, nature and significance. *New Forest.* 17: 25–36.

Fairhurst, T. and Härdter, R. (2003). *Oil Palm: Management for Large and Sustainable Yields*. Singapore: Potash & Phosphate Institute.

Fairhurst, T. and McLaughlin, D. (2009). Sustainable oil palm development on degraded land in Kalimantan.

Fargione, J., Hill, J., Tilman, D. et al. (2008). Land clearing and the biofuel carbon debt. *Science* 319 (80): 1235–1238. https://doi.org/10.1126/science.1152747.

Fitzherbert, E.B., Struebig, M.J., Morel, A. et al. (2008). How will oil palm expansion affect biodiversity? *Trends Ecol. Evol.* 23: 538–545. https://doi.org/10.1016/j.tree.2008.06.012.

Gaveau, D.L.A., Sheil, D., Husnayaen et al. (2016). Rapid conversions and avoided deforestation: examininge four decades of industrial plantation expansion in Borneo. *Sci. Rep.* 6: 32017. https://doi.org/10.1038/srep32017.

Gaveau, D.L.A., Pirard, R., Salim, M.A. et al. (2017). Overlapping land claims limit the use of satellites to monitor no-deforestation commitments and no-burning compliance. *Conserv. Lett.* 10: 257–264. https://doi.org/10.1111/conl.12256.

Germer, J. and Sauerborn, J. (2008). Estimation of the impact of oil palm plantation establishment on greenhouse gas balance. *Environ. Dev. Sustain.* 10: 697–716. https://doi.org/10.1007/s10668-006-9080-1.

Gharibreza, M., Raj, J.K., Yusoff, I. et al. (2013). Land use changes and soil redistribution estimation using 137Cs in the tropical Bera Lake catchment. *Malaysia. Soil Tillage Res.* 131: 1–10. https://doi.org/10.1016/j.still.2013.02.010.

Gilroy, J.J., Prescott, G.W., Cardenas, J.S. et al. (2015). Minimizing the biodiversity impact of neotropicaln oil palm development. *Glob. Chang. Biol.* 21: 1531–1540. https://doi.org/10.1111/gcb.12696.

Gingold, B., Rosenbarger, A., Muliastra, Y.I.K., et al. (2012). How to identify degraded land for sustainable palm oil in Indonesia.

Guillaume, T., Damris, M., and Kuzyakov, Y. (2015). Losses of soil carbon by converting tropical forest to plantations: erosion and decomposition estimated by δ13C. *Glob. Chang. Biol.* 21: 3548–3560. https://doi.org/10.1111/gcb.12907.

Guillaume, T., Holtkamp, A.M., Damris, M. et al. (2016a). Soil degradation in oil palm and rubber plantations under land resource scarcity. *Agric. Ecosyst. Environ.* 232: 110–118. https://doi.org/10.1016/j.agee.2016.07.002.

Guillaume, T., Maranguit, D., Murtilaksono, K., and Kuzyakov, Y. (2016b). Sensitivity and resistance of soil fertility indicators to land-use changes: new concept and examples from conversion of Indonesian rainforest to plantations. *Ecol. Indic.* 67: 49–57. https://doi.org/10.1016/j.ecolind.2016.02.039.

Halim, R.M. and Aziz, A.A. (2019). MPOB commitment towards clean air regulatory compliance. *Palm Oil Eng. Bull.* 130: 13–17.

Hoffmann, M.P., Castaneda Vera, A., van Wijk, M.T. et al. (2014). Simulating potential growth and yield of oil palm (Elaeis guineensis) with PALMSIM: modelm description, evaluation and application. *Agric. Syst.* 131: 1–10. https://doi.org/10.1016/j.agsy.2014.07.006.

Hooijer, A., Page, S., Canadell, J.G. et al. (2010). Current and future CO2 emissions from drained peatlands in Southeast Asia. *Biogeosciences* 7: 1505–1514. https://doi.org/10.5194/bg-7-1505-2010.

Jennings, S., Nussbaum, R., and Evans, T. (2003). *The High Conservation Value Forest Toolkit.* Oxford: ProForest.

Kaniapan, S., Hassan, S., Ya, H. et al. (2021). The utilisation of palm oil and oil palm residues and the related challenges as a sustainable alternative in biofuel, bioenergy, and transportation sector: a review. *Sustainability* 13: 3110. https://doi.org/10.3390/su13063110.

Kent, M., Bradshaw, A.D., and Chadwick, M.J. (1980). The restoration of land: thet ecology and reclamation of derelict and degraded land. *J. Biogeogr.* https://doi.org/10.2307/2844672.

Koh, L.P., Levang, P., and Ghazoul, J. (2009). Designer landscapes for sustainable biofuels. *Trends Ecol. Evol.* 24: 431–438. https://doi.org/10.1016/j.tree.2009.03.012.

Kudesia, R., Rani, M., and Pal, A. (2009). Reclaiming degraded land in India through the cultivation of medicinal plants. *Bot. Res. Int.* 2: 174–181.

Manning, F.C., Kho, L.K., Hill, T.C. et al. (2019). Carbon emissions from oil palm plantations on peat soil. *Front. For. Glob. Chang.* 2: 1–21. https://doi.org/10.3389/ffgc.2019.00037.

Montagnini, F. (2001). Strategies for the recovery of degraded ecosystems: experiencese from Latin America. *Interciencia* 26: 498–503.

Murniati, F. (2002). *From Imperata cylindrica Grasslands to Productive Agroforestry.* Netherland: Wageningen University.

Mutsaers, H.J.W. (2019). The challenge of the oil palm: usingu degraded land for its cultivation. *Outlook Agric.* 48: 190–197. https://doi.org/10.1177/0030727019858720.

Padmanaba, M. and Sheil, D. (2007). Finding and promoting a local conservation consensus in a globally important tropical forest landscape. *Biodivers. Conserv.* 16: 137–151. https://doi.org/10.1007/s10531-006-9009-x.

Page, S.E. and Hooijer, A. (2016). In the line of fire: the peatlands of Southeast Asia. *Philos. Trans. R. Soc. B Biol. Sci.* 371: 1–9. https://doi.org/10.1098/rstb.2015.0176.

Paoli, G.D., Gillespie, P., Wells, P.L. et al. (2013). *Oil Palm in Indonesia: Governance, Decision Making and Implications for Sustainable Development, Governance*. Jakarta, Indonesia: Decision Making & Implications for Sustainable Development.

Phalan, B., Fitzherbert, E.B., Rafflegeau, S. et al. (2009). Conservation in oil-palm landscapes. *Conserv. Biol.* 23: 244–245. https://doi.org/10.1111/j.1523-1739.2008.01151.x.

Pirard, R. and Irland, L.C. (2007). Missing links between timber scarcity and industrial overcapacity: lessons from the Indonesian pulp and paper expansion. *For. Policy Econ.* 9: 1056–1070. https://doi.org/10.1016/j.forpol.2006.10.003.

Rodríguez, J.C. and Sabogal, C. (2019). Restoring degraded forest land with native tree species: the experience of "bosquesb Amazónicos" in Ucayali, Peru. *Forests* 10: 851–869. https://doi.org/10.3390/f10100851.

RSPO (2007). RSPO Principles and Criteria for Sustainable Palm Oil Production. October. https://doi.org/10.1111/j.1523-1739.2008.01026.x

RSPO (2017). Guidance on Map Submission for Land Use Change Analysis (LUCA) for Independent Smallholders.

RSPO (2019). Reflecting on a Decade of Growth Impact Report.

Ruysschaert, D., Darsoyo, A., Zen, R. et al. (2011). *Developing Palm-Oil Production on Degraded Land: Technical, Economic, Biodiversity, Climate, Legal and Policy Implications.* PanEco–World Agroforestry Centre.PI.

Sarminah, S., Karyati, K., Simbolon, J., and Tambunan, E. (2018). Rehabilitation and soil conservation of degraded land using sengon (Falcataria moluccana) and peanut (Arachis hypogaea) agroforestry system. *Biodiversitas* 19: 222–228. https://doi.org/10.13057/biodiv/d190130.

Smit, H.H., Meijaard, E., van der Laan, C. et al. (2013). Breaking the link between environmental degradation and oil palm expansion: a method for enabling sustainable oil palm expansion. *PLoS One* 8: e68610. https://doi.org/10.1371/journal.pone.0068610.

Smith, D.R., Townsend, T.J., Choy, A.W.K. et al. (2012). Short-term soil carbon sink potential of oil palm plantations. *GCB Bioenergy* 4: 588–596. https://doi.org/10.1111/j.1757-1707.2012.01168.x.

Suhardi, Faridah, E., and Handojo, H.N. (2016). Rehabilitation of degraded forests in Indonesia. *World Bank Tech. Pap.* 270: 27–56.

Suprihatin, A. and Waluyo (2015). Nutrient needs of producing oil palm crops in sour dry land of South Sumatra (in Indonesian). In: *Prosiding Seminar Nasional Swasembada Pangan*, 337–342. Lampung, Indonesia: Politeknik Negeri Lampung.

Taheripour, F., Hertel, T.W., and Ramankutty, N. (2019). Market-mediated responses confound policies to limit deforestation from oil palm expansion in Malaysia and Indonesia. *Proc. Natl. Acad. Sci. U. S. A.* 116: 19193–19199. https://doi.org/10.1073/pnas.1903476116.

Tyrie, G.R. and Gunawan, A. (1998). The environmental framework as a basis for forest management planning. In: *Data Management and Modelling Using Remote Sensing and GIS for Tropical Forest Land Inventory* (ed. Y. Laumonier, B. King, C. Legg, and K. Rennolls). Jakarta, Indonesia: Rodeo International.

Van Straaten, O., Corre, M.D., Wolf, K. et al. (2015). Conversion of lowland tropical forests to tree cash crop plantations loses up to one-half of stored soil organic carbon. *Proc. Natl. Acad. Sci. U. S. A.* 112: 9956–9960. https://doi.org/10.1073/pnas.1504628112.

Vermeulen, S. and Goad, N. (2006). Towards better practicebp in smallholder palm oil productionspop. In: *Natural Resource Series*, vol. 5. International Institute for Environment and Development.

Wawan, Dini, I.R., and Hapsoh (2019). The effect of legume cover crop Mucuna bracteata on soil physical properties, runoff and erosion in three slopes of immature oil palm plantation. *IOP Conf. Ser. Earth Environ. Sci.* 250: 012021. https://doi.org/10.1088/1755-1315/250/1/012021.

Whitmore, T.C. (1988). *Tropical Rain Forests of the Far East*, 22e. Oxford Science Publications. Oxford University Press.

Wich, S.A., Garcia-Ulloa, J., Kühl, H.S. et al. (2014). Will oil palm's homecoming spell doom for Africa's great apes? *Curr. Biol.* 24: 1659–1663. https://doi.org/10.1016/j.cub.2014.05.077.

Woittiez, L.S., Sadikin, H., Turhina, S. et al. (2016). *Smallholder Oil Palm Handbook*. Wageningen, Netherland: Wageningen University & SNV Netherland.

Woittiez, L.S., van Wijk, M.T., Slingerland, M. et al. (2017). Yield gaps in oil palm: a quantitative review of contributing factors. *Eur. J. Agron.* 83: 57–77. https://doi.org/10.1016/j.eja.2016.11.002.

18

Reclaiming Urban Brownfields and Industrial Areas–Potentials for Agroecology

Petra Schneider[1], *Tino Fauk*[1], *and Florin-Constantin Mihai*[2]

[1] *Department Water Environment, Civil Engineering and Safety, Magdeburg-Stendal University of Applied Sciences, Breitsscheidstr. 2, D-39114, Magdeburg, Germany*
[2] *CERNESIM Environmental Research Center, Institute of Interdisciplinary Research "Alexandru Ioan Cuza" University of Iasi, Bulevardul Carol I 11, 700506, Iași, Romania*

18.1 Introduction

The countries in Europe are facing a particular situation: while fertile land is giving way to new construction, more and more land in cities and communities is lying fallow. Land use and urban sprawl are among the most pressing recent environmental problems. Other consequences include higher costs for the community and a loss of urban life quality. Between 2005 and 2015, a population increase of around 1.5 million was reported for 53 of 77 major cities in Germany (Beckmann and Dosch 2018). Another interesting aspect is the immigration of woman between 18 and 25 years in urban areas, which positively affected the birth rate (Beckmann and Dosch 2018). However, the other 24 large cities have a decreasing number of inhabitants (shrinking cities). This effect is not only noticeable in big cities as the demographic development is much more pronounced in rural areas.

For reasons of sustainability and to protect natural resources, the reuse of previously used and maybe even polluted areas must always be given priority over the use of greenfield sites. Land recycling is an important tool for reducing land use transformation from natural land cover or high nature value farmland to settlements and traffic systems. At the same time, the linking of polluted site remediation and land recycling leads to a much more effective, including the economic use of funds. The revitalization of fallow land has also become a focus of urban renewal and urban development in Europe in recent years Important elements of municipal development and infrastructure policy are the revitalization of commercial and industrial wasteland, military properties, and infrastructure fallows. In parallel, urban brownfields play a substantial role in the municipal context including in Eastern Europe (Costa et al. 2018) while rural areas could also benefit from such projects (Bostenaru Dan and Bostenaru-Dan 2021).

Agroecological Approaches for Sustainable Soil Management, First Edition.
Edited by Majeti Narasimha Vara Prasad and Chitranjan Kumar.

Even urban brownfields and industrial areas have been considered land resources for revitalization in order to establish new industries and living centers, such land resources are able to contribute to biodiversity, and also provide enormous potential for reforestation and the development of succession areas against the background of the current climate debate. Such "lost places" are therefore dynamic ecosystems whose development process is subject to natural succession. For central Europe, this means that a forest develops at the end of natural succession (Wittig 2002; Zerbe 2019). A later use of woodlands does not have to be ruled out. One hectare of forest (with around 1000 trees per hectare) can bind around 200 kg of CO_2 per tree, i.e., the most important thing needed for efficient forest development to bind CO_2 is space. Moreover, there is potential for agroforestry and land remediation through phytoremediation applications. In this way, urban brownfields and industrial areas may provide land resources for agroecological purposes, particularly forestry. Some successful urban forest projects are revealed in Germany as a proven nature-based solution for urban brownfields (Rink and Schmidt, 2021).

According to the Food and Agriculture Organization of the United Nations (FAO) agroecology is a holistic and integrated approach that simultaneously applies ecological and social concepts and principles to the design and management of sustainable agriculture and food systems. It seeks to optimize the interactions between plants, animals, humans, and the environment while also addressing the need for socially equitable food systems. FAO addresses 10 elements of agroecology: diversity, co-creation and sharing of knowledge, synergies, efficiency, recycling, resilience, human and social values, culture and food traditions, responsible governance, as well as circular and solidarity economy. Moreover, according to the Soil Association, agroecology is the application of ecological concepts and principals in farming (Soil Association 2022). In this way, it is part of ecological engineering, a discipline that uses ecology and engineering to predict, design, construct or restore, and manage ecosystems that integrate "human society with its natural environment for the benefit of both" (Mitsch and Jorgensen 1989).

The present contribution illustrates how urban brownfields and industrial areas can be reclaimed and what contribution can be provided by agroecology. Having in view the 2030 Forest Strategy and the Soil Strategy of the European Commission, there are substantial synergies in order to benefit both.

18.2 Characterizing Urban Brownfields and Industrial Areas

18.2.1 Overview on Urban Brownfields and Industrial Areas and Respective Hazards

Due to their history, urban brownfields and industrial areas often contain material, technical, physical, or social hazard potentials. This setting can involve physico-chemical factors such as the presence of surface burdens, e.g., from hazardous substances or munitions, but also the decay of buildings and the corresponding problems with unstable building fabric or potential social hazards such as desertification or criminal hotspots or dangerous "playground" for children. In case of huge dimensions, hazardous substances represent an obstacle to subsequent use. Typical challenges relate to polluted sites and harmful soil modifications as well as the presence of munition, especially on former military sites. The development of those types of lands requires a corresponding risk assessment and, if necessary, risk elimination, since potential risks can restrict the subsequent use, and can also lead to a reduction in the property value.

The term urban brownfields and industrial areas, which are to be recycled or reused, relate in particular to following land types:

- urban brownfield sites and polluted waste lands,
- conversion areas (abandoned military properties),
- abandoned roads and railway tracks,
- deserted areas (desolation including depopulation), including abandoned manors/estates, and
- abandoned industrial areas, including post-mining areas.

What all these areas have in common is that they have lost their previous function and use and are to be given a new use. As part of a circular economy approach, the type of subsequent use is in principle open-ended and is determined by the market situation, that is the demand for land. In structurally weak regions there is often a low land demand for revitalization projects. The reintegration of such properties into the natural cycle offers great future potential for the subsequent land use, including agroecological applications.

Urban brownfields and industrial areas may contain potential hazards. According to ISO/IEC Guide 51, a hazard describes a condition or a situation in which there is the possibility that an object to be protected (e.g., a person, but also a natural basis of life) can coincide spatially and/or temporally with a potential source of damage, i.e., the hazard reaches the potential receptor via an impact pathway and can course a damage. In the case of urban brownfields and industrial areas, the question of potential hazards is naturally linked to the presence of contaminated sites, ordnance, or warfare agents and/or geotechnical stability problems. In general, a distinction is made between acute and latent hazards. The hazard potential can be material (e.g., hazardous substances), technical (e.g., building remains, soil instability), physical (e.g., radioactivity), but also social (e.g., depopulation, mental stress situations) in nature. The term "fallow land" that is referring to urban brownfields denotes plots of land that are either not used at all or not used in accordance with their urban planning potential. This means that the areas are unused or underused. Brownfield sites are mostly abandoned business premises or business premises that are no longer needed by the company. Economic structural change and company closures are usually the cause of the emergence of fallow land. Table 18.1 shows a summary of the hazards derived for urban brownfields and industrial areas.

As a rule, urban brownfields and industrial areas are often considered problematic areas. The importance of the topic results from the economic, historical, and resource-related perspective: rehabilitation and development costs, availability of reserve areas for new structural uses, renewable energies, or nature conservation.

Table 18.1 Types of urban brownfields and industrial areas and respective hazards.

Type of land	Associated land types	Hazard potential				Hazard for after use
		Substances	Technical	Physical	Social	
Urban brownfields	Fallow land	Possibly metals, organic substances depending on former use	Possibly building remains, cavities	Pits, destroyed building fabric (threat to physical integrity)	criminal hotspot, dangerous playground (threat to psychological integrity)	Usually small
Conversion land (former military use)	Military training areas, drop places	Munitions, warfare agents, metals, organic pollutants	Building remains, cavities, possibly subsoil stability, possibly soil compaction	Possibly radioactive radiation	Desolation	Munitions, warfare agents, hazardous materials, dimension, desolation
	Air bases	Organic pollutants, de-icing agents	Remains of buildings, cavities, possibly subsoil stability		Desolation	Hazardous materials, desolation
	Legacy armaments	Munitions, explosives, metals, organic pollutants	Remains of buildings, cavities, possibly subsoil stability			Munitions, warfare agents, hazardous materials, dimension, desolation
	Ammunition dismantling plants	Munitions, explosives, metals, organic pollutants	Remains of buildings, cavities, possibly subsoil stability	Possibly radioactive radiation		Munitions, warfare agents, hazardous materials, dimension, desolation

Industrial brownfields	Old industrial sites	Depending on the industry: metals, organic and/or inorganic pollutants, industrial chemicals	Remains of buildings, cavities, possibly subsoil stability, possibly industrial tailings (sludge ponds), soil compaction			Dimension, hazardous materials
	Polluted waste lands	Depending on the industry: metals, organic and/or inorganic pollutants, industrial chemicals	Remains of buildings, cavities, possibly subsoil stability	Possibly radioactive radiation		Hazardous materials
	Former lignite mining sites	Metals, inorganic pollutants	Unstable slopes, industrial tailings (sludge ponds)			Acid mine drainage, hazardous substances, dimensions
	Former ore mining sites	Metals, inorganic pollutants	Unstable slopes, industrial tailings (sludge ponds)	Possibly radioactive radiation		Acid mine drainage, hazardous substances, dimensions
Transport browfields	Abandoned track systems	Metalloide, organische/ anorganische Schadstoffe, Pestizide	Remains of buildings, cavities, possibly subsoil stability, soil compaction			Hazardous materials
	Abandoned roads	Metals, organic/ inorganic pollutants	Soil compaction			Hazardous materials
	Abandoned air bases	Metals, organic/ inorganic pollutants	Remains of buildings, cavities, possibly subsoil stability, soil compaction			Hazardous materials
Deserted areas	Abandoned economic or agricultural areas	Metals, organic/ inorganic pollutants, possibly pesticides, manure	Remains of buildings, cavities, possibly subsoil stability		Desolation	Desolation
	Abandoned settlements	Hazardous substances only in case of accident	Remains of buildings, cavities, possibly subsoil stability		Desolation	desolation

18.2.2 Development Potentials of Urban Brownfields and Industrial Areas

Land is scarce in many regions of the world and is therefore subject to competing claims for use, resulting in so called land take, which practically means land consumption. Land consumption refers to the conversion of agricultural or natural areas in particular into "settlement and traffic areas." The expansion of settlement and traffic areas is accompanied by the loss of soil and natural soil functions as basis of terrestrial habitats and their biodiversity.

According to the German Soil Legislation, soil is the carrier of natural soil functions, but also has an archival function. Land use relates significantly to soil sealing, i.e., a more or less complete covering and replacement of the natural soil with building materials. When the natural soil is removed, the remaining soil loses its natural soil functions and becomes so called "soil material," hence waste in the sense of the waste legislation. This means that land consumption relates not only to soil sealing but also to soil loss. Land consumption is estimated to be one of the least well-known environmental issues globally. The conversion of suitable and available areas, such as brownfield sites, is therefore the order of the day. One instrument for reducing land consumption is land recycling as a tool for circular land use.

Land recycling was defined by the German Scientific-technical Association for Environmental Remediation and Brownfield Redevelopment (ITVA) as the "use-related reintegration of such properties into the economic and natural cycle that have lost their previous function and use–such as disused industrial and commercial areas, military properties, traffic areas, etc. by means of planning, environmental and economic policy measures" (German Scientific-technical Association for Environmental Remediation and Brownfield Redevelopment ITVA, 1998). The main technical difference between land recycling and the remediation of polluted sites is that the remediation of polluted sites is a risk prevention measure. Properties subject to land recycling need not necessarily to be polluted, however, if they are, they are called polluted sites. Land recycling, including the depollution, shall be part to a particular type of sustainable land management that is called "Circular land management."

Circular land management is a circular economy approach that has its roots in the recycling approach similar to other resources. Although originally the recycling approach does not primarily refer to land, but to waste resources which ideally becomes raw materials again, the basic recycling concept can be transferred also to land sustainable management. The principles are laid down in common recycling strategies:

- avoidance
- preparation for reuse
- recycling (material recycling)
- other recycling, in particular energetic recycling and backfilling
- elimination

and are re-interpreted in the field of circular land management with "avoid – use – compensate." For this purpose, the project "Experimental Housing and Urban Development (ExWoSt)" of the Federal Ministry of Transport, Building and Urban Development (BMVBS) and the Federal Office for Building and Regional Planning (BBR) coined the term circular land management.

Circular land management goes beyond conventional land recycling and is a holistic, strategic approach to (re)using fallow land and reducing land consumption, which also takes into account the requirements of climate protection, energy efficiency, remediation of polluted sites and urban development (Deutsches Institut für Urbanistik, 2001). On the one hand, existing, underused potential areas such as fallow land and vacant lots should be valued. Previously used areas that are not suitable for subsequent structural use are released from the real estate cycle by way of renaturation.

The principle of circular land management includes:

- a system of planning, use, abandonment of use, lying fallow and reintroduction of areas that is strategically planned.
- the premise of mobilizing existing space potential and prioritizing it for new uses,
- no longer the open space, but the fallow land, i.e., to consider the existing area as a resource for new uses,
- to develop a usage cycle of land provision, if necessary, development, use, fallow, and reuse.

Land recycling as part of circular land management is naturally a dynamic process. A key question in connection with land recycling is the decision criteria to find the optimal use in the frame of subsequent usage options and of brownfield development. Ultimately, this decision must always be the result of a site-related consideration process. In this context, restoration should also be understood as a potential solution, or as an interim solution until industrial infrastructure or residential development can give the area a new function. Nevertheless, there will always remain areas that are permanently excluded from the revitalization process. However, this does not mean, that these areas do not have any economic potential: they can, for example, be developed as an area pool for compensatory measures in nature conservation.

18.2.3 New Approaches to a Land Saving Management

In Germany, land trading certificates are intended to quantitatively control the use of land. The basic principle of trading of land trading certificates is comparable emission certificates trading and based on an environmental economic approach. The use of the protected asset, in this case the land (that is also a soil resource), is limited and divided into partial use claims. These are issued to interested parties as tradable rights (e.g., certificates) (German Federal Environment Agency 2018).

Trading land certificates was tested in practice for the first time from 2013 to 2017 in a supra-regional pilot project with municipalities (German Federal Environment Agency 2018). According to the German Federal Environment Agency (2018), the principle of land trading is based on the fact that the municipalities are assigned so-called land designation rights in the form of certificates, which they need if new building rights and that are to be created by a development plan in the outer area. The number of certificates must then correspond to the area dedicated for the first time for settlement and transport purposes (Federal Environment Agency 2018). If the area required outside is greater than the available certificates, additional certificates can be purchased from other municipalities. If municipalities do not need the certificates assigned to them themselves, they can sell the rights to other municipalities (German Federal Environment Agency 2018). An upper limit can be set for the sum of the allocated certificates. According to the Federal Environment Agency (2018), for example, 30 ha per day could be used in Germany which correlates with the goal for maximum land consumption per year in Germany.

The objective of the concept of land certificates is that the municipalities no longer pursue a policy of land take but tend to sell the certificates allocated to them on the market from an economic point of view (German Federal Environment Agency 2018). Potential buyers of area certificates are growth municipalities that have neither enough certificates nor sufficient interior space to implement their land use policy (German Federal Environment Agency 2018).

According to the information service of the German Economic Institute iwd (2019), shrinking regions benefit from trading in area certificates, since they can sell unused certificates and white certificates obtained through demolition to municipalities that want to designate a particularly large amount of building land, and the money raised can be invested in their own infrastructure and invest in the activation of brownfields and vacancies (IWD 2019). So far there has only been one pilot test for trading in area certificates, which runs from 2014 to 2028 and in which 87 German municipalities are involved. However, the result of the test has already shown that saving land can work in this way.

Based on the previous statements, the following equations are concluded. With focus on an annual maximum limit ($LTC_{max,a}$) in hectares per year for the transformation of other land use categories (LU_i) into the land use category settlement and transport maybe also industrial or other commercial areas (LU_{sat}), the maximum years of land trade (t_{maxLT}) can be determined in a simplified way with Eq. (18.1).

$$t_{maxLT} = \frac{\left[\left(\sum_i LU_i\right) - \left(LU_{sat} + \sum_i LU_{fix,i}\right)\right]}{LTC_{\max,a}} \tag{18.1}$$

The previous equation contains the summed area of community-specific land use ($\Sigma_i LU_i$) and the summed fixed (restricted) land use of categories i ($\Sigma_i LU_{fix,\,i}$) in hectares at current time. With the addition of a municipality-specific expected value or function for the forecast land use of settlement and transportation areas in the coming years (ΔLU_{sat}), the planning of annual land trade (LTC_a) for a legislative period in years (t_{lp}) can be derived with Eq. (18.2).

$$LTC_a = \frac{\left[\left(\sum_i LU_i\right) - \left(\Delta LU_{sat} + \sum_i LU_{fix,i}\right)\right]}{t_{lp}} \leq LTC_{\max,a} \tag{18.2}$$

The change of land use category takes place in the year of political decision-making, accordingly, these areas are to be transferred from, for example, the land use category of grassland to the land use category of settlements. Land use certificate trading is limited due to municipal area. Equation 18.2 may be useful to approximate the maximum area for certificate trading. For calculate the maximum area without into account the legislature period, one option is given in Eq. (18.3).

$$LTC_{\max} = \left(\Sigma_i\, LU_i\right) - \left(\Delta LU_{sat} + \Sigma_i\, LU_{fix,i}\right) \tag{18.3}$$

The most feasible solution for a sustainable and equitable land use is therefore certificate trading.

18.3 After Use Options for Urban Brownfields and Industrial Areas

18.3.1 General Options and Restrictions

Future area development potential of urban brownfields and industrial areas include the options of revitalization and renaturation, as well as smooth transitions between these. The real usage potentials are determined by primary boundary conditions, in particular the regional demographic development and the soil quality as well as the diversity of species and green structure elements. Both boundary conditions significantly influence the property quality and thus the property value.

Urban brownfields and industrial areas are usually characterized by use-restricting properties, which leads to a limited marketability of such areas in the short term. The marketability is essentially determined by the presence of obstacles and the quality of the property, which can be determined through the ABC model of land recycling (Ferber 1997). However, in the long term, urban brownfields and industrial areas represent valuable soil resources and area potential and thus a substantial potential for the generation of ecosystem services. Standard land value forecasts show that green spaces and residential areas with a high proportion of greenery are increasing in value.

In practice, the desire of the municipalities to revitalize the areas played a central role in connection with the demand for such areas. Not infrequently, the municipalities in demographically shrinking regions had to recognize that the areas that had been given up for the century in connection with industrial or residential use are no longer in demand due to structural regional change. Here it is important to examine the objectives and options for potential subsequent use and their boundary conditions in more detail. From the author's point of view, new options for use are opening up in connection with the European Union's Green Deal and the associated strategies, in particular:

- "EU Biodiversity Strategy for 2030: More room for nature in our lives" as of June 2021,
- New EU forest strategy up to 2030 as of July 2021,
- EU agricultural strategy "Farm to Fork" as of October 2021,
- "EU Soil Strategy for 2030 – Reaping the benefits of healthy soils for people, food, nature and climate" as of November 2021,
- EU Action Plan: "Toward Zero Pollution for Air, Water and Soil" as of May 2021,

and the proposal for the new EU Law on Nature Restoration as of June 2022. The proposed regulation sets EU-wide legally binding targets for the restoration of nature in different ecosystems, and states, among other things, that by 2030 at least 20% of the land and sea areas of the EU must be taken to restore nature and that all natural and semi-natural ecosystems must also be put on the path to recovery by 2030. Here, urban brownfields and industrial areas offer great potential, especially since many of the affected areas such as military training areas and deserted areas are already subject to natural succession.

Subsequent use as green infrastructure for climate protection and CO_2 binding purposes is also conceivable, for example by planting forests. Last but not least, the recreational function should be mentioned due to the nature conservation upgrading of the areas. For example, former landfill sites are transformed into functional urban green areas for recreational activities (Mihai et al. 2021). This also pays off economically, since the obligation to compensate for the impairment of nature and landscape resulting from an intervention is anchored for instance in the German Federal Nature Conservation Act. Last but not least, the renaturation of fallow land contributes to the development of biotope networks, which serves to implement the EU Biodiversity Strategy by 2030.

Both the existing and the future land use are the basis and decisive decision-making criteria for the options for future land development. The use of land in a municipality is usually laid down in the land use plan. The use of land thus forms the framework for other uses, such as commercial, residential, use of resources and infrastructure. Land use potentials describe the possible uses that can be derived from the quality properties and characteristics of the available areas.

In the frame of spatial development, natural and infrastructural features are usually checked in order to then determine which subsequent use options can be developed in the sense of, for example, use for residential or commercial purposes. This approach is primarily aimed at services and trade (revitalization) and requires the existence of a demand for land. If there is no land demand with the aim of revitalization, it is worth thinking about natural options for subsequent use (restoration) and perceiving nature as a potential service provider for the subsequent use and development of the area, i.e., establishing a connection to the ecosystem services of the area in question.

The assessment of the options for future area development and the weighing up between revitalization and restoration is naturally a fundamental task in land management, which is largely determined by structural boundary conditions. The following aspects can be considered as possible influencing factors:

Primary Boundary Conditions

- The regional demographic development and therefore resulting land demand, which largely determines the space requirements.
- The soil quality, which largely determines the property quality and thus the property value.
- Diversity, both at the species level and at the level of landscape structures in relation to current and future climatic conditions.

Secondary Boundary Conditions

- The existence of taboo zones (e.g., protected areas, natural risk areas, etc.), which can result in usage restrictions.
- The potential natural vegetation, which gives an indication of the vegetation development potential of the areas (for instance in Germany does already exist a complete mapping).
- The renaturation and habitat connectivity potential, which is of enormous importance for the long-term resilience of the urban landscape and human well-being (Schmidt 2020; Zerbe 2019).
- The potential of the area to generate ecosystem services.
- The infrastructural equipment, which is a demand feature, especially for commercial use.

The above classification of primary and secondary boundary conditions results from empirical values. In particular, the classification of the infrastructural equipment could be viewed differently by land development institutions. However, the evaluation of urban brownfields and industrial areas, has shown that the infrastructural equipment (all media are usually available) does not improve the development and marketing potential of such areas. The reason for this is often both the unfavorable location (conversion areas are often located in very remote regions) and the enormous size of the areas, which are no longer needed in this dimension as usable areas, especially in Eastern Germany. This is usually related to the demographic development and the resulting shrinking demand for space in the regions concerned. Nevertheless, the quality of the location plays a central role for the future development potential, albeit in a different way than in growing regions, where the infrastructural equipment can be assessed as a central feature.

In shrinking regions, the probability of revitalization as a development potential is significantly lower, which means that the option of restoration, including a possible use of the landscape by agriculture, is gaining in importance. Soil quality forms the basis for solid ecosystem development. Therefore, the regional demographic development, which largely determines the space requirement, and the soil quality, which largely determines the property quality and thus the property value, were identified as the primary boundary conditions for future space development potential.

18.3.2 Restoration and Green Infrastructure

Green infrastructure (GI) describes a strategically planned network of high-quality natural and semi-natural resources that provides ecosystem services and protects biodiversity in rural and urban areas (European Commission 2013). In addition, GI helps to understand the value of the benefits that nature offers to human society. It can sometimes provide an alternative or complement to gray infrastructure (European Commission 2013).

Ecosystem services refer to direct and indirect contributions of ecosystems to human well-being, i.e., services and goods that bring direct or indirect economic, material, health, or psychological benefits to humans (Millennium Ecosystem Assessment 2005). In the Millennium Ecosystem Assessment (2005), providing, regulating, supporting and cultural services were derived as ecosystem services. Biological diversity is the prerequisite for the healthy and natural development of all living things and ecosystems and creates many ecosystem services that are used every day, such as:

Provisioning ecosystem services: comprises the provision of resources like wood, medicinal plants, agricultural and horticultural products, minerals and drinking water (Zerbe 2019). These ecosystem services can only be provided by unsealed soil, sufficient water and soil conditions and the genetic resource in the form of animal and plant diversity.

Regulatory ecosystem services: are services such as micro- and mesoclimate, water balance, air and water purification, noise reduction, crop pollination, nutrient cycles, humus formation, carbon fixation and oxygen production (Zerbe 2019).

Supporting ecosystem services: basic ecosystem services that enable all other services, such as bio- and geochemical cycles, e.g., oxygen production and carbon sequestration, the maintenance of nutrient cycles or the water cycle.

Cultural ecosystem services: Ecosystems and species contribute to diverse landscapes. Cultural ecosystem services also consider spiritual, recreational, and cultural benefits such as making people happy and giving meaning to life.

In the European Union, the classification Common International Classification of Ecosystem Services (CICES) is also used, which emerged from the work on environmental accounting by the European Environment Agency (EEA). She supports her contribution to the revision of the Systems of Environmental-Economic Accounting (SEEA), currently led by the United Nations Statistical Division (UNSD) Schwaiger et al. (2015). In CICES, ecosystem services are defined as the contributions that ecosystems make to human well-being and are distinguished according to the goods and benefits that people derive from them. Green infrastructure thus provides ecosystem services that benefit people, e.g., in the form of CO_2 sequestration and oxygen production, but also heat buffering.

For the American Society of Landscape Architects (2019), green infrastructure can be at the heart of smart regional and urban planning, guaranteeing communities a livable environment with clean air and water for future generations. GI planning also includes the restoration of degraded ecosystems. Ecosystem restoration needs to be supported by predicting habitat evolution during restoration activities. Typically, such a type of model takes into account two steps: (i) the reference state and (ii) the ecological direction, i.e., the diversity of life forms present, the structure of the species to be created and the role that these species play in the ecological system (Gann & Lamb, 2006). Well-functioning ecosystems can provide a constant flow of ecosystem services that are central to human well-being. Remediation and ecosystem restoration often results in a stronger and more balanced flow of services, which is the main motivation for ecological remediation from an economic and social point of view.

The reintegration into the natural balance is aimed at restoring ecosystem structures and functions. There are several ways to restore ecosystems (Figure 18.1), namely (Paternoster et al. 2021):

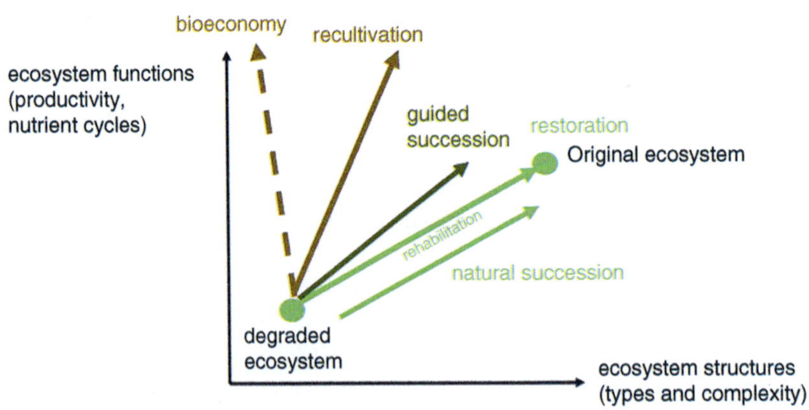

Figure 18.1 Restoring ecosystem structures and functions. *Source:* Wong and Bradshaw 2003; modified Schneider 2022.

Rehabilitation, which represents a real restoration service, i.e., the achievement of a more natural state, i.e., H. a state of lower intensity of use or intervention. In the event of abandonment, this is associated with allowing natural succession.

Natural succession, in which nature is given complete free rein, thus allowing the natural return of the plant, animal and fungal communities (biocenosis) typical of a site. Guided succession means minor management of the succession process (Connell und Slatyer, 1977).

Recultivation, i.e., the active return to a usable state (e.g., with regard to agricultural and forestry interests) or restoration after extremely intensive use or destruction. It is primarily about the creation of new habitats for which there is no direct reference in the natural and cultural landscape.

Ecosystem restoration supports the development or restoration of an ecosystem that has been more or less severely degraded or completely destroyed by humans toward a more natural state. In this way, certain ecosystem services and structures are restored against the background of current ecological, socio-economic and nature conservation framework conditions (Zerbe et al. 2009). Reference states can be defined for near-natural and semi-natural ecosystems, but in the case of irreversibly changed landscapes, such as urban industrial landscapes, new types of ecosystems can also arise for which there are no reference systems so far, but which are usually very well adapted to the special site conditions.

18.3.3 Revitalization Options

Revitalization refers to the rehabilitation of vacant buildings, derelict areas, and vacant lots as well as underused development areas indoors and outdoors. The revitalization of fallow land protects the environment and is a key concept for shaping structural changes. The revitalization of brownfield sites comprises the initialization and evaluation phase, the planning and development phase, the execution and project implementation phase, and the marketing and brand launch. It has a focus on commercial, industrial, residential, as well as real estate use. Generally, it can be considered a gray infrastructure approach that might be complemented with green infrastructure elements, particularly in residential areas. If revitalization can be implemented in reality, depends on the market demand (Pongratz and Vogelgesang 2016).

18.3.4 Market Demand, Barriers, and Requirements

The following criteria play a decisive role in the real estate marketing of areas to be used in a circular economy with regard to future usability and feasibility: market situation and financial viability, ownership structure, soil conditions, legal restrictions and existing statutes, degree of development, monument protection, existing buildings, and asking price of the owners.

The ITVA e.V. recently named the existing but largely non-harmonized requirements for material flow and soil management (German Scientific-technical Association for Environmental Remediation and Brownfield Redevelopment ITVA, 2019) as another major obstacle from a legal point of view.

In earlier studies on obstacles to the re-utilization of areas to be used in a circular economy, the primary problem identified was that no new user could be found (Alpha Research

Study Industriebrachen, 2003). This was followed by "unsuitable planning framework conditions" and "lack of start-up financing." Only in fourth place was "unfavorable situation" mentioned as an obstacle in this study. Ferber (1997) developed the ABC model for the classification of areas, in which the option of area recycling as a self-perpetuating factor is described as a question of marketability (Figure 18.2).

The marketability is essentially determined by the non-existence or presence of the above-mentioned obstacles and thus also the question of whether, depending on the quality of the property, it is a "self-perpetuating" (= A-property) or a reserve area (= B-property) due to high preparation costs and/or low demand (= C property). As a rule, C-areas can only be put to subsequent use within public planning, as they are of no interest to private investors. In between are so-called development areas (= B properties), e.g., areas whose development can be "pushed" by means of public-private partnership, e.g., that the public sector must create incentives to market these areas. Operator models that can be designed as

a) concession models, and
b) Public Private Partnership (PPP) business models.

Private "venture capitalists" only invest if the project is profitable in the long term (A and B properties, profitable property portfolio) and the risks are distributed fairly. PPP business models can be designed as a trustee model, developer model or agency model.

Figure 18.2 illustrates that B properties typically dominate the portfolio, since A properties can be marketed quickly. C-properties are practically "slot sellers." In Northern Germany, one of the reasons for the difficult marketability of areas that have been abandoned was the sometimes-immense size of the area (especially in the case of conversion and former mining areas), but also often the unfavorable location of the areas in regions that are structurally weak anyway.

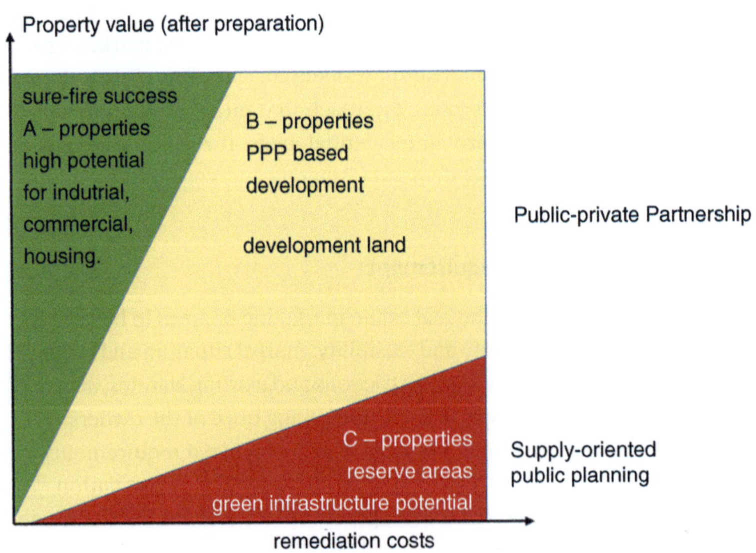

Figure 18.2 ABC model of land recycling – a question of marketability, *Source:* Ferber 1997; modified Schneider 2022.

18.3.5 Land Management

Land management represents an interdisciplinary field of action in the area of tension between land use planning, economic development, marketing, and financing (Deutsches Institut für Urbanistik 2001; Pongratz 2016) (Figure 18.3). So far, there is no academic institution that trains "land managers." Persons acting in this field typically come from land use planning, soil protection or the real estate industry. Land management was defined as follows in 2012 by the German Land Working Group of the Federal States: "...an instrument at municipal or regional level that makes it possible to deal with land and land efficiently." It thus serves to optimize land use and the provision of building land. However, this definition says nothing about the market situation, which naturally has a significant impact on the marketability of land, as has been impressively observed since 2015 when interest rates began to fall. Since then, the costs of cleaning up polluted sites have receded significantly, while the decisive criterion at the beginning of the construction boom was still the location, which no longer plays the decisive role after the return of the properties close to the city to the land cycle due to the persistently high demand. Anyhow, land management is an interdisciplinary field of action (Figure 18.3).

In preparation for the new use, the following is done:

- Technically: preparation of the land (dismantling of plants and buildings, stabilization of the subsoil, securing or decontamination of polluted sites, etc.)
- Legally: creating new building rights for the land (e.g., development plan),
- Economically: aligning the land potential and the investor's financial proposal.

Against the background of the still growing use of land, it can be assumed from an economic point of view that the value of land and property will continue to increase steadily. There is a

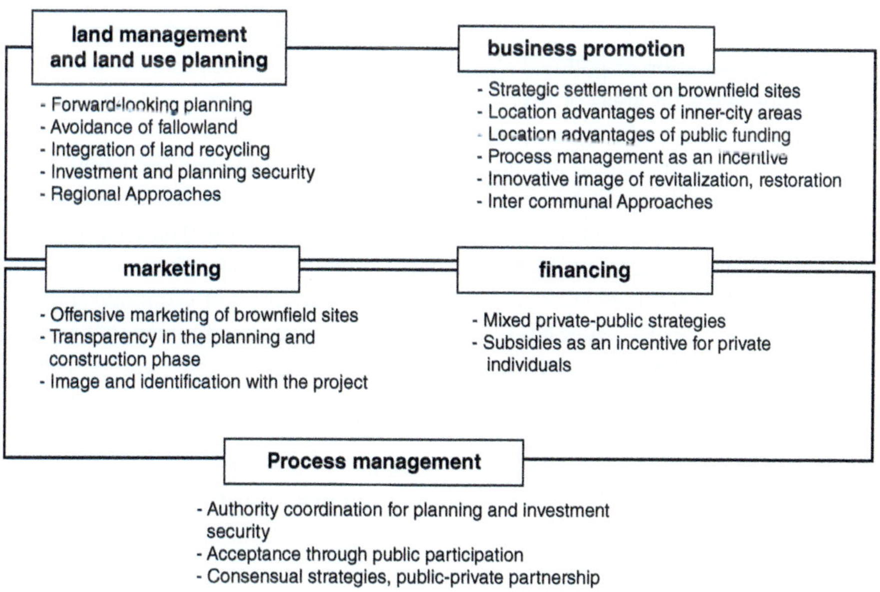

Figure 18.3 Interdisciplinary fields of action and links in land management and land recycling. *Source:* Deutsches Institut für Urbanistik 2001, modified Schneider 2022.

high probability that this development will also affect structurally weak regions in the future. The reason for this is, on the one hand, the increasing scarcity of resources, but also the insight into the need for an increasingly ecological orientation of the economy, e.g., bioeconomy, which uses bio-based renewable raw materials. The production of renewable raw materials requires land, a resource that urban brownfields and industrial areas can provide.

This means that the thematic orientation of demand is shifting from classic "grey" revitalization toward the production of bio-based raw materials and ecosystem services. According to the Millennium Ecosystem Assessment (2005), ecosystem services are services that natural capital provides for people and represents a new economic perspective on nature and its natural resources. The definition of ecosystem services is relatively young. In 1977, Walter Westmann published "How Much Are Nature's Services Worth?" describing "nature's services" (Westman 1977). Today's term, *ecosystem services*, comes from Ehrlich and Ehrlich in 1981 (Ehrlich and Ehrlich 1981). Urban ecosystem services are directly related to actions to improve the urban footprint through ecological open space planning or outreach (Kang et al. 2020). Meanwhile, green infrastructure was recognized an economic value also for properties.

18.4 Role of Soil Management

Soil management is a substantial part of the reclamation of urban brownfields and industrial areas, comprising two pillars: the handling of soil material and the circular management of land.

In case of urban brownfields and industrial areas, the greatest problem in connection with hazardous substances from historical uses is that these can result in harmful soil modifications, that is pollution or compaction, which often have to be removed before potential subsequent use (remediation of polluted sites). Depending on previous use, the spectrum of hazardous substances on areas of urban brownfields and industrial areas can contain metals, organic pollutants, or inorganic pollutants.

In contrast to conventional locations, however, urban brownfields and industrial areas have an unusually large dimension (e.g., brownfield sites, post-mining locations), which means that the removal of polluted sites causes exceptionally high costs. The social dimension is not even taken into account. Many of urban brownfields and industrial areas are located in structurally weak regions, where they used to be a center of economic activity, among other things. Since this center has broken away due to the abandonment of use, this results in regional social desolation, which as a rule cannot be compensated for with modern industrial activities (revitalization), even if this option for subsequent use is still preferred at the municipal level.

If the need for remediation is determined, a remediation plan must be drawn up and approved by the competent authority. The hazard elimination can therefore typically involve measures of:

- Remediation,
- explosive and munition clearance, and
- waste disposal.

Munition and waste (particularly hazardous waste) should generally be completely disposed of. In case of polluted sites, this does not necessarily have to be the case, the need for remediation results here from the result of the risk assessment. As part of the remediation plan, a remediation target value is specified which is based on the site conditions and the relevant hazard potential and until which remediation measures are to be carried out in the groundwater, in the soil air and/or the unsaturated zone.

Phytoremediation refers to the removal and possible metabolism of metals and organic pollutants from the soil by plants, microorganisms and fungi (Ansari et al. 2016; Alvernia and Soesilo 2019; Deutschmann 2018; Neilson and Rajakaruna 2015). On the one hand, this is a long-term effective process that does not allow rapid subsequent use, but on the other hand it has been tried and tested for areas that are very large and for which there is no short-term land use claim. This setting does not reduce the risk for owners of being sued by the competent authority for the implementation of investigation and remediation measures, but the probability increases that not only the land remediation costs can be reduced substantially, but also additional cost advantages through the generation of ecosystem services can be generated. Experience has shown that the investment risk in terms of building and waste legislation can also be reduced, since the probability or necessity of building work on a plot of land intended for renaturation is often lower. These would only be used as a remediation measure if there was a harmful change in the soil that necessitated an urgent need for remediation, e.g., B. in the sense of police law (imminent danger) or for reasons of acute endangerment of protected goods in the sense of the soil protection requirements.

18.5 Potentials for Agroecology

18.5.1 Dimensions of Potential Agroecological Applications

Agroecology can contribute to the rehabilitation of urban brownfields and industrial areas in different ways. It might (i) form a feasible after for brownfields, and (ii) serve the depollution of polluted sites. Agroecological principles apply at all scales, and therefore may be small scale or large-scale applications.

18.5.2 Small-Scale Applications

Small-scale applications are usually considered up to an extension of 1500 m^2, however the respective land might be even much smaller (Schneider et al. 2021). In practice, this dimension applies especially to urban brownfields as industrial waste lands and brownfields are usually much larger. In recent times, there is a strong development to use urban brownfields für urban gardening or urban farming purposes. According to Lohrberg and Timpe (2011), urban farming refers to agricultural activities in the urban realm that are market-oriented, professional, specialized, adaptable, area-related, with low media impact, and sustainable. Another form that is even more small scale is urban gardening, being subsistence-oriented (own use), civic, quality-of-life-oriented, location-based, neighborhood or neighborhood oriented, and receives a higher public awareness. Also, allotment gardens belong to this section, even they are usually not urban brownfields. However, in Eastern Germany there is a tendency to become so, because a lot of allotment gardens are abandoned due to demographic change and the respective shrinking of the population (Schneider and Fauk 2022). Schneider and Fauk (2022) described cases of 30% abandoned parcels in East German allotment garden associations.

A lot of interesting examples of urban gardening and urban farming activities have been implemented recently in Europe's cities on urban brownfields. A pilot project was implemented in the city of Andernach, Germany. "Edible City" is the name of the Andernach city administration's concept, with which the municipal green planning was sustainably redesigned in 2007 at no additional cost. Every citizen may harvest the fruits and vegetables grown in the city. In addition, a public permaculture facility was built on the outskirts of Andernach. The used spaces have been former urban brownfields.

A special form of agroecology is permaculture which particularly complies with the agroecological principles (Ferguson and Lovell 2014). Also, this activity can be practiced at urban brownfields. Permaculture offers people a holistic and low-threshold approach to ensuring food security and well-being (Aranya 2021). In its original form, the concept of permaculture arose with the aim of modeling agriculture on natural ecosystems (Mollison and Holmgren 1990). However, the application of permaculture also goes hand in hand with shaping individual living and working conditions and changing personal behavioral patterns, which, according to Dahlgren and Whitehead, represent different determinants of health (Dahlgren and Whitehead 1993; Hurrelmann et al. 2018). According to Ferguson and Lovell (2014), permaculture is the harmonious integration of landscape and people who provide their food, energy, shelter and other material and non-material needs in a sustainable way.

Figure 18.4 gives an impression of a permaculture garden at the Regional Garden Exhibition in Beelitz, Federal State Brandenburg in Germany in 2022. The used land for the Regional Garden Exhibition was an urban brownfield of a former wastewater treatment site.

Figure 18.4 Permaculture garden at the Regional Garden Exhibition in Beelitz, Federal State Brandenburg in Germany in 2022. Photo credit: Petra Schneider.

18.5.3 Large-Scale Applications

However, industrial, or military brownfields are usually much larger and can be subject to agricultural or horticultural use as well. This is for instance practiced in Eastern Germany in the former lignite mining areas. Starting from the point that agriculture is being deprived of its production base to a considerable extent due to the land consumption by mining, the restoration of appropriate areas is therefore part of the mining process. In East Germany this transformation is performed by the semi-governmental operator Lausitzer und Mitteldeutsche Bergbau-Verwaltungsgesellschaft mbH (LMBV). Use as a competitive agricultural production site places particularly high demands on the soil substrate and the entire recultivation process as only the most valuable soil substrates can be used (LMBV 2020a). Impressions from the LMBV mining land reclamation are given in Figure 18.5.

Since the soil substrates in East Germany usually have good pH values, soil amelioration with lime is only rarely necessary. On the other hand, the tendency to defective soil compaction is increased. Crucial to the success of agricultural recultivation is soil application by the large opencast mining machine (spreader) from a low fall height, as few passes as possible for leveling and area preparation and several years of soil-conserving cultivation in the recultivation phase (LMBV 2020a). A storage density of between 1.5 and 1.6 g/cm^3 and a coarse pore volume of at least 8% are aimed for. Location, size, area shape, and relief are other relevant aspects for the agricultural use of the dump sites. A slightly wavy surface with a gradient of at least 0.5 to a maximum of 7% ensures the drainage of surface water into the receiving water without the risk of erosion and meets the requirements for machine processing of the dump surfaces. Between the 20 and 30 ha fields, strips of wood are integrated to reduce wind erosion, evaporation and to develop productivity and a biotope network with the surrounding opencast mining areas (LMBV 2020a).

Figure 18.5 Impressions from the LMBV mining land reclamation, East Germany. Photo credit: Petra Schneider.

The accumulation of humus, the creation of structure in the soil, the development of soil life and the necessary nutrient cycles is a process that takes many years. The alternating cultivation of certain grasses (ryegrass, winter cereals) and legumes such as alfalfa and lupine have proven successful on young soils (LMBV 2020a). Depending on the location, such a recultivation crop rotation takes between 7 and 15 years. Subsequent cultivation of field crops, oilseeds and corn allows yields and further develops the soil. Well recultivated agricultural areas are at least equivalent to those of the surrounding landscape. However, they only reach their full yield potential after two to three decades (LMBV 2020a). The agricultural use might be combined with agroforestry.

The term *agroforestry* encompasses land use systems in which perennial woody plants are cultivated and reared together with crops and/or livestock on the same area (Nair 1993). The aim of these systems is always to bring about a positive interaction between the individual components, both in ecological and economic terms (Nair 1993). This may mean using trees to improve soil organic matter, water retention or biodiversity for environmental reasons. From an economic point of view, for example, the protective function of trees against the increasing occurrence of extreme weather events can be used to reduce harvest losses. The highest added value can always be achieved in case of the generation of income sources during the vegetation period or even further products might emerge from the harvest (Landgraf et al. 2018). Examples are beekeeping as a proven way to produce honey as by-product using tree species such as *Robinia pseudoacacia* for agroforestry. In addition to energy-rich wood, by-products that serve as supplementary income source for instance as poplar fluff and bark bast for fiber production, or poplar bark for the chemical industry and willow bark for the pharmaceutical industry (Landgraf et al. 2018).

Another combination with novel technologies is agrovoltaics, that denotes the approach of using agricultural land to simultaneously produce agricultural crops and generate photo voltaics electricity (Fraunhofer ISE 2020). According to Fraunhofer ISE (2020), the installed Agrivoltaics power increased exponentially from app. 5 MWp in 2012 to app. 2.9 GWp in 2018 to more than 14 GWp in 2021 with national funding programs in Japan (since 2013), China (ca. 2014), France (since 2017), the USA (since 2018) and most recently Korea. An agri-photovoltaic system enables the areas under the solar modules to be cultivated, and this is also applicable on former mining sites that are redeveloped as agricultural land (Figure 18.6).

Figure 18.6 Agri-photovoltaic system. Photo credit: Petra Schneider.

18.5.4 Forestry and Natural Succession

As discussed already above, forestry and nature near forms of woodlands are another feasible solution to be established on urban brownfields and industrial areas (Rink and Schmidt, 2021). Forest recultivation plays a key role in post-mining landscape development in the East German lignite mining areas (LMBV 2020b). On dumps of lignite opencast mining, especially in the opencast mines or opencast mining areas that were closed until 1999, initial afforestation is difficult due to the very low availability of nutrients (nitrogen, phosphorus, potassium), extensive biological inactivity and the lack of available organic matter in the raw soil (LMBV 2020b). The ecosystem processes, such as nutrient turnover, can only set in very slowly. The tree species are selected according to the location, with a high proportion of hardwood and great ecological diversity. Native woody species are preferred, and forest edges are designed in a targeted manner and their structure is adapted to the requirements of fire protection and the biotope network (LMBV 2020b).

Moreover, in some places natural succession is particularly permitted and fostered, like former military sites. An example of a succession area specifically developed as a tourist destination is the pilot project for the international nature exhibition on the Lieberoser Heide (Figure 18.7). With the Lieberose Succession Park, the Brandenburg Natural Landscape Foundation has created a vantage point and information point that makes it possible to experience the development of the wilderness on the former military training

Figure 18.7 Lieberose Succession Park, former military site, now administrated by the Brandenburg Natural Landscape Foundation. Photo credit: Petra Schneider.

area. From 1942 to 1992 the Lieberoser Heide was used for military purposes. After the armed forces withdrew in 1992, the Bundeswehr decided in 1994 against further use of the site and the areas of the Lieberose military training area were transferred to the Federal State of Brandenburg in Germany.

A particular case is illustrated in Figure 18.8, the re-use of the lost place Beelitz-Heilstätten, Federal State of Brandenburg, Germany. The workers' pulmonary sanatorium Beelitz-Heilstätten, built between 1898 and 1930 by the Berlin State Insurance Institution, is one of the largest hospital complexes in the Berlin area. It is a listed ensemble of 60 buildings on a total area of around 200 ha. During the first and second World Wars, the Beelitz sanatorium served as a military hospital and sanatorium for 17,500 sick and wounded soldiers. After the end of the World War II, in which the sanatoriums were severely damaged in some cases, the site was taken over by the Red Army in 1945. The sanatorium served as the Soviet/Russian Army's largest military hospital abroad until 1994. After years of standstill and further decay, the creditor banks managed to find a buyer for the area in March 2008, but only after repeated sales to private initiators did the gradual securing, renovation, and new use of a large part of the building begin in 2015. In 2015, the first treetop path in Brandenburg was opened on the site of the former women's lung sanatorium. It is 320 m long, up to 23 m high and crosses the tree-covered ruins of the so-called "Alpine House," which burned down in 1945. The place is a particular teaching site how natural succession on abandoned places takes place, as can be seen particularly through the natural roof top forest that established itself in the last 30 years.

Figure 18.8 Re-use of a Lost Place: Beelitz-Heilstätten, Federal State of Brandenburg, Germany. A former lung clinics, abandoned more than 30 years ago was refurbished into a touristic site to learn about the development of lost places and natural succession. Photo credit: Petra Schneider.

The so-called “Alpine House” is one of the extensions that were commissioned in 1907 to the sanatoriums opened in 1902, which originally served to treat 273 tubercular women (Rottenplaces 2020). Pioneer plants settled on the World War ruins. A roof forest sprouted, probably the largest “bonsai garden” in Germany, because without sufficient soil and hardly any water, the trees only thrived in small stature and yet hold the ceiling of the building with their roots. In the meantime, this biotope with its around 15 plant species is also home to animals, from sand lizards to dormouse to foxes and a family of raccoons. With the beginning of the extremely dry summer, since 2018, a new period in the history of the rooftop forest has also begun. Pine trees, some of which had been growing here for 70 years, dried up. However, it can already be observed that new plants are using this space (Rottenplaces 2020).

18.5.5 Agroecological Applications on Polluted Sites–Phytoremediation

The relevant phytoremediation methods are: phytoextraction, rhizofiltration, phytostabilization, rhizo- and phytodegradation, pump and tree, land farming, phytogassing, hydraulic control and others (Trapp 2000). The conventional phytoremediation approach is landfarming, means the technically enhanced approach to remediate polluted sites using basically agricultural technologies, and plants that remediate the pollution. Beside the environmental benefits, an efficient cleanup process of contaminated urban site using phytoremediation could lead also to social benefits to underdevelopment areas of cities with vulnerable inhabitants by creating more green spaces (Nissim and Labrecque 2021).

Gómez et al. (2019) pointed out that trees are getting into the focus in comparison to previously used herbaceous plants in field trials because of their greater phytoremediation effectiveness. Recent investigations underline that fast-growing trees commonly used in short-rotation coppicing are naturally endowed with key advantages to fight pollution (Gómez et al. 2019). Moreover, this kind of application offers additional benefit through the greenhouse gas binding potential and work as climate mitigation solution. The use of biomass for phytoremediation offers up further benefits, as the biomass may be used for energetical valorization.

18.6 Conclusions

The UN Sustainable Development Goal 11 (SDG-11), which calls for inclusive, safe, resilient and sustainable cities, covers the spatial aspect of urbanization through its indicator of land use (SDG indicator 11.3.1: “Ratio of land use rate to population growth rate”). The indicator posits that where the rate of land conversion due to urban development is disproportionate to population dynamics, such “growth violates every premise of sustainability by which an urban area might be measured” (UN Habitat 2021).

The options for the future area development urban brownfields and industrial areas offer potential for both revitalization and renaturation. The real usage potential for revitalization is determined by the regional demographic development and the associated infrastructural equipment and demand for goods. In connection with the utilization potential for renaturation, the soil quality and the climatic conditions play an important role in assessing the development of the possible vegetation types. The subsequent use of urban brownfields and industrial areas as green infrastructure opens up a number of subsequent use options that can be assigned to the use-related reintegration of such properties into the natural cycle. This enables new possibilities for innovative land use, which also includes economic valorization potential, especially against the background of the need for climate adaptation measures, but also the preservation and improvement of the state of biodiversity. In the long term, this facilitates "green marketing" of the areas. Depending on the planned subsequent use, the entire range of possible reintegration options into the natural balance is conceivable, i.e., both recultivation (e.g., production of arable or forest land, for arable land provided that it is free of pollutants if the arable crops are to be used for food), as well as restoration (nature conservation use) and the completely natural creation of the ecosystem through natural succession.

The assessment of the usability risk, e.g., the risk that parts of the property or the entire property cannot be used in the manner permitted by building law or customary in the area must be assessed in a differentiated manner. It can be assumed that restoration will be preferred anyway if there is no demand for space for revitalization. Therefore, the probability is small that building rights will be claimed at all, which means that competing claims for use are the exception. However, if these exist, there is also a usability risk. The assessment of the mercantile reduction in value also leads to a differentiated result. In principle, there is a risk that the "history" of the location could lead to a poor image of the property and thus to depreciation. On the other hand, what has already been stated for the risk of recourse in terms of soil protection law also applies here. If and therefore value deductions are to be applied in addition to the renovation costs or the additional expense.

It can therefore be said that the development of the value of the area after restoration can be positive and cost-reducing, especially in the case of harmful soil impacts that are not in acute need of remediation. How the marketability of the areas develops after renaturation must therefore be determined in relation to the location. The existence or non-existence of taboo zones (e.g., protected areas, natural risk areas, etc.) is relevant for both subsequent use options, as this can lead to restrictions on use. Subsequent use through renaturation naturally does not exclude the designation of nature conservation protection areas. In any case of there is a high potential for nature conservation use, but also agricultural or forestry use including productive urban forestry.

18.7 Outlook

Due to their history, urban brownfields and industrial areas often contain potential hazards that must be taken into account when developing the area, on the one hand because potential hazards can restrict the type of subsequent use, and on the other hand because potential hazards can lead to a reduction in the property value of the area. The determination of the real hazard potential requires, in addition to an exploration of polluted sites and, if necessary, a probing of explosive ordnance by a competent institution. Regained land resources can be used in a variety of ways in a range between revitalization and renaturation. They offer a high potential for climate mitigation through the establishment of greenergy and forests, and synergies might be used in case of phytoremediation of polluted soil.

References

Alpha Research Zürich (2003): Study Industriebrachen (Die brachliegende Schweiz - Entwicklungschancen im Herzen von Agglomerationen), performed by alpha research Zürich 2003, published by ARE/ Bundesamt für Umwelt BAFU, 2004. https://www.newsd.admin.ch/newsd/message/attachments/3796.pdf

Alvernia, P. and Soesilo, T.E.B. (2019). Phytoremediation as a sustainable way for land rehabilitation of heavy metal contamination. *J. Phys. Conf. Ser.* 1381: 012062.

Ansari, A.A., Gill, S.S., Gill, R. et al. (2016). *Phytoremediation: Management of Environmental Contaminants*, vol. 1. Springer. ISBN-13: 978-3319346342.

Aranya (2021). *Permaculture Design – Step by Step*. East Neon, Hampshire: Permanent Publications.

Beckmann, G. and Dosch, F. (2018). Monitoring der Siedlungsflächenentwicklung. In: *Flächeninanspruchnahme in Deutschland – Auf dem Wege zu einem besseren Verständnis der Siedlungs- und Verkehrsflächenentwicklung* (ed. M. Behnisch, O. Kretschmer, and G. Meinel), 3–24. Dresden: Springer Spektrum.

Bostenaru Dan, M. and Bostenaru-Dan, M.M. (2021). Greening the brownfields of thermal power plants in rural areas, an example from Romania, set in the context of developments in the industrialized country of Germany. *Sustainability* 13: 3800. https://doi.org/10.3390/su13073800.

Bundesministerium für Verkehr, Bau und Stadtentwicklung (BMVBS) und des Bundesamtes für Bauwesen und Raumordnung (BBR) (2007). *Kreislauf-wirtschaft in der städtischen/stadtregionalen Flächennutzung*, 51. Werkstatt: Praxis Heft.

Connell, J.H. and Slatyer, R.O. (1977). Mechanisms of succession in natural communities and their role in community stability and organization. *Am. Nat.* 111 (982): 1119–1144.

Costa, S.O., Kellecioglu, I., and Weber, R. (2018). Developing brownfields via public-private-people partnerships Lessons learned from Baltic Urban Lab. Nordregio Report.

Dahlgren, G. and Whitehead, M (1993). Tackling inequalities in health: what we can learn from what has been tried? Working Paper Prepared for the King's Fund International Seminar on Tackling Inequalities in Health. London: King's Fund.

Deutsches Institut für Urbanistik (2001). Aktive Baulandpolitik für Stadt und Umland. Vom kommunalen zum regionalen Bodenmanagement, Berlin (29–30 Oktober 2001). https://difu.de/veranstaltungen/2001-10-29/aktive-baulandpolitik-fuer-stadt-und-umland.html?embedded=5519 (accessed 29 December 2021).

Deutschmann, C. (2018). Phytoremediation von verunreinigten und kontaminierten Böden mit Hilfe von ausdauernden Pflanzen: Wie wird mit kontaminierten Brach-flächen im urbanen und suburbanen Bereich umgegangen? Bachelorarbeit Fachhochschule Weihenstephan.

Ehrlich, P. and Ehrlich, A. (1981). *Extinction: The Causes and Consequences of the Disappearance of Species*, 72–98. New York: Random House.

European Commission (2013). Communication from the Commission to the European Parliament, the Council, the European Economic and Social Committee and the Committee of the Regions - Green Infrastructure (GI)—Enhancing Europe's Natural Capital. Brussels: European Commission. https://eur-lex.europa.eu/resource.html?uri=cellar:d41348f2-01d5-4abe-b817-4c73e6f1b2df.0014.03/DOC_1&format=PDF (accessed 22 January 2021).

Ferber, U. (1997). Brachflächen-Revitalisierung: internationale Erfahrungen und mögliche Lösungskonzeptionen. Sächsisches Staatsministerium für Umwelt und Landesentwicklung.

Ferguson, R.S. and Lovell, S.T. (2014). Permaculture for agroecology: design, movement, practice, and worldview. A review. *Agron. Sustain. Dev.* 34 (2): 251–274. https://doi.org/10.1007/s13593-013-0181-6.

Fraunhofer ISE (2020). Agrivoltaics. https://www.ise.fraunhofer.de/en/key-topics/integrated-photovoltaics/agrivoltaics.html (accessed 1 August 2022)

Gann, G. and Lamb, D. (2006). Ecological restoration: a means of conserving biodiversity and sustaining livelihoods. Society for Ecological Restoration International, Tucson, Arizona.

German Federal Environmental Agency–Umweltbundesamt (2018). Handel mit Flächenzertifikaten. https://www.umweltbundesamt.de/themen/boden-landwirtschaft/flaechensparen-boeden-landschaften-erhalten/handel-flaechenzertifikaten#modellprojekt-handel-mit-flachenzertifikaten (accessed 31 January 2022).

German Federal Environmental Agency–Umweltbundesamt (2019). Siedlungs- und Verkehrsfläche. https://www.umweltbundesamt.de/daten/flaeche-boden-land-oekosysteme/flaeche/siedlungs-verkehrsflaeche#anhaltender-flachenverbrauch-fur-siedlungs-und-verkehrszwecke- (accessed 31 January 2022).

German Scientific-technical Association for Environmental Remediation and Brownfield Redevelopment (ITVA) – Ingenieurtechnischer Verband für Altlastenmanagement und Flächenrecycling e.V. (ITVA) (1998). Arbeitshilfe C5-1 Flächenrecycling, available online: https://web.archive.org/web/20070927202248/http://www.itv-altlasten.de/142.0.html, retrieved 29.11.2021

German Scientific-technical Association for Environmental Remediation and Brownfield Redevelopment (ITVA) – Ingenieurtechnischer Verband für Altlastenmanagement und Flächenrecycling e.V. (ITVA) (2019). Zeitenwende für das Flächenrecycling – Bedeutung und Potenziale Positionen und Empfehlungen des ITVA für eine nachhaltige Flächennutzung, Positionspapier des ITVA-Fachausschusses C5 Flächenrecycling. https://www.itv-altlasten.de/publikationen/positionspapiere (accessed 29 November 2021).

Gómez, l., Contreras, A., Bolonio, d. et al. (2019). Chapter Ten – Phytoremediation with trees. In: *Advances in Botanical Research*, vol. 89 (ed. F.M. Cánovas), 281–321. Academic Press. ISSN 0065-2296, ISBN 9780128154656. https://doi.org/10.1016/bs.abr.2018.11.010.

Hurrelmann, K., Richter, M., Klotz, T., and Stock, S. (ed.) (2018). *Referenzwerk Prävention und Gesundheitsförderung. Grundlagen, Konzepte und Umsetzungsstrategien, 5. vollst*, überarb.e. Bern: Hogrefe. ISBN 978-3-456-85590-5.

Iwd Informationsdienst des Instituts der deutschen Wirtschaft (2019). Mit Zertifika-ten gegen den Flächenverbrauch. https://www.iwd.de/artikel/mit-zertifikaten-gegen-den-flaechenverbrauch-447462 (accessed 31 January 2022).

Kang, W., Chon, J., and Kim, G. (2020). Urban ecosystem services: a review of the knowledge components and evolution in the 2010s. *Sustainability* 12 (23). https://doi.org/10.3390/su12239839.

Landgraf, D., Bärwolff, M., Burger, F. et al. (2018). Produktivität, Management und Nutzung von Agrarholz. In: *Agrarholz – Schnellwachsende Bäume in der Landwirtschaft – Biologie, Ökologie, Management* (ed. M. Veste and C. Böhm), 447–510. Stuttgart, Cottbus: Springer Spektrum.

Lausitzer und Mitteldeutsche Bergbau-Verwaltungsgesellschaft mbH (LMBV) (2020a). Landwirtschaftliche Rekultivierung. https://www.lmbv.de/aufgaben/rekultivierung/landwirtschaftliche-rekultivierung (accessed 1 August 2022).

Lausitzer und Mitteldeutsche Bergbau-Verwaltungsgesellschaft mbH (LMBV) (2020b). Forstliche Rekultivierung. https://www.lmbv.de/aufgaben/rekultivierung/forstliche-rekultivierung (accessed 1 August 2022).

Lohrberg, F. and Timpe, A. (2011). Urbane Agrikultur – Neue Formen der Primärproduktion in der Stadt, Fachmagazin "Planerin" 5/2, Berlin.

Mihai, F., Schneider, P., and Eva, M. (2021). Ecological engineering and green infrastructure in mitigating emerging urban environmental threats. In: *Majeti Narasimha Vara Prasad (ed.): Handbook of Ecological and Ecosystem Engineering*, 95–121. Wiley https://doi.org/10.1002/9781119678595.ch5.

Millennium Ecosystem Assessment Program (2005). *Ecosystems and Human Well-Being: Synthesis*. Island Press.

Mitsch, W.J. and Jorgensen, S.E. (1989). Introduction to ecological engineering. In: *Ecological Engineering: An Introduction to Ecotechnology* (ed. W.J. Mitsch and S.E. Jorgensen), 3–12. New York: Wiley.

Mollison, B. and Holmgren, D. (1990). *Permaculture One*. Tagari Publications.

Nair, P.K.R. (1993). An Introduction to Agroforestry. Dordrecht, The Netherlands: Kluwer Academic Publishers.

Neilson, S. and Rajakaruna, N. (2015). Phytoremediation of agricultural soils: using plants to clean metal-contaminated arable land. In: *Phytoremediation* (ed. A. Ansari, S. Gill, R. Gill, et al.). Cham: Springer. https://doi.org/10.1007/978-3-319-10395-2_11.

Nissim, W.G. and Labrecque, M. (2021). Reclamation of urban brownfields through phytoremediation: implications for building sustainable and resilient towns. *Urban For. Urban Greening* 65: 127364.

Paternoster, D., Danzinger, F., Koukal, T., et al. (2021). Strategischer Rahmen für die Restauration von Ökosystemen, Endbericht, im Auftrag der Umweltbundesamt GmbH Österreich. https://www.umweltbundesamt.at/fileadmin/site/publikationen/rep0741.pdf (accessed 25 July 2022).

Pongratz, P. (2016). Serviceorientierte Verwaltung und Wirtschaftsförderung. In: *Standortmanagement in der Wirtschaftsförderung. Wirtschaftsförderung in Lehre und Praxis* (ed. J. Stember, M. Vogelgesang, P. Pongratz, and A. Fink). Wiesbaden: Springer Gabler.

Pongratz, P. and Vogelgesang, M. (2016). Flächen- und Immobilienmanagement. In: Standortmanagement in der Wirtschaftsförderung. Wirtschaftsförderung in Lehre und Praxis. Springer Gabler, Wiesbaden.

Rink, D. and Schmidt, C. (2021). Afforestation of urban brownfields as a nature-based solution. 2021. Experiences from a project in Leipzig (Germany). *Land* 10. 893. https://doi.org/10.3390/land10090893.

Rottenplaces – Magazin rund um verfallene Bauwerke, Denkmalschutz und Industriekultur (2020). Dachwald als Biotop: 75 Jahre Weltkriegsruine „Alpenhaus" http://www.rottenplaces.de/main/dachwald-als-biotop-75-jahre-weltkriegsruine-alpenhaus-39824 (accessed 1 August 2022).

Schmidt, C. (2020). *Landschaftliche Resilienz – Grundlagen, Fallbeispiele, Praxisempfehlungen*. Dresden: Springer Spektrum.

Schneider, P. and Fauk, T. (2022). The role of allotment gardens for connecting nature and people. In: *Human-Nature Interactions* (ed. I. Misiune, D. Depellegrin, and L. Egarter Vigl). Cham: Springer. https://doi.org/10.1007/978-3-031-01980-7_21.

Schneider, P., Rochell, V., Plat, K., and Jaworski, A. (2021). *Circular Approaches in Small Scale Food Production, Circular Economy and Sustainability 1*, 1231–1255. Springer Nature. https://doi.org/10.1007/s43615-021-00129-7.

Schwaiger, E., Berthold, A., Gaugitsch, H., Götzl, M., and Milota, E. (2015). Wirtschaftliche Bedeutung von Ökosystemleistungen monetäre Bewertung—Risiken und Potenziale. Umweltbundesamt. http://hdl.handle.net/11159/78.

Soil Association (2022). Agroecology is Sustainable Farming that Works with Nature. https://www.soilassociation.org/causes-campaigns/a-ten-year-transition-to-agroecology/what-is-agroecology (accessed 14 June 2022).

The American Society of Landscape Architects (2019). Professional Practice – Green Infrastructure. https://www.asla.org/greeninfrastructure.aspx (accessed 22 January 2022).

Trapp, S. (2000). Aspekte der Phytoremediation organischer Schadstoffe. *UWSF - Z. Umweltchem. Ökotox.* 12: 246–255. https://doi.org/10.1065/uwsf2000.09.030.

UN Habitat (2021) UN Habitat Metadata Sheet on SDG Indicator 11.3.1. https://unstats.un.org/sdgs/metadata/files/Metadata-11-03-01.pdf (accessed 2 October 2021).

Westman, W.E. (1977). How much are nature's services worth? *Science* 197 (4307): 960–964. https://doi.org/10.1126/science.197.4307.960.

Wittig, R. (2002). *Siedlungsvegetation*. Frankfurt: Eugen-Ulmer.

Wong, M.H. and Bradshaw, A.D. (2003). *The Restoration and Management of Derelict Land: Modern Approaches*. World Scientific Publishing Co Pte Ltd.

Zerbe, S. (2019). *Renaturierung von Ökosystemen im Spannungsfeld von Mensch und Umwelt – Ein interdisziplinäres Fachbuch*. Bozen-Bolzano (Italy): Springer Spektrum.

Zerbe, S., Wiegleb, G., and Rosenthal, G. (2009). Einführung in die Renaturierungsökologie. In: *Renaturierung von Ökosystemen in Mitteleuropa* (ed. S. Zerbe and G. Wiegleb), S.1–S.21. Berlin Heidelberg: Springer-Verlag.

19

Plant Growth Promoting Rhizobacteria Sustaining Saline and Metal Contaminated Soils

Chitranjan Kumar[1], *Ajay Tomar*[1], *Sangeeta Pandey*[1], *and Majeti Narasimha Vara Prasad*[2]

[1] *Amity Institute of Organic Agriculture, Amity University, Uttar Pradesh, Noida, India*
[2] *School of Life Sciences, University of Hyderabad, Hyderabad, India*

19.1 Introduction

Traditional farming mostly depends on the applications of pesticides, such as organic and inorganic chemicals, for the elimination of phytopathogens and to increase crop production. The phytopathogen atmospheric problems have arisen due to the maximum application of chemicals in agriculture. Now there is an emerging interest in biological control leading to a promising alternative for phytopathogens, a complementary way to reduce the use of agrochemicals (Weston et al. 2012). A growing plant in soil indicates a complex microbe-plant-soil relationship (Backer et al. 2018). This relationship is being continuously degraded due to salinization and metal contamination through the continued recirculation and reuse of marginal quality water without any disposal of saline water and sewage treatment (Kumar et al. 2022). Halotolerant and halophilic bacteria have been successfully utilized in saline soils (Kannauj, Auraiya, Raebareli, Unnao, Manipuri, and Pratapgarh districts of Uttar Pradesh, India) to alleviate salt stress (Orhana 2016). The salinity and sodic surface mostly depend on irrigation water quality, and irrigation with alkaline water has caused a decline in soil fertility in the topsoil (Paul and Nair 2008). The improvement in agricultural sustainability in saline soils is required for better agriculture as in India and in developing countries, most of the population depends on farms for their livelihood. Several physiological and biochemical abnormalities adversely affect plant growth (Paul and Lade 2014). The plants maintain a higher cytosolic K^+/Na^+ ratio via ion homeostasis but this ionic equilibrium is disturbed in salinity. The similar radius of Na^+ and K^+, makes transporting channels difficult to distinguish between these two ions.

Agroecological Approaches for Sustainable Soil Management, First Edition.
Edited by Majeti Narasimha Vara Prasad and Chitranjan Kumar

19.2 PGPR: Modes of Action to Improve Plant Growth

The microorganisms present in close association with plants show various adaptations to face the adverse environment. The high osmolality causes the water to move out from their cell which reduces turgidity and causes dehydration of cell. Several PGPRs assist in plant growth and development under saline conditions (Tewari and Arora 2013). The molecular characterization using 16s rRNA ribotyping has identified several strains belonging to the species *Bacillus subtilis* (PS-1), *Pseudomonas aeruginosa* (PS-2), *B. cereus* (PS-3), *B. subtilis* (PS-4), *Micrococcus luteus* (PS-5), *Serratia marcescens* (PS-6), *Enterobacter hormaechei* (PS-7), *Pseudomonas putida* (PS-8), *B. aerius* (PS-9), *B. pumilus* (PS-10), *B. pumilus* (PS-11), *Pantoea ananatis* (PS-12), *Staphylococcus aureus* (PS-13), *Microbacterium aurum* (PS-14), and *B. sonorensis* (PS-15). Some of these strains have shown positive results for ammonium production, solubilization of Ca, Mg, Zn, P, K, and production of growth hormones, siderophores, and HCN. The PGPR role in crop improvement is also observed in several crops like rice, wheat, and barley (Mishra et al. 2016; Singh and Jha 2015; and Sen and Chandrasekhar 2014). PGPRs have been considered a recent boon to sustainable agriculture. They have been seen to positively affect several parameters of plant growth. Some of them are studied and listed here. Root proliferation or root growth is enhanced by increasing the root surface by absorbing more water and nutrients (Chen et al. 2016; Paul and Lade 2014). The chlorophyll content in maize has been increased by inoculating *Azotobacter* strains in promoting plant growth (Yaish et al. 2015), and almost similar observations are seen in maize and canola (Paul and Lade 2014; Paul and Sarma 2006). Dual inoculation of PGPR consortium comprising *Pseudomonas moraviensis* and *Bacillus cereus* increased maize growth parameters like protein, sugar, proline contents, and yield by 10–15% higher over single inoculation (Hassan and Bano 2016). Similarly, the dual inoculation with *Azotobacter* and phosphorus solubilizing bacteria caused significant improvement in growth and yield parameters of wheat (Singh et al. 2016). The nutritional requirements of plants should be fulfilled as it is a very important factor in combating stress. Nutritional deficiency is caused by different factors like salinity competition in nutrient uptake. This nutritional deficiency caused due to salinity in soil affects plant growth parameters. Isolation and characterization of several halotolerant PGPR and their effects on the wheat crop were conducted in the Kutch region of Gujarat. The isolated bacteria showed a positive response for IAA and their significant role in phosphate solubilization, HCN production, and N-fixing capacity (Marakana and Sharma 2018). The secondary metabolites released from microorganisms are also seen to improve plant growth (Mishra et al. 2016). Several research has been conducted on saline-affected crops like wheat, rice, barley, etc. and the positive effect of PGPR was observed (Singh and Jha 2015; Sen and Chandrasekhar 2014; Yu et al. 2014a).

19.3 Molecular Characterization of PGPRs

The NCBI GeneBank database reveals the significance of several strains viz. *B. subtilis* (MZ448473), *P. aeruginosa* (MZ448508), *B. cereus* (MZ448512), *B. subtilis* (MZ448514), *M. luteus* (MZ448513); *S. marcescens* (MZ448578); *E. hormaechei* (MZ448579); *P. putida* (MZ448586), *B. aerius* (MZ448638), *B. pumilus* (MZ448637), *B. pumilus* (MZ449219),

P. ananatis (MZ449224), *S. aureus* (MZ449233), *M. aurum* (MZ449235), *B. sonorensis* (MZ449312), and others. Some endophytic bacteria from different plants such as *Bacillus* species from the coniferous roots, *Pseudomonas* sp. from tomato, *Micrococcus* sp. from rice seed, *Pantoea* sp. from eucalyptus seed; *Staphylococcus* sp. from tomato and wheat and *Microbacterium* from rape roots have been characterized (Tian et al. 2017; Shahzad et al. 2017; Kruh et al. 2019). Various studies asserted that PGPR strains have the ability to improve plant performance by producing various metabolites (Li et al. 2016; Khan et al. 2020).

19.4 PGPR: A Competent, Facultative, and Intracellular Microorganism

PGPRs are identified as free-living endophytes that can be used to increase the yield of crops in salt-affected areas where agriculture is the prime source of living for people. The PGPR strains are converted to bio-fertilizers are eco-friendly, healthy, and safe alternatives to the present chemical fertilizers. This review will help readers to convert barren land to fertile soils and will help to lessen the use of chemical fertilizers, and thus will further help to save the environment. Thus, the most efficient carrier materials are mixed with the PGPR strains and the resulting biofertilizer is checked on crops at the field level. Some PGPRs and Rhizobacteria (RB) beneficial for alleviating salt stress are presented in Table 19.1.

Table 19.1 Alleviation of salt stress in rice using PGPR and RB – Salient research findings.

S.No.	Outcome of research	Authors	Year
1.	Inoculation of PGPR *Bacillus megaterium* improved crop response in rice KDML-105 in salt-affected soils	Sapsirisopa et al.	2009
2.	PGPR *Bacillus amyloliquefaciens* inoculated rice produced 29% higher yield as compared to non-inoculated rice under saline soils	Putaporntip et al.	2014
3.	Identified four actinomycetes (*Streptomyces jiujiangensis, Nocardiopsis yanglingensis, Pseudonocardia oroxyli Streptomyces psammoticus*) that increased root and shoot growth of rice under saline conditions	Suksaard et al.	2017
4.	*Streptomyces* sp. GMKU-336 produced ACC-Deaminase enzyme, reduced ethylene production, and enhanced salt tolerance in rice (cv KDML-105)	Jaemsaeng et al.	2018
5.	Rice production (cv KDML-105) increased by an ACC-Deaminase producing *Streptomyces venezuelae* [ATCC-10712] through modulating salt tolerance mechanism	Yoolong et al.	2019
6.	PGPR *Sinomonas* sp. (strain OR 15–23) developed aromatic 2-Acetyl-1-pyrroline in rice under saline conditions	Chinachanta et al.	2021

19.5 Signal Exchange between PGPRs and Root Hairs

Soil organic amendments and biochar have promoted soil microbial biomass in soils. These organic amendments are also used as organic carriers for soil microbes (Nadeem et al. 2017). Plant growth-promoting rhizobacteria (PGPRs) are the focus of research as plant growth parameters are regulated by several phytohormones. These phytohormones act as molecular signals between PGPRs and root hairs of plants (Backer et al. 2018). In the past few decades, innovative techniques have been relying on the application of endophytes as a non-toxic, natural, and biological source against environmental stresses (Berg et al. 2016). An enduring and intimate relationship between plant roots and endophytic bacteria can be developed without damaging the plant tissues. PGPRs have displayed huge diversity in bacterial taxa (Bacon and Hinton 2007). These bacteria possess the ability to upsurge plant performance and yield via activating mechanisms by stimulating phytohormone syntheses such as indole-3 acetic acid (IAA), gibberellin and cytokinin, solubilization of P, K, and Zn, N-fixation, production of siderophores, and reduction in plant ethylene level by producing 1-amino-cyclo-propane-1-carboxylate-deaminase to alleviate the abiotic stress (Iqbal et al. 2010; Backer et al. 2018). According to published data, plant growth-promoting endophytes have the potential to increase plant growth by improving nutrient status through decaying complex substrates via lytic enzymes. *Pinus sylvestris* (commonly called Scotch or Scots pine) is a common, widely distributed gymnosperm tree in Europe, North America, Scotland, and North Asia. It is mostly found in sunny to somewhat shady areas and can withstand frost and drought conditions. It can effortlessly grow in nutrient-poor soil, sandy soil, peat bogs, and rocky ridges. It requires a low temperature to break the dormancy of seeds and begins to flourish in the spring. It is commercially used as timber for furniture, construction, and buildings, whereas it is also used to prepare ointments, pain relievers, and medication related to the respiratory tract. This research was aimed at isolating and screening the endophytic bacteria having plant growth-enhancing properties from the roots of *P. sylvestris*. The other purpose of the work was to molecularly characterize the isolated strains by using 16S rRNA ribotyping. These PGPRs also regulate abscisic acid (ABA) pathways in plants, especially under water-stress conditions (Zhou et al. 2016). There exist multitrophic interactions in the agroecosystem within microbiomes comprising biotic and abiotic factors (presented in Figure 19.1) qualitatively and quantitatively affect soil sustainability.

PGPR-Arabidopsis interactions have been considered useful to study signaling pathways in plant development (Desbrosses et al. 2009). PGPRs can secrete IAA and ABA hormones which are responsible for the depolarization and hyperpolarization of the leaf membrane resulting in the opening and closing of stomata (Figure 19.2). Kin channels are activated by membrane hyperpolarization in response to H+-ATPase. Shaker channels KAT1, KAT2, AKT1, AKT2, and AtKC1 are found in guard cells and these channels form heterodimers AKT1-AtKC1, KAT1-KAT2, and AKT2-KAT2. Extracellular acidification results in activation of Kin channels whereas extracellular Ca^{2+} negatively affects K-weak channels. KAT1 currents are furthermore modulated by intracellular pH, ATP, cGMP, 14–3-3 proteins. K-out channels, GORK get activated by membrane depolarization by activating anion channels and deactivating H+-ATPases. GORK currents are positively influenced by extracellular and intracellular alkalinization (Sharma et al. 2013).

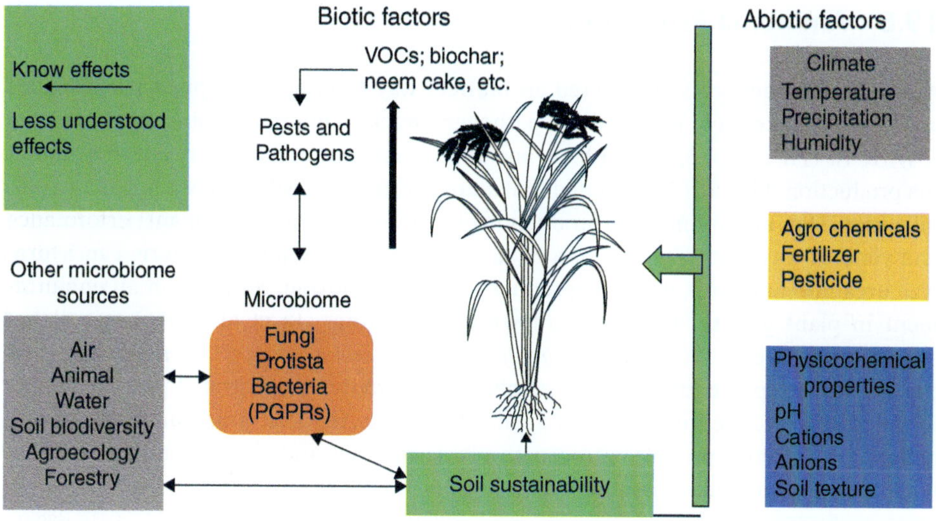

Figure 19.1 Multitrophic interactions in the agroecosystem within microbiomes affect soil sustainability.

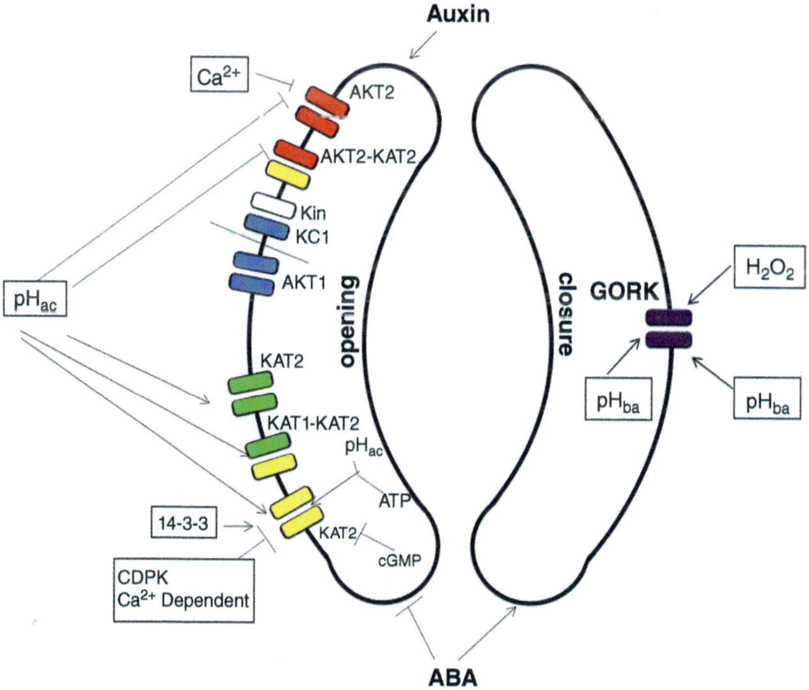

Figure 19.2 Potassium channels in guard cells. *Source:* Adapted from Sharma et al. (2013); *Note:* The phytohormones Auxins and ABA secreted by PGPRs have been reported for the opening and closing of stomata; Abbreviations: pHac, acidification; pHba, alkalinisation.

19.6 Ammonia Production

Ammonia production is considered an indirect way to progress in plant performance and can play a key role in inhibiting pathogen attacks. Different strains obtained from *P. sylvestris* (PS-1, 2, 3, 6, 9, 11, 12, 13, and 15) have shown noticeable results for ammonia production. Li et al. (2016) stated that among different mechanisms, phosphate solubilization, IAA, and ammonium production play key role in enhancing plant performance and yield. It was described by Khan et al. (2020) those endophytic bacteria can hydrolyze urea into ammonia with a byproduct i.e. carbon dioxide, resulting in an enhancement in plant growth. In agreement with our findings, Li et al. (2016) stated that endophytic bacteria for instance *Bacillus* spp., *Enterobacter* spp., *Pantoea* spp., and *Sphingomonas* spp., extracted from elephant-grass roots demonstrated positive results for ammonia production. Marques et al. (2010) asserted that ammonium producing endophytic bacteria provide nitrogen to plants for root and shoot elongation, which leads to improvement in plant biomass.

19.7 Production of IAA and HCN

It has been noticed that stain PS-6 possesses IAA-producing ability. The IAA is considered as an important hormone that helps in the improvement in plant growth, root hair formation, adventitious root formation, metabolite biosynthesis and cotyledon cell elongation (Strader et al. 2010) and can be produced by plants and microbes (like fungi and bacteria). Ali et al. (2017) asserted that the application of IAA-producing bacteria is a promising approach to increase plant yield, height, root length and root surface. Strader et al. (2010) stated that the production of IAA concentration varies from species to species, and it may be affected by plant-microbe interaction. Various researchers have reported the enhancement in IAA production by plant-associated endophytes such as *Bacillus, Micrococcus Pseudomonas, Brevibacterium, Acinetobacter,* and *Pantoea* (Makar et al. 2021). Pearson correlation analysis exhibited that IAA has a positive correlation with Ca, a negative correlation with K and P, and a weak positive correlation with Mg and Zn.

PGPRs have the potential to yield HCN also. HCN production plays a vital role in controlling pathogens' attacks on plants. Saharan and Nehra (2011) reported that the host plant is not affected by the HCN production but keeps the plants away from competition or predation. Strains like PS-1, PS-3, PS-7,8,9, PS-11, and PS-14 have shown visible results for HCN production. Tamilarasi et al. (2007) reported *Bacillus* as a major genus isolated from *Coriandrum sativum* roots and possessed the potential to suppress disease. HCN can improve Fe and P availability to plants, which leads to an improvement in plant growth-promoting traits. Studies have reported that bacterial genera belonging to *Rhizobium, Bacillus, Pseudomonas, Erwinia, Klebsilla, Enterobacter,* and Alcaligenes can suppress the pathogen attach in plants by stimulating the HCN production mechanism.

19.8 Solubilization of Nutrients (P, K, Ca, Zn, and Mg)

PGPRs are known to have the potential to increase the availability of the insoluble P to host plants. Strains PS-3, PS-6, PS-8, and PS-9 have shown considerable phosphate solubilization potential. Pearson correlation analysis exhibited that P had positive correlation with K. Several researchers have found that *Bacillus, Enterobacter, Pseudomonas,* and *Serratia* could solubilize unavailable forms of P and increase plant growth by stimulating hormone production, which ultimately improves the morphological and physiological plant attributes (Hameeda et al. 2008). According to Chen et al. (2017), there is an inverse link between soluble P concentration and pH. Phosphate solubilizing endophytic bacteria play a significant role in agricultural ecologies to improve plant yield and growth by rising phosphorus availability. Passari et al. (2016) screened 84.6% phosphate solubilizing bacteria from the *Clerodendrum colebrookianum* roots and demonstrated that organic acid produced by bacteria could play a vital part in substrate acidification, resulting in improvement in P solubilization.

PGPRs have shown K solubilization potential in soils. Besides nitrogen and P, K is thought as a vital plant nutrient that provides an ionic atmosphere for various metabolic activities. About 95–99% of K is found in an insoluble form, thus unavailable to plants for uptake but the application of K solubilizing bacteria could increase its availability to plants. Several studies have revealed that bacterial genera belonging to *Agrobacterium*, *Bacillus*, *Erwinia*, *Klebsilla*, *Pseudomonas,* and *Rhizobium* can solubilize K from the soil and make it available to plants for uptake (Shrivastava et al. 2018). PS-12 strain has shown a higher K solubilization potential. Chen et al. (2017) stated that bacteria secrete various organic acids, which lead to a decline in rhizosphere pH, resulting in the dissolution of the insoluble potassium. Various researchers reported K solubilizing endophytic bacteria for instance *Enterobacter* sp., *Alcaligenes* sp., and *Bacillus* sp. from bamboo root and *Microbacterium* sp. from tobacco rhizosphere (Yuan et al. 2015).

Several PGPRs have shown Ca, Mg, and Zn solubilization potential. The strain PS-5 has shown a higher Ca solubilization. Pearson correlation analysis exhibited that Ca has positive correlation with IAA and negative correlation with K. The strain PS-3 has shown a higher Mg solubilization. Lad et al. (2015) showed that *Serratia* sp. had the potential to solubilize Mg on agar plates supplemented with magnesium trisilicate. Besides Ca and Mg, PS strains have shown positive results for Zn solubilization. The strains showing nutrient solubilizing potential can be used as biofertilizers.

19.9 Siderophore Production

Bacteria could increase the plant yield and suppress diseases by producing siderophores. The strains PS-2, PS-5, 6, 7, PS-9, 10 and PS-13, 14, 15 have shown positive responses for siderophore production. Plant growth-promoting bacteria could improve plant performance by solubilizing phosphate, fixing nitrogen, and producing HCN, IAA and siderophores (Yuan et al. 2015). It was reported by Ali and Vidhale (2013) that iron is limiting to plants for uptake because of the low solubility of ferric ions and siderophore helps in the chelation of iron via an energy-dependent mechanism and makes it available to plants for uptake. Siderophore production also protects the plant from attack by soil-borne pathogens through chelating iron. Pahari et al. (2014) isolated four bacteria from the rhizosphere and one strain showed more siderophore-producing ability compared to others. They asserted that *Bacillus* produced maximum siderophores after 48 hours of incubation. Studies have reported that bacterial genera belonging to *Sphingomonas*, *Enterobacter*, *Burkholderia*, *Bacillus*, *Pseudomonas*, *Klebsilla* and Pantoea can increase the availability of iron to plants by producing siderophores (Li et al. 2016). In natural plant communities, positive and synergistic interactions between higher plant and associated microbes are observed that help plants to thrive under all odd conditions, like in Scot's pine forest. Such microbial associations indicate a long-term evolutionary history where symbiotic associations between two different types of organisms turned into mutualism.

19.10 The Phenomenon of Antagonism and Hyperparasitism

Plant diseases are eliminated by a different bio-agents by a direct mechanism such as competition, parasitism, hyperparasitisam, and indirect mechanism triggered by Induced Systemic Resistance (ISR) and Systemic Acquired Resistance (SAR) in plants against the phytopathogens. Several bio-agents like *Rhizobium*, *Pseudomonas fluorescens*, *Azospirillium*, and *Azotobactor* are presents in the rhizosphere which behave like bio-control agents, and they also increase the plant disease resistance against biotic and abiotic stresses (Meena et al. 2020). *Pseudomonas flourescens* can perform its biocontrol activity by siderophore production which is defined as "low molecular weight biosynthetic compounds, produced by microbial cells under iron limiting conditions, chelate the ferric ion (Fe^{+++}) with high specific activity, and serve as vehicles for the transport of iron (III) into the cell." The other mechanism of pathogen control is "hyper-parasitism." When one microorganism parasitizes on other microorganisms and completely inhibits the growth and multiplication of other microorganisms, such action is called hyper-parasitism. The percent disease incidence of root rot reduced year after year might be due to the continuous application of *Trichoderma viride* and *P. fluorescens* Wettable Powder (WP) formulation. Different strains of *P. fluorescens* exhibited differential growth-promoting responses against different crop plants. This differential growth-promoting response of *P. fluorescens* was observed even at the cultivar level in rice (Meliani and Bensoltane 2016; Meliani et al. 2017). Seed priming with PGPRs (like Trichoderma) and soil amendment with neem cake have minimized soil-borne pathogens like *Fusarium oxysporum*.

Various elements act as impediments that can be acknowledged to influence the commercialization of microbial bio-manipulate marketers (an important step for enhancing bio-manipulate efficiency). One of those impediments is the dearth of the era for cost-effective mass manufacturing of bio-manipulate marketers and their inadequate sturdiness during garage with enough stage of powerful population. Microbial manipulation of soil living pests and pathogens relies upon the hit status quo and survival of microbial inoculums in soil and the rhizosphere. It is being noticed that microbial biocontrol marketers are frequently powerful withinside the laboratory, however, the stage of pest manipulation performed withinside the area is now and again disappointing and unpredictable. Some of those screw-ups may be attributed to insufficient status quo and survival of microbial inoculums in soil (Sivasakthi et al. 2013). Microbial or bio-manipulate of crop sicknesses have emerged as an essential element of integrated ailment control and thereby for natural crop cultivation. Biological manipulation in large part relies upon the use of a microorganism (Fungi, Bacteria, Viruses, Nematodes, etc.) for suppressing the ailment-inflicting microorganisms i.e., plant pathogens. Bacterial bio-manipulate marketers can consist of *P. fluorescens, P. putida, P. aeruginosa, B. subtilis*, and different *Bacillus spp*. Seed priming with a bioagent *T. viride* @ 4 g/kg seed and soil amendment with neem cake @ 250 kg/ha minimized wilt incidence in tomatoes. This treatment also resulted in a higher number of branches per plant (15.0), increased number of flowers per plant (119.8), plant height (82.3 cm), and number of fruits per plant (32.6), over control (Barnwal et al. 2011). Thus, PGPRs are capable of controlling severe soil-borne diseases (such as Fusarium wilt) in agriculture.

19.11 Alleviation of Metal Stress

Heavy metals are elements that enter into their surrounding environments at low amounts through natural sources, including the erosion of soil and the earth's crust (Jaishankar et al. 2014). But, in recent decades, human activities like mining, discharge of untreated industrial effluents and urban sewage, and use of chemical pesticides and fertilizers have raised dramatically metals concentration to the dangerous doses for biological functions in all living organisms (Jaishankar et al. 2014; Jan et al. 2015). Among the heavy metals, which can cause environmental pollutions, lead, cadmium, and arsenic have become an important global problem, because they have no essential role in biological reactions (Shen et al. 2019). Heavy metals have harmful effects on the soil texture and its fertility, the number and activity of soil microbial population, and physiological and morphological features of plants that lead to crop yield reduction. Also, heavy metals can easily be uptake by the plant's root and subsequently arrive at the food chain of animals and humans (Jiwan and Ajah 2011; Tiwari and Lata 2018). Metals can cause severe diseases in humans through damage to vital biological substances, such as lipids, proteins, and nucleic acids through the production of free radicals (Jan et al. 2015). Unlike organic contaminants in the environment, heavy metals are persistent to accumulate in plants and biological systems because they cannot be broken down to non-toxic elements gradually. Thus, it is necessary to use suitable techniques for cleaning up heavy metals from contaminated environments to reduce their toxicity (Khashei et al. 2018). Numerous chemo-physical methods which are used for metal remediation have various deficiencies such as high expense, loss of soil fertility, and generation of secondary pollutants (Khashei et al. 2018; Ali et al. 2013). However, the limitation of plant growth, due to heavy metal stress, decreases the effectiveness of phytoremediation (Fan et al. 2018). To overcome this limitation, heavy metal-resistant rhizobacteria with plant growth-promoting (PGP) properties can be applied to enhance plant biomass production and phytoremediation output (Yu et al. 2014a,b). These bacteria stimulate plant growth by different mechanisms such as increasing the availability of essential nutrients through phosphorus solubilization, nitrogen fixation, siderophore production for iron uptake. Also, they regulate many plant developmental processes by the secretion of phytohormones like indole-3-acetic acid (IAA) and protect plants against deleterious effects of phytopathogens by the synthesis of antibiotics, HCN, and degradative enzymes (Ahemad and Kibret 2014; Olanrewaju et al. 2017; Bhattacharyya and Jha 2012). These bacteria can preserve plants against metal toxicity by numerous mechanisms (Banerjee et al. 2018). So, plant-microbe-combined remediation has gained more attention currently. We need more studies in future to isolate new heavy metal-resistant bacterial strains with plant growth-promoting properties for enhancing phytoremediation efficiency in contaminated soils.

19.12 Assessment of Plant Growth-Promoting Activities

Phytoremediation is considered a novel and efficient technique for the remediation of heavy metal polluted soil. It has gained more attention because it is a low-cost, durable, and environmentally friendly alternative solution among the various traditional methods used for the rehabilitation of polluted areas. Metal hyperaccumulator plants can be used as efficient tools in this process due to their high metal accumulating potency. However, most of these plants have slow growth and low biomass at high concentrations of metals, so their optimum performance can be greatly reduced in these conditions. These restrictions can be overcome by the inoculation of plants with PGPBs. These microorganisms can improve phytoremediation efficacy by providing essential nutrients and increasing the production of plant biomass via different processes such as phosphate solubilization, fixation of atmospheric nitrogen, and secretion of phytohormones like IAA (Tirry et al. 2018; Yahaghi et al. 2018; He and Yang 2007). Several new bacterial strains have been identified which can increase plant growth as well as remove heavy metals from polluted soils. Several rhizospheric strains have been isolated from the heavy metal-polluted soils for bioremediation that have shown enhanced plant growth. An outline of recent advances in the field of PGPR for sustainable soil management is presented in Table 19.2.

Table 19.2 Recent advances in PGPR for sustainable soil management.

S.No.	The outcome of PGPR research	Authors	Year
1.	The *in vitro* generated root exudates of plants *Arabidopsis thaliana* and *Medicago truncatula* enhanced fungal population and diversity in soils	Broeckling et al.	2008
2.	PGPR and AMF enhanced the uptake of nutrients in plants and thereby increased the efficiency of nutrient use or fertilizer use	Adesemoye and Kloepper	2009
3.	The combined culture of Fe and S-oxidizing bacteria (*Acidithiobacillus ferrooxidans* and *Acidithiobacillus thiooxidans*) dissolved tri-calcium phosphate in the inoculated leachates	Bhatti and Yawar	2010
4.	The Gram-positive Actinomycetes and fungi improved the priming of SOM utilizing the labile substrate	Garcia-Pausas and Paterson	2011

Table 19.2 (Continued)

S.No.	The outcome of PGPR research	Authors	Year
5.	Treatment of seeds with *T. harzianum* T22 enhanced plant growth and biochemical parameters in maize	Akladious and Abbas	2012
6.	Composts of kitchen wastes enriched with *T. harzianum* T22 improved the production and nutritional quality of tomatoes	Molla et al.	2012
7.	Salicylic acid and Gamma amino butyric acids (GABA) applied to soils as root exudates of Arabidopsis revealed that phenolic compounds develop a specific group of soil microbiome	Badri et al.	2013
8.	PGPRs mobilized zinc in cereals and were recommended for cereals biofortification	Hafeez et al.	2013
9.	PSB releases plant growth regulators that increase bioleaching, biosoption, metal solubilization, and bio-precipitation to enhance phytoremediation in metal-polluted soils	Ahemad	2015
10.	Developed and commercialized formulations for Azospirillum spp.	Bashan and de-Bashan	2015
11.	*Serratia nematodiphila* (PEJ1011) can grow actively at a low temperature (5 °C) and produce gibberellins. PEJ1011 can regulate abscisic, jasmonic, gibberellins, and salicylic acids and protect *Capsicum annuum* L under low-temperature stress.	Kang et al.	2015
12.	PGPR can control soil-borne pathogens and regulate microbiome (microhabitat interactions), and successfully develop a new niche of microbial communities.	Adam et al.	2016
13.	*Dietzia natronolimnaea* (STR1) can induce salinity tolerance in wheat through a complex mechanism, upregulating TaABARE and TaOPR1 (ABA pathway), modulating SOS1 and SOS2 (SOS pathways), transporting TaNHX1, TaHAK, and TaHKT1, and enhanced antioxidant gene expressions.	Bharti et al.	2016
14.	*Bacillus amyloliquefaciens* (SQR9) counteracts the increase of ABA inducing salt tolerance in maize.	Chen et al.	2016
15.	*Paenibacillus lentimorbus* (B-30488) can produce ACC-deaminase. It is used to control the Southern blight disease of tomato.	Dixit et al.	2016
16.	Biochar inoculated with a pseudomonad is alternatively used for peat-moss for sustainable corn production.	Głodowska et al.	2016
17.	*Enterobacter sp.* (UPMR18) produced ACC-Deaminase that increased antioxidants (SOD, APX, and CAT activities) in Okra. The PGPR increased salt tolerance in plants.	Habib et al.	2016
18.	PGPR inoculated *Camelina sativa* produced ACC-Deaminase that triggered gene expressions (acdS) to increase salt tolerance in plants.	Heydarian et al.	2016
19.	Identified chitin oligomers Myc-COs in AMF exudates that activated Ca^{++} in the root epidermis of rice. Established the role of a bifunctional Lysin-motif (LysM)-receptor OsCERK1 in AMF signaling.	Carotenuto et al.	2017
20.	Application of *Pseudomonas fluorescens*, biochar, and compost synergistically enhanced cucumber growth under water stress.	Nadeem et al.	2017
21.	Established *Pseudomonas entomophila* (23S) as a biocontrol agent against *Clavibacter michiganensis*. The PGPR was used for controlling the bacterial Canker disease in tomatoes.	Takishita et al.	2018

(*Continued*)

Table 19.2 (Continued)

S.No.	The outcome of PGPR research	Authors	Year
22.	A PGPR *Pseudomonas aeruginosa* was used for controlling the weed *Phelipanche aegyptiaca* in tomatoes.	Kruh et al.	2019
23.	Introduced major determinants of PGPR that induce systemic resistance in plants. Outlined lipopolysaccharides, siderophores, lipopeptides, etc. as major determinants which alter the growth of pathogens in plants.	Meena et al.	2020
24.	*Pantoea* sp. and *Bacillus* sp. Synthesized indole-related compounds and Improved production and nutritional quality of wheat.	Makar et al.	2021
25.	The combined application of PGPR and composts promoted antioxidants, cytokinin, exopolysaccharides, and pH regulation to alleviate drought, salt, and heavy metal stress in plants.	Malik et al.	2022

19.13 Assessment of Bacterial Reactions to Heavy Metals

Heavy metal-resistant bacteria with plant growth-promoting traits can increase metal accumulation capacity and its translocation in plants by several mechanisms, including detoxification, accumulation, transformation, and sequestration of metals. Also, they can enhance metal solubility by various processes, including the release of the proton, siderophores, and solubilization of phosphate. Therefore, these bacteria reduce the metal phytotoxicity as well as increase the phytoavailability of heavy metals in polluted soil, and thereby improving metal uptake by plants (Yahaghi et al. 2018; He and Yang 2007). Isolated bacteria from metal-polluted areas are usually more resistant to high concentrations of heavy metals than those isolated from intact sites (Tirry et al. 2018). The microbial diversity of *Aeromonas, Pseudomonas, Bacillus, Lavobacteria, Corynebacteria, Acinetobacter, Mycobacteria, Streptomyces, Arthrobacter*, and *Cyanobacteria* have shown bioremediation potentials in the heavy metal-polluted soils (Mani and Kumar 2014). Among fungi, *Aspergillus niger, Cladosporium resinae, Ganoderma lucidum, Penicillium spp, Rhizopus arrhizus, Trametes versicolor* and *Pleurotus ostreatus* can recover heavy metals from the polluted soil-water systems (Kumar and Mani 2012). Several studies have also demonstrated that metal-resistant bacteria were able to promote plant growth and phytoremediation efficacy (Tirry et al. 2018; Sheng et al. 2008). *Pseudomonas taiwanensis, Enterobacter cloacae, Pantoea agglomerans*, and *Enterobacter ludwigi* have been found novel and efficient in alleviating metal stress in plants. *P. agglomerans* have exhibited higher solubilization of insoluble phosphate and IAA production. *E. ludwigi* has shown a higher MIC value in the case of Cd^{2+}. *P. taiwanensis, P. agglomerans*, and *E. ludwigi* have exhibited a higher MIC value (>7 mM) in the case of Pb^{2+}. *P. agglomerans* and *E. ludwigi* have shown a higher MBC of Pb^{2+} (>120 mM), and a higher MBC of Cd^{2+} (>25 mM) belonging to *P. agglomerans*. Thus, based on PGPR characteristics and metal resistance, strains HO, RA, SV, and MR have been suggested as new enhancers of phytoremediation efficiency in contaminated soils with heavy metals (Tirry et al. 2018).

The direct application of the bacteria isolated from vermicompost could significantly increase crop production and plant growth several folds viz. application of *Pseudomonas* sp. and *Bacillus* sp. boosted development and production of fennel (Darzi et al. 2008), lemon grass (Ratti et al. 2001), black cumin, borage (Shaalan 2005a,b), basil (Rashmi et al. 2008), and other crops (Kohler et al. 2007; Vaz-Moreira et al. 2009; Pandya et al. 2017). Besides phenotypic identity, 16S rRNA sequence homology-based phylogenetic analysis of *B. subtilis* has confirmed the phenotypic identity of the bacteria, a well-known Gram-positive rod-shaped bacterium, which can survive in extreme environmental conditions by forming endospores. It is commonly found in topmost layer of soil along with *Bacillus megaterium, B. cereus, B. subtilis, Bacillus licheniformis, B. benzoevoran., B. pumilus, B. macroides,* etc. in vermicompost and many of them has been recognized as PGPR (Vaz-Moreira et al. 2009). *B. subtilis* can increase the shoot dry weight of lettuce (Kohler et al. 2007), promote growth, and act as control agents against *Macrophomina phaseolina, Rhizoctonia solani* (Pandya et al. 2017), and root rot fungus in *Pinus roxburghii* (Singh et al. 2008). Besides, it could produce the growth regulators like auxins (IAA), gibberellins, and cytokinins like other *Bacillus* spp., in vermicompost and increase nutrient content in the rhizospheric soil which ultimately leads to acting as effective PGPR and improve production and growth of paddy and other crops.

19.14 Conclusion

Plant growth-promoting rhizobacteria, especially genera *Bacillus, Pseudomonas, Micrococcus, Pantoea, Staphylococcus, Serratia,* and *Enterobacter* have shown several positive characteristics that promote the growth of the plants and their production several folds by increasing nutrient contents in soils, or by controlling plant and soil-pathogens, or by controlling plant stress element, or by producing plant growth hormones. They could be used as a biocontrol agent and biofertilizer for good growth and sustainable production of food and cereal crops. PGPRs can be inoculated with the desired crop and seedlings to improve plant growth. PGPRs are capable of developing stress tolerance mechanisms in plants. The potential of different halophytes and metal-hyperaccumulating plants has been established as an environmentally friendly and low-cost solution for alleviating salt and metal stress from salt and metal-contaminated soils. The metal-resistant bacteria with PGP traits are in high demand for the restoration of several types of salt-affected and metal-polluted soils. Multi-metal tolerant rhizobacteria and plant root exudate have wider applications in the field of agricultural sustainability. Strains of *P. taiwanensis, E. cloacae, P. agglomerans,* and *Enterobacter ludwigi* have the potential to promote plant growth by preparing essential nutrients through the production of IAA, fixation of nitrogen, and solubilization of insoluble phosphate. Also, they are being utilized for removing metals from the environment. These PGPRs should be utilized as candidates for application in developing salt tolerance and restoring bacteria-assisted remediation of degraded soils. In addition, biocontrol of crop diseases has become a vital component of integrated disease management and thereby for organic crop cultivation. Biological control largely depends upon the use of a microorganism (Fungi, Bacteria, Viruses, Nematodes, etc.) for suppressing the disease-causing microorganisms i.e., plant pathogens.

References

Adam, E., Groenenboom, A.E., Kurm, V. et al. (2016). Controlling the microbiome: microhabitat adjustments for successful biocontrol strategies in soil and human gut. *Front. Microbiol.* 7: 1079. https://doi.org/10.3389/fmicb.2016.01079.

Adesemoye, A.O. and Kloepper, J.W. (2009). Plant–microbes interactions in enhanced fertilizer-use efficiency. *Appl. Microbiol. Biotechnol.* 85: 1–12. https://doi.org/10.1007/s00253-009-2196-0.

Ahemad, M. (2015). Phosphate-solubilizing bacteria-assisted phytoremediation of metalliferous soils: a review. *3 Biotech* 5: 111–121. https://doi.org/10.1007/s13205-014-0206-0.

Ahemad, M. and Kibret, M. (2014). Mechanisms and applications of plant growth promoting rhizobacteria: current perspective. *J. King Saud Univ. Sci.* 26: 1–20.

Akladious, S.A. and Abbas, S.M. (2012). Application of Trichoderma harziunum T22 as a biofertilizer supporting maize growth. *Afr. J. Biotechnol.* 11: 8672–8683.

Ali, S.S. and Vidhale, N.N. (2013). Bacterial siderophore and their application: a review. *Int. J. Curr. Microbiol. Appl. Sci.* 2 (12): 303–312.

Ali, H., Khan, E., and Sajad, M.A. (2013). Phytoremediation of heavy metals—concepts and applications. *Chemosphere* 91: 869–881.

Ali, S., Charles, T.C., and Glick, B.R. (2017). Endophytic phytohormones and their role in plant growth promotion. In: *Functional Importance of the Plant Microbiome*, 89–105. Cham: Springer.

Backer, R., Rokem, J.S., Ilangumaran, G. et al. (2018). Plant growth-promoting rhizobacteria: context, mechanisms of action, and roadmap to commercialization of biostimulants for sustainable agriculture. *Front. Plant Sci.* 9: 1473. https://doi.org/10.3389/fpls.2018.01473.

Bacon, C.W. and Hinton, D.M. (2007). Bacterial endophytes: the endophytic niche, its occupants, and its utility. In: *Plant Associated Bacteria*, 155–194. Dordrecht: Springer.

Badri, D.V., Chaparro, J.M., Zhang, R. et al. (2013). Application of natural blends of phytochemicals derived from the root exudates of Arabidopsis to the soil reveal that phenolic-related compounds predominantly modulate the soil microbiome. *J. Biol. Chem.* 288: 4502–4512. https://doi.org/10.1074/jbc.M112.433300.

Banerjee, A., Jhariya, M., Yadav, D., and Raj, A. (2018). *Micro-Remediation of Metals: A New Frontier in Bioremediation*. Handbook of Environmental Materials Management. Springer.

Barnwal, M.K., Maiti, D., and Pandey, A.C. (2011). Management of tomato wilt caused by *Fusarium oxysporum* f. sp. lycopersici by using bioagents and organic amendments. *Indian Phytopathol.* 64 (2): 194–196.

Bashan, Y. and de-Bashan, L.E. (2015). Inoculant preparation and formulations for azospirillum spp. In: *Handbook for Azospirillum* (ed. F.D. Cassán, Y. Okon, and C.M. Creus), 469–485. Berlin: Springer.

Berg, G., Rybakova, D., Grube, M., and Köberl, M. (2016). The plant microbiome explored: implications for experimental botany. *J. Exp. Bot.* 67 (4): 995–1002.

Bharti, N., Pandey, S.S., Barnawal, D. et al. (2016). Plant growth promoting rhizobacteria *Dietzia natronolimnaea* modulates the expression of stress responsive genes providing protection of wheat from salinity stress. *Sci. Rep.* 6: 34768. https://doi.org/10.1038/srep34768.

Bhattacharyya, P.N. and Jha, D.K. (2012). Plant growth-promoting rhizobacteria (PGPR): emergence in agriculture. *World J. Microbiol. Biotechnol.* 28: 1327–1350.

Bhatti, T.M. and Yawar, W. (2010). Bacterial solubilization of phosphorus from phosphate rock containing sulfur-mud. *Hydrometallurgy* 103: 54–59. https://doi.org/10.1016/j.hydromet.2010.02.019.

Broeckling, C.D., Broz, A.K., and Bergelson, J. (2008). Root exudates regulate soil fungal community composition and diversity. *Appl. Environ. Microbiol.* 74: 738–744. https://doi.org/10.1128/AEM.02188-07.

Carotenuto, G., Chabaud, M., Miyata, K. et al. (2017). The rice LysM receptor-like kinase OsCERK1 is required for the perception of short-chain chitin oligomers in arbuscular mycorrhizal signaling. *New Phytol.* 214: 1440–1446. https://doi.org/10.1111/nph.14539.

Chen Luo, S., Wu, Y., Ye, J. et al. (2017). The effects of the endophytic bacterium *Pseudomonas fluorescens* Sasm05 and IAA on the plant growth and cadmium uptake of *Sedum alfredii* hance. *Front. Microbiol.* 8: 2538.

Chen, L., Liu, Y.P., Wu, G.W. et al. (2016). Induced maize salt tolerance by rhizosphere inoculation of Bacillus amyloliquefaciens SQR9. *Physiol. Plant.* 158: 34–44. https://doi.org/10.1111/ppl.12441.

Chinachanta, K., Shutsrirung, A., Herrmann, L. et al. (2021). Enhancement of the Aroma Compound 2-Acetyl-1-pyrroline in Thai Jasmine Rice (*Oryza sativa*) by Rhizobacteria under Salt Stress. Biology 10: 1065. https://doi.org/10.3390/biology10101065.

Darzi, M.T., Ghalavand, A., and Rejali, F. (2008). Effect of mycorrhiza, vermicompost and phosphate biofertilizer application on flowering, biological yield and root colonization in fennel (*Foeniculum vulgare* mill.). *Iran. J. Crop. Sci.* 10 (1): 88–109.

Desbrosses, G., Contesto, C., Varoquaux, F. et al. (2009). PGPR-Arabidopsis interactions is a useful system to study signaling pathways involved in plant developmental control. *Plant Signal. Behav.* 4 (4): 321–323. https://doi.org/10.4161/psb.4.4.8106.

Dixit, R., Agrawal, L., Gupta, S. et al. (2016). Southern blight disease of tomato control by 1-aminocyclopropane- 1-carboxylate (ACC) deaminase producing *Paenibacillus lentimorbus* B-30488. *Plant Signal. Behav.* 11: e1113363. https://doi.org/10.1080/15592324.2015.1113363.

Fan, M., Liu, Z., Nan, L. et al. (2018). Isolation, characterization, and selection of heavy metal-resistant and plant growth-promoting endophytic bacteria from root nodules of Robinia pseudoacacia in a pb/Zn mining area. *Microbiol. Res.* 217: 51–59.

Garcia-Pausas, J. and Paterson, E. (2011). Microbial community abundance and structure are determinants of soil organic matter mineralisation in the presence of labile carbon. *Soil Biol. Biochem.* 43: 1705–1713. https://doi.org/10.1016/j.soilbio.2011.04.016.

Głodowska, M., Husk, B., Schwinghamer, T., and Smith, D. (2016). Biochar is a growth-promoting alternative to peat moss for the inoculation of corn with a pseudomonad. *Agron. Sustain. Dev.* 36: 1–10. https://doi.org/10.1007/s13593-016-0356-z.

Habib, S.H., Kausar, H., and Saud, H.M. (2016). Plant growth-promoting rhizobacteria enhance salinity stress tolerance in okra through ROS-scavenging enzymes. *Biomed. Res. Int.* 2016: 6284547. https://doi.org/10.1155/2016/6284547.

Hafeez, F.Y., Abaid-Ullah, M., and Hassan, M.N. (2013). Plant growth-promoting rhizobacteria as zinc mobilizers: a promising approach for cereals biofortification. In: *Bacteria in Agrobiology: Crop Productivity* (ed. D.K. Maheshwari, M. Saraf, and A. Aeron), 217–235. Berlin: Springer.

Hameeda, B., Harini, G., Rupela, O.P. et al. (2008). Growth promotion of maize by phosphate-solubilizing bacteria isolated from composts and macrofauna. *Microbiol. Res.* 163 (2): 234–242.

Hassan, T. and Bano, A. (2016). Biofertilizer: a novel formulation for improving wheat growth, physiology and yield. *Pak. J. Bot.* 48 (6): 2233–2241.

He, Z. and Xe, Y. (2007). Role of soil rhizobacteria in phytoremediation of heavy metal contaminated soils. *J. Zhejiang Univ. Sci. B* 8: 192–207.

Heydarian, Z., Yu, M., Gruber, M. et al. (2016). Inoculation of soil with plant growth promoting bacteria producing 1-aminocyclopropane-1-carboxylate deaminase or expression of the corresponding acds gene in transgenic plants increases salinity tolerance in camelina sativa. *Front. Microbiol.* 7: 1966. https://doi.org/10.3389/fmicb.2016.01966.

Iqbal, U., Jamil, N., Ali, I., and Hasnain, S. (2010). Effect of zinc-phosphate-solubilizing bacterial isolates on growth of *Vigna radiata*. *Ann. Microbiol.* 60 (2): 243–248.

Jaishankar, M., Tseten, T., Anbalagan, N. et al. (2014). Toxicity, mechanism and health effects of some heavy metals. *Interdiscip. Toxicol.* 60–72.

Jan, A.T., Azam, M., Siddiqui, K. et al. (2015). Heavy metals and human health: mechanistic insight into toxicity and counter defense system of antioxidants. *Int. J. Mol. Sci.* 16: 29592–29630.

Jaemsaeng, R., Jantasuriyarat, C., and Thamchaipenet, A. (2018). Positive role of 1-aminocyclopropane-1-carboxylate deaminase-producing endophytic *Streptomyces* sp. GMKU 336 on flooding resistance of mung bean. *Agric. Nat. Resour.* 52(4): 330–334. https://doi.org/10.1016/j.anres.2018.09.008.

Jiwan, S. and Ajah, K. (2011). Effects of heavy metals on soil, plants, human health and aquatic life. *Res. J. Chem. Environ.* 1: 15–21.

Kang, S.M., Khan, A.L., Waqas, M. et al. (2015). Gibberellin-producing *Serratia nematodiphila* PEJ1011 ameliorates low temperature stress in *Capsicum annuum* L. *Eur. J. Soil Biol.* 68: 85–93. https://doi.org/10.1016/j.ejsobi.2015.02.005.

Khan, N., Bano, A., and Curá, J.A. (2020). Role of bfeneficial microorganisms and salicylic acid in improving rainfed agriculture and future food safety. *Microorganisms* 8 (7): 1018.

Khashei, S., Etemadifar, Z., and Rahmani, H.R. (2018). Immobilization of Pseudomonas putida PT in resistant matrices to environmental stresses: a strategy for continuous removal of heavy metals under extreme conditions. *Ann. Microbiol.* 68: 931–942.

Kohler, J.F., Caravaca, L., and Carrasco, R.A. (2007). Interactions between a plant growth-promoting rhizobacterium, an AM fungus and a phosphate-solubilising fungus in the rhizosphere of *Lactuca sativa*. *Appl. Soil Ecol.* 35: 480–487. https://doi.org/10.1016/j.apsoil.2006.10.006.

Kruh, L.I., Abu-Nassar, J., Lidor, O., and Aly, R. (2019). An endophytic bacteria (*Pseudomonas aeruginosa*) strain translocated and protected tomato (*Solanum lycopersicum* L.) plants from its parasitic weed *Phelipanche aegyptiaca*. *BioRxiv* 1: 842757.

Kumar, C. and Mani, D. (2012). *Advances in Bioremediation of Heavy Metals: A Tool for Environmental Restoration*, 245. Saarbrucken, Germany: Academic Publishing AV Akademikerverlag GmbH & Co. KG. ISBN No. 978-3-659-29951-3.

Kumar, C., Ramawat, N., and Verma, A.K. (2022). Organic fertigation system in saline-sodic soils: a new paradigm for the restoration of soil health. *Agron. J.* (ISSN 0002-1962) 2022: 114: 317–330.

Lad, S.D., Shanware, A., and Trivedi, M. (2015). Embedding potential bioinoculant strain for effective soil conditioner production. *World J. Pharm. Res.* 4 (10): 1395–1405.

Li, X., Geng, X., Xie, R. et al. (2016). The endophytic bacteria isolated from elephant grass (*Pennisetum purpureum Schumach*) promote plant growth and enhance salt tolerance of hybrid Pennisetum. *Biotechnol. Biofuels* 9 (1): 1–2.

Makar, O., Kuźniar, A., Patsula, O. et al. (2021). Bacterial endophytes of spring wheat grains and the potential to acquire Fe, cu, and Zn under their low soil bioavailability. *Biology* 10 (5): 409.

Malik, L., Sanaullah, M., Mahmood, F. et al. (2022). Unlocking the potential of co-applied biochar and plant growth-promoting rhizobacteria (PGPR) for sustainable agriculture under stress conditions. *Chem. Biol. Technol. Agric.* 9: 58. https://doi.org/10.1186/s40538-022-00327-x.

Mani, D. and Kumar, C. (2014). Biotechnological advances in bioremediation of heavy metals contaminated ecosystems: an overview with special reference to phytoremediation. *Int. J. Environ. Sci. Technol.* 11: 843–872.

Marakana, T. and Sharma, M. (2018). Isolation and characterization of halotolerant bacteria and its effects on wheat plant as PGPR. *J. Pharm. Innov.* 7 (7): 102–111.

Marques, A.P., Pires, C., Moreira, H. et al. (2010). Assessment of the plant growth promotion abilities of six bacterial isolates using *Zea mays* as indicator plant. *Soil Biol. Biochem.* 42 (8): 1229–1235.

Meena, M., Swapnil, P., Divyanshu, K. et al. (2020). PGPR-mediated induction of systemic resistance and physiochemical alterations in plants against the pathogens: current perspectives. *J. Basic Microbiol.* 60 (10): 828–861.

Meliani, A. and Bensoltane, A. (2016). Biofilm-mediated heavy metals bioremediation in PGPR Pseudomonas. *J. Bioremed. Biodegr.* 7 (370): 2.

Meliani, A., Bensoltane, A., Benidire, L., and Oufdou, K. (2017). Plant growth-promotion and IAA secretion with *Pseudomonas fluorescens* and *Pseudomonas putida*. *Res. Rev. J. Bot. Sci.* 6 (2): 16–24.

Mishra, J., Arora, N.K., and Fatima, T. (2016). *Role of Secondary Metabolites from Plant Growth-Promoting Rhizobacteria in Combating Salinity Stress.*

Molla, A.H., Haque, M.M., Haque, M.A., and Ilias, G. (2012). Trichoderma-enriched biofertilizer enhances production and nutritional quality of tomato (*Lycopersicon esculentum* mill.) and minimizes NPK fertilizer use. *Agric. Res.* 1: 265–272. https://doi.org/10.1007/s40003-012-0025-7.

Nadeem, S.M., Imran, M., Naveed, M. et al. (2017). Synergistic use of biochar, compost and plant growth-promoting rhizobacteria for enhancing cucumber growth under water deficit conditions. *J. Sci. Food Agric.* 97: 5139–5145. https://doi.org/10.1002/jsfa.8393.

Olanrewaju, O.S., Glick, B.R., and Babalola, O.O. (2017). Mechanisms of action of plant growth promoting bacteria. *World J. Microbiol. Biotechnol.* 33: 197.

Orhana, F. (2016). Alleviation of salt stress by halotolerant and halophilic plant growth-promoting bacteria in wheat (*Triticum aestivum*). *Braz. J. Microbiol.* 47 (3): 621–627.

Pahari, A., Dangar, T.K., and Mishra, B.B. (2014, 2017). Siderophore quantification of bacteria from sundarban and its effect on growth of brinjal (*Solanum melongena* L.). *Education.*

Pandya, U., Prakash, S., Shende, K. et al. (2017). Multifarious allelochemicals exhibiting antifungal activity from *Bacillus subtilis* MBCU5. *3 Biotech* 7 (3): 175. https://doi.org/10.1007/s13205-017-0827-1.

Passari, A.K., Mishra, V.K., Leo, V.V. et al. (2016). Phytohormone production endowed with antagonistic potential and plant growth promoting abilities of culturable endophytic bacteria isolated from *Clerodendrum colebrookianum* Walp. *Microbiol. Res.* 193: 57–73.

Paul, D. and Lade, H. (2014). Plant-growth-promoting rhizobacteria to improve crop growth in saline soils: a review. Agronomy for sustainable development. *EDP Sci.* 34 (4): 737–752.

Paul, D. and Nair, S. (2008). Stress adaptations in a plant growth promoting rhizobacterium (PGPR) with increasing salinity in the coastal agricultural soils. *J. Basic Microbiol.* 48 (5): 378–384.

Paul, D. and Sarma, Y.R. (2006). Antagonistic effects of metabolites of *Pseudomonas fluorescens* strains on the different growth phases of *Phytophthora capsici*, foot rot pathogen of black pepper (*Piper nigrum* L.). *Arch. Phytopathol. Plant Protect.* 39 (0): 1–6.

Putaporntip, C., Miao, J., Kuamsab, N. et al. (2014). The *Plasmodium vivax* merozoite surface protein 3β sequence reveals contrasting parasite populations in Southern and Northwestern Thailand. *PLoS Negl. Trop. Dis.* 8:e3336. https://doi.org/10.1371/journal.pntd.0003336.

Rashmi, K.R., Earanna, N., and Vasundhara, M. (2008). Influence of biofertilizers on growth, biomass and biochemical constituents of *Ocimum gratissimum*. L. *Biomed* 3 (2): 123–130.

Ratti, N., Kumar, S., Verma, H.N., and Gautam, S.P. (2001). Improvement in bioavailability of tricalcium phosphate to *Cymbopo gonmartinii* var. *motia* by rhizobacteria, AMF and *Azospirillum* inoculation. *Microbiol. Res.* 156: 145–149. https://doi.org/10.1078/0944-5013-00095.

Saharan, B.S. and Nehra, V. (2011). Plant growth promoting rhizobacteria: a critical review. *Life Sci. Medi. Res.* 21 (1): 30.

Sapsirisopa, S., Chookietwattana, K., Maneewan, K., and Khaengkhan, P. (2009). Effect of salt-tolerant *Bacillus* inoculum on rice KDML 105 cultivated in saline soil. Asian J. Food Agro-Industry 2 (Special Issue): S69–S74.

Sen, S. and Chandrasekhar, N.C. (2014). Effect of PGPR on growth promotion of rice (*Oryza sativa* L.) under salt stress. *Asian J. Plant Sci.* 4 (5): 62–67.

Shaalan, M.N. (2005a). Effect of compost and different sources of biofertilizers, on borage plants (*Borago officinalis* L.). Egypt. *J. Agric. Res.* 83 (1): 271–284.

Shaalan, M.N. (2005b). Influence of biofertilizers and chicken manure on growth, yield and seeds quality of (*Nigella sativa*, L.) plants. Egypt. *J. Agric. Res.* 83 (2): 811–828.

Shahzad, R., Waqas, M., Khan, A.L. et al. (2017). Indoleacetic acid production and plant growth promoting potential of bacterial endophytes isolated from rice (*Oryza sativa* L.) seeds. *Acta Biol. Hung.* 68 (2): 175–186.

Sharma, T., Dreyer, I., and Riedelsberger, J. (2013). The role of K^+ channels in uptake and redistribution of potassium in the model plant *Arabidopsis thaliana*. *Front. Plant Sci.* 2013 (4): 224. https://doi.org/10.3389/fpls.2013.00224.

Shen, X., Chi, Y., and Xiong, K. (2019). The effect of heavy metal contamination on humans and animals in the vicinity of a zinc smelting facility. *PLoS One* 14.

Sheng, X.F., Xia, J.J., Jiang, C.Y. et al. (2008). Characterization of heavy metal-resistant endophytic bacteria from rape (Brassica napus) roots and their potential in promoting the growth and lead accumulation of rape. *Environ. Pollut.* 156: 1164–1170.

Shrivastava, M., Srivastava, P.C., and D'souza, S.F. (2018). *Phosphate-Solubilizing Microbes: Diversity and Phosphates Solubilization Mechanism*, 137–165. Singapore: InRole of rhizospheric microbes in soil, Springer.

Singh, R.P. and Jha, P.N. (2015). The plant growth promoting bacterium *Klebsiella* sp.SBP- 8 confers induced systemic tolerance in wheat (*Triticum aestivicum*) under salt stress. *J. Plant Physiol.* 184: 57–67.

Singh, N., Pandey, P., Dubey, R.C., and Maheshwari, D.K. (2008). Biological control of root rot fungus *Macrophomina phaseolina* and growth enhancement of *Pinus roxburghii* by rhizosphere competent *Bacillus subtilis* BN1. *World J. Microbiol. Biotechnol.* 24: 1669–1679. https://doi.org/10.1007/s11274-008-9680-z.

Singh, M.P., Kumar, P., Kumar, A. et al. (2016). Effect of NPK with biofertilizers on growth, yield and nutrient up take of wheat (*Triticum aestivum* L.) in Western Uttar Pradesh condition. *Soc. Recent Dev. Agric. Prog. Agric.* 16 (1): 83–87.

Sivasakthi, S., Kanchana, D., Usharani, G., and Saranraj, P. (2013). Production of plant growth promoting substance by *Pseudomonas fluorescens* and Bacillus subtilis isolates from paddy rhizosphere soil of Cuddalore District, Tamil Nadu, India. *Int. J. Microbiol. Res.* 4 (3): 227–233.

Strader, L.C., Culler, A.H., Cohen, J.D., and Bartel, B. (2010). Conversion of endogenous indole-3-butyric acid to indole-3-acetic acid drives cell expansion in Arabidopsis seedlings. *Plant Physiol.* 153 (4): 1577–1586.

Suksaard, P., Pathom-Aree, W., and Duangmal, K. (2017). Diversity and plant growth promoting activities of actinomycetes from mangroves. *Chiang Mai J. Sci.* 44(4): 1210–1223.

Takishita, Y., Charron, J., and Smith, D.L. (2018). (Pseudomonas Entomophila 23S, a PGPR with Potential for Control of Bacterial Canker Disease in Tomato (*Solanum lycopersicum* L.) Against *Clavibacter Michiganensis* Subsp. Michiganensis. Doctoral thesis. McGill University, Montreal, QC.

Tamilarasi, S., Nanthakumar, K., Karthikeyan, K., and Lakshmanaperumalsamy, P. (2007). Diversity of root associated microorganisms of selected medicinal plants and influence of rhizomicroorganisms on the antimicrobial property of Coriandrum sativum. *J. Environ. Biol.* 29 (1): 127.

Tewari, S. and Arora, N.K. (2013). Plant growth promoting rhizobacteria for ameliorating abiotic stresses triggered due to climatic variability. *Clim. Chang. Environ. Sustain.* 1 (2).

Tian, B., Zhang, C., Ye, Y. et al. (2017). Beneficial traits of bacterial endophytes belonging to the core communities of the tomato root microbiome. *Agric. Ecosyst. Environ.* 247: 149–156.

Tirry, N., Joutey, N.T., Sayel, H. et al. (2018). Screening of plant growth promoting traits in heavy metals resistant bacteria: prospects in phytoremediation. *JGEB* 16: 613–619.

Tiwari, S. and Lata, C. (2018). Heavy metal stress, signaling, and tolerance due to plant-associated microbes: an overview. *Plant Sci.* 9: 452.

Vaz-Moreira, I., Faria, C., Lopes, A.R. et al. (2009). *Sphingobium vermicomposti* sp. nov., isolated from vermicompost. *Int. J. Syst. Evol. Microbiol.* 59 (12): 3145–3149. https://doi.org/10.1099/ijs.0.006163-0.

Weston, D.J., Pelletier, D.A., Morrell-Falvey, J.L. et al. (2012). *Pseudomonas fluorescens* induces strain-dependent and strain-independent host plant responses in defense networks, primary metabolism, photosynthesis, and fitness. *Mol. Plant-Microbe Interact.* 25 (6): 765–778.

Yahaghi, Z., Shirvani, M., Nourbakhsh, F. et al. (2018). Isolation and characterization of pb-solubilizing bacteria and their effects on pb uptake by Brassica juncea: implications for microbe-assisted phytoremediation. *J. Microbiol. Biotechnol.* 28: 1156–1167.

Yaish, M.W., Antony, I., and Glick, B.R. (2015). Isolation and characterization of endophytic plant growth promoting bacteria. *Antonie Van Leeuwenhoek* 107(6): 1519–1532. https://doi.org/10.1007/s10482-015-0445-z

Yoolong, S., Kruasuwan, W., Thanh Phạm, H.T. et al. (2019). Modulation of salt tolerance in Thai jasmine rice (*Oryza sativa* L. cv. KDML105) by *Streptomyces venezuelae* ATCC 10712 expressing ACC deaminase. *Sci Rep.* 9(1): 1275. https://doi.org/10.1038/s41598-018-37987-5.

Yu, X., Bernard, R., Greenberg, B.M. et al. (2014a). Plant growth-promoting bacteria facilitate the growth of barley and oats in salt-impacted soil implications for phytoremediation of saline soils. *Int. J. Phytoremediation* 16: 1133–1147.

Yu, X., Li, Y., Zhang, C. et al. (2014b). *Culturable Heavy Metal-Resistant and Plant Growth Promoting bacteria in V-Ti Magnetite Mine Tailing Soil from Panzhihua*, 9. PloS One: China.

Yuan, Z.S., Liu, F., and Zhang, G.F. (2015). Isolation of culturable endophytic bacteria from Moso bamboo (*Phyllostachys edulis*) and 16S rDNA diversity analysis. *Arch. Biol. Sci.* 67 (3): 1001–1008.

Zhou, C., Ma, Z.V., Zhu, I. et al. (2016). Rhizobacterial strain *Bacillus megaterium* bofc15 induces cellular polyamine changes that improve plant growth and drought resistance. *Int. J. Mol. Sci.* 17: 976. https://doi.org/10.3390/ijms17060976.

20

Internet of Things (IoT) in Soil Management for Achieving Smart Agriculture

Amir Parnian[1], Mehdi Mahbod[2], Chanchal K. Mitra[3], Hossein Beyrami[1], and Majeti Narasimha Vara Prasad[3]

[1] *National Salinity Research Center (NSRC), Agricultural Research, Education and Extension Organization (AREEO), Yazd, 8917357676, Iran*
[2] *Department of Water Sciences and Engineering, College of Agriculture, Jahrom University, Jahrom, Iran*
[3] *School of Life Sciences, University of Hyderabad, Hyderabad, India*

20.1 Introduction

Until recently access to the Internet was limited via devices like the desktop, tablet, or smartphone, but now with the Internet of Things (IoT) practically all appliances can be connected to the Internet and monitored remotely. IoT is a system of interconnected devices that transfer and exchange data over a wireless network without any human intervention. The things on the Internet of Things could be the doorbell, thermostats, smoke detectors, water heaters, security alarms, etc. A device is a piece of equipment that should have capabilities of communication and could have capabilities of actuation, sensing, data capture, data processing, and data storage (Ahmad and Nabi 2021).

IoT devices (Figures 20.1 and 20.2) are connected to the network with the help of gateways or processing nodes, which process the information collected from the sensors and transfer it to the cloud. Functioning as the storage and processing unit, the cloud is where all actions are performed, and where the collected data are used for further learning and inferences. Figure 20.1 shows three layers that are the basic architecture of IoT. The specific uses and layer characterizations of the Internet for data communication and connecting "things" occur according to the work (i.e., collection and transmission) performed on data from its source to the physical world (Kumar et al. 2019; Lee and Lee 2015). Regarding IoT, a "thing" could be a physical thing or a virtual thing. A physical thing is an object of the physical world, and a virtual thing is information. Things are capable of being identified and integrated into communication networks (Lombardi et al. 2021). A network should have a specialized communication protocol with processing power to handle that communication. The "thing" in IoT may be able to perform certain actions once it is commanded, but for that purpose, it must recognize the command and confirm its completion. Routers, switches, and gateways are considered part of the network, but

Agroecological Approaches for Sustainable Soil Management, First Edition.
Edited by Majeti Narasimha Vara Prasad and Chitranjan Kumar.

Figure 20.1 Basic architecture of IoT.

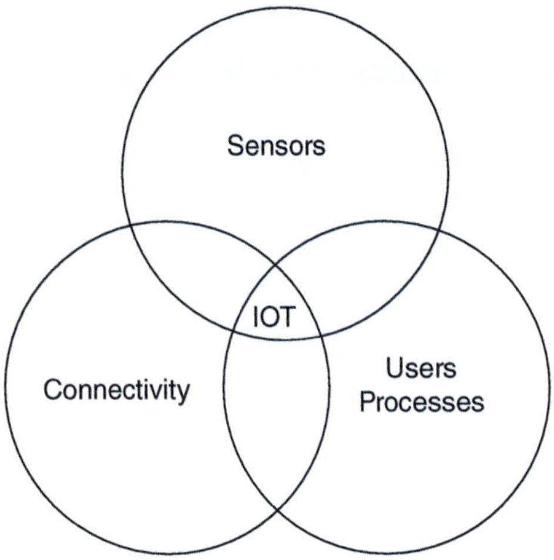

Figure 20.2 The main components of IoT.

they may also be classified as things (Ahmad and Nabi 2021; Li et al. 2015). With improved response monitoring and analytical capabilities, IoT is being adopted in almost all industries and domains, opening doors to countless applications. Both public and private sectors, e.g., manufacturing, transportation, logistics, retail, and healthcare are readily deploying IoT with such a wide spectrum of applications (Asghari et al. 2019). The future of IoT looks more promising than ever before. The integration of IoT with other technologies, e.g., cloud computing, machine learning (ML), and artificial intelligence (AI) is paving the way for many new and exciting innovations (Aghdam et al. 2021; Golpîra et al. 2021; Zhou et al. 2021).

The IoT describes physical objects (or groups of such objects) with sensors, processing ability, software, and other technologies that connect and exchange data with other devices and systems over the Internet or other communications networks. As shown in Figure 20.1, IoT is collaborative interaction of sensors, connectivity, and user processes. Meanwhile, this technology is reducing the workforce as human-to-human and human-to-machine interactions are eliminated in IoT, as data collection, transmission, and processing are carried out by machines (Ahmad and Nabi 2021). The IoT could be summarized into four stages (Pattnaik et al. 2020):

Stage 1: As the primary stage of this architecture, sensors and actuators receive "data or information" from the "things, which is digitized for further processing."

Stage 2: Data acquisition systems and Internet portals that perform important functions are activated during this phase. This is achieved by collecting all kinds of data from different input sources and transforming them into a suitable form for further processing. This stage is also responsible for A/D (analog to digital) metering and control. The IoT gateway is part of this phase that makes communication possible; Therefore, all contact data must pass from here.

Stage 3: Referred to as the "edge," this is the stage where "data or information" is received from the second stage through various gateways either through wired or wireless transmission. The edge typically receives data over wireless modes such as Wi-Fi, LoRa, or ZigBee. An important feature of a good architecture is to process data as close to the ends as possible; thus, pre-analytical processing takes place here to support subsequent processing.

Stage 4: In the last phase of this architecture, the actual processing occurs and converges in a cloud. Data analysis, data management, and some data archiving are the main functions of this stage. After processing the data, a purpose-based analysis is carried out and valuable results have been obtained that help in control, management, and decision-making.

For a better understanding of IoT connectivity and responses in the matter of agriculture and natural resources management, four questions are answered below:

20.1.1 What Is a Network?

We generally describe a network as a set of connected (inter-connected) devices that can exchange (some are transmitted only-for example a computer keyboard mostly transmits information- while many are both receive-transmit). Perhaps many homes today have more than one PC/laptop and a printer (or a multi-function device) and scanner. Modern cellphones (smartphones) are almost as powerful as a laptop computer a decade old. We show the few components of a network in a small office-home office (SOHO) in the diagram below (Figure 20.3).

The above diagram is a stand-alone local network, but it can be connected to the outside (Internet) via the telephone (may be dial-up or a broadband connection) or the Internet connection can be supported by a mobile phone data service. It is the router (shown in Figure 20.3 as item M) that distributes one external Internet connection to several devices internally. In this context, we use a hub, switch and router synonymously although there are subtle differences. For a large local network (for example a large office or a university), a direct fiber optic connection from the ISP (Internet service provider) may be considered.

20.1.2 How Does the IoT Work?

The heart of the IoT devices is a microprocessor (also called microcontroller) first pioneered by Intel (Intel 2022). The year was 1971 when "That's when the Intel® 4004 became the first general-purpose programmable processor on the market-a "building block" that engineers could purchase and then customize with software to perform different functions in a wide variety of electronic devices." and in 2022, microcontrollers are still around with greater functions and powers. A schematic block diagram of a modern microprocessor is shown in Figure 20.4.

A common IoT device consists of a microcontroller, a chip (integrated circuit) for Wi-Fi (wireless network capability), several sensors that are electronically connected with the microcontroller, and a small connection that can be used to program the microcontroller (initial programming can be done before the chip is installed on the board and future

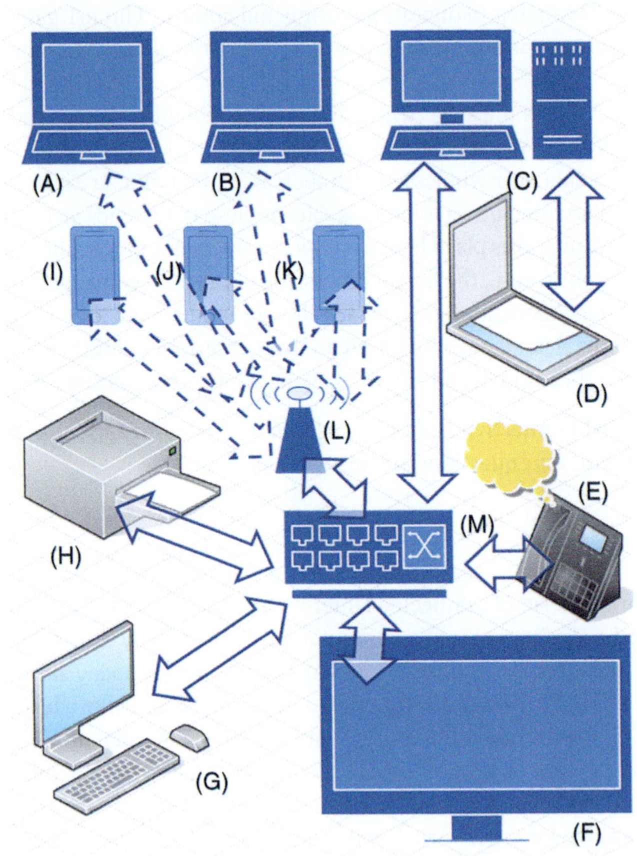

Figure 20.3 A small office/ home office (soho) setup with networking. A typical setup can accommodate 2–4 laptops (A, B above), a couple of desktops (C and G), a number of smartphones (I, J, and K), a scanner, and a printer (D and H: can be replaced with a multifunction device with a telephone connection), an ethernet router (or a hub or a switch; M), a wireless router (L) and broadband device (E: connecting to the internet access). The wireless connections are shown with dotted double arrows and the ethernet connections are shown with solid double arrows. This setup (shown above) can serve a small office with shared computer access. Also present is a large display suitable for presentations and group discussions (F). A somewhat larger setup may demand a dedicated local server. The above setup can be spread over two to three rooms (may need extra wifi routers). Each network is controlled by a router that has a local name (name of the network)- called the SSID (Service Set IDentifier) of the router (of the network). When in range, the Wi-Fi card can scan and list all the names (SSIDs) of the networks it can connect to. However, this (the SSID of a given router) can be hidden so that it does not show up in a public list.

updates can be made via Wi-Fi connection). The device receives information from the network and sends sensor data over the network.

The nature of the data IoT sends down the network depends on the nature of the sensors attached to it. For example, the IoT device can send the ground temperature data every hour or every minute (depending on the program inside the microcontroller) to the network and there is another computer on the same network that will collect the data and process it at regular intervals. The major task of this computer is actionable

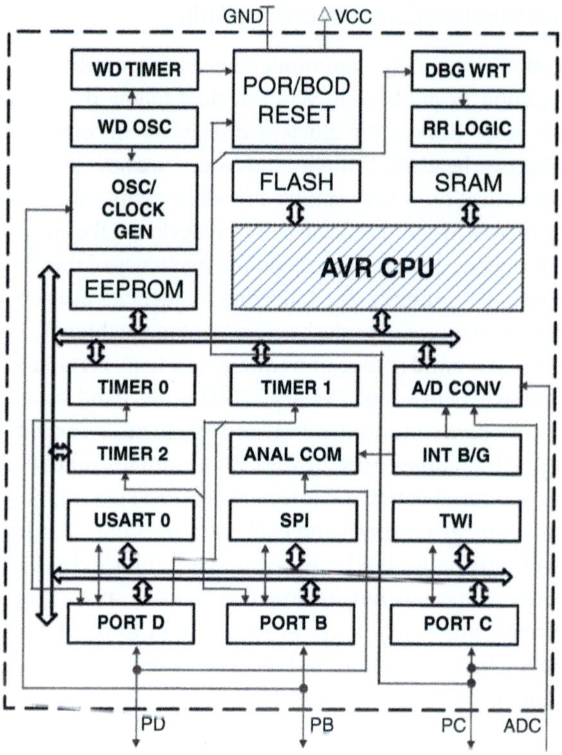

Figure 20.4 Block diagram of a modern microprocessor adapted from ATmega328P datasheet. This microcontroller is widely used in the Arduino series of microcontroller boards (Arduino UNO) and is very popular with hobbyists. The microcontroller is simple yet has 32 KB flash memory, 2 KB SRAM, 1 KB EEPROM, Watchdog timer, three internal timers (two 8 bit and one 16 bit), an internal oscillator, built-in ADC, and a host of other desirable features.

information (for example, if the temperature is high, the water sprinkler may be activated; but usually simple tasks like this are best left at the local level) from the set of different types of sensors.

20.1.3 How Does the Network Work?

Most common computer networks can be very broadly divided into either (i) wired or (ii) wireless. The most common wired connections today use the ethernet protocol (Ethernet Alliance 2022). Although wired connections are inherently more secure compared to wireless connections, they are vulnerable to the common man-in-the-middle attack (Martens 2022). However, using encryption of the user data can reduce the risks involved. For the local area wireless network, the interception of the data is rather easy and hence encryption is rather essential. For the ethernet networks, the common encryption too is SSL (secure socket layer) which is software based. For the wireless network, the encryption is partly hardware-based and commercial integrated circuits are available that provide the most common functionalities with the minimum hassle (Figure 20.4).

20.1.4 What Is Wi-Fi and How Does Wireless Communication Work?

With the advent of mobile communication technology, communications using wireless systems (Wi-Fi devices) have become common in public use. Wi-Fi devices follow the IEEE 802.11 standard for compatibility with other devices for efficient transfer of information. Although the technology has considerable overhead, its flexibility and ease of use have become a star attraction. In brief, wireless communication is very similar to mobile communication technology in the sense that it uses radio waves to communicate between devices. Like an AM or a FM radio, the digital signal is used to encode a carrier signal and the frequency of this carrier signal is specified on the Wi-Fi device. For example, the following frequencies are common today (sequence numbers 2, 3, and 4 are common for desktop and laptop machines):

The frequency mentioned in Table 20.1 is a band (range of frequencies); (e.g., 2.4 GHz in Table 20.1 represents a band of frequencies covering 2300–2400 MHz; the band covers 100 MHz but only a part of it is available for data communication) and this band can accommodate 4 non-overlapping channels (each channel uses 20 MHz plus a small gap called guard band of 5 MHz that is left empty or unused to prevent interference) or 11 overlapping channels (not all are available for use at any given time) and each card uses one channel (sometimes more than one) of fixed width (usually 20 MHz plus one spacer of 5 MHz). The actual frequency allocation varies between nations but for low-power local applications (very limited range) most of the channels are available (local rules are in force). Lower frequency operations (2.4 GHz frequency band) are preferred if the network is spanning several rooms (5 and 6 GHz signals cannot cross masonry walls).

Each Wi-Fi device is associated with a transmitter and a receiver. For user security, transmitted data is usually encrypted using established protocols. Most WLAN (wireless local area network; used interchangeably with Wi-Fi network) support WEP (Wired Equivalent Privacy) and WPA (Wi-Fi Protected Access), and modified WPA2 and WPA3 are common (circa 2022). Digital data received for transmission is encrypted first using the selected protocol. The encrypted data is next coded for transmission using a codec (codec = coder and decoder; coder is used for transmission and the decoder is used for reception). This is next

Table 20.1 Common Wi-Fi card band designations (based on IEEE standard)[a]

No	Name (designation)	Frequency	Max speed
1	802.11a	5 GHz	54 Mbps
2	802.11b	2.4 GHz	11 Mbps
3	802.11g	2.4 GHz	54 Mbps
4	80211n	2.4 GHz	150 Mbps
5	802.11ac	5 GHz	433 Mbps
6	802.11ac	5 GHz	866 Mbps
7	802.11ax	2.4; 5 GHz	1200 Mbps
8	802.11axe	6 GHz	1200 Mbps

Wi-Fi is a registered trademark of the Wi-Fi alliance. Ghz = 10^9 Hz; Mbps = million bits per sec.
[a] This table is not an exhaustive list.

put on a carrier using a modem (modem = modulator and demodulator; modulator is used for transmission and the demodulator is used for reception). Further, this signal is next amplified and put out on the antenna. The same antenna is used to receive information from the router. Although, the information transmitted by the router is seen and received by all the devices sharing the same channel, only one device (that matches the MAC, Media Access Control address of the network card, this is a hexadecimal number of 12 hex digits that uniquely identified each network card) can respond. However, the MAC address can be spoofed and that is widely used by hackers to listen to unencrypted communication between the router and a device. A modern network chip (integrated circuit) is itself a powerful microprocessor that can combine all the above tasks in a single chip (often the antenna is external to the device). A block diagram of a widely used chip (ESP32) used in Wi-Fi is shown below. Note that the most common Internet traffic protocol (http and https: hypertext transfer protocol/secure) may or may not be (depending on whether http: or https: being used) encrypted using the SSL (secure socket layer) protocol, but the security offered by the Wi-Fi transmission is over and above (and is not restricted to http or https) other forms of data protection. Further, it is relatively easy to sniff Wi-Fi data packets off the air and therefore data security is rather important although the data usually travels a short distance.

ESP32 chip can be used for both Wi-Fi and Bluetooth (Figure 20.5). This chip is available in both single-core and dual-core versions and also supports wireless modes 802.11 b/g/n.

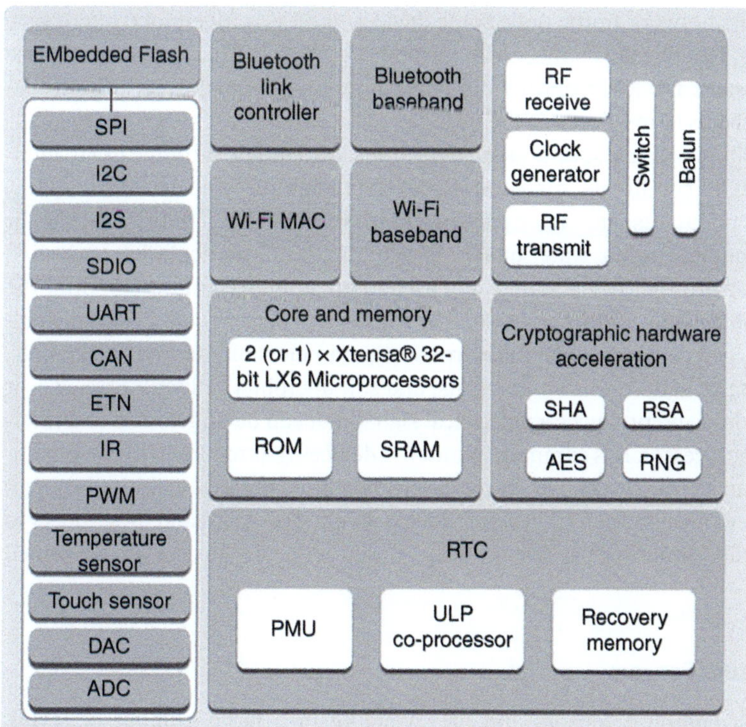

Figure 20.5 Block diagram of ESP32 chip – widely used in low-power Wi-Fi devices (*Source:* taken from the data sheet at https://www.espressif.com/sites/default/files/documentation/esp32_datasheet_en.pdf).

The data encryption is carried out in hardware. The chip has built-in support for a large number of communication protocols (SPI, I2C, I2S, SDIO, UART, Ethernet, and IR). The chip supports SHA (Secure Hash Algorithm), RSA (Developed by Ron Rivest, Adi Shamir, and Leonard Adleman), and AES (Advanced Encryption Standard) in hardware. The chip also has hardware, Random Number Generator (RNG). For technical details, the full datasheet can be consulted.

20.2 Sensors and Data in IoT-Based Systems

20.2.1 The Sensors

Most common IoT-based systems use a microcontroller, a wireless device (that can connect to a nearby hub or switch), and a host of sensors. These sensors are responsible for collecting information about the environment and transmitting them to the wirelessly connected hub for onward transmission. Most commonly, the hub uses the POTS (Plain Old Telephone System) with the help of a SIM (subscriber identity module) card to connect to a remote server. In a domestic setup, the IoT devices may collect information about a host of regular but diverse items (for example, television, refrigerator, microwave ovens, kitchen appliances, and voice-enabled assistants (Alexa with Amazon and Siri with Apple are omnipresent) by analyzing the network traffic only. Some devices like TV, refrigerators, microwave ovens, etc. may be IoT enabled to connect with the manufacturer and may report on the use and abuse of the system. Device errors are routinely communicated to the manufacturer to better understand the uses and needs of the user.

Besides these built-in systems, several standalone units are coming into wide use to monitor the environmental conditions, for example, the temperature, wind speed, humidity, various kind of air pollutants, and other parameters to get a reliable indication of the air quality. Similarly monitoring the water quality is also important, particularly concerning water pollution. Recently, IoT devices are being used in greenhouse setups to monitor the overall health of the system. All this work done by the devices depends on sensor technology, a small and rugged device that can measure some parameters of interest (for example, temperature, moisture, light level, soil nutrients, and soil microbial activity). Each sensor performs one task only and produces an electrical signal that can be understood and interpreted by the microprocessor. It is common to find IoT devices that report up to five parameters at a time (it can be fitted with five sensors). The data is locally used (within the farm) and handheld devices are common (that can reduce network overhead) and they appear to be more user-friendly. Some typical sensors widely used with IoT devices are presented below paragraphs.

20.2.2 Temperature Sensors

They are the most common sensors and are available in a large number of styles. However, the most common are the ones that are cheap and can measure ambient temperatures (−55 to 125 °C are common) with reasonable accuracy (1 °C). They are rugged, cheap, and need minimal interface electronics (or drivers). One such sensor is seen in Figure 20.6. They can be directly connected to a microprocessor. If higher temperatures are needed to be measured, there are different types (for example, a platinum resistance thermometer can be used to measure temperatures up to 600 °C). This particular device (on left) has an integrated circuit built into the sensor and several sensors can be connected to monitor temperatures at different points. These devices are widely used in air-conditioning systems, room heating systems, body temperature measurements, industrial device control systems, etc. With a reduced range, they can provide higher resolution (0.5 °C) but when higher accuracy (0.1 °C or better) is needed, other devices can be used.

20.2.3 Humidity Sensors

These devices measure humidity in the ambient air and are widely used in humidity control systems. These are simple oxide-based sensors that can read both temperatures (±2C accuracy) and humidity (±5% RH) a sample device can be seen in Figure 20.7. They are damaged in a highly contaminated environment, and high temperatures and have a slow response. Like the temperature sensor (Figure 20.6) this is also a three-wire device (one of the pins has no connection) and has a built-in chip in the body that controls the sensing and data handling. Mainly because it can handle both temperature and humidity, it is very popular with the electronic display of ambient temperature and humidity and for air handling units where a close regulation of the humidity is needed. The humidity is reported as RH (relative humidity; has a range of 0–100 but effective measurement range is around 30–80 with about 5% accuracy). This device is not suitable for soil moisture measurements for which a different type of sensor is used.

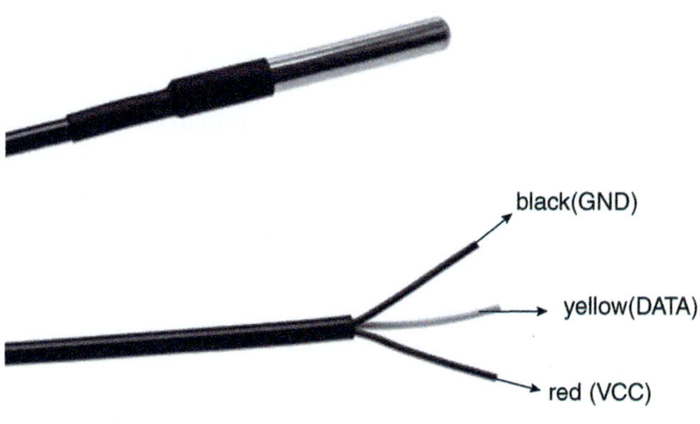

Figure 20.6 A typical temperature sensor. The stainless steel sheath contains an integrated circuit and other electronics. The device can be directly connected to a microcontroller (*Source:* Taken from Aliexpress.com).

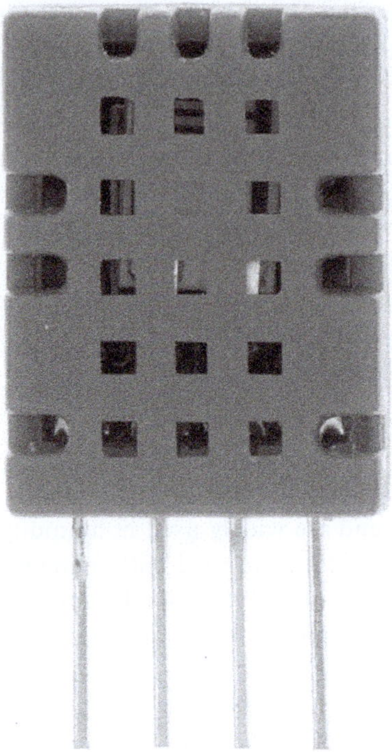

Figure 20.7 A typical humidity sensor. The device can measure both temperature and humidity. It can be directly connected to a microcontroller (*Source:* Taken from Aliexpress.com).

20.2.4 Sensors for Soil Moisture

These are devices that measure moisture content in common soils and there are several types in use. The capacitative types are popular because they can be left in the field and may be connected to a handheld device when reading is needed. Alternatively, they can be connected to a sprinkler system that can be automated to run when the soil gets dry beyond some threshold. These devices are ultra-simple and very robust. To determine the moisture content at a given place, the probe is simply pushed into the soil and a small reader is connected to the probe. Soil moisture is classified into surface moisture (top 10 cm of the surface) and root zone region moisture (top 200 cm from the surface) (NOAA 2022). Most common sensors measure the surface soil moisture (Figure 20.8) but sensors are available that can be inserted deep into the soil.

20.2.5 Sensors for pH and Dissolved Solids

A large number of sensors are available today that can measure the acidity (or alkalinity) of soil. The common state of the soil is slightly alkaline (pH 7–8) but soils can turn acidic under various environmental conditions. Moreover, the acidity (the concentration of the H^+ ions measured on a logarithmic scale) or alkalinity is expressed in the pH scale (https://www.usgs.gov/media/images/ph-scale; accessed August 04, 2022). Soils with high acidity or alkalinity may be considered damaged and can affect the underlying water

Figure 20.8 Common soil moisture sensor which able to examine the surface moisture (*Source:* Taken from Aliexpress.com).

quality. Similarly, we have sensors that can measure the total amount of dissolved solids in a given soil. This is measured by the electrical conductivity (EC) of the soil sample and cannot distinguish different chemical substances present in the soil. Most common fertilizers are inorganic salt in nature, and they are not completely utilized by the plants. Periodic measurements of the soil conductivity can tell us about the residual salinity in the soil and high salinity is inhibitory for most plants.

20.3 The Data

For a small IoT network of several sensors, the analysis is rudimentary and is performed manually. However, when the network contains many devices (say more than 100) manual analyses are not easy. The devices (IoT devices that belong to one unique network) form a set that can be considered one unit. There may be several hundred similar devices each identified by a computer ID. Each device ID has location information physically built into a table (map of the devices). While transmitting information, each device sends the device ID by default, and this information is translated into the physical map (location information) by the recipient. The data (a surface map of the sensor values) may be considered a time series over a long period (say one year) and need to be treated by statistical tools.

20.4 IoT in Agriculture

Extreme weather conditions climatic changes and environmental impacts are some of the challenges faced by the agriculture industry in meeting the growing food demands (Hasegawa et al. 2021) and IoT has been helping to tackle these challenges (Raj et al. 2021). Smart farming is a high-tech efficient and sustainable approach to agriculture (Boursianis et al. 2022). This is achieved by applying connected devices and innovative technologies within the agriculture process from seeding to harvesting crops. IoT provides valuable real-time insight to farmers (Idoje et al. 2021). Devices can mean anything from pump sheds and tractors to weather stations (Gómez-Chabla et al. 2019). The IoT system can detect real-time weather conditions like humidity rainfall temperature and wind conditions very accurately. The collected data helps select the right crops to grow and maintain, by allowing farmers to act quickly when weather conditions turn unfavorable (Keswani et al. 2019). IoT has a wide scope in agriculture, as it helps the tedious task of collecting, processing, and analyzing data to take appropriate decisions by minimizing human effort, human error, waste of resources, etc. Remote monitoring of almost all of the parameters is now practically possible. Thus, the goal of automation in farming methodologies has become a reality thanks to IoT on a larger scale. Therefore, IoT has boosted smart farming (Kim et al. 2020). In addition, IoT applications are also being used for precise farming, livestock monitoring, vehicle tracking, field observation, and inventory monitoring. They enable farmers to analyze information and take informed and quick decisions. Farmers can also analyze soil water and nutrient conditions to increase farming effectiveness and efficiency. IoT applications in the agricultural industry has made farming more efficient, effective, and sustainable (Mekala and Viswanathan 2020).

Some of the recent and common applications of IoT in agricultural-specific domains are (i) Solar-Powered Automated Drip Irrigation systems, (ii) Agricultural Drones, (iii) Precision Farming (iv) Smart Cultivation (v) Green House Production (vi) Monitoring Activities (vii) Crop Management (viii) Agricultural Machinery, and (ix) Disease and Pest Control (Ahmad and Nabi 2021). IoT helps to reduce water loss and at the same time helps to apply fertilizers. Drones can be used for various tasks such as field surveillance, monitoring crop growth, precision spraying, investigation of ground and soil, chlorophyll content determination, nitrogen content determination, and high-resolution imaging for various analyses. Drone-based yield monitoring can be further processed and analyzed to provide useful insights. The most important application of IoT is precision farming, as the principles of precision farming have been brought to reality. These cost-effective IoT systems are available on the market, helping this emerging trend of precision farming further (Nayak et al. 2020). By calculating the adequate amount of input requirements, IoT could minimize wastage in cultivation. The harvest field can be monitored, and smart systems in IoT help the farmer by reducing manpower. With IoT in action, greenhouse shortcomings have been greatly mitigated. The environmental conditions are detected and transmitted to a storage location through wireless sensor networks. The data are then analyzed using a myriad of sophisticated tools and analytical models (Maraveas et al. 2022). In order to obtain expert advice from agronomists, it is necessary to continuously monitor various parameters of the field and crop. With manual methods, it is impossible to oversee things in such a way. IoT, therefore, offers the advantage of monitoring all aspects of agriculture and making this data available to professionals in order to provide farmers with reliable advice (Nayak et al. 2020).

Crop management is an important task to improve crop yield and quality. To this end, historical data plays a key role in crop management. Precision agriculture requires knowledge gained from deeper data analysis to make every action with maximum efficiency. Examples of these include the amount of water required, the timing of irrigation, the rate and timing of fertilizer and pesticide application, etc. Under sophisticated IoT-based management technology, real-time information reaches farmers either by email or SMS in a convenient way, taking crop management to the next level (Vitali et al. 2021).

The scenario of machine applications in agriculture has changed entirely with the implementation of IoT (Nayyar et al. 2020). Machines could serve directly or indirectly. Some are used for surveillance, replacing humans in performing agricultural practices and in monitoring various parameters. IoT makes machines more powerful than traditional alternatives. For example, seeding devices controlled by technologies like GIS and GPS connected to IoT networks offer higher seeding accuracy (Marcu et al. 2019).

There are diverse benefits of IoT implementation, e.g., disease and pest control (Parsons et al. 2020). Approximately 37% of crop production would be lost each year (Ahmad and Nabi 2021). The data from the IoT system is stored in a cloud, which is subsequently analyzed. Data is shared directly with experts or rendered on ML platforms that yield results. Forecasts are based on computer analysis. Both ML and deep learning are integrated using different algorithms. Hence, diseases and pest attacks can be forecasted (Kanuru et al. 2021). Convolutional neural networks are also used to train the system (Bhoi et al. 2021a). IoT requires monitoring of main agricultural parameters like irrigation and water quality and amount, weather, and soil. Sustainable use of irrigation water is a key concern in agriculture. This issue is not possible until there is an accurate assessment of soil moisture content.

20.5 IoT in Soil Science

Moisture is crucial for crop growth, as nutrients are supplied to the crop through the water. Moisture in the soil is efficiently determined by some types of sensors, which used to require manpower. In addition to reducing water wastage, an automated irrigation system can also be used to identify the macronutrients (i.e., potassium, phosphorus, and nitrogen). It is a tedious process for farmers to perform this test manually. IoT enables farmers to obtain information about soil conditions and remotely monitor their fields with useful suggestions on soil requirements and corresponding application rates. Examples of sensors that are involved in soil monitoring are (i) High-accuracy sensors of location for determining coordination (i.e., altitude, longitude, and latitude). (ii) Sensors are used to measure the soil properties like clay content, the quantity of organic matter, and soil water content optically using different spectra of light (e.g., infrared, polarized, etc.). (iii) Ion detection sensors are used to determine pH and nutrient levels in soil. (iv) Mechanical sensors that help to determine soil resistance (Ahmad and Nabi 2021).

20.6 IoT Parts: Soil Sensors and Parameter Monitoring with IoT-Linked Sensors

Soil monitoring/ surveillance with IoT-based tech allows farmers and producers to achieve maximum yield, reduce disease defects and optimize resource efficiency. IoT sensors in the matter of soil monitoring can measure moisture, temperature, macronutrients (e.g., N, P, and K), pH, water potential, EC, and air quality (e.g., oxygen and carbon dioxide levels). Output data from IoT-linked sensors are collected and transmitted/ transferred to a central point/cloud system (cloud computing technology) for analysis, visualization, and analysis (Rehman et al. 2022; Hughes 2022).

These results and data- obtained from IoT-linked sensors- then can be used to optimize many proposes in farming operations, identify the trends and lead to make subtle decisions for adjustments to changing conditions to reach the maximum crop yield and high quality in a commercial approach. Furthermore, the use of IoT in agriculture is also known as Smart Agriculture (or Smart Farming), and IoT is the skeleton and one of the central components of Precision Farming (Vikranth 2021).

Profitable and efficient cultivation depends on several soil parameters. Various soil sensors like pH, moisture, nutrients (soil N, P, and K sensors), temperature and humidity, salinity/EC, etc. would be utilized for this kind of IoT base measurement and data collection. In these IoT-based systems, obtained data from the sensors of soil moisture, soil temperature, temperature, and humidity could be analyzed and integrated for managing an effective irrigation system (Grimblatt et al. 2021), which could help to preserve water and soil properly.

Devise equipped with salinity/EC sensors can be used to detect saltwater instantly. Many of these soil salinity sensors are available as soil-medium-connected (Figure 20.9) or none-connected (Figure 20.10). Soil pH sensor reveals the acidity or alkalinity contained in the soils. Soil salinity, pH, and nutrients IoT-linked sensors can be used to perform in fertilizer/soil fertility management systems.

20.6.1 Soil Temperature

Soil temperature is an effecting factor in the matter of plant growth and activity, which have a direct impact on roots, soil microorganism's respiration, and soil organic components decomposition and nutrient mineralization in soil. IoT-linked sensors can evaluate soil temperature by measuring air temperature in association with other soil factors; anyway, using a buried probe is the most accurate measurement of soil temperature. Depending on the monitored root morphology and plant age, multiple probes can be

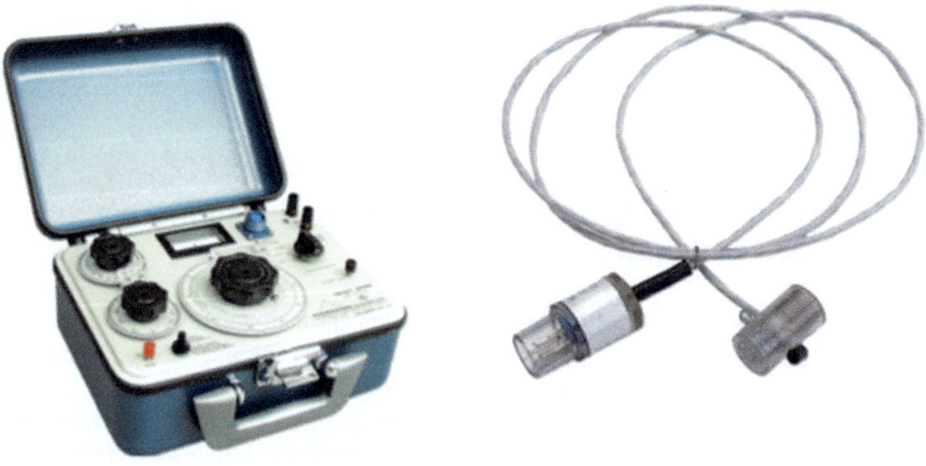

Figure 20.9 5500 Salinity Bridge devise and its sensor, Soilmoisture Equipment Corp., www.soilmoisture.com.

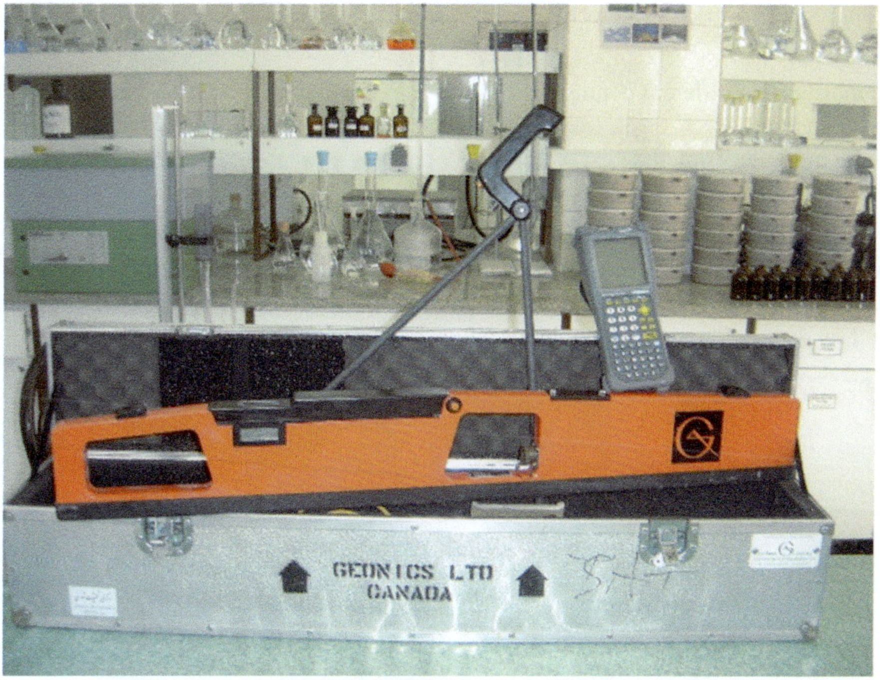

Figure 20.10 EM38 instrument, which provides measurements of soil conductivity and magnetic susceptibility without any soil-instrument contact, Geonics Limited, www.geonics.com.Smart Agriculture is focused on soil monitoring/management, weather, and crop conditions. Given the importance of weather and irrigation, many Smart Farming solutions are coupled with Smart Environment (Air Quality) and Smart Water (Pollution, Turbidity, Nutrients) for a complete solution (Rehman et al. 2022). The most common IoT sensors are listed and briefly explained in the below paragraphs.

installed at separate depths to achieve a better evaluation of the soil conditions. Different technology types of IoT-linked sensors (like IR sensors) can be used to monitor the surface soil temperature. It should be mentioned; mostly soil environment temperature is related to humidity and soil moisture, which is the most commonly measured data (Na et al. 2017).

20.6.2 Soil Moisture

The soil moisture content is the most considerable factor in IoT-based soil management systems. This is mostly monitored using buried probes with electrodes. Soil moisture content is a factor, which plays a vital role in hydrology, soil science, and agricultural purposes. Soil moisture content is an essential factor utilized for many reasons (Placidi et al. 2020):

a) Water in the soil serves as a critical nutrient for all crops and plants, water is an essential component of photosynthesis,
b) Crop yield is heavily influenced by the availability of water in the soil, soil water is an important carrier of soluble food nutrients for plant growth
c) Soil water helps regulate soil temperatures.

IoT and Smart Agriculture technology in some cases can measure the following altogether:

a) Soil moisture (in different depths),
b) Conductivity,
c) Volumetric water content
d) Soil water potential.

The smart lysimeter of the National Salinity Research Center of Iran (IR-NSRC), which is located in their base in Yazd city is a good example of IoT application in agricultural research. The device measures water content and temperature in the soil. Further, it collects and sends the data remotely to the IR-NSRC researcher's office (Figure 20.11).

Figure 20.11 IoT-based lysimeter in IR-NSRC, Yazd, Iran.

20.6.3 Solar Radiation

IoT-linked sensors can measure different types of solar radiation that would play a key role in plant photosynthesis. More than basic light measurement levels (mostly determined in lux units), IoT technology can measure the followings (Ghosh et al. 2020):

a) Solar – photo synthetically active radiation,
b) Solar – UV,
c) Solar – shortwave.

Solar radiation factors have an undeniable impact on plant growth and productivity, which IoT related technologies allowed users to monitor solar radiation levels to understand trends/correlations.

20.6.4 Weather

Rainfall/precipitation, wind, humidity, and atmospheric pressure all play an essential role in plant growth. Some smart agriculture systems support several advanced weather stations. IoT weather stations can measure the following (Bhoi et al. 2021a):

a) Precipitation,
b) Temperature,
c) Humidity,
d) Air pressure, and
e) Wind speed and direction.

20.6.5 Fertilizer

Soil macronutrient sensors (like nitrogen, phosphorous, and potassium sensors) are relatively new and advanced. However, these new techs provide a reliable method in which these key soil nutrients can be measured using IoT-linked sensors. Fertilizer IoT-linked sensors use complex various technologies. Nowadays, advanced combined fertilizer sensors can measure nitrogen, phosphorous, potassium, pH, EC, temperature, and moisture almost together easily (Lova Raju and Vijayaraghavan 2020).

20.7 A Better Understanding of Soil Conditions (Fertility, Degradation, Irrigation, Detection of Soil-Borne Diseases, etc.)

Environment monitoring (EM) represents an adjusted plan and management to control and better respond to environmental changes and stresses. EM in agriculture and terrestrial natural resource management mostly deals with water, soil, solar radiation, climate, and biological stress parameters. The sources of these environmental challenges are caused by several factors, whether man-made/man-controllable or the ones due to natural causes which are unable to manage properly by human acts/decisions (Ullo and Sinha 2020).

Obviously, in most agricultural and terrestrial natural resource ecosystems, the soil is the base skeleton, which could be the base for land management. On the other hand, land management in that concept is mostly considered equal to soil management; besides, knowing and understanding soil conditions is fundamental for soil/land management (Al Mamun and Yuce 2019). IoT lined sensors, concerning their role in EM, would accurately address the challenges for to land-managers for better soil management, which these main challenges are fertility parameters, soil degradation and loss, soil moisture condition (irrigation demands), detection of soil-borne diseases (determination related of soil physicochemical properties).

Furthermore, EM in soil management helps land managers to have a better understanding of soil conditions, but data analysis in some cases part of data collection is up to land managers. More importantly, decision-making and acting in response to challenges/changes is 100% reliant on land managers. In fact, land managers only receive warnings from the system and they are alone in decision making which increases the risk of human errors (Raut et al. 2018). For example, when EM sends some data to a land manager about soil moisture which shows it falls to the level of permanent wilting point (PWP), he/she has to these following actions:

1) Collecting and checking the data,
2) Analyzing the data,
3) Deciding irrigation or do nothing,
4) If the demand for irrigation is recognized by the land manager, he/she has to consider the volume of water demand for this individual soil, and
5) Afterward, acting for irrigation is the available step for the land manager.

EM relay on IoT-linked sensors and data collection of these sensors may be achieved through local area network (LAN) cables. Bluetooth wireless technology, Wi-Fi In this context, the current work presents three different IoT-based wireless sensors for soil condition monitoring: one employing User Datagram Protocol (UDP)-based Wi-Fi communication, one communicating through Wi-Fi and Hypertext Transfer Protocol (HTTP), and a third one using Bluetooth Smart (Ullo and Sinha 2020).

As it is obvious, EM is a helping tool for land-manager in the matter of soil conditions understanding. Recently, with some signs of progress in AI technology, especially in the terms of ML, EM utilizing these advanced technologies developed and upgraded to smart environment monitoring (SEM) system. SEM because the technology has enabled EM methods to accurately survey the important/impacting factors, with optimal control of elements/pollutants/environmental stresses and their effects/impacts. In near future, due to many reasons, especially the demand for efficient/effective resource management, the design of smart farming systems is taking the place of old and traditional methods to create and plan farm/ garden/greenhouse environments (Ullo and Sinha 2020; Zhang et al. 2018). In the matter of agriculture, as a key point for the sustainability and development of nations, it is easy to underline how the SEM approach in land management can play a noteworthy function by granting "smart/green agriculture."

20.8 The Future Role of IoT in Smart Agriculture

Recent achievements, such as IoT, cloud computing technology, and cyber-physical systems (CPS), grant us to transmit and manage huge amounts of data after observation of agricultural-related parameters. As mentioned previously, IoT parts related to soil (Soil sensors and soil sample equipment) could be the main skeleton for smart agriculture. But the worldwide spreading of smart agriculture will happen faster and become cheaper when the availability of the Internet, could technology and CPSs have become at a high level for most economical sectors of land-managers (Friha et al. 2021).

The appearance of IoT has shown a fascinating enlightened direction of innovative research in the matter of agriculture. Additionally, IoT-based devices accelerate research and make the research project for scientists easier to manage and more comfortable (an IoT-based device developed and used for agricultural and environmental research showed in Figure 20.12).

IoT needs to be experimented with and studied in all aspects, before being widely/fully applied in agricultural applications, e.g., farming, horticulture, greenhouses, etc. There are various potential applications for IoT in agriculture, and many specific issues and challenges may be associated with IoT deployment for better resource performance in farming and the meantime, for economical efficient widespread smart farming (Ray 2017).

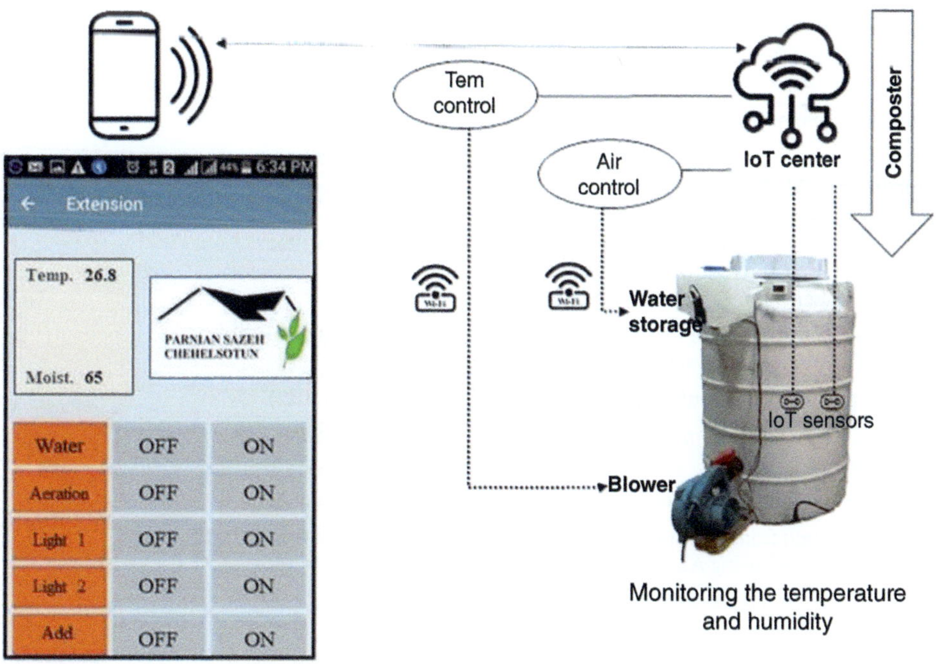

Figure 20.12 The smart composting reactor, which is a smart IoT-based device for research in the matter of composting and bioremediation (designed and provided by Dr. A. Parnian and Parnian Sazeh Chehelsotoun Co. Ltd.).

IoT applications for smart agriculture can classify into seven categories (Friha et al. 2021):

1) Smart monitoring
2) Smart water management
3) Agrochemicals applications
4) Disease management
5) Smart harvesting
6) Supply chain management
7) Smart agricultural practices

In addition, supply chain management IoT uses in agriculture is a challenge, which is the blockchain assisting tech mentioned as a potential solution (Friha et al. 2021). Moreover, this approach still has several challenging areas like (i) hardware boards, (ii) interoperability of systems, (iii) networking and energy management, (iv) security and privacy threats, (v) hardware and software costs, and (vi) education challenges, which should be developed before the widespread use of IoTs in agriculture.

20.9 Technology in Advanced Farming

In the fields of AI, IoT, and robotics, farming has predicted a huge boom. We have split our focus among AI, IoT, and Robotics in this area to offer a full understanding of the issues in order to determine various implantation and scopes of these disciplines as they arise, which some recent applications of IoT in agriculture mentioned in Table 20.2.

Table 20.2 Some examples for recent applications of IoT in agriculture.

Applications of IoT in agriculture	Authors (Year)
Solar fertigation: A sustainable and smart IoT-based irrigation and fertilization system for efficient water and nutrient management.	Ahmad et al. (2022)
IoT and GIS management for direct sowing and high efficiency food production.	Alcoberro et al. (2022)
IoT for precision in agricultural system.	Babu et al. (2022)
IoT in the agriculture domain with future directions.	Bansal and Goyal (2022)
Machine learning approaches for agro IoT systems.	Dhivyaa et al. (2022)
Rice cultivation from plowing–harvesting to increase the yield using AI.	Dulhare and Gouse (2022)
A BIM-GIS-IoT-based system for excavated soil recycling.	Huang et al. (2022)
Machine learning approaches for Agro-IoT systems.	Jadhav and Lal (2022)

Table 20.2 (Continued)

Applications of IoT in agriculture	Authors (Year)
Effect of different land-use types on soil properties in South-Central Ethiopia.	Kebebew et al. (2022)
Agriculture-IoT-based sprinkler system for water and fertilizer conservation and management.	Kumar and Choudhury (2022)
IoT enabled technologies in smart farming and challenges for adoption	Kumar et al. (2022)
Inhibitory effects of soil amendment coupled with water management on the accumulation of Cd and Pb in double-cropping rice.	Li et al. (2022)
Applications of IoT for optimized greenhouse environment and resources management. Computers and electronics in agriculture	Maraveas et al. (2022)
Developing an IoT-based data analytics system for predicting soil nutrient degradation level.	Najeeb Ahmed and Kamalakkannan (2022)
Agricultural science with IoT.	Pant (2022)
Overview of IoT in the agroecosystem.	Parmar and Kumar (2022)
Application of IoT and cloud computing in automation of agriculture irrigation.	Phasinam et al. (2022)
Exploiting IoT and its enabled technologies for irrigation	Ramachandran et al. (2022)
AI based yield prediction and smart irrigation.	Ramdinthara et al. (2022)
Vertical agriculture monitoring system	Rani et al. (2022)
A review of crop husbandry and soil management practices using meta-analysis studies: toward soil-improving cropping systems.	Rietra et al. (2022)
IOT based smart irrigation management system for environmental sustainability in India.	Sangeetha et al. (2022)
IoT-based soil nutrient analyzer using Gaussian process regression.	Senevirathne et al. (2022)
LoRa based intelligent soil and weather condition monitoring with Internet of Things for precision agriculture in smart cities.	Singh et al. (2022)
Farm management and resource optimization using IoT.	Srinivas et al. (2022)
Engineering IoT systems in the convergence between agronomic and computer sciences.	Tabuenca et al. (2022)
Automated smart irrigation system using IoT with sensor parameter	Tephila et al. (2022)
Corn rhizosphere microbial community in different long term soil management systems.	Vendruscolo et al. (2022)
Increasing productivity of rice plants based on IoT	Wicaksono et al. (2022)

(*Continued*)

Table 20.2 (Continued)

Applications of IoT in agriculture	Authors (Year)
Artificial intelligence and machine learning for the green development of agriculture in the emerging manufacturing industry in the IoT platform.	Zhou et al. (2022)
IoT technologies for precision agriculture:	Pyingkodi et al. (2022)
Mathematical and visual understanding of a deep learning model toward m-agriculture for disease diagnosis.	Francis and Deisy (2021)
A predictive machine learning application in agriculture: Cassava disease detection and classification with imbalanced dataset using convolutional neural networks.	Sambasivam and Opiyo (2021)
Reflections on the future of swarm robotics.	Dorigo et al. (2020)
Smart sprayer for precision weed control using artificial intelligence	Partel et al. (2020)
Detection of sunn pest-damaged wheat grains using artificial bee colony intelligence techniques.	Sabanci (2020)
Artificial intelligence approach to estimate discharge of drip tape irrigation based on temperature and pressure	Seyedzadeh et al. (2020)
Zero-and few-shot learning for diseases recognition of *Citrus aurantium* L. using conditional adversarial autoencoders	Zhong et al. (2020)
The role of geospatial technology with IoT for precision agriculture	Bhanumathi and Kalaivanan (2019)
Combined life cycle assessment and artificial intelligence for prediction of output energy and environmental impacts of sugarcane production.	Kaab et al. (2019)
Evolution of IoT and its significant impact in the field of precision agriculture.	Khanna and Kaur (2019)
Development and evaluation of a low-cost and smart technology for precision weed management utilizing artificial intelligence.	Partel et al. (2019)
-crop plant disease classification over cell phone acquired images taken on real field	Picon et al. (2019)
Designing future precision agriculture: detection of seeds germination using artificial intelligence	Shadrin et al. (2019a)
Enabling precision agriculture through embedded sensing with artificial intelligence.	Shadrin et al. (2019b)
The newest agricultural technologies.	Ozguven (2018)
Research and development in agricultural robotics: a perspective of digital farming.	Shamshiri et al. (2018)
Evaluation of artificial intelligence models for actual crop evapotranspiration modeling in mulched and non-mulched maize croplands.	Tang et al. (2018)

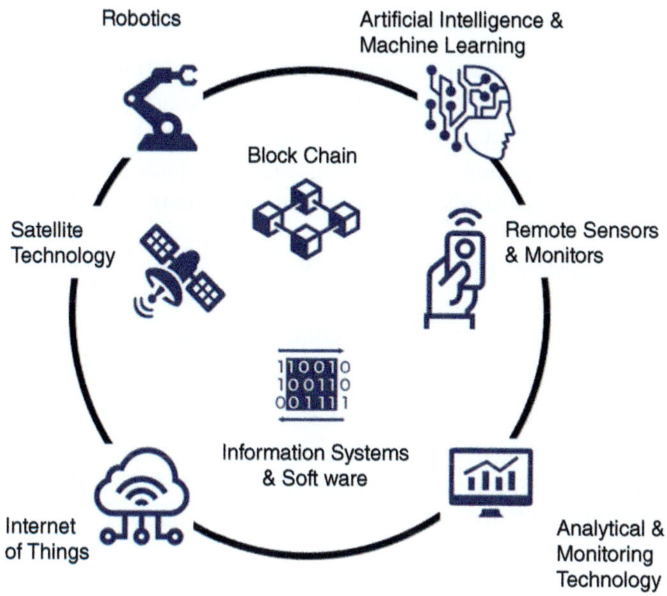

Figure 20.13 Integration of AI, Robotics and IoT in soil management and smart agriculture, a conceptual frame.

Figure 20.13 shows a conceptual framework for integrating AI, robotics, and the Internet of Things with soil management and smart agriculture.

20.10 Risks of IoT in Land Management and Food Security

IoT could be a prosperous tool to meet the needs of the human communities, further, it produces plenty of benefits in resource preservation and management. IoT really visualized a situation whereby many physical objects are connected/related to the Internet, which able to transfer data, communicate, and identify themselves to other devices. This would generate/produce a huge amount of data to process to making decisions in consequence. However, risks regarding the adoption of IoT also occur at all levels (Brou and Janssen 2015).

There are many threats in IoT uses in agriculture. A brief categorization of the main causes of low security in IoT based systems according to (Demestichas et al. 2020) are "a) Unpatched firmware and/or extended use of default passwords, which allow for compromising the devices within an IoT network; b) Limited computational resources of smart devices, which hinders the implementation of complex cryptographic algorithms, due to the efforts of vendors to reduce the costs of their products in a competitive environment; c) Vulnerabilities in the communication protocols used by smart devices (e.g., ZigBee, Bluetooth); d) Low security level of the old version of the WPA protocol, which is still used in many cases; e) Search engines that can be used for executing passive vulnerability detection; f) The danger of organizing millions of smart devices in a powerful botnet (e.g., Mirai), due to the relatively easy detection of vulnerabilities by means of Internet scanning; g) General lack of attention to the security of smart devices."

In addition, IoT-based systems could be vulnerable (Demestichas et al. 2020):

a) Application (Data Thefts, Access Control Attacks, Service Interruption Attacks, Malicious Code Injection Attacks, Sniffing Attacks, and Reprogram Attacks)
b) Middleware (Man-In-the Middle Attack [MITM], SQL Injection Attack, Signature Wrapping Attack, Cloud Malware Injection, and Flooding Attack in Cloud)
c) Internet (Phishing Site Attack, Access Attack, DDoS/DoS Attack, Data transit Attacks, and Routing Attack)
d) Access Gateway (Secure on-Boarding, Extra Interfaces, End-to-End Encryption, and Firmware Updates)
e) Edge Technology (Node Capturing, Malicious Code Injection attack, False data Injection Attack, Side-Channel Attacks, Eavesdropping and Interference, Sleep Deprivation Attacks, and Booting Attacks)

Any disruption in IoT-based land management would lead to resource loss and low productivity. Which in large-scale use of IoTs, it means a huge lake of food or food loss. In matter of food security there is a high risk that can tremendously affect a nation in wars or political challenges.

20.11 Conclusion

IoT is a new promoting tech in natural resource and agricultural management. In addition, this progressing technology is very helpful in soil monitoring and management. Nowadays, many applied IoT devices are available in the matter of soil monitoring in the market, and a variety of IoT soil management systems are developing by scientists and engineers around the world. This manner pledge vast infiltration and uses of IoT-based technology in soil management, which might be a game-changing point in future globalization of smart agriculture and food security calculations for governments. IoT rising uses in soil management also have many risks that could shape a weak point in the food security of human societies. In collusion, IoT technology and smart agriculture bases are ready to rise in near future for soil management issues but it felt that huge fields of R&D remain for more accurate, reliable, and secure devices, networks, and AIs.

References

Aghdam, Z.N., Rahmani, A.M., and Hosseinzadeh, M. (2021). The role of the internet of things in healthcare: future trends and challenges. *Comput. Methods Programs Biomed.* 199: 105903. https://doi.org/10.1016/J.CMPB.2020.105903.

Ahmad, L. and Nabi, F. (2021). *Agriculture 5.0: Artificial Intelligence, IoT and Machine Learning*. CRC Press. https://www.routledge.com/Agriculture-50-Artificial-Intelligence-IoT-and-Machine-Learning/Ahmad-Nabi/p/book/9780367646080.

Ahmad, U., Alvino, A., and Marino, S. (2022). Solar fertigation: a sustainable and smart IoT-based irrigation and fertilization system for efficient water and nutrient management. *Agronomy* 12 (5): 1012.

Al Mamun, M.A. and Yuce, M.R. (2019). Sensors and systems for wearable environmental monitoring toward IoT-enabled applications: a review. *IEEE Sens. J.* 19 (18): 7771–7788.

Alcoberro, R., Bauer, J., and Durakbasa, N.M. (2022). Direct-drive integrated motor/controller with IoT and GIS management for direct sowing and high efficiency food production. In: *Digitizing Production Systems. Lecture Notes in Mechanical Engineering* (ed. N.M. Durakbasa and M.G. Gençyılmaz), 271–280. Cham: Springer. https://doi.org/10.1007/978-3-030-90421-0_23.

Asghari, P., Rahmani, A.M., and Javadi, H.H.S. (2019). Internet of things applications: a systematic review. *Comput. Netw.* 148: 241–261. https://doi.org/10.1016/J.COMNET.2018.12.008.

Babu, R.G., Lavanya, D., Raja, L. et al. (2022). IoT for precision in agricultural system. In: *AIP Conference Proceedings*, vol. 2405 (ed. R. Kalidoss, K. George, and M. Vinayagam), 030008. AIP Publishing LLC. https://doi.org/10.1063/5.0072865.

Bansal, R. and Goyal, N. (2022). A study on IoT in the agriculture domain with future directions. In: *Real-Life Applications of the Internet of Things, Challenges, Applications, and Advances* (ed. M. Mangla, A. Kumar, V. Mehta, et al.), 139–160. Apple Academic Press. https://doi.org/10.1201/9781003277460-8.

Bhanumathi, V. and Kalaivanan, K. (2019). The role of geospatial technology with IoT for precision agriculture. In: *Cloud Computing for Geospatial Big Data Analytics. Studies in Big Data*, vol. 49 (ed. H. Das, R. Barik, H. Dubey, and D. Roy), 225–250. Cham: Springer. https://doi.org/10.1007/978-3-030-03359-0_11.

Bhoi, A., Nayak, R.P., Bhoi, S.K. et al. (2021a). IoT-IIRS: internet of things based intelligent-irrigation recommendation system using machine learning approach for efficient water usage. *Peer J. Comput. Sci.* 7: e578.

Bhoi, S.K., Jena, K.K., Panda, S.K. et al. (2021b). An internet of things assisted unmanned aerial vehicle based artificial intelligence model for rice pest detection. *Microprocess. Microsyst.* 80: 103607. https://doi.org/10.1016/J.MICPRO.2020.103607.

Boursianis, A.D., Papadopoulou, M.S., Diamantoulakis, P. et al. (2022). Internet of things (IoT) and agricultural unmanned aerial vehicles (UAVs) in smart farming: a comprehensive review. *Internet Things* 18: 100187. https://doi.org/10.1016/J.IOT.2020.100187.

Brou, P. and Janssen, M. (2015). Effects of the Internet of Things (IoT): a systematic review of the benefits and risks. *ICEB 2015 Proceedings*, Hong Kong, SAR China, p. 48. https://aisel.aisnet.org/iceb2015/48.

Demestichas, K., Peppes, N., and Alexakis, T. (2020). Survey on security threats in agricultural IoT and smart farming. *Sensors* 20 (22): 6458.

Dhivyaa, C.R., Anbukkarasi, S., and Saravanan, K. (2022). Machine learning approaches for agro IoT systems. In: *Internet of Things and Analytics for Agriculture*, vol. 3 (ed. P.K. Pattnaik, R. Kumar, and S. Pal), 93–111. Singapore: Springer. https://doi.org/10.1007/978-981-16-6210-2_5.

Dorigo, M., Theraulaz, G., and Trianni, V. (2020). Reflections on the future of swarm robotics. *Sci. Rob.* 5 (49). https://doi.org/10.1126/scirobotics.abe4385.

Dulhare, U.N. and Gouse, S. (2022). Automation of rice cultivation from ploughing–harvesting with diseases, pests and weeds to increase the yield using AI. *ICCCE 2021*, 505–513. Singapore: Springer.

Ethernet Alliance (2022). Ethernet Roadmap 2022. https://ethernetalliance.org/technology/ethernet-roadmap (accessed 20 March 2023).

Francis, M. and Deisy, C. (2021). Mathematical and visual understanding of a deep learning model towards m-agriculture for disease diagnosis. *Arch. Comput. Methods Eng.* 28 (3): 1129–1145.

Friha, O., Ferrag, M.A., Shu, L. et al. (2021). Internet of things for the future of smart agriculture: a comprehensive survey of emerging technologies. *IEEE/CAA J. Autom. Sin.* 8 (4): 718–752.

Ghosh, S., Sarkar, A., Mitra, A., and Das, A. (2020). Smart cropping based on predicted solar radiation using IoT and machine learning. *2020 IEEE International Conference on Advent Trends in Multidisciplinary Research and Innovation (ICATMRI)*, 1–5. IEEE.

Golpîra, H., Khan, S.A.R., and Safaeipour, S. (2021). A review of logistics internet-of-things: current trends and scope for future research. *J. Ind. Inf. Integr.* 22: 100194. https://doi.org/10.1016/J.JII.2020.100194.

Gómez-Chabla, R., Real-Avilés, K., Morán, C. et al. (2019). IoT applications in agriculture: a systematic literature review. *Adv. Intell. Syst. Comput.* 901: 68–76. https://doi.org/10.1007/978-3-030-10728-4_8/COVER.

Grimblatt, V., Jégo, C., Ferré, G., and Rivet, F. (2021). How to feed a growing population—an iot approach to crop health and growth. *IEEE J. Emerging Sel. Top. Circuits Syst.* 11 (3): 435–448.

Hasegawa, T., Sakurai, G., Fujimori, S. et al. (2021). Extreme climate events increase risk of global food insecurity and adaptation needs. *Nat. Food* 2 (8): 587–595. https://doi.org/10.1038/s43016-021-00335-4.

Huang, T., Kou, S., Liu, D. et al. (2022). A BIM-GIS-IoT-based system for excavated soil recycling. *Buildings* 12 (4): 457.

Hughes, J. (2022). Soil Monitoring with IoT – Smart Agriculture. https://www.manxtechgroup.com/soil-monitoring-with-iot-smart-agriculture (accessed 20 March 2023).

Idoje, G., Dagiuklas, T., and Iqbal, M. (2021). Survey for smart farming technologies: challenges and issues. *Comput. Electr. Eng.* 92: 107104. https://doi.org/10.1016/J.COMPELECENG.2021.107104.

Intel (2022). The Story of the Intel® 4004. Intel. https://www.intel.com/content/www/us/en/history/museum-story-of-intel-4004.html (accessed 20 March 2023).

Jadhav, S.A. and Lal, A. (2022). Machine learning approaches for agro-IoT systems. In: *Cloud IoT Systems for Smart Agricultural Engineering* (ed. S. Krishnan, J.B.R. Rose, N.R. Rajalakshmi, and N. Prasanth), 89–108. Chapman and Hall/CRC. https://doi.org/10.1201/9781003185413-6.

Kaab, A., Sharifi, M., Mobli, H. et al. (2019). Combined life cycle assessment and artificial intelligence for prediction of output energy and environmental impacts of sugarcane production. *Sci. Total Environ.* 664: 1005–1019.

Kanuru, L., Tyagi, A.K., Aswathy, S. et al. (2021). Prediction of pesticides and fertilizers using machine learning and internet of things. *2021 International Conference on Computer Communication and Informatics, ICCCI 2021*. https://doi.org/10.1109/ICCCI50826.2021.9402536.

Kebebew, S., Bedadi, B., Erkossa, T. et al. (2022). Effect of different land-use types on soil properties in Cheha District, south-Central Ethiopia. *Sustainability* 14 (3): 1323.

Keswani, B., Mohapatra, A.G., Mohanty, A. et al. (2019). Adapting weather conditions based IoT enabled smart irrigation technique in precision agriculture mechanisms. *Neural Comput. Applic.* 31 (1): 277–292. https://doi.org/10.1007/S00521-018-3737-1/TABLES/5.

Khanna, A. and Kaur, S. (2019). Evolution of internet of things (IoT) and its significant impact in the field of precision agriculture. *Comput. Electron. Agric.* 157: 218–231.

Kim, W.S., Lee, W.S., and Kim, Y.J. (2020). A review of the applications of the internet of things (IoT) for agricultural automation. *J. Biosyst. Eng.* 45 (4): 385–400. https://doi.org/10.1007/S42853-020-00078-3.

Kumar, D. and Choudhury, U. (2022). *Agriculture-IoT-Based Sprinkler System for Water and Fertilizer Conservation and Management*, 229–244. Design and Development of Efficient Energy Systems.

Kumar, S., Tiwari, P., and Zymbler, M. (2019). Internet of things is a revolutionary approach for future technology enhancement: a review. *J. Big Data* 6 (1): 1–21. https://doi.org/10.1186/S40537-019-0268-2/FIGURES/9.

Kumar, R., Sinwar, D., Pandey, A. et al. (2022). IoT enabled technologies in smart farming and challenges for adoption. In: *Internet of Things and Analytics for Agriculture*, vol. 3 (ed. P.K. Pattnaik, R. Kumar, and S. Pal), 141–164. Singapore: Springer. https://doi.org/10.1007/978-981-16-6210-2_7.

Lee, I. and Lee, K. (2015). The internet of things (IoT): applications, investments, and challenges for enterprises. *Bus. Horiz.* 58 (4): 431–440. https://doi.org/10.1016/J.BUSHOR.2015.03.008.

Li, S., Da Xu, L., and Zhao, S. (2015). The Internet of Things: a survey. *Inf. Syst. Front.* 17 (2): 243–259. https://doi.org/10.1007/S10796-014-9492-7/FIGURES/7.

Li, L.F., Wang, Y.H., Li, Y.C. et al. (2022). Inhibitory effects of soil amendment coupled with water management on the accumulation of Cd and Pb in double-cropping rice. *Huan Jing ke Xue=Huanjing Kexue* 43 (1): 472–480.

Lombardi, M., Pascale, F., and Santaniello, D. (2021). Internet of things: a general overview between architectures, protocols and applications. *Information* 12: 87. https://doi.org/10.3390/INFO12020087.

Lova Raju, K. and Vijayaraghavan, V. (2020). IoT technologies in agricultural environment: a survey. In: *Wireless Pers. Commun.*, vol. 113(4), 2415–2446. US: Springer. https://doi.org/10.1007/s11277-020-07334-x.

Maraveas, C., Piromalis, D., Arvanitis, K.G. et al. (2022). Applications of IoT for optimized greenhouse environment and resources management. *Comput. Electron. Agric.* 198: 106993. https://doi.org/10.1016/J.COMPAG.2022.106993.

Marcu, I.M., Suciu, G., Balaceanu, C.M., and Banaru, A. (2019). IoT based system for smart agriculture. *Proceedings of the 11th International Conference on Electronics, Computers and Artificial Intelligence, ECAI 2019*. https://doi.org/10.1109/ECAI46879.2019.9041952.

Martens, B. (2022). Man-in-the-Middle Attack. https://www.safetydetectives.com/blog/avoiding-the-man-in-the-middle-preventing-a-common-cyberattack (accessed 20 March 2023).

Mekala, M.S. and Viswanathan, P. (2020). (t,n): Sensor stipulation with THAM index for smart agriculture decision-making IoT system. *Wireless Pers. Commun.* 111 (3): 1909–1940. https://doi.org/10.1007/S11277-019-06964-0/FIGURES/15.

Na, A., Isaac, W., Varshney, S., and Khan, E. (2017). An IoT based system for remote monitoring of soil characteristics. *2016 International Conference on Information Technology, InCITe 2016 – The Next Generation IT Summit on the Theme – Internet of Things: Connect Your Worlds*, 316–320. https://doi.org/10.1109/INCITE.2016.7857638.

Najeeb Ahmed, G. and Kamalakkannan, S. (2022). Developing an IoT-based data analytics system for predicting soil nutrient degradation level. In: *Expert Clouds and Applications* (ed. I. Jeena Jacob, F.M. Gonzalez-Longatt, S. Kolandapalayam Shanmugam, and I. Izonin), 125–137. Singapore: Springer. https://doi.org/10.1007/978-981-16-2126-0_12.

Nayak, P., Kavitha, K., and Mallikarjuna Rao, C. (2020). *IoT-Enabled Agricultural System Applications, Challenges and Security Issues*, 139–163. http://dx.doi.org/10.1007/978-981-13-9177-4_7.

Nayyar, A., Le Nguyen, B., and Nguyen, N.G. (2020). The internet of drone things (Iodt): future envision of smart drones. *Adv. Intell. Syst. Comput.* 1045: 563–580. https://doi.org/10.1007/978-981-15-0029-9_45/COVER.

NOAA. (2022). Soil moisture. *The 2022 National Soil Moisture Workshop*. https://www.drought.gov/events/2022-national-soil-moisture-workshop (accessed 20 March 2023).

Ozguven, M.M. (2018). The newest agricultural technologies. *Curr. Invest. Agric. Curr. Res.* 5 (1): 573–580.

Pant, Y. (2022). Agricultural science with IoT. In: *Internet of Things for Agriculture 4.0: Impact and Challenges* (ed. R. Singh, A.K. Thakur, A. Gehlot, and A.K. Kaviti), 111–130. CRC Press. https://doi.org/10.1201/9781003161097-5.

Parmar, M. and Kumar, R. (2022). Overview of IoT in the agroecosystem. In: *Agri-Food 4.0* (ed. R.S. Mor, D. Kumar, and A. Singh), 111–122. Emerald Publishing Limited. https://doi.org/10.1108/S1877-636120220000027008.

Parsons, L., Ross, R., and Robert, K. (2020). A survey on wireless sensor network technologies in pest management applications. *SN Appl. Sci.* 2 (1): 1–12. https://doi.org/10.1007/S42452-019-1834-0/TABLES/1.

Partel, V., Kakarla, S.C., and Ampatzidis, Y. (2019). Development and evaluation of a low-cost and smart technology for precision weed management utilizing artificial intelligence. *Comput. Electron. Agric.* 157: 339–350.

Partel, V., Kim, J., Costa, L., Pardalos, P.M., and Ampatzidis, Y. (2020). Smart sprayer for precision weed control using artificial intelligence: comparison of deep learning frameworks. *ISAIM*.

Pattnaik, P.K., Kumar, R., and Pal, S. (ed.) (2020). *Internet of Things and Analytics for Agriculture*, vol. 2, 67. https://doi.org/10.1007/978-981-15-0663-5.

Phasinam, K., Kassanuk, T., Shinde, P.P. et al. (2022). Application of IoT and cloud computing in automation of agriculture irrigation. *J. Food Qual.* 2022: 8285969.

Picon, A., Seitz, M., Alvarez-Gila, A. et al. (2019). Crop conditional convolutional neural networks for massive multi-crop plant disease classification over cell phone acquired images taken on real field conditions. *Comput. Electron. Agric.* 167: 105093.

Placidi, P., Gasperini, L., Grassi, A. et al. (2020). Characterization of low-cost capacitive soil moisture sensors for IoT networks. *Sensors* 20 (12): 3585.

Pyingkodi, M., Thenmozhi, K., Karthikeyan, M., et al. (2022). IoT technologies for precision agriculture: a survey. *2022 6th International Conference on Computing Methodologies and Communication (ICCMC)*, 372–376. IEEE.

Raj, M., Gupta, S., Chamola, V. et al. (2021). A survey on the role of internet of things for adopting and promoting agriculture 4.0. *J. Netw. Comput. Appl.* 187: 103107. https://doi.org/10.1016/J.JNCA.2021.103107.

Ramachandran, V., Ramalakshmi, R., Kavin, B.P. et al. (2022). Exploiting IoT and its enabled technologies for irrigation needs in agriculture. *Water* 14 (5): 719.

Ramdinthara, I.Z., Bala, P.S., and Gowri, A.S. (2022). AI-based yield prediction and smart irrigation. In: *Internet of Things and Analytics for Agriculture*, vol. 3 (ed. P.K. Pattnaik, R. Kumar, and S. Pal), 113–140. Singapore: Springer. https://doi.org/10.1007/978-981-16-6210-2_6.

Rani, T.P., Jasmine, J.A., Janani, M. et al. (2022). Vertical agriculture monitoring system in IoT. *Elementary Educ. Online* 20 (1): 5220–5220.

Raut, R., Varma, H., Mulla, C., and Pawar, V.R. (2018). Soil monitoring, fertigation, and irrigation system using IoT for agricultural application. In: *Intelligent Communication and Computational Technologies. Lecture Notes in Networks and Systems*, vol. 19 (ed. Y.C. Hu, S. Tiwari, K. Mishra, and M. Trivedi), 67–73. Singapore: Springer. https://doi.org/10.1007/978-981-10-5523-2_7.

Ray, P.P. (2017). Internet of things for smart agriculture: technologies, practices and future direction. *J. Ambient Intell. Smart Environ.* 9 (4): 395–420.

Rehman, A., Saba, T., Kashif, M. et al. (2022). A revisit of Internet of Things technologies for monitoring and control strategies in smart agriculture. *Agronomy* 12 (1): 127.

Rietra, R., Heinen, M., and Oenema, O. (2022). A review of crop husbandry and soil management practices using meta-analysis studies: towards soil-improving cropping systems. *Land* 11 (2): 255.

Sabanci, K. (2020). Detection of sunn pest-damaged wheat grains using artificial bee colony optimization-based artificial intelligence techniques. *J. Sci. Food Agric.* 100 (2): 817–824.

Sambasivam, G. and Opiyo, G.D. (2021). A predictive machine learning application in agriculture: cassava disease detection and classification with imbalanced dataset using convolutional neural networks. *Egypt. Inf. J.* 22 (1): 27–34.

Sangeetha, B.P., Kumar, N., Ambalgi, A.P. et al. (2022). IOT based smart irrigation management system for environmental sustainability in India. *Sustainable Energy Technol. Assess.* 52: 101973.

Senevirathne, I., Ambegoda, T., Wijesena, R. and Perera, I. (2022). IoT-based soil nutrient analyser using Gaussian process regression. *2022 2nd International Conference on Advanced Research in Computing (ICARC)* (7–12). IEEE.

Seyedzadeh, A., Maroufpoor, S., Maroufpoor, E. et al. (2020). Artificial intelligence approach to estimate discharge of drip tape irrigation based on temperature and pressure. *Agric. Water Manag.* 228: 105905.

Shadrin, D., Menshchikov, A., Ermilov, D., and Somov, A. (2019a). Designing future precision agriculture: detection of seeds germination using artificial intelligence on a low- power embedded system. *IEEE Sens. J.* 19 (23): 11573–11582.

Shadrin, D., Menshchikov, A., Somov, A. et al. (2019b). Enabling precision agriculture through embedded sensing with artificial intelligence. *IEEE Trans. Instrum. Meas.* 69 (7): 4103–4113.

Shamshiri, R., Weltzien, C., Hameed, I.A., et al. (2018). Research and development in agricultural robotics: a perspective of digital farming.

Singh, D.K., Sobti, R., Jain, A. et al. (2022). LoRa based intelligent soil and weather condition monitoring with Internet of Things for precision agriculture in smart cities. *IET Commun.* 16 (5): 604–618.

Srinivas, P., Swami Das, M., and Malathi Latha, Y.L. (2022). Farm management and resource optimization using IoT. *ICDSMLA 2020*, 1527–1538. Singapore: Springer.

Tabuenca, B., García-Alcántara, V., Gilarranz-Casado, C., et al. (2022). Engineering IoT systems in the convergence between agronomic and computer sciences. *2022 IEEE Global Engineering Education Conference (EDUCON)*, 2084–2087. IEEE.

Tang, D., Feng, Y., Gong, D. et al. (2018). Evaluation of artificial intelligence models for actual crop evapotranspiration modeling in mulched and non-mulched maize croplands. *Comput. Electron. Agric.* 152: 375–384.

Tephila, M.B., Sri, R.A., Abinaya, R., Lakshmi, J.A. and Divya, V. (2022). Automated smart irrigation system using IoT with sensor parameter. *2022 International Conference on Electronics and Renewable Systems (ICEARS)*, 543–549. IEEE.

Ullo, S.L. and Sinha, G.R. (2020). Advances in smart environment monitoring systems using IoT and sensors. *Sensors* 20 (11): 3113.

Vendruscolo, E.C.G., Mesa, D., and de Souza, E.M. (2022). Corn rhizosphere microbial community in different long term soil management systems. *Appl. Soil Ecol.* 172: 104339.

Vikranth, K. (2021). An implementation of IoT and data analytics in smart agricultural system–a systematic literature review. *Int. J. Manage. Technol. Soc. Sci. (IJMTS)* 6 (1): 41–70.

Vitali, G., Francia, M., Golfarelli, M., and Canavari, M. (2021). Crop management with the IoT: an interdisciplinary survey. *Agronomy* 11: 181. https://doi.org/10.3390/AGRONOMY11010181.

Wicaksono, M.G.S., Suryani, E., and Hendrawan, R.A. (2022). Increasing productivity of rice plants based on IoT (internet of things) to realize smart agriculture using system thinking approach. *Procedia Comput. Sci.* 197: 607–616.

Zhang, L., Dabipi, I.K., and Brown Jr., W.L. (2018). Internet of things applications for agriculture. *Internet of Things A to Z: Technologies and Applications*, 507–528.

Zhong, F., Chen, Z., Zhang, Y., and Xia, F. (2020). Zero-and few-shot learning for diseases recognition of Citrus aurantium L. using conditional adversarial autoencoders. *Comput. Electron. Agric.* 179: 105828.

Zhou, I., Makhdoom, I., Shariati, N. et al. (2021). Internet of things 2.0: concepts, applications, and future directions. *IEEE Access* 9: 70961–71012. https://doi.org/10.1109/ACCESS.2021.3078549.

Zhou, Y., Xia, Q., Zhang, Z. et al. (2022). Artificial intelligence and machine learning for the green development of agriculture in the emerging manufacturing industry in the IoT platform. *Acta Agric. Scand. Sec. B—Soil Plant Sci.* 72 (1): 284–299.

Index

Agroecological Approaches for Sustainable Soil Management, First Edition.
Edited by Majeti Narasimha Vara Prasad and Chitranjan Kumar.

k

l

r

t